Optics

Tenth Edition

M. H. Freeman PhD, MPhil, BSc, MInstP, Hon CGIA
Optics and Vision Limited, Denbigh, Clwyd; Visiting Professor, Department of
Optometry, City University, London

Butterworths
London   Boston   Singapore   Sydney   Toronto   Wellington

 PART OF REED INTERNATIONAL P.L.C.

First published 1934
Second edition 1936
Third edition 1939
Fourth edition 1942
Fifth edition 1945
Reprinted 1947, 1949
Sixth edition 1951
Reprinted 1954, 1956, 1959

Seventh edition 1965
Reprinted 1969, 1972
Eighth edition 1974
Reprinted 1976, 1977, 1978, 1979 (twice)
Ninth edition 1980
Reprinted 1981
Tenth edition 1990

British Library Cataloguing in Publication Data

Freeman, M. H. (Michael Harold)
Optics – 10th ed.
1. Optics
I. Title II. Fincham, W. H. A. (Walter Henry Angel).
Optics.
535

ISBN 0-407-00530-7

Library of Congress Cataloging-in-Publication Data

Freeman, M. H. (Michael Harold)
Optics/M. H. Freeman – 10th ed.
p. cm.
Rev. ed. of: Optics/W. H. A. Fincham, M. H. Freeman.
9th ed. 1980
Includes bibliographical references.
ISBN 0-407-00530-7
1. Optics. I. Fincham, W. H. A. Optics. II. Title
[DNLM: 1. Optics. QC 355.2 F855o]
QC355.2.F74 1990
535 – dc20
                                        89-22363

Composition by Genesis Typesetting, Laser Quay, Rochester, Kent
Printed in Great Britain at the Alden Press,
Bound by Hartnolls Ltd, Bodmin, Cornwall

Preface to the tenth edition

The last revision of this book was carried out in the late Seventies and, while the basic science of optics remains the same, technology has advanced considerably. This is particularly evident in the greater use of lasers and fibre optics and the development of diffractive (holographic) optical elements. Related technological developments are in micro-positioning control and personal computers. The general availability of the latter impinges in two ways on the author of a textbook on optics.

In the first place it is generally accepted that a large part of the learning process in geometrical optics is engendered by the longhand calculations of conjugate images, locations and magnifications for optical elements and systems. While these can readily be simulated on a personal computer with a teaching program, the author maintains that the optimum *learning* conditions comprise a text book, a calculator with a reciprocal function and a writing pad for sketches and results.

In the second place, however, it is recognized that the personal computer now has the computing power to run optical design and analysis programs with speed and precision. This has an inverse effect on a textbook such as this which places no great mathematical demands on the reader. In particular, the specific techniques of matrix algebra and complex exponent trigonometry have been avoided. The principles of refraction and reflection, conjugate images and optical aberrations as well as wave motion phenomena such as interference and diffraction may be understood without advanced mathematics. While the further application of these principles to lens design and diffractive systems including image formation and optical transfer functions does demand a high level of mathematics, these technologies are now generally available via well-constructed computer programs for personal computers, the intelligent use of which primarily requires an understanding of the optical principles described in this book. This is an important consideration for students of professions where optical knowledge is a necessary but relatively small part of their professional skills. This book therefore is a basic text in optical science as related to visible light.

The principles of visible optics are applicable to a wider range of the electro-magnetic spectrum even though different sources, detectors and materials are usually required. This restriction to the visible spectrum allows a compact and cohesive treatment of the subject. This book should meet the specific needs for optical knowledge experienced by optometrists, ophthalmologists and visual scientists as well as scientists and engineers with a particular interest in optical systems for visual use, and provides an introduction for those involved in non-visual applications.

For this new edition, the chapters on geometrical optics have been entirely re-written and provide a steady progression from simple refraction and reflection through to instrument design principles and geometrical aberrations including non-spherical optical elements. The chapters on physical optics have been rearranged to provide a basic introduction to light as a wave motion, sources, detectors and optical materials; the chapters on interference and diffraction have been largely retained, with the latter now covering diffractive optical elements. The final chapters on image analysis and the eye as an optical instrument have been extended.

It has been the current author's purpose that the tenth edition remains the essentially basic textbook that W. H. A. Fincham intended.

M. H. F.

Preface to the first edition

During recent years considerable progress has been made in bringing the teaching of optics in line with practical requirements, and a number of text books dealing with Applied Optics have been published. None of these, however, caters for the elementary student who has no previous knowledge of the subject and who, moreover, is frequently at the same time studying the mathematics required.

This book, which is based on lectures given in the Applied Optics Department of the Northampton Polytechnic Institute, London, is intended to cover the work required by a student up to the stage at which he commences to specialize in such subjects as ophthalmic optics, optical instruments and lens design. It includes also the work required by students of Light for the Intermediate examinations of the Universities.

The first eleven chapters deal with elementary geometrical optics, Chapters XII to XVI with physical optics, and the last three with geometrical optics of a rather more advanced character.

The system of nomenclature and sign convention adopted is that in use at the Imperial College of Science and the Northampton Polytechnic Institute, London. The sign convention is founded on the requirement that a converging lens shall have a positive focal length measured from the lens to the second principal focus. This is easily understood by the elementary student and is the convention commonly used throughout the optical industry. In ophthalmic optics – the most extensive branch of optical work at the present time – lenses are always expressed in terms of focal power; this idea has been introduced quite early and used throughout the work.

The solution of exercises plays an important part in the study of a subject such as Optics, and it is hoped that the extensive set of exercises with answers will be found useful. Typical examples from the examination papers of the London University, the Worshipful Company of Spectacle Makers and the British Optical Association are included by permission of these bodies.

My best thanks are due to my colleagues Messrs H. T. Davey and E. F. Fincham for their valuable assistance in the preparation of the diagrams, which, together with the photographs, have been specially made for the book. I wish particularly to express my indebtedness to Mr H. H. Emsley, Head of the Applied Optics Department, Northampton Polytechnic Institute, for reading the manuscript and for his very valuable help and suggestions given during the whole of the work.

January, 1934

W.H.A.F.
London

Contents

Preface to the tenth edition

Preface to the first edition

1 The basics of light and optical surfaces   1

2 Reflection and refraction at plane surfaces   25

3 Refraction and reflection at spherical surfaces   54

4 Thin lenses   84

5 Thick lenses and systems of lenses   115

6 Principles of optical instruments   170

7 Aberrations and ray tracing   230

8 Non-spherical and segmented optical surfaces   270

9 Light sources and the nature of light   299

10 Photometry and detectors   329

11 Optical materials: interaction of light with matter   357

12 Interference and optical films   397

13 Diffraction: wavefronts and images   422

14 Optical design: forming a good image   464

15 The eye as an optical system   492

Answers to exercises   515

Index   521

Colour plates 1 to 12 between pages 408 and 409

Chapter 1

The basics of light and optical surfaces

1.1 Light and optics

The branch of science known as Optics is mainly about light and vision. It also includes the study of other radiations very similar to light but not seen by the human eye.

Light is a form of radiant energy. Radiant heat, radio waves and X-rays are other forms of radiant energy. Light is sent out through space by **luminous sources**. Most of these are able to emit light because they are very hot. The high temperature of these sources means that their constituent atoms are in a state of considerable agitation, the effects of which are transmitted outwards from the source in all directions.

A piece of dark metal, when cold, emits no radiation that we can see. When gradually heated, it sets up a disturbance in the form of vibrations or **waves** in the surrounding medium, which radiate outwards at very high speed. At some distance away we can feel the effect as heat; we detect this form of radiation by our sense of touch. As the temperature rises and the vibrations become faster, the metal is seen to glow red; the radiation is such that we can see its effects; it is in the form of light. We detect this form of radiation by our sense of sight, the eye acting as a detector. With a further rise in temperature the metal passes to a yellow and then to a white heat.

The exact nature of light is not completely known, but a working idea of these 'wave motions' can be found in the ripples that occur when the calm surface of water is disturbed by dropping a stone into it. The important characteristics of the disturbance are: the speed or **velocity** at which it travels outwards; the distance between the wave crests, called the **wavelength**; and the **frequency** or rate of the rise and fall of the water surface. These will be studied in more detail in Chapter 9 where it will be shown that the velocity is equal to the frequency multiplied by the wavelength.

In the case of light, heat and radio waves, the velocity has been found by experiment to be 186 000 miles per second or 300 million metres per second. We find that the frequency of the vibration is greater for light than for heat and, as the increasing temperature of metal showed, the vibrations are faster for yellow light than for red light (which appeared first). As the velocity (in vacuum) is the same for all colours, it follows that the wavelength is longer for red light and shorter for yellow light and blue light. Table 1.1 gives typical wavelengths and frequency values from each type of radiation. A more complete description is given in Section 9.4.

Table 1.1

Radiation	Velocity (m s^{-1})	Frequency (Hz)	Wavelength (nm)
Radio	3×10^8	1×10^6	3×10^{11}
Heat			
Thermal infrared	3×10^8	30×10^{12}	10 000
Near infrared	3×10^8	300×10^{12}	1 000
Light			
Red	3×10^8	395×10^{12}	759
Yellow	3×10^8	509×10^{12}	589
Violet	3×10^8	764×10^{12}	393
Ultraviolet	3×10^8	1000×10^{12}	300
X-rays	3×10^8	3×10^{18}	0.1

When the metal is white hot it is emitting all colours equally. The nature of white light was investigated by Sir Isaac Newton (1642–1727) when he passed a narrow beam of white light through a glass prism and found that a white patch on a screen was now broadened out into a spectrum with the colours the same as those seen in the rainbow. The white light is said to be *dispersed* into its constituent colours and the effects of these on the eye and with different optical materials are studied in Chapter 13.

1.2 Rectilinear propagation of light (or: Light travels in straight lines)

Any space through which light travels is an optical medium. Most optical media have the same properties in all directions and are said to be isotropic. Some exceptions are described in Chapter 11. Also, most optical media have these same properties throughout their mass and are said to be homogeneous. Again, there are some exceptions and these are described in Chapter 11.

When light starts out from a point source B (Figure 1.1) in an isotropic homogeneous medium, it spreads out uniformly at the same speed in all directions; the position it is in at any given moment will be a sphere (such as C) having the source at its centre. Such imaginary spherical surfaces will be called light fronts or **wavefronts**. In the case of the water ripples the disturbance is propagated in one plane only and the wavefronts are circular.

Huygens (1629–1695), usually considered to be the founder of the wave theory of light, assumed that any point on a wave surface acted as a new source from which spherical 'secondary waves' or 'wavelets' spread out in a forward direction. The surface DAE, which touches or envelops all these wavelets, forms a new wavefront which is again a sphere with its centre at B. This action can be repeated as long as the medium remains isotropic and homogeneous. Huygens refined this idea by assuming that the wavelets travel out in a forward direction only and that their effect is limited to that part which touches the enveloping new wavefront. This concept is known as **Huygens' principle** and, with developments by Fresnel (1788–1827) and Kirchhoff (1824–1887) is used to explain the effects of interference and diffraction in Chapters 12 and 13. The basic construction of wavelets defining the new wavefront enables us to find the change in the form or

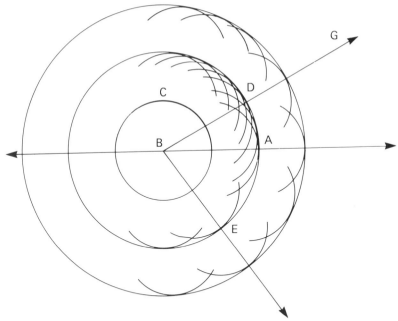

Figure 1.1 Huygens' construction for the propagation of light using wavefronts and wavelets

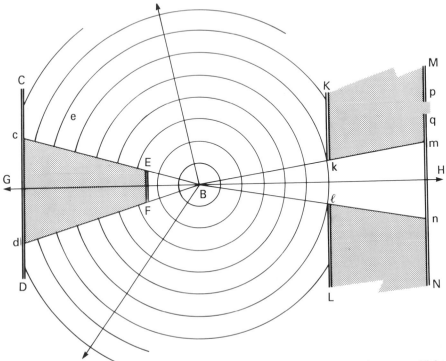

Figure 1.2 Rectilinear propagation of light shown using four opaque screens, a point source of light at B and apertures at kl and pq

the direction of the wavefront on passing to a new medium or on reflection at a surface.

From Figure 1.1 it is clear that the path, such as BDG, travelled by any part of the disturbance, is a straight line perpendicular to the wavefront. For a large number of optical effects it is enough to assume that this is how light travels. We speak of such a line along which light travels as a **ray**. Neither wavelets nor rays actually exist but both provide a useful way of dealing with the actions of light.

A luminous source, such as B in Figure 1.2, will be seen only if the light from it enters the eye. Thus, an eye in the region between C and K, or between D and L, of the two screens CD and KL will see B because the light from B reaches these regions without interruption. An eye located between c and d will not see the object B because the opaque screen EF shields the region cEFd which is said to be in shadow. B will be visible to an eye at e just beyond the limits of the shadow region.

If we have two opaque screens KL and MN, with circular holes at kl and pq, the screen MN will be illuminated over a circular patch mn but will be dark everywhere else, and an eye placed at the aperture pq will not see the source B.

The fact that B is invisible to any eye placed in the region cEFd and visible to one placed in the region kmnl shows, as was seen from Huygens' principle, that the light effect at any point is due to that part of the light which has travelled along the straight line joining the point to the source.

If the shadow cd and the bright patch mn were studied more closely it would be found that they did not have sharp edges even when the source B was a very small point. The light waves, in passing the edges of these apertures, bend round into the space behind them in the same way as water waves may be seen curving round the end of a jetty or breakwater. Light waves are so very small that this bending or **diffraction** is a very small effect and special methods have to be used to see it at all (Chapter 13).

1.3 Pencils and beams

The light from a point source which passes through a limiting aperture, such as kl of Figure 1.2, forms a small group of rays which is called a **pencil** of light. Sometimes the word **bundle** of rays is used to mean the same thing. The term ray bundles is most often used in optical design work with computers and in the calculation of aberrations (Chapters 17 and 14). We will use the word pencil for most of this book. Pencils of rays can also be thought of as coming from any point on a large extended source or illuminated object. The aperture that defines the pencil may be an actual hole in an opaque screen or the edges of a lens, mirror or window.

In Figure 1.2, the pencil formed by kl becomes larger as it gets further away from B and the limiting aperture. We say that the light in this pencil is **divergent**. Sometimes the pencil is **convergent**, getting smaller along the direction of the light. This is the case with a convex lens when the light pencil converges to a point or **focus**. This focus is the **image** of the object point from which the light started. Beyond the focus the pencil diverges. When the object point or the focus is a very large distance away, the rays in the pencil will be almost parallel. The pencil will be parallel when the object or the image is at infinity. Divergent, parallel and convergent pencils, with their corresponding wavefronts, are shown in Figure 1.3. The ray passing through the centre of the limiting aperture is called the **principal ray** or **chief ray** of the pencil.

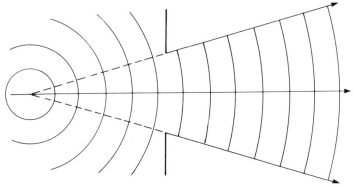

(a) Diverging

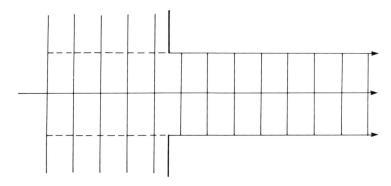

(b) Parallel

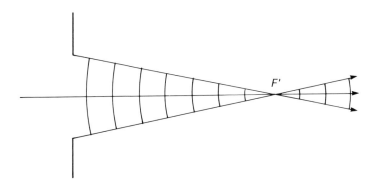

(c) Convergent (up to F')

Figure 1.3 Light pencils and wavefronts

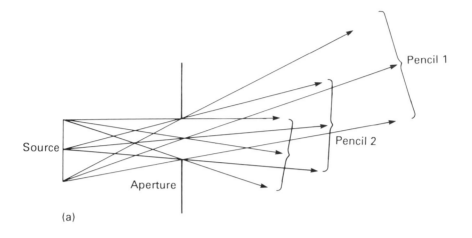

(a)

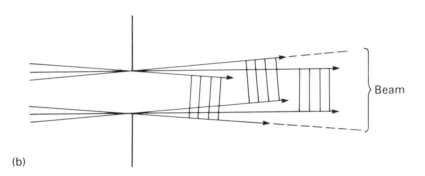

(b)

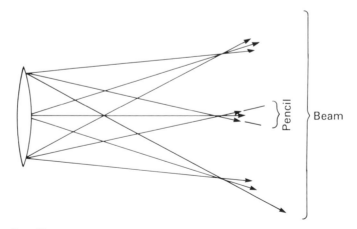

(c)

Figure 1.4 Pencils and beams

The pencils of light described above start out at a point source of light. If the source of light or the illuminated object is larger than a point, we imagine that such extended sources or extended objects comprise a large number of point sources. When an aperture is restricting the light from the extended source, each point on the source has a pencil of rays going through the aperture (Figure 1.4(a)). This collection of pencils is called a **beam** of light. The edges of the beam may be diverging or converging independently of the pencils of light that form it. Thus, if light from the sun passes through an aperture (Figure 1.4(b)), the individual pencils from each point on the sun will be parallel but the pencils are not parallel to each other and the edges of the beam are divergent. The beam from a lens (Figure 1.4(c)) may be divergent while the pencils of light in it are convergent. In this book the terms divergent, convergent and parallel light refer to *the form of the pencils* and *not to that of the beam*. We will study beams more in Chapter 7.

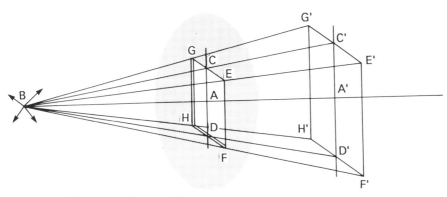

Figure 1.5 Width of a bundle or pencil

In Figure 1.5, BC'D' is the section of a pencil diverging from a luminous point B, and limited by an aperture CD at position A. The width of the rectangular aperture is GE and FH and a rectangular patch of light is formed on a screen at position A'. The patch has the height C'D' and the corners E', F', G' and H'. In the similar triangles ABC and A'BC',

$$\frac{C'A'}{CA} = \frac{BA'}{BA} \tag{1.1}$$

Because the light spreads out uniformly in all directions, all the other dimensions on the light patch are in the same proportion to their corresponding dimensions of the aperture.

Thus,

$$\frac{E'F'}{EF} = \frac{BA'}{BA}$$

$$\frac{E'G'}{EG} = \frac{BA'}{BA} \quad \text{etc.}$$

and

$$\frac{\text{Area } E'F'H'G'}{\text{Area } EFHG} = \frac{E'F' \times E'G'}{EF \times EG} = \frac{(BA')^2}{(BA)^2} \tag{1.2}$$

This means that the area of the cross-section of a pencil varies as the square of its distance from the source, since the amount of light in a pencil depends only on the amount given out by the source. This effect is very important in the science of photometry and will be studied in Chapter 10. If the source is large compared with the aperture a different effect is found.

In this chapter and most chapters of this book we are more concerned with the actual size of images and shadows rather than their area. This means that Equation 1.1 forms the main basis for our calculations.

1.4 The pinhole camera

The action of a pinhole camera shows that light travels in straight lines and therefore this device provides a verification of the law of rectilinear propagation of light. As its name suggests, the pinhole camera uses a small hole to form images. The light from each point on an illuminated object, on passing through a small aperture in an opaque screen, forms a narrow pencil and, if the light is received on a second screen at some distance from and parallel to the first screen containing the aperture, each pencil produces a patch of light of the same shape as the aperture. Because the light travels in straight lines, the patches of light on the screen are in similar relative positions to those of the corresponding points on the object. The illuminated area of the screen is similar in shape to that of the original object but is turned upside down or **inverted** (Figure 1.6(a)).

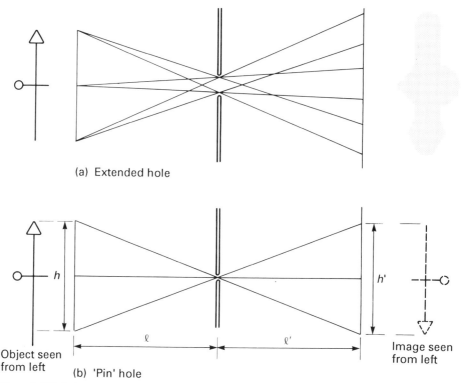

Figure 1.6 Pinhole camera

If the aperture is made small (a pinhole), the individual patches of light will overlap only to a small extent and a fairly well defined picture or image of the object is formed. As can be seen in Figure 1.6(b), the size of the inverted image of any object will depend on the distances of object and image from the aperture. By using the mathematics of similar triangles again, we find that

$$\frac{h'}{h} = \frac{l'}{l} \tag{1.3}$$

The degree of sharpness of the image formed by a pinhole can never be very good because if the diameter of the hole is made very small the effects of diffraction (Chapter 13) begin to blur the image. Also, the illumination of the image is very low compared with that of a camera having a lens, which will have a larger aperture allowing more light to pass through. The pinhole camera image is free from distortion (Chapter 7) and has large depth of focus; that is, the images of objects at greatly varying distance are all reasonably sharp at any screen position.

1.5 Shadows and eclipses

The formation of shadows is also explained by the law of rectilinear propagation of light. If the source of light is a point, the shadow of any object will be sharp, neglecting the very small effects of diffraction. The size of a shadow cast onto a screen is calculated in just the same way as the size of the patch of light in Figure 1.5 was calculated. Now we imagine EFHG to be an object rather than an aperture in a screen. The shadow it casts due to source B is E′F′H′G′. If the object is tilted, we must calculate the shadow size with a different value of A for each corner and the shadow will appear distorted. However, it remains the same shape as the object appears when viewed from the position of the source.

Usually, the light is coming from an extended source and we must consider the shadow as formed from different points on the source. In Figure 1.7(a), B_1B_2 is an extended disc source, CD an opaque disc object and LP a screen. No light from the source can enter the space CDNM and on the screen there is a circular area of diameter MN in total shadow. This is called the **umbra**. Surrounding the umbra is a space, CLM, DNP, in partial shadow which is called the **penumbra**. In the penumbra the illumination gradually increases from total darkness at the edge of the umbra to full illumination beyond the outer edge of the penumbra. For example, the point m is receiving light from the portion B_1b of the source. Because light travels in straight lines, the diameters of the umbra and penumbra may be found using similar triangles.

When the source is larger than the opaque object (Figure 1.7(b)), the umbra is a cone having the object as its base. At a certain distance from the object, the umbra disappears and the shadow is completely penumbral, as shown. All shadows in sunlight are like this because nothing on earth is bigger than the sun.

In Figure 1.7(b):

$$\frac{AO}{BO} = \frac{CD}{B_1B_2} \quad \text{or} \quad AO = CD\left(\frac{BO}{B_1B_2}\right) \tag{1.4}$$

As the sun is 1.4 million km in diameter and 150 million km away, it gives

$$\frac{BO}{B_1B_2} = 107$$

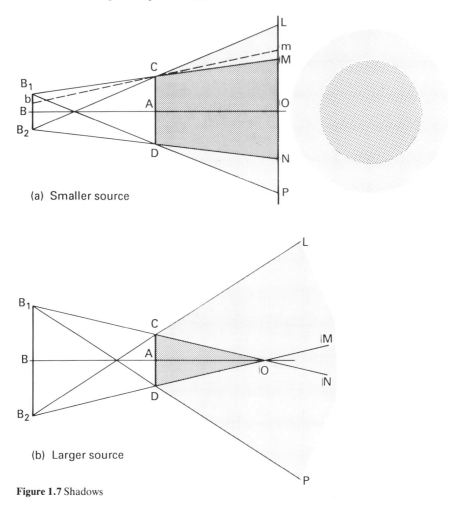

(a) Smaller source

(b) Larger source

Figure 1.7 Shadows

All shadows due to sunlight of objects on the earth have an umbra which is 107 times larger than the object. The largest object close to the earth is the moon. This 'object' is 3500 km in diameter and 380 000 km from the earth. These values are also approximately in the ratio 1 to 107. Thus, when the moon passes between the sun and the earth, we have the situation represented by Figure 1.7(b) where B_1B_2 is the sun, CD is the moon, and the earth's surface is at O. This is called a solar eclipse. An observer at O sees the moon just covering the sun and can then clearly see the flares (coronae) which are emitted from the sun's surface. When the earth passes between the moon and the sun we have a lunar eclipse. The moon passes into the earth's umbra but is still partly visible because the earth's atmosphere scatters sunlight into the umbral region.

Shadows play an important part in the way we see solid objects. A distant white sphere, evenly illuminated, cannot be distinguished from a flat white disc. However, if the illumination gives some shadowing on the sphere, so that it is immediately seen as a solid object.

1.6 Optical surfaces

The light from a source in an isotropic homogeneous medium continues to travel
out in straight rays and expanding wavefronts until it meets the surface of some
other medium. At this surface a number of effects can occur, depending on the
nature of the two media and of the surface between them. In all cases, some of the
light is sent back or reflected. The remainder passes into the new medium, where
some will be changed into some other form of energy or absorbed and some will
continue travelling on through the new medium and is said to be transmitted.

These three effects always take place but, often, one of them accounts for almost
all the light energy. If the screen LK in Figure 1.2, for example, were of black card,
most of the light would be absorbed. The screen MN in Figure 1.2 would probably
be white card so that most of the light would be reflected and we would be able to
see the patch MN. If MN were a mirror, the light would also be reflected but in a
different way, as explained in Section 1.7. If the screen MN were clear glass, most
of the light would be transmitted.

In the following sections and chapters, these three effects will be considered
separately, but it must be remembered that all three are always taking place at an
optical surface. More unusual effects can also occur, such as the absorption and
re-emission of the light with a different wavelength. Different effects can occur with
surfaces having very fine structure. However, the main effects are reflection,
absorption and transmission.

1.7 Reflection – specular and diffuse

When a surface is polished nearly all the reflected light travels back in definite
directions, as though coming from a source placed in some new position. The
polished surface then acts as an aperture and limits the reflected beam. An
observer in this reflected beam sees the image of the original source of the light
without being aware of the reflecting surface if this is well polished. Such a
reflection is said to be **regular** or **specular**. Figure 1.8 shows the effect on the ray
directions and the beam of light. Once the position of the image of the object is
found, the reflected rays can be drawn in by regarding the reflecting surface as an
aperture. The calculations needed to find the image position will be developed in
Chapter 2.

When a surface is not polished, every irregularity of the surface will reflect light
in a different direction and the light will not return as a definite beam. The surface
itself will be visible from all directions, as the light is spreading out in all directions
from each point on the surface. This type of reflection is said to be **irregular** or
diffuse and the light is **diffused** or **scattered**, although the actual reflection at each
minute portion of the surface will be regular. All illuminated objects are visible
because of the light irregularly reflected at their surfaces.

The amount of light reflected at a surface will vary greatly with different media.
For example, light diffusely reflected from white, grey and black paper will be
about 80%, 25% and 8% respectively of the incident light, the light arriving at the
surface. Only a few materials have reflections greater than 90% and it is also quite
difficult to obtain materials that reflect very little light. Special black paints have
been formulated for use inside optical instruments where any scattered light must
be absorbed as quickly as possible (see Table 10.2, p. 340).

For specular reflection, a lot depends on the quality of polish. Mirrors made of steel do not reflect as much light as those made of silver. The best polished surfaces are obtained with glass and most mirrors are made by putting onto polished glass a thin layer of metal or other reflecting material. The ratio of reflected light to incident light is known as the reflectance of the mirror and values above 99% can be achieved as described in Chapter 12. Polished surfaces are not perfectly smooth, because of the molecular structure of materials. However, once the irregularities are less than the wavelength of the incident light, the reflection becomes more and more specular. This means that quite rough surfaces can reflect heat radiation and radar waves. Even visible light can be reflected from unpolished surfaces if the incident light is nearly along the surface. A flat white sheet of paper can be seen to reflect specularly if the eye is placed at the edge of a flat sheet and looks across the sheet to objects beyond the other edge.

In the case of transparent substances, such as glass, water, etc., the specular reflectance from a surface can be found using Fresnel's equations (Chapter 12). It will depend not only on the angle at which the incident light meets the surface, but also on the difference in the refractive indices (Section 1.10) of the two media. For example, if some pieces of spectacle glass are placed in glycerine or, better still, canada balsam, media which have almost the same refractive index as the glass, almost no light is reflected at the surface of the glass and the pieces become nearly invisible. A similar effect is seen in the case of a grease spot on paper. The grease filling the pores of the paper has a refractive index more nearly that of the paper fibres than the air has, and so less light is reflected. Thus, the spot appears darker than the surrounding paper.

The light reflected by some materials is coloured. This is because, of the various colours in white light, only some are reflected and the others are absorbed by the substance. Such reflection is said to be **selective**. A red object when illuminated with white light reflects only the red, the other colours passing into the material and being absorbed. If the red object is illuminated with green light it appears black because there is no red light to reflect.

1.8 The law of reflection

In Figure 1.8 a ray of light BA is incident on an optically smooth surface DE, and is regularly reflected to form the ray AF. This ray is one of a collection of rays leaving the object and being reflected along the whole length of the mirror surface DE to form the image which appears to be behind the mirror. In the geometry of the light ray BAF, BA and AF are straight lines while DE is a surface. This can cause confusion and the mathematics is easier to use (and to explain) if we deal with three straight lines. This can be the case if, instead of the surface itself, we use a line (AN) perpendicular to the surface at the point of incidence A. This line is called the **normal to the surface at the point of incidence** and is a unique line. No other line is perpendicular and no other surface through the point A has this line for its normal. The plane containing the incident ray BA and the normal AN is called the **plane of incidence**. The angle BAN which the incident ray makes with the normal is the **angle of incidence**, i. The light after reflection will travel along AF and the angle FAN is the **angle of reflection**, r.

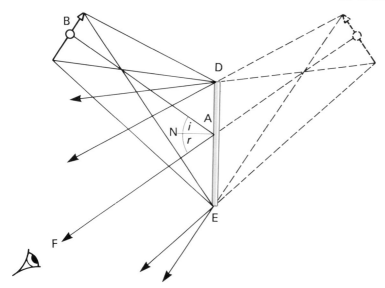

Figure 1.8 Specular reflection

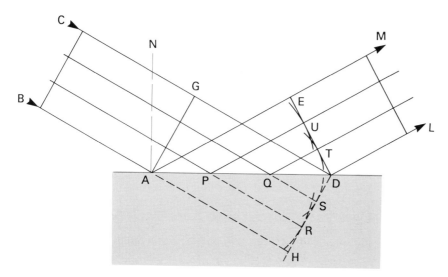

Figure 1.9 Reflection at a plane surface – Huygens' principle

The law of reflection may be stated as:

The incident and reflected rays and the normal to the surface lie in one plane and the incident and reflected rays make equal angles with the normal on opposite sides of it.

This law of reflection may be found by experiment and also by applying Huygens' principle. In Figure 1.9, BCGA represents a parallel pencil of light incident on a plane reflecting surface AD. AG is the position of the wavefront at one particular instant. If the reflecting surface had not been there, the new wavefront, after a period of time, would have reached the position HD. If we assume that the speed of the light is the same before and after reflection, then, in a given time, each part of the reflected light will have travelled back from the surface a distance equal to that which it would have travelled forward if there had been no reflection. Each point on the surface between A and D becomes, in succession, the centre of a reflected wavelet as the incident wavefront meets it. For example, the reflected wavelets from the points A, P and Q will have travelled out distances AE, PU and QT equal to AH, PR and QS respectively, while the edge of the wavefront at G is reaching the surface at D. From Huygens' principle, the envelope ED of these wavelets will be the reflected wavefront. Because we have used a parallel pencil of rays, the length GD = AH and we have made AE = AH. Thus GD and AE are equal and AD common. Both angles AGD and AED are right angles and so

angle GAD = angle EDA   (1.5)

Remembering that NAD is a right angle because AN is the normal to the surface, we have angle BAN equal to angle GAD and angle NAE equal to angle EDA. Using Equation 1.5 we have the angle of incidence equal to the angle of reflection.

1.9 Absorption and transmission

After reflection, the remainder of the light enters the new medium. Its behaviour is now dependent on the properties of the new medium. In most cases all the light, after travelling a short distance, is changed into some other form of energy, such as heat or chemical action, and the light is said to be absorbed. Materials through which light cannot pass are said to be **opaque**. The amount of absorption will depend on the properties of the material and on its thickness. Opaque metals such as gold and aluminium transmit very well when they are very thin (see Chapter 12).

When all colours are equally absorbed, which is usually the case when the material has a reasonable thickness, the absorption is said to be **neutral**. The reflection may be selective, as with red paint, but the absorption of the remaining light is neutral. Some materials, such as coloured glasses, dyes, etc., have **selective absorption**; some colours are absorbed and others are transmitted. This can be thought of as **selective transmission**, as discussed below. Some materials absorb light and then re-radiate it at a different wavelength. This is **fluorescence** and is dealt with in Chapter 9.

Some materials absorb very little light and very nearly all the light is transmitted. These materials are **transparent** and are very important in optics. If the surfaces of such transparent materials are polished, the light passes through as a definite beam. No material is perfectly transparent and some fraction of the light is always absorbed. Good quality glass absorbs about 0.5% of the incident light when it passes through a thickness of 10 mm. Special glasses have been developed for use in optical fibres which absorb less than 10% of the incident light through lengths of over 1 km. Other materials which contain fine particles of different optical properties from their surroundings transmit the light but diffuse or scatter it and objects cannot be seen clearly through them. These media, such as opal glass, thin paper, smoke and fog, are said to be **translucent**. Even when the individual particles are transparent, the many reflections from the mass of particles scatter the light. The same effect is obtained with a transparent material having an unpolished surface, such as 'frosted' or 'ground' glass. Screens made of translucent materials are often used to diffuse light from a source, or as surfaces on which to see images.

As stated above, the absorption and therefore the transmission of some materials is selective with respect to the colour or wavelength of the light absorbed or transmitted. Thus, when white light is incident on a plate of green glass, some of the light is reflected at the polished surface and this will still be white. Of that which passes into the glass only the green light is transmitted; the other colours making up white light are absorbed. All coloured glasses and dyes behave in this way. A good clear glass is transparent to all colours of visible light, but it is quite opaque to light in the ultraviolet of wavelength shorter than the so-called cut-off wavelength of the glass. In high-index glasses (Chapter 11) this cut-off wavelength may encroach into the blue light and these glasses appear yellow. Other glasses are opaque to infrared light to provide protection for the eyes of people looking into furnaces and fires. The action of filters is dealt with in Chapter 10.

1.10 Refraction – refractive index

When a ray of light enters a new medium, its direction is usually changed. This apparent break in the ray path is called 'refraction', from the same Latin word as fracture. The law of refraction, which describes the ray directions, is more complicated than the law of reflection. This means that, although the action of refraction of light rays at a polished surface gives an image of the object providing those rays, the image is clear only for narrow pencils of rays. In Figure 1.10 the narrow ray bundles from A and B, after refraction, appear to be coming from A' and B', but the ray BC at a greatly different angle of incidence refracts to become CD, which does not appear to be coming from B'.

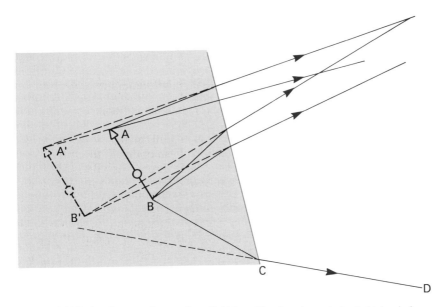

Figure 1.10 Refraction at a plane surface: light travelling from lower index to higher index

For the refraction of a single ray at a plane polished surface, we use a normal to the surface, as in the case of reflection. However, as Figure 1.11 shows, the normal must be shown right through the surface. The angle of incidence is shown as i and the angle of refraction as i'. We find that the relationship between these two angles depends on the difference in the velocity of the light in the two media. Foucault, in 1850, found by experiment that the speed of light in air was greater than in water. This result supported the wave theory of light suggested by Huygens and justifies our use of Huygens' construction to find the law of refraction (Section 1.11). The action of all lenses and prisms is entirely due to this change in velocity of light in different media.

The velocity of light in space (or in a vacuum) is internationally recognized as $299\,792.5\,\mathrm{km\,s^{-1}}$. In all other optical media the velocity is less than this.

We establish an absolute refractive index, which is the ratio of the velocity in vacuum divided by the velocity in the medium. Thus:

$$\text{Absolute refractive index of water} \ = \ \frac{\text{Velocity of light in vacuum}}{\text{Velocity of light in water}} \ = \ n_\mathrm{w} \qquad (1.6)$$

When light passes from one optical medium to another, such as from air to glass, glass to water, etc., the refractive effect is due to the ratio of the velocity of light in one medium, divided by the velocity of light in the other medium. This is called the relative refractive index. Thus the relative refractive index (water to glass), $_\mathrm{w}n_\mathrm{g}$, is given by:

$$_\mathrm{w}n_\mathrm{g} \ = \ \frac{\text{Velocity in water}}{\text{Velocity in glass}}$$

$$= \ \frac{n_\mathrm{g}}{n_\mathrm{w}} \qquad (1.7)$$

This relative refractive index can also be found by dividing the absolute refractive index for glass by the absolute refractive index for water. Therefore, there is no need to keep tables of relative refractive index for every possible *pair* of materials. Neither is there any need to measure the absolute refractive index of materials because the relative refractive index between two materials can be found using the relative refractive index values between each of the two materials and a third common material such as air. Thus:

$$_\mathrm{w}n_\mathrm{g} \ = \ \frac{_\mathrm{a}n_\mathrm{g}}{_\mathrm{a}n_\mathrm{w}} \qquad (1.8)$$

Because of this, it is usual to quote for the refractive index, n, of a given material the relative refractive index with respect to air. In any case, this is only slightly different from the absolute refractive index, because the absolute refractive index of air is $1.000\,29$. The terms 'rare' and 'dense' are often used in a comparative sense when referring to media of low and high index respectively.

A further complication is the fact that in most materials the refractive index changes with the wavelength of the light used. This is useful when we want to separate different colours of light, as with spectrometer prisms (Chapter 11) but becomes a problem when we want very good images from lenses (Chapter 14). In stating the refractive index of a material, it is necessary to quote the wavelength of the light used. Generally, a mean value of refractive index is taken as that for

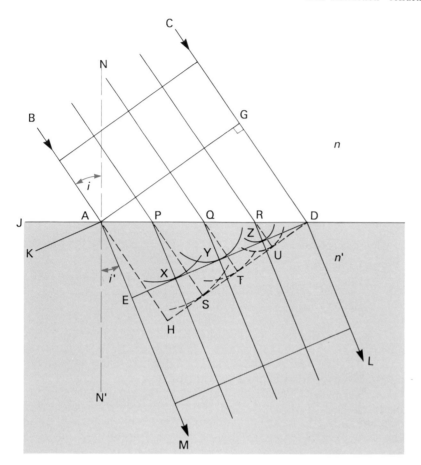

Figure 1.11 Refraction at a plane surface – Huygens' principle

yellow light. This used to be the yellow light from sodium vapour as used in many street lamps, but is now that from either helium vapour or mercury vapour – although the sodium light values are still commonly quoted. Table 1.2 gives the mean refractive index of some optical materials.

Table 14.2 (p. 470) gives more values and the subject is discussed more thoroughly in Chapters 11 and 14.

Table 1.2

Optical material	Refractive index
Diamond	2.4173
Water	1.3334
Optical plastics	1.46–1.7
Optical glass	1.46–1.95
Spectacle crown glass	1.523
'High index' spectacle glass	1.7

1.11 The law of refraction

The new direction taken by light when it passes through the surface between two media of different refractive index can be found by applying Huygens' principle in a way similar to that used for the law of reflection. In Figure 1.11, a parallel pencil of light BCDA is incident on the surface AD between two media of refractive index n and n'; n' is greater than n. AG is a plane wavefront, one edge of which has just reached the surface at A. If the surface had not been there, this wavefront would continue moving at velocity V and reach the position HD a short time later. As the incident wavefront meets the surface, each point on the surface, such as A, P, Q, R and D, becomes in succession the centre of wavelets travelling in the new medium. These wavelets travel more slowly in the second medium, at the velocity V', where

$$\frac{V'}{V} = \frac{n}{n'}$$

according to our definition of refractive index. At this slower speed, by the time the light at G has travelled to D in the first medium, the radii of the wavelets in the second medium will be:

$$AE = AH\left(\frac{n}{n'}\right) \qquad PX = PS\left(\frac{n}{n'}\right) \qquad QY = QT\left(\frac{n}{n'}\right) \qquad \text{etc.}$$

The envelope of all these wavelets will be the refracted wavefront ED. The refracted wavefront is still a plane surface and it is refracted *towards the normal* NN'. This is because the light travelled from a rare to a dense medium; n' is greater than n. If the light had moved from a dense to a rare medium, we would take n' to be less than n, the new wavelets would have radii longer than in the first medium, and a similar diagram would show that the light is refracted *away from the normal*.

The angle BAN in Figure 1.11 is the angle of incidence, i, and the angle MAN' is the angle of refraction, i'. From the definition of refractive index (Section 1.10):

$$nDG = n'AE$$

and so

$$n\frac{GD}{AD} = n'\frac{AE}{AD}$$

which is $n \sin DAG = n' \sin ADE$.   (1.9)

If we rotate the line AG through 90° it lies on AB, because the wavefront is at right angles to the ray. If we rotate the line AD through 90° it lies on AN, because the normal is at right angles to the surface. Thus, it is shown that angle GAD is the same size as angle BAN, the angle of incidence, i.

If we draw the line AK parallel to DE, the new wavefront, we have angle ADE equal to angle JAK. Rotating AJ and AK through 90° shows that angle JAK is equal to the angle of refraction, i'.

Thus, from Equation 1.9 we have:

$$n \sin i = n' \sin i'$$ (1.10)

Sometimes this equation is expressed as:

$$\frac{\sin i}{\sin i'} = \frac{n'}{n}$$ (1.10a)

but this is a bad way to remember it – there is less chance of error with Equation 1.10. However, the law of refraction may still be stated as:

The incident and refracted rays and the normal to the surface at the point of incidence lie in the one plane on opposite sides of the normal and the ratio of the sine of the angle of incidence to the sine of the angle of refraction is a constant for any two media for light of any one wavelength.

This law, which was first stated, but in a somewhat different form, by Willebrord Snell (1591–1626), is often known as **Snell's law**. (Snell stated the law of refraction as a constant ratio of the cosecants of the angles. The statement in terms of sines was first given by Descartes in 1637.)

Refraction at a plane surface and the law of refraction may be demonstrated by sighting at two pins seen through a block of glass having two parallel polished sides. The glass block is placed on a piece of paper with its polished sides vertical. Two pins (Figure 1.12) are placed a few centimetres apart to represent the path of a narrow beam of light incident obliquely on one surface. On looking through the glass, we can see that the pins will be in the apparent positions A′B′. Two further pins, C and D, can be placed in line with A′ and B′ seen through the glass. A line can then be drawn through A and B to the surface of the glass block and another line through CD to the other surface.

Then the line AEFD represents the path of a ray through the glass. By drawing lines to represent the ray paths and the surfaces, the normals at the points of

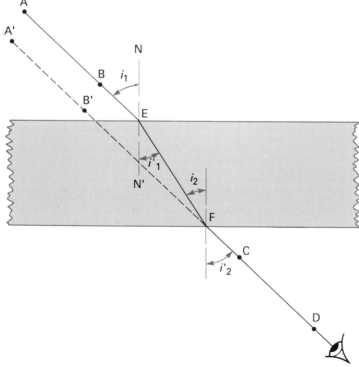

Figure 1.12 Experiment to verify the law of refraction

incidence can be drawn and the angles of incidence and refraction measured. If the experiment is done a number of times at different angles of incidence, the law of refraction may be experimentally verified and the refractive index of the glass obtained.

The law of refraction (and of reflection) may be briefly stated in a general law known as **Fermat's principle of least time**:

The actual path travelled by light in going from one point to another is that which, under the given conditions, requires the least time.

Sometimes, when the optical surface is curved, the time may be a maximum rather than a minimum but it is always one or the other. In mathematics, maxima and minima are known as stationary points. This principle defines a *unique* ray path for the light. In Figure 1.13(a) a ray from B reflected at A to reach C appears to come from B'. As proved earlier, B' is as far behind the mirror as B is in front. Clearly, when going from B' to C, the direct route B'AC is always shorter than any such as B'XC because the straight line is the shortest distance between two points.

With refraction, the shortest time may not be the shortest distance because of the different velocities in the two media. The velocity of light in different materials is related to their refractive indices.

From Equation 1.6, we can see that:

$$\text{Velocity of light in water} = \frac{\text{Velocity of light in vacuum}}{\text{Absolute refractive index of water}}$$

For any medium:

$$\text{Velocity of light in the medium} = V = \frac{V_0}{n}$$

where V_0 is the velocity of light in vacuum and n the refractive index of the medium. Thus, the time taken by light to travel a distance l in vacuum is l/V_0, while the time taken to travel the same distance in a medium of refractive index n is nl/V_0. The term nl is called the **optical path** or **optical thickness** and is used in Chapter 12 when computing interference effects.

In Figure 1.13(b) the path BAC is that satisfying the law of refraction. Although, for path BYC, BY is longer in the faster medium and YC is shorter in the slower medium, the extra optical length $n(LY)$ is longer than $n'(AM)$ even though n' is greater than n. For the path BXC the extra optical length $n'(XQ)$ is greater than the optical length $n(PA)$ even though XQ is dimensionally shorter than PA.

The ray paths defined by Fermat's principle and, therefore, by the laws of reflection and refraction, take no account of the direction of the light. The principle holds for light going from B to C in Figure 1.13, as shown, and for light going from C to B. This **principle of reversibility of optical path** is true for light passing through almost every optical system. Sometimes the calculation of ray paths is simplified by assuming that the light is travelling in the opposite direction (see Section 2.8).

1.12 The fundamental laws of geometrical optics

The study of optics has, for many years, been divided into two parts:

1. Study of the wave nature of light, usually called **physical optics**.
2. Study of the way in which the directions in which light travels are modified by prisms, mirrors and lenses, known as **geometrical optics**.

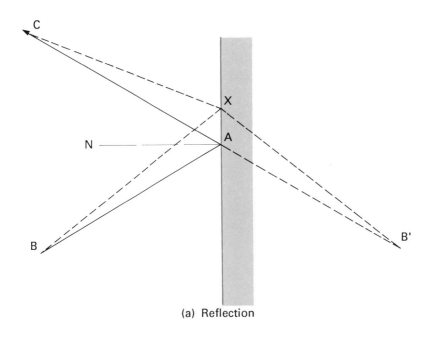

(a) Reflection

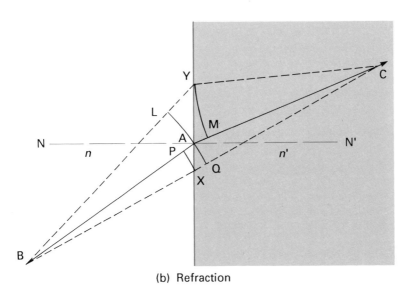

(b) Refraction

Figure 1.13 Fermat's principle

Recently, it has become possible to make optical elements with very fine structures which interact with the wave nature of light and so change its direction. This is known as **diffractive optics**. The term **electro-optics** has come into use to describe the study and use of the interaction between light and electrical fields. To add to the confusion, the term **opto-electronics** is used to describe work where the light is converted to electrons and electrons to light.

In this book, we are mainly concerned with geometrical optics (Chapters 1–8) and physical optics (Chapters 9–13). However, aspects of opto-electronics are included in Chapter 10 and electro-optics in Chapter 9. Ideas on diffractive optics are raised in Chapter 13.

For the early chapters, we can say that geometrical optics, ignoring diffractive effects due to the wave nature of light, assumes:

1. Neighbouring rays of light are independent of each other.
2. The propagation of light is rectilinear. That is, light travels in straight lines.
3. The law of reflection.
4. The law of refraction.

Exercises

1.1   A person holding a tube 6 in long and 1 in in diameter in front of one eye just sees the whole of a tree through the tube. What is the apparent (angular) height of the tree, and what is its distance if its actual height is 40 ft?

1.2   Explain the terms 'convergent', 'divergent' and 'parallel' light. Illustrate your answer with diagrams showing the form of the wavefronts in each case.

1.3   What are meant by pencils and beams of light? Illustrate your answer with diagrams. A spherical source of 5 cm diameter is placed 30 cm from a circular aperture of 8 cm diameter in an opaque screen; find the size and nature of the patch of light on a white screen 50 cm from and parallel to the plane of the aperture.

1.4   A small source of light is 30 cm from a rectangular aperture 18 cm × 8 cm. A screen is placed at 105 cm from and parallel to the plane of the aperture; what will be the area of the illuminated patch on the screen?

1.5   A pencil of light on passing through a lens of 40 mm diameter converges to a point 60 cm from the lens. Find the areas of the cross-sections of the pencil at 20, 50 and 75 cm from the lens.

1.6   Explain carefully the formation of the image in a pinhole camera. How does the character and size of the image depend on:
(a) the size of the aperture,
(b) the shape of the aperture,
(c) the distance of the aperture from the screen,
(d) the distance from the aperture to the object?

1.7   If the distance between an object and its image formed by a pinhole is 5 ft, what will be the position of the pinhole for the image to be one-tenth the size of the object?

1.8   A pinhole camera produces an image of 2.25 in diameter of a circular object, and when the screen is withdrawn 3 in further from the pinhole the diameter of the image increases to 2.75 in. What was the original distance between the pinhole and the screen?

1.9   What is the height of a tower that casts a shadow 65 ft 3 in in length on the ground, if the shadow of the observer who is 6 ft tall is 7 ft 6 in in length at the same time?

1.10   A circular opaque object of 3 in diameter is placed 12 in from and parallel to a circular source of 5 in diameter. Find the nature and size of the shadow on a screen perpendicular to the line passing through the centres of the source and the object and 3 ft from the object.

1.11 Explain with diagrams the way in which total and partial eclipses of (a) the sun and (b) the moon are produced.

1.12 Describe and give reasons for the appearances of:
(a) grey paper
(b) finely crushed glass
(c) paraffin wax, solid and molten

1.13 Why is a lump of sugar practically opaque, while the small separate crystals of which the lump is composed are themselves transparent?

1.14 Describe Huygens' principle and show how it may be used to explain the laws of reflection and refraction, on the wave theory.

1.15 Using Huygens' principle, construct the reflected wavefront when a parallel pencil of light is incident on a plane surface (a) normally, (b) at 30°, (c) at 70°.

1.16 A poster has red letters printed on white paper; describe and explain its appearance when illuminated (a) with red light and (b) with green light.

1.17 The velocity of yellow light in carbon disulphide is $183.8 \times 10^6 \, \mathrm{m\,s^{-1}}$; in ether it is $221.5 \times 10^8 \, \mathrm{m\,s^{-1}}$. Find the relative refractive index for refraction (a) from ether into carbon disulphide, (b) from carbon disulphide into ether.

1.18 The refractive indices of a few substances are:

Water   1.33
Dense flint glass   1.65
Crown glass   1.50
Diamond   2.42

Find the relative refractive index for the case of light being refracted from:
(a) Water into crown glass
(b) Water into diamond
(c) Diamond into water
(d) Dense flint glass into air
(e) Crown glass into dense flint glass

1.19 Given that the speed of light is $3 \times 10^{10} \, \mathrm{cm\,s^{-1}}$ in air and $2.25 \times 10^{10} \, \mathrm{cm\,s^{-1}}$ in water, construct the emergent wavefront from a small source situated in the water 10 cm below the surface, and from this determine the position of the image.

1.20 State the law of refraction and describe a method of measuring the refractive index of a block of glass having plane parallel faces.

1.21 Show that if a ray of light passes from a medium in which its velocity is V_1 to another in which it is V_2, $V_2 \sin i = V_1 \sin i'$ where i and i' are the angles of incidence and refraction respectively. What is the ratio V_1/V_2 called?

1.22 A parallel pencil is incident at an angle of 35° on the plane surface of a block of glass of refractive index 1.523. Find the angle between the light reflected from the surface and that refracted into the glass.

1.23 State the fundamental laws of geometrical optics and describe an experimental verification of each.

1.24 Give diagrams showing the change in direction of the wavefronts of a parallel pencil on passing from air to water when the pencil is incident at (a) 30°, (b) 60° to the normal (n of water = $4/3$).

1.25 A parallel pencil of light is incident at 45° to the upper surface of oil ($n = 1.47$) floating on water ($n = 1.33$). Trace the path of the light in the oil and the water.

1.26 A white stone lies on the bottom of a pond. Its edges are generally observed to be fringed with colour, blue and orange. Explain this and state which is the blue edge.

1.27 A ray incident at any point at an angle of incidence of 60° enters a glass sphere of refractive index $\sqrt{3}$ and is reflected and refracted at the further surface of the sphere. Prove that the angle between the reflected and refracted rays at this surface is a right angle. Trace the subsequent paths of the reflected ray.

1.28 A pinhole camera has a screen 100 mm in height. When it is 20 metres from a tree, the image of the tree just fits on the screen. When the camera is moved 2 metres closer to the tree, 1 metre of the tree cannot be imaged.

    (a) Calculate the distance within the pinhole camera from the pinhole to the screen.

    (b) Calculate the height of the tree.

    (c) If the camera were filled with water in the second position would the image now fit onto the screen and if so calculate an approximate size assuming the refractive index of the water is ⁴⁄₃ (and ignoring the leak from the pinhole!)

Chapter 2

Reflection and refraction at plane surfaces

2.1 Imaging by reflection

When we look at a plane reflecting surface, such as an ordinary mirror, we see –
apparently behind the mirror – images of objects which are in front of the mirror.
Some of the light from each object point is reflected at the mirror surface and
enters the eye *as though* it is coming from a point behind the mirror. These
non-existent points behind the reflecting surface form images which are called
virtual images. We can see them but we cannot get to them. The light from the
original object is diverging and continues to diverge after reflection at the plane
surface.

Sometimes, reflection (or refraction) at a curved surface makes the light
converge to a point again and images are formed which actually exist at the position
to which the light converges. These images, which can be received on a screen, are
called **real images**. Examples of these are the image formed on the film by a camera
lens and the image formed on a screen by a slide projector.

The position of any image formed by a plane reflecting surface can be found by
using the law of reflection or by using Huygens' principle to find the form of the
reflected wavefront. Using Huygens' method first, we see in Figure 2.1 point object
B in front of a plane mirror DE. Spherical wavefronts are moving out from B and,
if the mirror had not been there, FLH would have been the wavefront at some
moment. However, as the incident wavefront meets the mirror, each point on the
mirror becomes a series of wavelets travelling back at the same speed as the
incident light was travelling forward. Thus, the part of the wavefront meeting the
mirror at A forms wavelets which travel back to M in the same time that the
incident light would have travelled to L. Because the speed is the same, AM is the
same distance as AL.

The same thing happens at G and at every point on the mirror surface between R
and S. Because RS is a straight line and every point on SMR is the same distance
from the mirror as it would have been on SLR, the curvature of SMR is the same as
SLR, except that SLR is centred on B and SMR is centred on B', where AB' =
BA.

The same result can be found using the law of reflection. In Figure 2.1 again, the
ray BS is incident on the mirror at S and is reflected to form the ray ST which
appears to come from B'. NS is the normal to the mirror surface at S and so angle
TSN equals angle NSB ($i = r$) by the law of reflection. If point A is chosen so that
the normal from A passes through B, then the ray BA is reflected back along AB

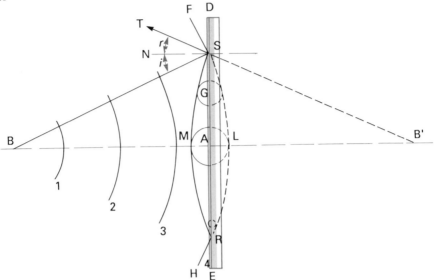

Figure 2.1 Image formed by reflection at a plane surface – Huygens' construction

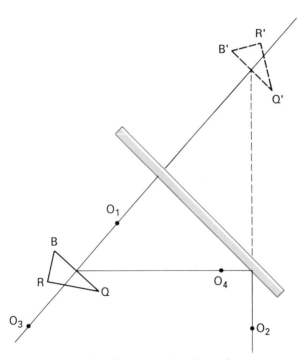

Figure 2.2 Reversion of images seen in a plane mirror

and appears to come from B', both the angle of incidence and the angle of reflection being equal to zero. For a plane reflecting surface, the normal at S is parallel to the normal at A. Thus, the angle SBA equals i and angle AB'S equals r. Because these angles are equal and the distance AS is common to the two triangles, ABS and AB'S, these triangles are the same size and so AB' = BA. We can say:

> The image formed by a plane mirror is as far behind the mirror as the object is in front, and the line joining the object and image is perpendicular to the mirror surface.

It follows from this that the shape and size of the image will be an exact replica of the object. The point on the object nearest the mirror will be represented by the point on the image nearest the mirror. The top and bottom of the object will correspond to the top and bottom of the image. We say that the image is **erect**. It is often difficult to decide whether the image is **reverted**; that is, reversed left for right. This depends on the position and posture of the observer in viewing the object and the image. In Figure 2.2, an observer at O_1 looking at the image sees B' on the left but, on turning round to view the object, he sees B on the right. He therefore describes the image as reverted. If he goes behind the object to O_3 he can then see B' and Q' on the same sides as B and Q but, if R is the rear of the object, he sees the front of it in the image. Mirrors are often used in viewing instruments to turn light through 90°. An observer viewing the image from O_2 sees it reverted compared with his direct view of the object from, say, O_4. Methods for reverting, rotating and inverting images are discussed in Section 2.10.

2.2 Field of view of plane mirrors

The extent of an image in a plane mirror that can be seen by an eye in any position will depend on the size of the mirror and the position of the eye. This extent of the image seen is called the **field of view** of the mirror. In Figure 2.3, the observer has his eye at O looking into the plane mirror DE. OD and OE will be the limiting directions along which the eye can see images reflected in the mirror. The reflected rays DO and EO correspond to rays incident on the mirror from points Q and R on the object. The distance QR is therefore the extent of the object of which the image Q'R' can be seen from the position O. From the figure, we have

$$\frac{Q'R'}{OB'} = \frac{DE}{OA} \quad \text{and} \quad \frac{QR}{BO'} = \frac{DE}{AO'}$$

O' is the image of the eye position and so OA = AO'. BA = AB' and therefore BO' = OB'.

The size Q'R' of the image that can be seen is equal to the size QR of the object that can be seen, and both are given by

$$QR = Q'R' = \frac{DE \times OB'}{DA}$$

$$= \frac{DE \times BO'}{AO'} \tag{2.1}$$

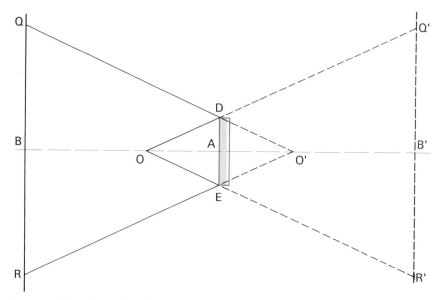

Figure 2.3 Field of view of a plane mirror

It is often needed to find how big and in what position a mirror needs to be, so that the whole of an object may be seen from a given eye position by reflection in the mirror. These problems can be solved, either by treating the mirror as an aperture of the same size and looking through it at the image, or by taking the reflected eye position and looking at the object, again treating the mirror as an aperture.

There are many applications of plane mirrors. The usual sight test for defective vision involves the patient looking at a letter chart from a distance of 6 m. The required distance can be reduced by placing the test chart beside or above the patient and placing a plane mirror 3 m in front of him. To the patient, the letters appear 6 m away. The actual letters must be reverted so that in the image they appear the right way round.

Another application of plane mirrors is in military aircraft where a partially reflecting mirror is positioned in front of the pilot, so that he can see through it to the outside world and also see, reflected by it, bright figures and symbols which appear superimposed on the outside world. This makes it unnecessary for him to look down to read the instruments and the system is called a 'heads-up display'.

2.3 Deviation of light by reflection; moving mirrors

When light is reflected from a plane mirror, there is a change in the direction of the beam of light. The angle through which the beam is turned – **the angle of deviation** – depends on the angle of incidence. In Figure 2.4 the beam BA is incident on the mirror DE and would have continued to C if it had not been reflected along AF. The angle v is the angle of deviation:

$$v = 180 - 2i \qquad\qquad (2.2)$$

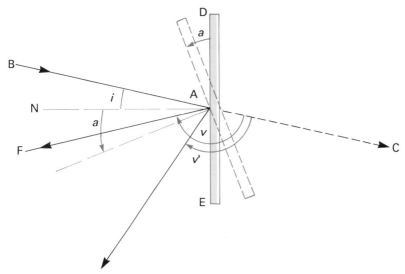

Figure 2.4 Rotating mirror. Reflected beam turns through twice the angle turned through by the mirror

Suppose now that the mirror is rotated through an angle a. The new angle of incidence is $i + a$. Thus, the new angle of deviation v' is given by:

$$v' = 180 - 2(i + a) \tag{2.2a}$$

The difference between v and v' is given by:

$$v - v' = 2a \tag{2.3}$$

Thus, the reflected light beam rotates through twice the angle that the mirror rotates. This fact must be taken into account when, for example, designing scanning systems for use with laser beams. It is common practice to use a polygon of mirrors which rotates at high speed. If the polygon has eight sides, it will rotate by 45° before each mirror is in the same position as the previous one. The reflected beam, however, will scan through an angle of 90°. Measuring instruments, in the days before electronics, often used a mirror attached to the coil of an electrical meter to measure small currents. A light beam reflected from this gave a useful amplification of 2, due to this fact of reflection.

2.4 Reflection at two mirrors

When light is reflected from two mirrors in succession, then, for the ray in the same plane as the normals of the mirrors, the total deviation it experiences depends only on the angle between the mirrors and not on the angle of incidence. Because of this, any rotation of both mirrors in that plane produces no change in the final direction of the light provided both mirrors are moved together.

In Figure 2.5, the total deviation of the ray is the sum of the deviations occurring at each reflection:

$$v = v_1 + v_2 \tag{2.4}$$

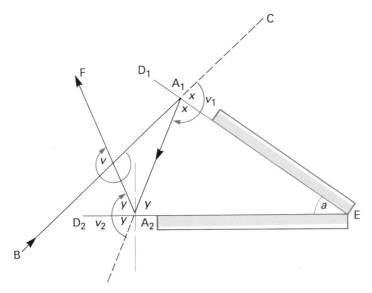

Figure 2.5 Deviation of light by reflection at two mirrors

Because the angle of incidence equals the angle of reflection at both reflections, v_1 is equal to $2x$ and v_2 is equal to $2y$, as shown in the figure. These are angles in the triangle A_1A_2E, which also contains a, the angle between the mirrors. By the usual rule in triangles:

$$x + y = 180° - a$$

We therefore have:

$$v = v_1 + v_2$$
$$= 2x + 2y$$
$$= 2(180 - a) \tag{2.5}$$

This equation has no dependence on angles of incidence, the total deviation depending only on a, the angle between the mirrors. Provided this can be kept fixed, and this is usually done by making the mirrors the sides of a prism (see Section 2.10), small errors in mounting the prism will not affect the total deviation.

If an object is placed between two plane mirrors, more than two images will be formed because some of the light reflected at one mirror may be reflected again at the other mirror. The number of such images that will be seen depends on the angle between the mirrors, the position of the object, and, to some extent, the position of the eye viewing the images. The positions of all the images may be found by finding any images formed by direct reflection and then considering these images as objects if they are in front of either of the mirrors.

In Figure 2.6, light from a luminous point B is reflected in mirror D_1E giving rise to the image B_1' as far behind this mirror as the object is in front. This image is in front of the mirror D_2E and some of the light from it is reflected in D_2E to form the image $B_{1.2}'$ as far behind this mirror as the image B_1' is in front. In the same way, light from the object reflected first by mirror D_2E forms the image B_2' and this in

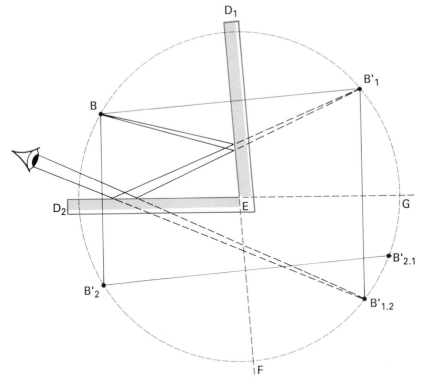

Figure 2.6 Multiple images formed by reflections at two mirrors

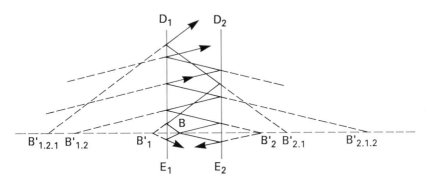

Figure 2.7 Multiple images with parallel mirrors

turn is reflected by D_1E to form the image $B'_{2.1}$. As both $B'_{1.2}$ and $B'_{2.1}$ are behind both mirrors, no further images are formed. The object and all the images lie on a circle centred on E, the intersection of the mirrors. The path of the light by which an eye at O sees the image at $B'_{1.2}$ is shown in the figure.

As the angle, a, between the mirrors is reduced, the number of images normally increases. For example, if the angle a in Figure 2.6 is made exactly 90°, then the images $B'_{1.2}$ and $B'_{2.1}$ coincide to give three images, plus the object, symmetrically disposed about the circle but, if the angle is reduced to 60°, five images plus the object can be seen. This is the arrangement used in the kaleidoscope. At 45°, seven images plus the object may be seen. At these values of a, which exactly divide into 180°, the number of images plus object is given by $2(180/a)$. At intermediate angles, the images seen depend on object and eye position.

When the angle a is made very small, the number of images increases dramatically. In the case where a is zero and the mirrors are parallel to each other, the number of images is theoretically infinity but, as some light is lost at each reflection, the more distant images shown in Figure 2.7 are growing increasingly faint.

2.5 Imaging by refraction

The law of refraction, developed in Chapter 1, was found to be more complex than the law of reflection. The images formed by plane refracting surfaces are therefore more complicated than images formed by plane reflecting surfaces. As shown in Section 1.11, the equation for refraction at a single surface may be written:

$$n \sin i = n' \sin i' \tag{2.6}$$

The mathematical expression sine (normally shortened to sin) has a *non-linear* relationship to the angle. Figure 2.8 shows what this means. If the value of sin i had a linear relationship to i, a straight line could be drawn so that any value of i such as i_1 could then be projected up to the line and the value of sin i_1 read off, using the straight line. However, the relationship is non-linear, as shown by the *curved* line, and the values obtained are therefore different. One important effect of this is that when, for example, the value of i_1 is doubled to i_2 the value sin i_2 is *not* twice that of sin i_1. The curved line describing the relationship actually reaches a maximum value of one when $i = 90°$ and then begins to decline.

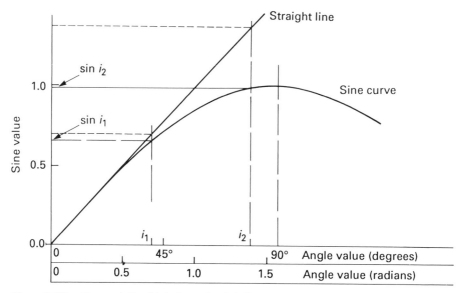

Figure 2.8 Non-linear relationship between i and sin i

It will be seen in Chapter 9 that this non-linear relationship is due to the way in which circles and lines interact. For the present, we can use circles in a *graphical construction* which automatically calculates the direction of the refracted ray for any angle of incident ray. The principle is shown in Figure 2.9. With their centres on A (the point of incidence on the plane refracting surface, DE), two circles are drawn whose radii are proportional to the refractive indices, n and n', of the media each side of the surface. The incident ray cuts the circle corresponding to n at the point B. From B a line is drawn parallel to the normal. This cuts the other circle at G. The refracted ray is drawn through A as if coming from G.

From the definition of sine:

$$\frac{BM}{AB} = \sin i \qquad \frac{GL}{AG} = \sin i'$$

However, BM = GL because the line GB has been drawn parallel to the normal, NN′, and so

$$AB \sin i = AG \sin i'$$

But AB and AG are the radii which we have made proportional to n and n' respectively, so that we can write the refraction equation $n \sin i = n' \sin i'$.

Thus, the construction provides a non-mathematical way of obtaining the direction of the refracted ray. This has been done in Figure 2.10 to obtain the refracted rays for light coming from an object B in glass of refractive index 1.5, and emerging through a plane surface into air of refractive index 1. The incident rays now use the larger circles and the refracted rays use the smaller circles, the

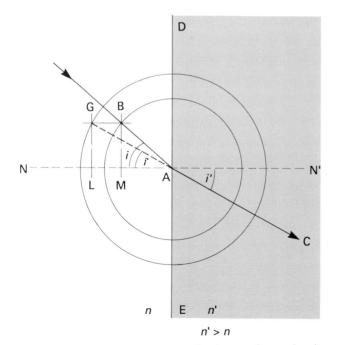

Figure 2.9 Graphical construction for refraction at a plane surface (for curved surfaces see Chapter 3)

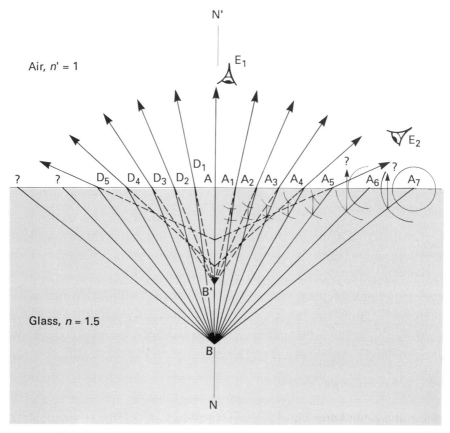

Figure 2.10 Light from an object point in glass refracted into air at a plane surface. Angles obtained by accurate graphical construction

construction line being drawn parallel to the normal as before. For accuracy, the circles should be drawn much larger than shown, keeping the same ratio between them. For clearness in the diagram, only the necessary part of each circle has been drawn. However, two features are very clear. First, the refractive rays do not appear to come from a single point as they did in the case of reflection at a plane surface. Secondly, for the larger angles of incidence, the construction lines do not cut the smaller circle and no refracted rays can be drawn. The only option open to the light is reflection. This is considered in Section 2.6.

The lack of a single image point for the refracted rays is an unfortunately common feature of optical systems. Perfect imaging, such as is obtained by reflection at a plane surface, is very rare. However, in Figure 2.10 it may be seen that the rays emerging through points A_1 and A_2 do appear to come from the single point B. The same is so for D_1 and D_2. Rays through D_3 and A_3 are not far away. It is the more extreme rays that show the larger error. Imperfect images such as this are known as **aberrated images** and the theory of aberrations is described in Chapter 7.

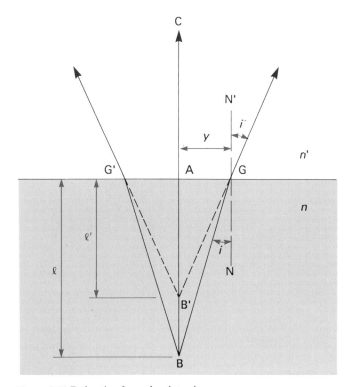

Figure 2.11 Refraction from glass into air

If these aberrations are very small, they can be ignored. One way to reduce the size of the aberrations is to reduce the aperture of the system. If rays between D_1 and A_1 only are used, the refractive image will be good. An eye, E_1, looking at point B will see a clear image at B′ because, in normal conditions, the pupil of the eye will restrict the rays used to a very small area. Even an eye at E_2 will see a reasonably clear image, but displaced upwards from B′. The upward displacement is related to the refractive index of the material and, as the index is greater for blue light compared with that for red light, the image seen will be displaced more for a blue object than for a red object. With a white point object a short spectrum may be seen, indicating chromatic aberration (described in Chapter 11).

The image at B′ seen by the eye at E_1 has a location with a simple relationship to the refractive index of the glass. In Figure 2.11, the rays forming the image at B′ are drawn with the distance G′G exaggerated. If we make the distance AG equal to y, we can see what happens to the mathematics as y becomes very small. In the figure, the distances AB and AB′ are given as l and l' respectively and, in the triangles ABG and ABG′, the hypotenuse lengths are $\sqrt{(l^2 + y^2)}$ and $\sqrt{(l'^2 + y^2)}$. Now, because the line BC is parallel to the normal NN′, the angle i is equal to the angle ABG and the angle i' is equal to the angle AB′G.

Thus

$$n \sin i = \frac{ny}{\sqrt{(l^2 + y^2)}} \tag{2.7a}$$

and

$$n' \sin i' = \frac{n'y}{\sqrt{(l'^2 + y^2)}} \tag{2.7b}$$

Using the equation of refraction (2.6), we have:

$$\frac{ny}{\sqrt{(l^2 + y^2)}} = \frac{n'y}{\sqrt{(l'^2 + y^2)}}$$

or

$$\frac{n}{n'} = \frac{\sqrt{(l^2 + y^2)}}{\sqrt{(l'^2 + y^2)}} = \sqrt{\left(\frac{l^2 + y^2}{l'^2 + y^2}\right)} \tag{2.8}$$

When y is zero, the right-hand side of this equation becomes l/l' and we say that it tends towards this value when y is very small compared to l and l'. Thus, for small y, Equation 2.8 becomes

$$\frac{n}{n'} = \frac{l}{l'} \tag{2.9a}$$

or

$$\frac{n}{l} = \frac{n'}{l'} \tag{2.9b}$$

This second form is the better equation to remember, and will be seen to be part of a larger equation for lens calculations. These equations show that, when objects in a higher index medium are viewed from a lower index medium, they appear closer to the surface than they really are; as is well known in the case of objects under water and bubbles in a block of glass. When the surrounding medium is air, $n' = 1$ and the glass block has a thickness l, the **apparent thickness** from Equation 2.9 is given by

$$l' = \frac{l}{n} \tag{2.9c}$$

The refractive index of a block of transparent material can be found by measuring the real thickness, l, and the apparent thickness using a microscope.

In Figure 2.11, the distance between B and B′ is found by subtracting l' from l. Using Equation 2.9a we have:

$$l - l' = l - \frac{n'l}{n}$$

$$= l\left(1 - \frac{n'}{n}\right) \tag{2.10}$$

This is important when viewing through plates of glass and this equation is used in Section 2.7.

2.6 Critical angle and total internal reflection

In Figure 2.10 it was seen that light rays from a point source inside glass were refracted away from the normal when emerging through a plane surface into air. It was also seen that beyond a certain angle of incidence no refracted ray could be

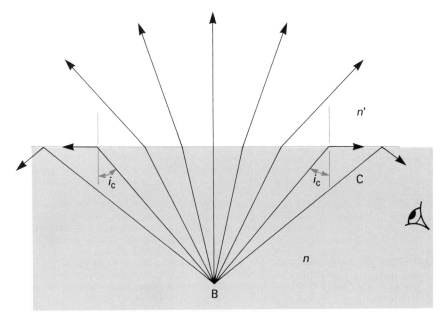

Figure 2.12 Critical angle and total internal reflection

found by the graphical method, because the construction line did not cut the smaller circles, as shown at A_6. The greatest possible angle of refraction is 90° and this occurs when the construction line just touches the smaller circle somewhere between A_5 and A_6. The angle of incidence at which this occurs is called the **critical angle**, i_c, and the value of this can be obtained from the refraction equation by putting i' equal to 90°, as shown in Figure 2.12.

Because $\sin 90° = 1$, the refraction equation becomes:

$$n \sin i_c = n' \sin 90° = n'$$

writing i_c for i. Therefore,

$$\sin i_c = \frac{n'}{n} \tag{2.11}$$

The sine of any angle is never greater than one and so this equation only gives values for i_c when n' is less than n. The refractive indices and, therefore, the critical angle between two media will depend on the colour of the light. Normally, the critical angle will be smaller for blue light than for red.

Table 2.1 gives the critical angle for yellow light of a few transparent substances when in contact with air and water.

When the angle of incidence is greater than the critical angle there is no way that the light can emerge from the higher-index material. Under these conditions all the light is reflected at the surface, obeying the law of reflection. This is called **total internal reflection** or sometimes **total reflection**. It must be remembered that some light is always reflected at the surface between two media of different refractive index and the amount reflected increases as the angle of incidence increases (see Figure 11.29). It is not an abrupt change in conditions at the critical angle,

Table 2.1 Critical angle and refractive index of some transparent substances

Material	Refractive index, n_d	Critical angle	
		In contact with air	In contact with water
Water	1.334	48.6	–
Polymethyl methacrylate*	1.49	42.2	63.5
Crown glass†	1.523	41.0	61.2
Light flint glass†	1.581	39.2	57.5
Dense flint glass†	1.706	35.9	51.4
Special flint glass†	1.952	30.8	43.1
Diamond	2.417	24.4	33.5

* Tradenames Perspex and Lucite
† Typical values

therefore, but the increase in the amount of reflected light appears quite sharp. The eye shown in Figure 2.12 looking at an extended source in the region of B would see a sharp line in the direction C on the one side of which the source would appear much brighter than on the other side. This can be used to measure the refractive index of materials.

The characteristic brilliancy of a diamond is due to its very high refractive index and, therefore, small critical angle. The material is cut in such a way that a very large part of the light entering it is totally reflected and passes out through the *table*, the large plane surface of the jewel.

Total internal reflection has many practical uses, particularly when replacing a silvered surface, the reflectivity of which is always less than total. Because the angle of incidence must be well away from normal incidence (to exceed the critical angle) most of these applications involve prisms and are described under Reflecting prisms (Section 2.10).

2.7 Refraction through parallel plates

It follows from the law of refraction that, if light is incident normally on a plane surface separating two different media, it continues in the new medium without deviation. At a second surface parallel to the first surface, the same conditions will apply and no change in direction will occur. Thus, light is undeviated when passing normally through a plane parallel plate of material of different index from the surrounding medium, such as a parallel plate of glass in air.

When the light is incident on such a plate at an angle, it will be deviated both on entering and on leaving the plate. The refraction equation can be used to find the directions. In Figure 2.13 light from a point source B is shown passing through the plate $D_1E_1D_2E_2$. While the ray BAFH passes through normally without deviation, the ray BA_1A_2C is deviated at both surfaces. If the refractive index of the plate is n_1' = n_2, that of the medium in front is n_1 and that of the medium behind is n_2'; then BA is the incident direction, i_1 is the angle of incidence and i_1' is the angle of refraction at the first surface. The angles of incidence and refraction at the second surface are i_2 and i_2'. We use the equation of refraction at each surface:

$$n_1 \sin i_1 = n_1' \sin i_1' \tag{2.12a}$$

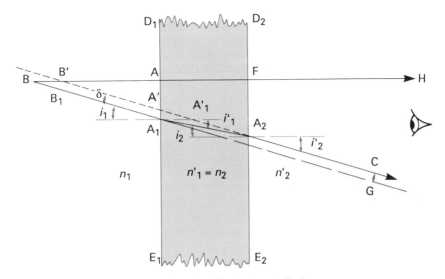

Figure 2.13 Refraction through a glass plate showing lateral displacement

$$n_2 \sin i_2 = n_2' \sin i_2' \tag{2.12b}$$

With the surfaces of the plate parallel with each other, we have, inside the glass,

$$i_1' = i_2 \tag{2.12c}$$

and we made n_1' and n_2 equal to the refractive index of the plate, so that

$$n_1' \sin i_1' = n_2 \sin i_2$$

and so, from Equations 2.12a and 2.12b, we have

$$n_1 \sin i_1 = n_2' \sin i_2' \tag{2.13}$$

When the media on each side of the plate are the same, as in the case of a plate of glass in air,

$$n_1 = n_2'$$

and so

$$i_1 = i_2'$$

Thus the light emerges *parallel to its original direction*. It has, however, been *laterally displaced* by the amount GC shown in Figure 2.13. The size of this displacement depends on the thickness of the plate, its refractive index, and the angle of incidence at the first surface. Because the light emerges parallel to its original direction, there will be no change in the apparent position of a *distant* object seen through the plate. The displacement effect for *near* objects is dealt with below.

When the light travels through a number of different parallel plates, as in Figure 2.14, the equation of refraction applies each time.

$$n_1 \sin i_1 = n_1' \sin i_1'$$
$$n_2 \sin i_2 = n_2' \sin i_2'$$
$$n_3 \sin i_3 = n_3' \sin i_3'$$
$$n_4 \sin i_4 = n_4' \sin i_4'$$

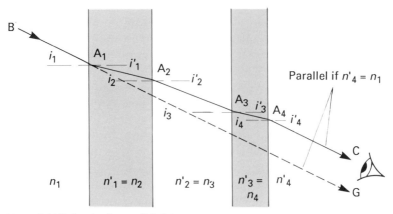

Figure 2.14 Refraction by parallel plates

If the surfaces are parallel, $i_1' = i_2$, $i_2' = i_3$, etc., and remembering that $n_1' = n_2$, $n_2' = n_3$, etc., we see from the equations above:

$$n_1 \sin i_1 = n_2 \sin i_2 = n_3 \sin i_3 = n_4 \sin i_4 = n_4' \sin i_4'$$

This shows that the final direction of the light is the same as if it had been refracted once at a single surface separating the first and last media. If the first and last media have the same refractive index, the final direction of the light is parallel to its original direction. There is, of course, a lateral displacement.

As stated above, this lateral displacement of the light causes an apparent lateral displacement of a *near* image seen by an eye looking along CA_4. To understand this better, the final direction has been traced back towards the object in Figure 2.13. This cuts the emerging light FH at B' and so, for an eye in the region CH, B' is where the object B appears to be. The *longitudinal* displacement distance, BB', is related to the apparent depth effect described in Section 2.5. If the ray BC had started at A rather than B, then the image of A_1 would be at A_1'. The distance A_1A_1', found using Equation 2.10, is:

$$A_1A_1' = t\left(1 - \frac{n_2'}{n_2}\right)$$

where t is the thickness of the plate and n_2, n_2' are the refractive indices as shown in Figure 2.13. So $n_2' = 1$ if the plate is in air.

The line $B'A'C$ is parallel to the line BA_1G and so the distance BB' is equal to A_1A_1' because they too are parallel, being aligned with normals to the plate surfaces. Thus, the longitudinal object to image shift due to a parallel plate of index n and thickness t is along a normal to the plate and given by:

$$\text{Image shift} = t\left(1 - \frac{1}{n}\right) \tag{2.14}$$

no matter where the object is behind the plate.

If this image is viewed along the ray A_2C it is seen at B' as shown in Figure 2.13. This, however, is the laterally displaced image because the plate is effectively tilted through an angle equal to i_1. If the plate were rotated to the normal to BA_1, the image of B would appear to be at B_1', where BB_1' is equal to BB' because the thickness and index of the plate remain constant. The lateral displacement due to a plate tilt of i_1 is, therefore, $B'B_1'$ and this is given *approximately* by

$$\delta = BB' \tan i_1 = t \left(1 - \frac{1}{n}\right) \tan i_1 \tag{2.15}$$

This approximation is accurate within 1% of the true value for values of i_1 up to nearly 45°. For larger angles, the error rapidly increases. However, when t is made small by using a thin plate, the apparent image position can be adjusted to very high accuracy by tilting the plate. Such a device is called a **parallel plate microcometer**. Two plates at an angle to each other were used by Helmholtz as a doubling device in his keratometer used for measuring the curvature of the cornea. An excellent description of the principle is given by Dr D. B. Henson in his book *Optometric Instrumentation* (Butterworths, London), p. 102.

2.8 Refracting prisms

In contrast to the study of parallel plates, we now consider refraction through a **prism** which is a plate of optical material where the two surfaces are *not* parallel. Refracting prisms always change the direction of the light even when the indices of the media on each side of the prism are the same. This change in direction or **deviation** of the light is usually the main purpose of a refracting prism. The angle between the surfaces of the prism may be very small or very large, as shown in Figure 2.15. Small angled prisms which deviate the light through small angles are mainly used in optometry, and are described in Section 2.9. Large-angled prisms trap the light by total internal reflection and so the third surface is used in reflecting prisms which are described in Section 2.10.

The mid-size prism in Figure 2.15 has refracting surfaces $A'C'C''A''$ and $A'E'E''A''$. The line of intersection, $A'A''$, of these two surfaces is the **refracting edge** or **apex** of the prism. Any section through the prism perpendicular to the edge, such as ACE, is a **principal section** and the angle CAE of such a section is called the refracting or apical angle, a.

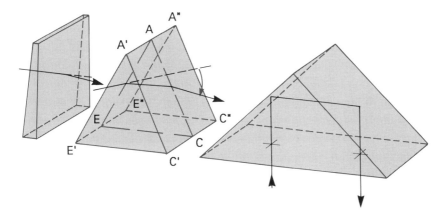

Figure 2.15 Prisms

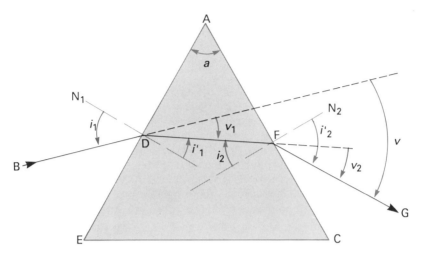

Figure 2.16 Refraction through a prism

A ray BD, in Figure 2.16, incident in the same plane as the principal section ACE will remain in the same plane after refraction and its path may be found using the equation of refraction, 1.10. We shall consider only prisms of refractive index n, surrounded by air. The equation of refraction becomes: at the first surface,

$$\sin i_1' = \frac{\sin i_1}{n} \tag{2.16a}$$

at the second surface,

$$\sin i_2' = n \sin i_2 \tag{2.16b}$$

In order to link these equations, we have, by adding the angles in triangle ADF,

$$(90 - i_1') + (90 - i_2) + a = 180°$$

or

$$a = i_1' + i_2 \tag{2.16c}$$

where a is the refracting angle of the prism.

The light is deviated through an angle v_1 at the first surface and v_2 at the second surface. From Figure 2.16,

$$v_1 = i_1 - i_1' \quad \text{and} \quad v_2 = i_2' - i_2$$

The total deviation, v, is given by

$$v = v_1 + v_2$$
$$= i_1 - i_1' + i_2' - i_2$$
$$= i_1 + i_2' - a \quad \text{(using Equation 2.16c)} \tag{2.16d}$$

It is clear from the figure that when the prism is of higher refractive index than the surrounding medium, as with a prism in air, the light is always deviated away from the refracting angle towards the base, CE. The amount of deviation, v, depends on the refractive index of the prism, its apical angle and the initial angle of incidence.

This value can be calculated using Equations 2.16. Figure 2.17 shows the paths of light rays incident at different angles through a prism of apical angle 60° and refractive index 1.51. The incident angles start at 15° to the normal and increase in 15° steps. It can be seen that the ray A travels into the prism but cannot emerge from the second surface due to total internal reflection. Ray B at 30° emerges as ray B′ at the angle shown. The other rays are similarly refracted up to the ray F which, at 90°, has the largest possible angle of incidence. The deviation of each ray is measured between the original directions shown by the short lines F_1, E_1, D_1, etc., and the emergent rays F′, E′, D′, etc.

The lines drawn in this diagram are merely mathematical calculations – the angles used satisfy Snell's law of refraction at each surface and the law is satisfied no matter whether the light travels from B to B′ or B′ to B. This is the principle of reversibility described in Section 1.11. Thus, Figure 2.17(b) can be drawn as the reverse of Figure 2.17(a), with the rays B′C′D′, etc., incident on the prism at the same angles at which they had emerged. The new emergent angles will be regularly spaced at the 15° intervals chosen for the first diagram. Because the angles of deviation are measured between the incident and emergent rays, they have the same values whichever way the light is going. It does not matter that the rays are at different *places* on the prism surfaces; it is the angles that matter.

Table 2.2 Deviation (all angles in degrees)

Ray	Angle of incidence	Angle of emergence	Angle of deviation	
F′–F	27.9	90	57.9	
B–B′	30	77.1	47.1	
E′–E	30.7	75	45.7	
D′–D	38.9	60	38.9	
C–C′	45	52.4	37.4	
				Minimum deviation
C′–C	52.4	45	37.4	
D–D′	60	38.9	38.9	
E–E′	75	30.7	45.7	
B′–B	77.1	30	47.1	
F–F′	90	27.9	57.9	

When all the rays are listed in order of angle of incidence, as in Table 2.2, it is seen that the angle of deviation starts off at a large value (57.9°) for the smallest angle of incidence and then reduces before increasing again back to the same large value for the largest angle of incidence. Thus, the largest deviations occur when the ray is either emerging or incident at grazing angle. The least deviation or **minimum angle of deviation** can be seen to occur somewhere between the incidence angles 45° and 52.4°. By doing many more ray calculations or by experiment, it can be verified that there is only one angle of incidence at which the deviation is a minimum. From this it can be deduced that the minimum angle of deviation occurs when the incident and emergent angles are equal. If this occurred when the angles of incidence and emergence were not equal, it would also occur when the light direction is reversed, giving a second minimum at a new angle of incidence. Thus, one minimum value can be obtained only if it occurs when the light passes symmetrically through the prism. This symmetry allows us to set:

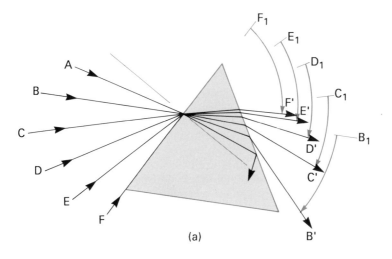

(a)

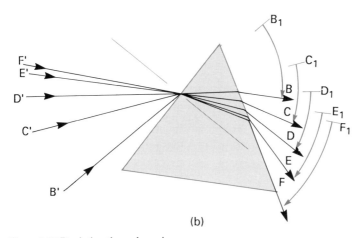

(b)

Figure 2.17 Deviation through a prism

$$i_1 = i_2' \quad \text{and} \quad i_1' = i_2$$

Thus, from Equation 2.16,

$$i_1' = \frac{a}{2} \tag{2.17a}$$

and, from Equation 2.16d,

$$i_1 = \frac{a + v_m}{2} \tag{2.17b}$$

where a is the apical angle of the prism and v_m is the angle of minimum deviation.

Using Equation 2.16a with Equations 2.17, we have

$$\sin\left(\frac{a}{2}\right) = \frac{\sin\left(\dfrac{a + v_m}{2}\right)}{n}$$

or

$$n = \frac{\sin\left(\dfrac{a + v_m}{2}\right)}{\sin\left(\dfrac{a}{2}\right)} \tag{2.18}$$

where n is the refractive index of the prism material.

If Equation 2.18 is used to calculate the angle of minimum deviation for a 60° prism of index 1.5, the value of v_m found is 37.2°. Equation 2.17b then gives i_1 at 48.6°, which is between 45° and 52.4° as predicted by the table.

The measurement of the minimum angle of deviation is important in practical science not only because it has one unique value but also because it hardly changes even if the angle of incidence is not precisely set at the symmetrical condition. Equation 2.18 shows that only the apical angle of the prism and the angle of minimum deviation need to be measured to find the refractive index of the prism material. Methods to measure refractive index are described in Chapter 11.

Prisms with apical angles of 60° are most common but other angles can be used. When the apical angle is reduced the deviation is reduced. This makes the calculation of n (using Equation 2.18) less accurate. When the apical angle is increased, the range of allowed angles of incidence is restricted. From Figure 2.16 it is seen that the minimum incidence angle (for grazing emergence) is 27.9° for a prism of 60° apical angle and 1.5 index. For an apical angle of 70° and the same refractive index, the minimum angle of incidence is 45.1°; and for an apical angle of 80° only light incident at grazing angle gets through the prism and emerges at grazing angle; for this prism the minimum and only deviation is 96.4°. Above this limiting apical angle, the prism cannot be used to refract light.

For prisms of higher refractive index, the limiting apical angle is reduced. At index 1.6 the angle is 77.4°; at index 1.8 it is 67.5°. Prisms made of most optical materials will not refract light if their apical angle is 90°. Such prisms can be used to reflect light if their third surface is polished (see Section 2.10).

The main purpose of the refracting prisms described in this section is the **dispersion** of the light passing through them either to measure the refractive index of the material of the prism as it changes with the wavelength of light or to separate white light into its spectrum so that a part of this can be selected and used for some other purpose. In both cases the incident light is in the form of a parallel beam, as described in Chapter 11. Prisms for looking through are described in the next section.

2.9 Ophthalmic prisms

When prisms are used for visual purposes, the refraction of the light produces aberrated images. Figure 2.18(a) shows light rays from a near object, B, reaching the prism at different angles of incidence. After two refractions, it is clear that the

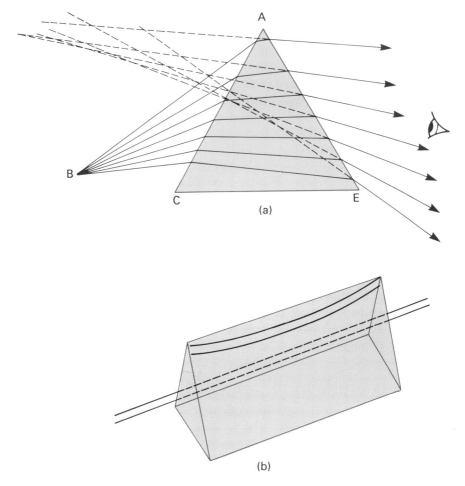

Figure 2.18 Viewing objects through a prism

deviations are also different and the emergent rays do not appear to come from a single point although an eye looking through the prism uses a very small range of rays, so that the image will look reasonably sharp. The best image will be seen when the used rays are near minimum deviation, because then the deviation is changing very little with angle of incidence.

The deviation produced by all prisms is towards the base, so the eye sees the image apparently displaced towards the apex. Unless the light from the object is monochromatic, that is of one colour only, the image will show the dispersion of the prism with the blue light deviated most and the red least. The white line shown in Figure 2.18(b) will therefore have a blue fringe at the top and a red fringe below when viewed through the prism.

When the eye looks at an extended line object like this, only one point on it is viewed through the principal section of the prism. Other parts of the line are viewed obliquely and, as the effective refracting angle in an oblique section of the prism is greater than the apical angle of the principal section, the deviation is

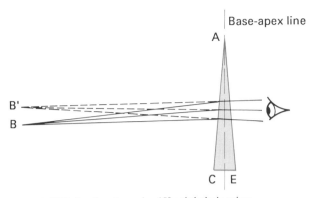

Figure 2.19 Refraction through a 10° ophthalmic prism

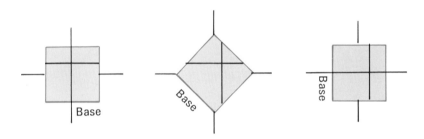

Figure 2.20 Appearance of crossline through an ophthalmic prism

greater for the other parts of the line. Thus the line in Figure 2.18(b) is distorted into a curved image as shown.

When prisms are used for the relief and correction of abnormal muscular conditions in the eyes, these three defects are present in the images seen. For this reason ophthalmic prisms have apical angles which rarely exceed 15° otherwise the images are noticeably blurred. Figure 2.19 shows a 10° prism with the image displaced in the direction of the base–apex line as shown. The appearance of this is shown in Figure 2.20.

This deviation may be measured in degrees or in units which give the apparent lateral displacement of an object at a distance of 1 m. This unit is the **prism dioptre**, Δ, and is the power of a prism which deviates the light passing through it so that an object at 1 m appears to be displaced by a distance of 1 cm. Thus, a prism of power 2.5 Δ will displace an object at 1 m by 2.5 cm and a prism of power $P\,\Delta$ will displace it by P cm. The angular deviation, v_p, of such a prism is given by the equation

$$\tan v_p = \frac{P}{100} \tag{2.19}$$

Angular deviations of all prisms are given by:

$$v = i_1 + i_2' - a \tag{2.16d}$$

but for prisms of small apical angle used almost normal to the incident light, the Equations 2.16a and 2.16b can be approximated to

$$i_1' = \frac{i_1}{n} \quad \text{and} \quad i_2'' = ni_2$$

so that Equation 2.16d becomes:

$$v = ni_1' + ni_2 - a$$

and, using Equation 2.16c, we find

$$v = n(a) - a = (n-1)a \tag{2.20}$$

Table 2.3

Apical angle (degrees)	Deviation (degrees)		
			Approximate (using Equation 2.20)
10	5.3	5.26	5.23
20	11.4	10.8	10.6
30	19.6	16.6	17.7

With all approximations, it is important to know what size of error they introduce. In Table 2.3, the angular deviation has been accurately calculated for two orientations of the prism (with the first surface and with the second surface normal to the light from the object) and then calculated using the approximate formula (Equation 2.20). A refractive index of 1.523, spectacle crown glass is assumed. It can be seen that the approximate value is generally smaller than the precise values and that the case where the second surface is normal to the incoming light (roughly minimum deviation condition) is the one best represented by the approximate calculation. Ophthalmic prisms are described in more detail in Chapter 7 of M. Jalie's book *The Principles of Ophthalmic Lenses* (Association of British Dispensing Opticians, London).

2.10 Reflecting prisms

The previous sections describe prisms that deviate light by refraction. This has the disadvantage that images are blurred and coloured unless the extent of the deviation is small. Deviation by reflection is better because the law of reflection is linear and there is no dispersion. The four prisms shown in Figure 2.21 deviate light by the amounts shown. The surfaces marked T must reflect by total internal reflection because they are also used as transmitting surfaces; those marked S must be coated with silver or aluminium. In all these prisms the angle between the reflecting surfaces is fixed and so the deviation is also fixed (Section 2.4).

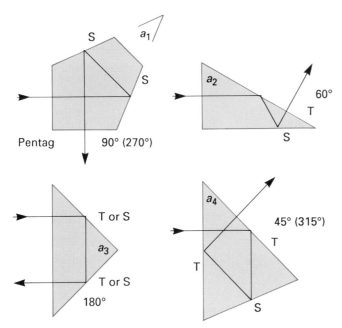

Figure 2.21 Prisms that deviate light by reflection: a_1, 45°; a_2, 60°; a_3, 90°; a_4, 45°

Often the angle over which these prisms can be used is very small due partly to their geometry and partly to the need to reflect and transmit on the appropriate sides of the critical angle. In many optical instruments, the field angles are sufficiently small and prisms like these are used to deviate the light and so bend the beam to the required path.

Sometimes it is required to rotate or invert the information in the beam rather than change its direction.

Image rotation is done by a combination of prisms which do not deviate the beam but reflect it an odd number of times. This means that the image is seen reverted compared to the object. When the prism combination is rotated about an axis parallel with the beam, the image rotates twice. For this reason the prism combinations shown in Figure 2.22 are sometimes called **half-speed rotators**. The **Dove prism** uses one reflection plus two refractions and is generally restricted to parallel beams and monochromatic (single-colour) light.

In the so-called **Abbe prism** the same thing is done with three reflections, but this takes up a lot of space when the prism rotates around the axis CC'. The **folded Abbe prism** system uses five reflections but is much more compact. Many other combinations have been invented (some many times over), each having different advantages and disadvantages.

Inverting prism systems also have many variants. The main requirement with these is an even number of reflections, but at least two of them must be arranged at right angles to the others. The most common is the **Porro prism** shown in Figure 2.23(a). This uses two 90° prisms to bring the light back to its original direction, but with the rays inverted vertically and horizontally. The emerging beam is displaced and the alternative arrangement shown in the figure gives a smaller displacement.

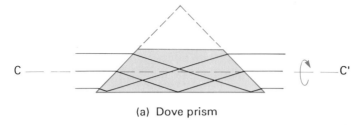

C

C'

(a) Dove prism

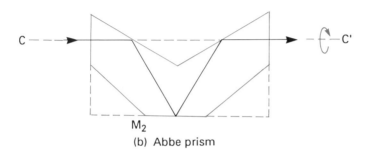

C

C'

M₂

(b) Abbe prism

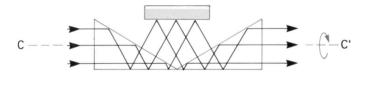

C

C'

(c) Folded Abbe system

Figure 2.22 Image rotators

   Porro prisms are used extensively in binoculars to bring the image the correct way up, and where the displacement can usefully allow bigger objective lenses (see Section 6.5). When a straight-through image inverter is needed without any displacement, an image rotator can sometimes be used with one of its reflecting surfaces converted to a **roof edge**. This is a reflecting prism with a 90° apical angle and its apical edge in the same plane as the axis of the light beam. Thus, the Abbe rotator shown in Figure 2.22 can be converted to an inverter by replacing M with a roof prism. This is the basis of the **Abbe roof prism** shown in Figure 2.23(b).

   More compact arrangements may be designed and these are used in the more expensive straight-through binoculars because the roof-edge prism must be very accurately made. The roof principle uses the $B'_{1.2}$ and $B'_{2.1}$ images in Figure 2.6. These images coincide only when the apical angle a is sufficiently precisely equal to 90°. A comprehensive survey of prisms is given by Dr D. W. Swift in *Optical Prisms* (Adam Hilger, Bristol).

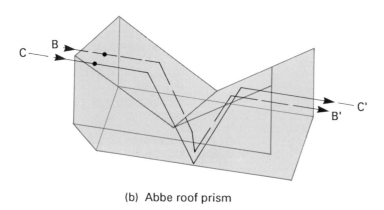

(a) Porro prism and variant

(b) Abbe roof prism

Figure 2.23 Inverting prism systems

Exercises

2.1 State the law of reflection. Show that the line joining an object and its image formed by a plane mirror is bisected at right angles by the mirror surface.
2.2 Describe, with a diagram, an experiment to prove that the image formed by a plane mirror is as far behind the mirror as the object is in front.
2.3 A small source of light is situated 50 cm from a plane mirror, 20 cm × 15 cm in size, and on a normal to its middle point. What will be the size of the patch of reflected light received on a screen 2 m from and parallel to the mirror?
2.4 A plane mirror 2 ft high is fixed on a wall of a room with its lower edge 3 ft above the floor. If a man, whose eyes are 6 ft above the floor, stands directly in line with the centre of the mirror and 3 ft from it, what will be the length of floor that he can see by reflection?

2.5   A sight-testing chart measuring 1.2 m × 0.8 m, the longer dimension being vertical, is to be viewed by reflection in a plane mirror. Find the smallest size of mirror that can be used if the chart is 3.25 m and the observer 2.75 m from the mirror. If the observer's eye is 1.2 m and the lower edge of the chart 1.5 m above the ground, what must be the height of the bottom edge of the mirror?

2.6   What will be the rotation of a mirror reflecting a spot of light on to a straight scale 1 m from the mirror if the spot of light moves through 5 cm?

2.7   It is required to turn a beam of light through 80° by means of mirrors; explain, giving diagrams, three ways in which this may be done. State the advantages and disadvantages of each method.

2.8   Two vertical plane mirrors are fixed at right angles to one another with their edges touching, and a person walks past the combination on a straight path at 45° to each mirror. Show how the image moves.

2.9   Two plane mirrors are placed 1 m apart with faces parallel. An object placed between them is situated 40 cm from one mirror; find the positions of the first six images formed by the mirrors.

2.10   Two plane mirrors are inclined at an angle of 60° to one another, and a small object is placed between them and 1 in from one mirror. Show on a diagram the position of each of the images and the path of the light by which an eye sees the last image formed.

2.11   A tank having a plane glass bottom 3 cm thick is filled with water to a depth of 5 cm. Trace the path of a narrow parallel beam of light, showing the angles at each surface, when the beam is (a) normal, (b) at an angle of incidence of 45° at the upper surface of the water. (Refractive index of the glass is 1.62.)

2.12   A narrow beam of light is incident on one surface of a plane parallel plate of glass (n = 1.62) 5 cm thick at an angle of 45° to the normal. Trace the path of the light through the plate and determine the lateral displacement of the beam on emergence.
       What will be the direction of the emergent light if the second surface of the glass plate is immersed in oil of refractive index 1.49?

2.13   An object held against a plate glass mirror of refractive index 1.54 appears to be 0.375 in from its image. What is the thickness of the mirror?

2.14   A 4 in cube of glass of refractive index 1.66 has a small object at its centre; show that, if a disc of opaque paper exceeding a certain diameter is pasted on each face, it will not be possible to see the object from anywhere outside the cube. Find the minimum diameter of disc required.

2.15   Find the critical angle for light passing from glass of refractive index 1.523 to (a) air, (b) water, (c) oil of refractive index 1.47.

2.16   A piece of glass (refractive index 1.5) 45 mm thick is placed against a picture. An observer views the picture from a position 500 mm from the *front* of the glass. Calculate the apparent distance of the picture from the observer.
       The picture and observer remain stationary while the piece of glass is repositioned 250 mm from the observer. Calculate the new apparent distance of the picture.

2.17   A prism of 30° refracting angle is made of glass having a refractive index of 1.6. Find the angles of emergence and deviation for each of the following cases:
       (a) Angle of incidence at first face = 24° 28'
       (b) Angle of incidence at first face = 53° 8'
       (c) Incident light normal to first face

2.18   A parallel beam of yellow light is incident at an angle of 40° to the normal to the first surface of a 60° prism of refractive index 1.53. Trace the path of the light through the prism and find the deviation produced on the beam. What will be the minimum deviation produced by this prism?

2.19   A prism has a refracting angle of 60° and a refractive index of 1.5. Plot a graph showing the relation of the total deviation to the angle of incidence of the light on the first surface and find the angle of incidence for which the deviation is a minimum. Show that this angle of incidence is equal to the angle of emergence.

2.20 A thin prism is tested by observing through it a scale 80 cm away, the scale divisions being 1 cm apart. The deviation is found to be 1.75 scale divisions. Express the deviation in prism dioptres.

   A ray of light falls normally on one face of this prism, is reflected by the second face, and emerges from the first face making an acute angle of 7.5° with its original direction. Find the apical angle and refractive index of the prism.

2.21 A convergent pencil of light is intercepted before reaching its focus by a right-angled isosceles prism. The angle between the edges of the pencil is 18° and the central ray meets one of the short faces of the prism normally. What must be the minimum refractive index of the prism in order that the whole of the pencil may be totally reflected at the hypotenuse face?

2.22 A pencil of parallel light is incident on the face PQ of a right-angled isosceles prism (right angle at P) in a direction parallel to the hypotenuse face QR. The width of the pencil is equal to one-quarter of the length of face QR.

   The prism is such as *just* to permit the whole width of the pencil to be reflected at face QR and then to emerge into air from face PR. Find the refractive index of the prism.

Chapter 3

Refraction and reflection at spherical surfaces

3.1 Spherical curvature and its measurement

A plane surface may be described as a surface where the normals at all points on it are parallel to each other. A spherical surface is one where all the normals pass through a single point. A tennis ball, a soap bubble, and the earth are all examples of spheres and each has a spherical surface. A person standing upright on the earth's surface represents the normal to the surface at that point. If a line is drawn from this person's head, through the feet and downwards into the earth, it will eventually meet, at the centre of the earth, lines drawn through other people standing at other points on the earth's surface. Clearly, mountains and valleys represent local differences from a true spherical surface and, in fact, the earth is flatter at the north and south poles than if it were a perfect sphere.

In optics nearly perfect spherical surfaces are very common because two surfaces rubbed together in a general way wear each other down to form two spherical surfaces, one like a ball and the other like a cup that fits the ball. This is the arrangement that allows them to move most easily and our bones have this form at hip joints and elbows, etc., where movement is needed.

The section through any spherical surface is a circle, or part of a circle, as shown in Figure 3.1. Each circle has a **centre of curvature**, C; and **radius of curvature**, r, defined as the distance from the surface to the centre of curvature. On each surface two normals are shown separated by the same distance along the surface. Although the distance between the normals is the same, the angle between the normals is very different.

We define the **curvature** of the surface as *the angle through which the surface turns in a unit length of arc*. In each of the cases shown in Figure 3.1, an object moving round the circle from P_1 to P_2 starts in the direction P_1D and ends in the direction DP_2 having turned through the angle a. Because each surface direction is at right angles to the normal at that point, the angle turned through is the same as the angle between the normals. If this angle is measured in radians* its value is

---

* *Radian measure of angles*

The size of any angle can be defined as the arc length, p, divided by the radius of that arc, r. Any radius can be used because the ratio p/r remains the same. These ratios are in dimensionless units called radians and 1 rad is the angle that gives p equal to r. In degrees this is about 57°. A right angle is $\pi/2$ rad and a full 360° is 2π rad, the circumference ratio of a circle.

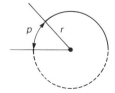

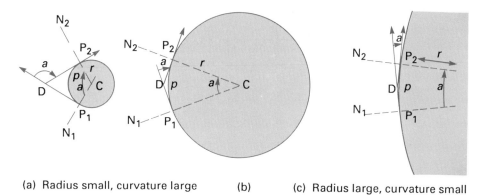

(a) Radius small, curvature large   (b)   (c) Radius large, curvature small

Figure 3.1 Curvature and radius of curvature

given by

$$a \text{ (in radians)} = \frac{p}{r}$$

where p is the arc length and r is the radius of the circle.
   Curvature is the value of a when p is one, so

$$\text{Curvature } (R) = \frac{a}{p} = \frac{1}{r} \tag{3.1}$$

Thus, the curvature of a surface is equal to the reciprocal of its radius and the common units of curvature are reciprocal metres (m^{-1}). To obtain the correct value for R in reciprocal metres the radius, r, must be measured in metres.
   Most optical surfaces are small parts of complete spheres as in Figure 3.1(c). To measure the curvature of an existing optical surface, a method is needed which does not require the position of the centre to be known. Optical methods, suitable for polished surfaces, are described in Chapter 12. Very high accuracy can be obtained, however, by the precise measurement of two dimensions. Figure 3.2 shows a circle, with centre C, representing a spherical surface. The points D, A and E lie on this circle and, most importantly, A is precisely mid-way between D and E. When the line is drawn between D and E (a chord of the circle) and the radius is drawn between C and A, these intersect at right angles at N as shown. The distance NA is called the **sag**, s, of the surface. The distances CD, CA and CE are all equal to the radius, r, and the half-chord NE is y.
   In the triangle CNE,

$$r^2 = y^2 + (r - s)^2$$
$$= y^2 + r^2 - 2rs + s^2$$

From this,

$$r = \frac{y^2}{2s} + \frac{s}{2} \tag{3.2}$$

If y and s are carefully measured, the radius of curvature, r, may be calculated. Usually s is very small and must be measured very precisely. For example, if r is 1 m

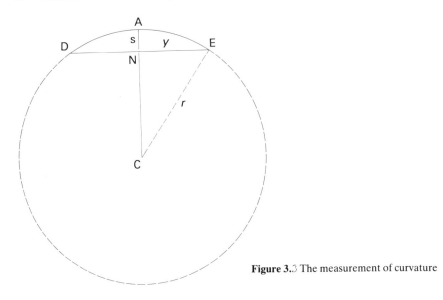

Figure 3.3 The measurement of curvature

and the whole chord is 8 cm, then y is 0.04 m and $y^2 = 0.0016\,\mathrm{m}^2$ so that s is 0.0008 m which is less than 1 mm. The term $s/2$ is, therefore, less than 0.5 mm and has been ignored in this calculation. If y is less than 5% of r, as is usually the case, then $s/2$ is less than 0.1% of r and the approximate forms of 3.02 are sufficient:

$$r = \frac{y^2}{2s} \quad \text{(approx.)} \tag{3.3a}$$

or

$$R\,\text{(curvature)} = \frac{2s}{y^2} \quad \text{(approx.)} \tag{3.3b}$$

Both y and s must be measured in metres if R is to be obtained correctly in reciprocal metres.

Instruments that measure curvature by measuring y and s are called **spherometers**. Old versions (Figure 3.3) have three outer legs so that, when they are on the surface to be measured, the central moving leg is approaching the surface along a normal. The value of y is the distance from the central leg to any of the outer three. The central leg has a fine-pitch thread by which it can be raised or lowered until it just touches the surface. Judging this is very difficult, the measurement of s is very crude and the pointed legs can easily scratch an optical surface.

If the accuracy required is not very high, it is sufficient to use two outer legs and a spring-loaded central leg which reads directly on to a dial gauge, all the legs having some non-scratch polymer material at their tips. This is the design used for the **optician's lens measure** or **lens clock**, shown in Figure 3.4. These are often calibrated directly in the power of lens surfaces and the normal tolerance for (low-power) spectacle lenses requires a curvature measurement accurate to $\pm 0.11\,\mathrm{m}^{-1}$ as shown in Section 3.5. If the outer legs are 40 mm apart $(2y)$, Equation 3.3b may be used to calculate the accuracy needed for the value of s.

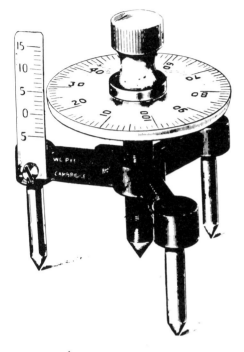

Figure 3.3 Spherometer – old style

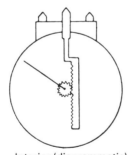

Front   Interior (diagrammatic)   **Figure 3.4** Optician's lens measure

$$S_{\text{Tolerance}} = \frac{y^2 R_{\text{Tolerance}}}{2}$$

$$= \frac{0.02^2 \times (\pm 0.11)}{2}$$

$$= \pm 0.000\,022 \text{ metres}$$

Thus, even with this relatively slack tolerance, s has to be measured to within 22 μm.

In the manufacture of precision optical elements, the quality and accuracy of their curvatures are measured using interference between the elements and accurate test plates (Section 12.6). The curvatures of the test plates are measured on highly accurate spherometers. The sag is measured to 1 μm and the value of y

kept as large as possible by having a range of supports with different diameters. For any given element the largest allowable support is used, although sometimes inaccuracies occur because the element sags under its own weight! The supporting structure may be a precision ground ring with its inner and outer diameters carefully measured, or a precision ground groove in which a non-contacting spacer holds three small steel balls of precisely measured diameter. With the heavier test plates the weight on each ball is often sufficient to deform them so that the continuous ring is now the preferred support.

The measurement of the sag is made using a probe which rises to touch the surface along a line precisely at the centre of the supporting ring. The actual measurement is made in two stages. With an accurate optical flat on the support the probe height is recorded. The flat is replaced by the test plate and the new probe height recorded. The difference gives the sag.

Figure 3.5 shows a typical instrument where the central probe is raised by a counter-weight. The accuracy of contact may be assessed by looking through the test plate at the interference fringes formed between the surface and the probe. As shown, the value of y for concave surfaces is different from the value for convex surfaces. The instrument and the test plate must be at a known temperature. With these precautions a curvature accuracy of 0.05% can be obtained.

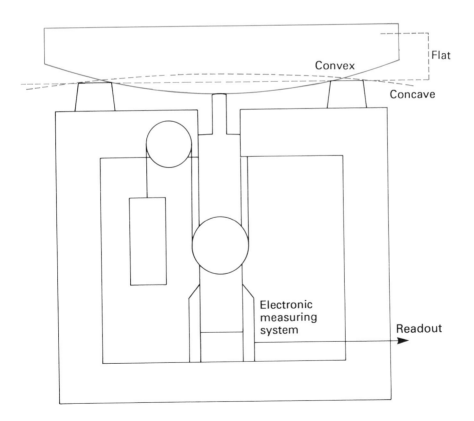

Figure 3.5 Precision spherometer

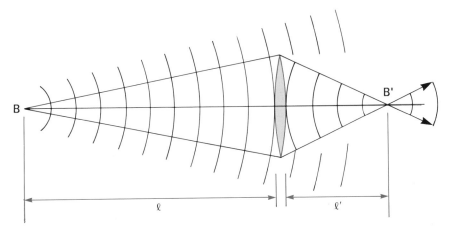

Figure 3.6 Curvature of wavefronts

3.2 Curvature of wavefronts

Optical elements such as lenses have an effect on light which changes the curvature of the wavefront as well as changing the direction of individual rays. Figure 3.6 shows a simple lens forming an image B′ of an object B. The diverging light from B arrives at the lens as a wavefront having one curvature and the lens changes this to a different curvature which converges to B′. The curvature of wavefronts can be stated in the same way as the curvature of spherical surfaces; the reciprocal of the distance to the centre of curvature.

In the case of refraction at a plane surface we may use the mathematics of curvature to find the effect on the light. In Figure 3.7 DAE is a plane surface between two different media of refractive index n and n'. Light is diverging from the object point B and DME is the section of the wavefront if the light had continued to travel in the first medium. Assuming that n' is less than n, the actual

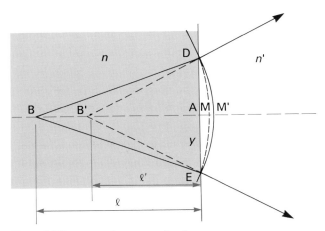

Figure 3.7 Curvature change on refraction

wavefront in the new medium is DM'E where AM' is greater than AM because the light is now travelling faster. The curvature of the wavefront has increased because the light at D and E has only just reached the new medium and the new wavefront is still coincident, at these points, with the original wavefront. Using Equation 3.1, the curvatures of the wavefronts are:

$$R = \frac{1}{l} \quad \text{and} \quad R' = \frac{1}{l'} \tag{3.4}$$

The light travels the distance AM' in medium n' in the same time as it would have travelled AM in medium n. From the definition of refractive index (Section 1.10) we have:

$$n\text{AM} = n'\text{AM}'$$

or

$$ns = n's'$$

where s and s' are the respective sag values of the wavefronts. Using the approximate formula, 3.3, on both sides we obtain

$$\frac{nRy^2}{2} = \frac{n'R'y^2}{2}$$

When Equations 3.4 are put into this, we have

$$\frac{n}{l} = \frac{n'}{l'} \tag{3.5}$$

which is the same as Equation 2.9.

This equation can also be written, very simply, as

$$L = L' \quad \text{(for plane surfaces)} \tag{3.6}$$

where L ($= n/l$) and L' ($= n'/l'$) are the **vergences** of the incidence and refracted light. When the light is travelling in air so that the refractive index is equal to one, its vergence is also equal to the curvature of the wavefront. Equation 3.6 shows that plane surfaces do not cause a change in the vergence of the light. We will see later in this chapter that curved surfaces and lenses do cause a change in vergence and their effect on light can be calculated most simply by using the concept of vergence. The unit of vergence is the **dioptre**. The vergences L and L' above are measured in dioptres when the values of l and l' are given in metres. Because lenses and curved optical surfaces cause a change in vergence, we can measure their power to do this in units of dioptres.

The reflection of light at a plane surface can also be described by Equation 3.6. It can be seen in Figure 3.8 that the reflected wavefront has the same curvature as the incident wavefront would have had because AM' is equal to AM, the light having the same velocity after reflection as before but in the reverse direction. We show this reversal of direction by using a negative value of refractive index to describe reflection. Thus, a plane mirror in air will have $n = 1$ and $n' = -1$. Equation 3.5 will then become:

$$\frac{1}{l} = \frac{-1}{l'}$$

which shows that $1/l$ and $1/l'$ are equal and opposite. The usefulness of this

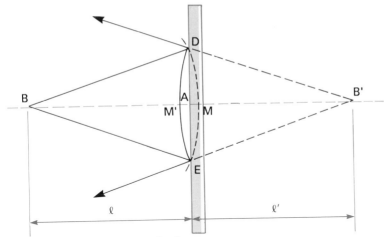

Figure 3.8 Curvature change on reflection

approach is that Equation 3.6 and all other equations using vergence remain the same for refraction and reflection. Very careful use of negative signs is needed and a systematic approach to this is explained in Section 3.4.

3.3 Graphical constructions

When a light ray strikes a curved surface it is refracted according to the law of refraction. The normal to a spherical surface at the point of incidence is the line joining that point to the centre of curvature. The construction used in Section 2.7 can be used at each point on the surface, but a general construction put forward by Thomas Young in 1807 gives the refracted ray for any ray incident on the surface. In Figure 3.9 the spherical surface DAE separates a medium or low index on the

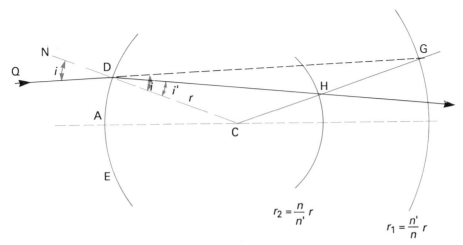

$$r_2 = \frac{n}{n'} r$$

$$r_1 = \frac{n'}{n} r$$

Figure 3.9 Young's construction

left from one of higher index on the right; n' is greater than n. For the construction, two circles are drawn with radii r_1 and r_2 related to the radius r of the surface by:

$$r_1 = \frac{n'}{n} r \quad \text{and} \quad r_2 = \frac{n}{n'} r$$

The ray QD is incident on the surface at D and, when extended, cuts the circle of radius r_1 at G. A line is drawn from G to C which cuts the other circle at H. DH is the line of the refracted ray. This is because we have made triangles CDG and CHD into similar triangles even though one is reverted with respect to the other. The distance CD is r, CG is r_1 and CH is r_2. By the equations above, we have

$$\frac{CG}{CD} = \frac{CD}{CH} = \frac{n'}{n}$$

which, together with the angle DCH being common to both triangles, is the condition for similar triangles.

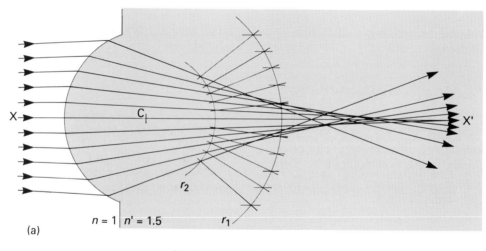

(a)

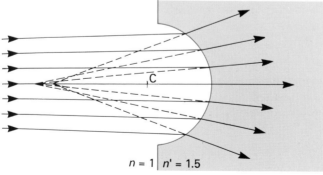

(b)

Figure 3.10 Refraction at a near-hemispherical surface

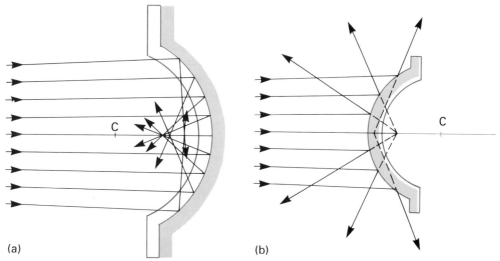

Figure 3.11 Reflection at a near-hemispherical surface

Therefore, the angle CHD is equal to i and, using the sine rule in triangle CHD, we have:

$$\frac{r_2}{\sin i'} = \frac{r}{\sin i} \quad \text{or} \quad \frac{\frac{n}{n'}r}{\sin i'} = \frac{r}{\sin i}$$

and so $n \sin i = n' \sin i'$, the law of refraction.

This construction is used in Figure 3.10 for a number of rays coming from a point object, on the axis XX' but some distance away, and incident on near-hemispherical surfaces separating air from glass of refractive index n' equal to 1.5. It can be seen that all these rays are refracted towards the axis but the more *peripheral* or *marginal* rays are not deviated to the same point as rays closer to the axis. No sharp image is formed.

The same construction cannot be used in the case of reflection because we cannot draw circles of negative radius to represent the $-n$ of reflection. However, the simpler law of reflection where the angles of incidence and reflection are equal may be used directly by drawing the normals and making the angles equal as in Figure 3.11.

The same rays from an axial object at the same distance as in Figure 3.11 are incident on a near-hemispherical reflecting surface having its centre at C. It can be seen that the peripheral rays are again deviated too much to form a good image even though we are using reflection rather than refraction. This lack of perfect imagery was discovered first with spherical mirrors. Because plane mirrors were known to give perfect images, Roger Bacon (1215–1294) called the defect **spherical aberration**. It is now known that other aberrations affect lenses and mirrors and these are described in Chapter 7.

The surface shown in Figure 3.10(a), which has its centre of curvature buried inside the material, is called a **convex** surface. Figure 3.11(b) shows a convex mirror with its centre of curvature behind the reflecting surface. The surfaces shown in Figures 3.10(b) and 3.11(a) are called **concave** surfaces.

3.4 Paraxial optics and the sign convention

Modern optical design has evolved methods for correcting the aberrations referred to above by using arrangements of lenses as described in Chapter 14. In particular, spherical aberration can be corrected by doublet lenses or surfaces with non-spherical curves as shown in Chapter 8. However, the simplest cure for spherical aberration is to restrict the aperture of the curved surface so that only rays near the axis are forming the image. This restriction is known as the **paraxial region** and these rays are known as **paraxial rays**. For such rays the angles of incidence and refraction will be small. As can be seen from Figure 2.8, the difference between the sine values and a straight line is only significant for large values of angle. When the angles are measured in radians rather than in degrees (see the footnote in Section 3.1), the actual value of the sine is found to be very close to the angle value in radians. Indeed the angle value in radians is between the sine and the tangent values for that angle, as Table 3.1 shows.

Table 3.1

Angle (degrees)	Sine value	Angle (rad)	Tangent value	Error (%)
0.3	0.005 236	0.005 236	0.005 236	<0.001
1.0	0.017 452	0.017 453	0.017 455	<0.01
3.0	0.052 336	0.052 360	0.052 408	<0.1
10.0	0.173 648	0.174 533	0.176 327	<1
30.0	0.500 000	0.523 599	0.577 350	<10

It is therefore possible, for paraxial rays, to calculate the positions of images and the directions of rays using the *approximate* expressions:

$$\sin i = i \quad \text{and} \quad \sin i' = i' \tag{3.7}$$

without incurring too much error. This is called the **paraxial approximation** and was developed mainly by Johann Karl Friedrich Gauss (1777–1855). Paraxial optics is sometimes known as **Gaussian optics**. For reasons developed in Chapter 7, it is also called **first-order optics**.

The paraxial approximation also includes the use of the approximate Equation 3.3 for relating curvature and sag. Unless otherwise stated, the paraxial approximation will be used from here on until the end of Chapter 6. It assumes that all object points are on or near the axis, that the apertures of all optical surfaces are small and that point images are formed of point objects. This is not as restricting as might be thought because the results obtained are found to apply for large apertures and objects (and images) away from the axis when the aberrations are corrected by the methods described in Chapter 14.

In order to calculate the positions of objects and images, it is necessary to adopt a standard way of measuring distances and angles. It is important to distinguish between objects and images which are in front of the lens or mirror from those which are behind. This is done by the use of negative and positive signs. The standard convention used in this book is described below, referring to Figure 3.12 in which a spherical refracting surface, S, is receiving light from an object O. In the

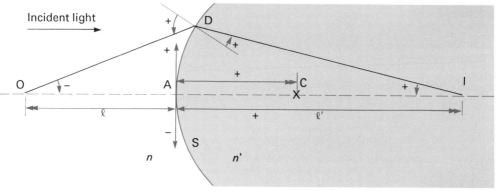

Figure 3.12 Sign convention

paraxial approximation, the image will be formed somewhere on the line through O and C, the centre of curvature of the surface. This line is called the **optical axis**. This axis is a normal to the surface, and the point at which it passes through the surface is called the **pole** or **vertex**, A, of the surface. With a single spherical surface like this, an axis can be drawn for any point object. As soon as *two* spherical surfaces are used, as in a lens, the optical axis is the line joining the two centres of curvature. In most optical systems, the surfaces have a common axis and the system is **centred** and **symmetrical** about this axis.

The following definitions apply to single surfaces and symmetrical systems.

Distances along the axis
All distances are measured *from* the lens or mirror. Those in the same direction as the incident light have positive values. Those in the direction against the incident light have negative values. Thus, in Figure 3.12, the object distance *from* the surface S *to* the point O is against the direction of the incident light and therefore has a negative value. The radius of curvature of the surface, measured *from* S to C has a positive value. In most diagrams in this book, the incident light will be travelling from left to right.

Distances perpendicular to the axis
All distances are measured *from* the axis. Those above have positive values and those below have negative values. Thus, the ray in Figure 3.12 strikes the surface S above the axis at a height having a positive value.

Angles
Angles measured anticlockwise have positive values; angles measured clockwise have negative values:

Slope of ray
This is defined as the acute angle between the direction of the ray *going to* the direction of the axis. Thus, the slope of the incident ray leaving O has a negative value.

Angles of incidence, refraction and reflection
These angles are measured *from* the normal *to* the ray. Thus, the angle of incidence in Figure 3.12 has a positive value.

An arrow head or extra arrow head shown on a dimension line for a distance or angle indicates the direction in which the measurement is made. The sign convention is fundamentally simple but great care must be taken and students often find it confusing. There is a strong temptation to look at Figure 3.12 and consider *l* as 'being negative' and to insert −*l* into the formulae of paraxial optics. *This is wrong and leads to incorrect answers.* In Figure 3.12, the distance *l* equals a negative value, say −50 mm. The minus sign is part of the value and not part of *l*. Thus −30 mm would be a different value for *l* and +50 mm would just be another different value for *l*. Whatever the value of *l*, it is put into the formula without changing any of the signs of the formula. Thus, if *l* = −50 mm and *f* = +30 mm, the formula

$$\frac{1}{l'} = \frac{1}{l} + \frac{1}{f}$$

is used by writing

$$\frac{1}{l'} = \frac{1}{(-50)} + \frac{1}{(+30)}$$

and calculating from there on.

3.5 Paraxial refraction at spherical surfaces

3.5.1 Change in ray path

Figure 3.13 shows a refracting surface similar to that in Figure 3.12, but the incident ray is now directed towards the axis at a point beyond the surface. The effect of this change is that all the distances and angles now have *positive values*. These

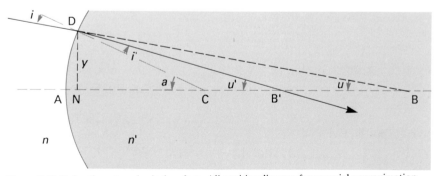

Figure 3.13 Refraction at a spherical surface. All-positive diagram for paraxial approximation

'all-positive' diagrams together with the paraxial approximation make the mathematics easier. To make the diagrams clear, the height y (and all the angles) are made *much bigger than the limits of the paraxial region*.

Using the law of refraction in Figure 3.13, we have:

$$n \sin i = n' \sin i'$$

and, in the paraxial approximation,

$$ni = n'i' \tag{3.8}$$

Again, in the diagram,

$$\tan u = \frac{y}{AB}$$

but we approximate to

$$u = \frac{y}{AB}$$

as AB is greater than NB and u is always smaller than tan u. This is not a bad approximation. In the same way,

$$u' = \frac{y}{AB'} \quad \text{and} \quad a = \frac{y}{AC}$$

In triangles DNC and DNB:

$$i = (90 - u) - (90 - a)$$
$$= a - u$$

and in triangles DNC and DNB':

$$i' = (90 - u') - (90 - a)$$
$$= a - u'$$

From the approximate law of refraction given above, we have:

$$n(a - u) = n'(a - u')$$

which gives:

$$n'u' = nu + (n' - n)a$$

Substituting the above expressions for u, u' and a, and dividing by y, we get:

$$\frac{n'}{AB'} = \frac{n}{AB} + \frac{n' - n}{AC}$$

As AB' = l'; AB = l and AC = r, we have:

$$\frac{n}{l'} = \frac{n}{l} + \frac{n' - n}{r} \tag{3.9}$$

3.5.2 Change in vergence

Equation 3.9 can also be obtained using the ideas on vergence developed in Section 3.2. Once again, we use an all-positive diagram (Figure 3.14). Light, converging

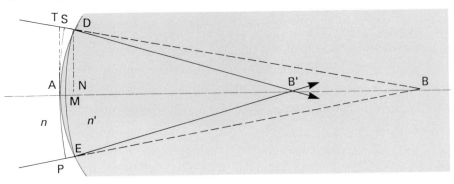

Figure 3.14 Refraction at a spherical surface. All-positive diagram for paraxial approximation

towards a point B, meets a spherical surface DAE separating media of refractive indices n and n'. The incident wavefront, SAP, is just meeting the surface at A. If n' is greater than n, the centre of this wavefront now begins to travel more slowly than the outer parts at S and P. By the time the outer parts reach D and E the centre has travelled the distance from A to M. The distance AM is less than SD because n' is greater than n. From the definition of refractive index (Section 1.10), we have:

$$nSD = n'AM \tag{3.10}$$

The original wavefront is spherical with its centre of curvature at B. In the paraxial region, it can be assumed that the refracted wavefront is also spherical with its centre of curvature at B'. In Figure 3.14, the sizes of the wavefronts are drawn much larger than the paraxial region, so that distances AM, SD, etc., can be seen clearly. The spherical wavefront, DME, has a sag, MN, as shown in Figure 3.2. The sag of the original wavefront, SAP, is TS; instead of drawing from S to P a line perpendicular to the axis, another perpendicular line has been drawn from A to T. The sag of the actual refracting surface is AN. In the paraxial region, TD is virtually parallel with the axis and is also equal to the sag of the refracting surface.

Developing Equation 3.10, we have:

$$n(TD - TS) = n'(AN - MN)$$

and so the new sag associated with the refracted light is given by:

$$n'(MN) = nTS + n'AN - nTD$$

All these distances are sag values and, using Equation 3.3, we can write

$$n'\left(\frac{y^2}{2l'}\right) = n\left(\frac{y^2}{2l}\right) + (n' - n)\left(\frac{y^2}{2r}\right)$$

which simplifies to the same equation as 3.9:

$$\frac{n'}{l'} = \frac{n}{l} + \frac{(n' - n)}{r} \tag{3.11}$$

where l' is the image distance, l the object distance and are therefore the radii of curvature of the refracted and incident wavefronts respectively, while r is the radius of curvature of the surface.

As n'/l' and n/l are the vergences defined for Equation 3.6, Equation 3.11 can be written:

$$L' = L + \frac{n' - n}{r}$$

$$= L + (n' - n)R \tag{3.12}$$

where R is the curvature of the refracting surface. The term $(n' - n)R$ is the amount by which L has to be changed to give L'. This power to change the vergence of the incident light is variously called the focal power or dioptric power, F.

In this case we shall use the term **surface power**, reserving other terms for lenses and systems. We define the surface power, F, as

$$F = (n' - n)R \tag{3.13}$$

This power is thus a constant for a given surface having a particular curvature and separating two media of given refractive index. Such a surface will always produce the same total change in vergence no matter what value is chosen for L, the incident vergence. Surface power is measured in dioptres and will be positive for a surface that tends to produce convergent light and negative for a surface tending to produce divergence.

A *convex* refracting surface of an optical medium of index greater than its surroundings will be a positive surface producing convergence. With reflection a *concave* surface produces convergence and is, therefore, positive.

When Equations 3.13 and 3.12 are put together, we obtain:

$$L' = L + F \tag{3.14}$$

This is the **fundamental paraxial equation** and can be applied to lenses and mirrors as well as single refracting surfaces.

The power of a surface remains the same whatever the direction of the light. In air ($n = 1$), the power of a spherical surface on glass of refractive index n_g is given by:

$$F = (n_g - 1)R \tag{3.15}$$

The power of a refractive surface as given by Equation 3.15 depends on the refractive index of the glass and the curvature. The optician's lens measure described in Section 3.1 assumes that the lens is made of ophthalmic crown glass of index 1.523. In recent years this has become less likely as more and more lenses are made of organic polymers such as CR39 (1.504) and polycarbonate (1.585). High index glasses (1.7 and 1.8) have also become more common.

The accepted tolerance for low-power spectacle lenses is $\pm 0.09\,\text{D}$ and, for ophthalmic crown glass, Equation 3.15 shows that this value gives a curvature tolerance of:

$$R_{\text{Tolerance}} = \frac{F_{\text{Tolerance}}}{(n_g - 1)}$$

$$= \frac{\pm 0.09\ \text{D}}{0.523}$$

$$= \pm 0.17\,\text{m}^{-1}$$

This is the value used in the analysis in Section 3.1.

3.6 Focal properties of spherical refracting surfaces

For a spherical refracting surface it follows from Equation 3.11 that for any object point B there is *one* corresponding image point B′. The distance of B′ from the vertex of the surface can be calculated using this equation and inserting the values for refractive index, object distance and radius of curvature. As the path of the light is reversible, the positions of the object and image are interchangeable. Thus, in Figure 3.14, light from B′ *in the reverse direction* will, on passing through the surface, appear to come from B. In this case B′ would be a real object and B a virtual image. In the case of Figure 3.14 *as drawn*, the object B is a virtual object and B′ a real image which gives the all-positive situation.

Pairs of point objects and images such as B and B′ are known as **conjugate points** or **conjugate foci**. Two of these pairs of conjugate points are of special importance. When the object is at an infinite distance from the surface the incident wavefront is plane and the rays are parallel. The vergence of this beam is zero. The light refracted at a surface of positive power will be converging to a real image at the point F′ in Figure 3.15. In the case of a surface of negative power the refracted light will appear to be coming from a virtual image at F′ in Figure 3.16. *In both cases*, this image, the conjugate of an infinitely distant object, is called the **second principal focus**, F′, of the surface and its distance *from* the surface is called the **second focal length**, *f*′.

Because *L* is zero under these conditions and *l*′ is redefined as *f*′, we have, from Equation 3.15,

$$L' = \frac{n'}{f'} = F$$

With a surface of positive power *f*′ has a positive value, and with a surface of negative power *f*′ has a negative value.

The other important condition is when the image is at an infinite distance. Under these conditions *L*′ is zero and the location at which the object must be placed is called the **first principal focus**, F. Figure 3.15 shows the situation with a positive surface where the real object must be located in front of the surface. In the case of a

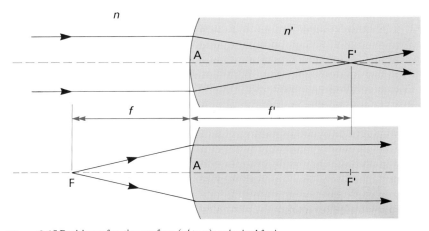

Figure 3.15 Positive refracting surface ($n' > n$): principal foci

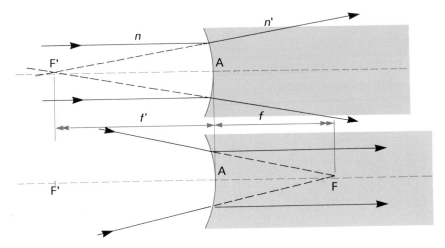

Figure 3.16 Negative refracting surface ($n' > n$): principal foci

negative surface, a virtual object must be provided, as shown in Figure 3.16. *In both cases* the distance of the first principal focus *from* the surface is called the **first focal length**, f.

Because L' is zero under these circumstances and l has been redefined as f, we have, from Equation 3.15,

$$-L = \frac{-n}{f} = F$$

With a surface of positive power f has a *negative* value, and with a surface of negative power f has a positive value.

It will be seen that both principal foci are real with a positive surface and virtual with a negative surface. With either surface the two focal lengths have values of opposite sign; the sign of the second is that of the surface power.

Combining the equations above,

$$\frac{-n}{f} = \frac{n'}{f'} = F = (n' - n)R \tag{3.16}$$

and

$$\frac{f'}{f} = -\frac{n'}{n}$$

We also have

$$f' + f = r$$

and

$$CF' = -f \qquad CF = -f'$$

The planes perpendicular to the axis and passing through the principal foci are known as the **focal planes**. Any incident ray parallel to the axis will, after refraction at a spherical surface, pass through or diverge from the second principal focus; and any incident ray passing through or travelling towards the first principal focus will, after refraction, travel parallel to the axis. These facts are useful in graphical constructions (Section 3.7).

3.7 Imaging by spherical surface refraction

Most objects are larger than a single point. For these extended objects only one point on them can lie on the axis and so it is necessary to consider off-axis points. The term **extra-axial points** has been used to describe these, but the more common term in optical design work is **field points** – that is, point objects within the field of view of the lens or instrument as described in Chapter 6. In this section we will assume that the field points are sufficiently near the axis for the paraxial approximation to be valid. In the diagrams, the off-axis heights are exaggerated for clarity.

In Figure 3.17, the spherical refracting surface, DAE, is receiving light from all points of the object BQ. The point B is on the axis of the surface and a ray from B travelling along this axis is incident normally on the surface at A and passes undeviated through C to B'. From the field point Q there is one ray which is also incident normally on the surface. This is also undeviated and must be the ray which passes through the centre of curvature, C. In the case of single refracting surfaces, we call this ray the **principal ray** from the field point Q, and C, the axial point through which no deviation occurs, is called the **nodal point**. Slightly different definitions are used for lenses and systems.

In the paraxial region, we assume that the image of B must be somewhere on the axis and the image of Q must be somewhere on this principal ray. If the image of B formed by the surface DAE in Figure 3.17 is at B', the same distances can be applied on the principal ray because this too can be considered an axis of the surface. If Bb is an arc of a circle drawn with C as centre, the image of b is b' on the arc B'b' drawn with C as centre. The image of the perpendicular BQ is a line *more* curved than B'b' because Q is further from the refracting surface than b and its image point is therefore closer to the surface than b'. This image curvature effect is discussed in Chapter 7.

When BQ is small, we can take B'Q' as the image, and in the paraxial region we shall assume that, for a plane object perpendicular to the axis, the image is also in a plane perpendicular to the axis. Object and image are said to be in **conjugate planes** and are drawn so in Figure 3.17.

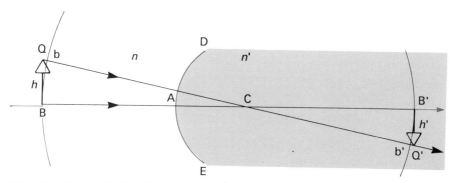

Figure 3.17 Imaging of field point, curvature of field

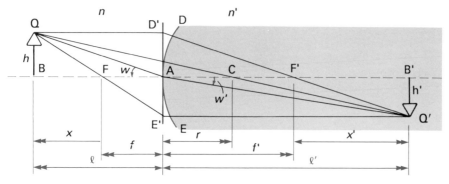

Figure 3.18 Image formation by positive refracting surface – paraxial region

In Figure 3.18, BQ is an object in front of a spherical refracting surface DAE. The position of the image B'Q' can be found by using the fundamental paraxial equation. It is also possible to find this position by drawing constructional rays on a scale diagram if the positions of the principal foci are known. In Figure 3.18 the distances above and below the axis are exaggerated for clarity, but it must be remembered that the construction applies only in the paraxial region. The central part of the surface DAE, when exaggerated, becomes a shallow curve and we can approximate this to the straight line D'AE'.

The position of the image Q' of field point Q can be found by drawing two rays. The principal ray through C is undeviated as shown. Secondly, a ray QD' drawn parallel to the axis is the same as a ray from a distant axial object and is therefore refracted through the second principal focus, F'. This ray intersects the principal ray at Q', thereby locating the position of the image of Q and, by making B'Q' perpendicular to the axis, locating the image of B. A third ray can also be used, as an alternative or a check. This is the ray from Q through F to the surface at E'. This ray could have started from F and is, therefore, refracted parallel to the axis and meets both the other rays at Q'.

As the principal ray QCQ' is undeviated, it is possible to relate the object height h to the image height h' by

$$\frac{h'}{h} = \frac{\text{CB}'}{\text{CB}} = \frac{l' - r}{l - r} = m \tag{3.17}$$

This ratio of the size of the image to the size of object is called the **lateral magnification**, m, of the image.

The ray QAQ' drawn through the vertex of the surface is refracted according to the law of refraction. The angles w and w' are small and so

$$\frac{n}{n'} = \frac{\sin w'}{\sin w} = \frac{\tan w'}{\tan w}$$

then

$$\frac{h'}{h} = \frac{l' \tan w'}{l \tan w} = \frac{nl'}{n'l} = \frac{L}{L'} = m \tag{3.18}$$

The lateral magnification will be positive when the image is erect and negative when the image is inverted. Images will be erect when they lie on the same side of the surface as the object and inverted when on the opposite side.

In Figure 3.18 it can be seen that the lateral magnification m can be shown as:

$$m = \frac{B'Q'}{AD'} = \frac{-x'}{f'}$$

and

$$m = \frac{AE'}{BQ} = \frac{-f'}{x}$$

Therefore,

$$\frac{x'}{f'} = \frac{x}{f}$$

and

$$xx' = ff' \tag{3.19}$$

where x and x' are the distances of the object and image from the first and second principal foci respectively. This is known as **Newton's relation** and can be useful when the object is near, but not at, the principal focus.

Another relationship uses the actual heights of object and image. From Equation 3.18, we have

$$\frac{h'}{h} = \frac{nl'}{n'l}$$

giving

$$n'h'l = nhl'$$

In the paraxial region, the angles u and u' (Figure 3.18) are small and so

$$u = \frac{AD'}{l} \quad \text{and} \quad u' = \frac{AD'}{l'}$$

We therefore have $n'h'u' = nhu$ and, for any other surfaces in an optical system,

$$H = nhu = n'h'u' = n''h''u'' \quad \text{etc.} \tag{3.20}$$

Each term has factors relating only to that surface and therefore H is an invariant throughout the system. This is very useful for aberration analysis in Chapter 7. It is usually known as the **Lagrange invariant**, H.

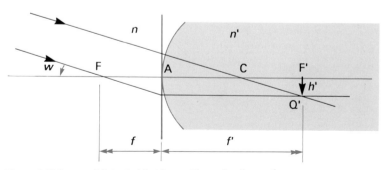

Figure 3.19 Image of distant object by positive refracting surface

When the object is large and at a distance from the surface the rays from the field point Q are parallel on reaching the surface but at an angle to the axis. These rays are imaged on the second focal plane of the surface as shown in Figure 3.19. The location of Q' on this plane is defined by the principal undeviated ray through C.

If the size of distant object is BQ where B is on the axis, then the **apparent size** is w, the angle subtended in the neighbourhood of the refracting surface. The size of the image F'Q' is

$$h' = f \tan w = fw \tag{3.21}$$

when w is small and measured in radians and f is the first focal length.

This gives another definition of the first focal length:

$$f = \frac{h'}{w}$$

$$= \frac{\text{Size of image in focal plane}}{\text{Apparent size of object}} \tag{3.22}$$

3.8 Paraxial reflection at spherical surfaces

The effect of spherical aberration at a spherical mirror surface is described in Section 3.3. A feature of spherical mirrors is that light from an object at the centre of curvature of a concave mirror (Figure 3.11(a)) is incident normally at the mirror surface and is therefore reflected back to the centre. Object and image are coincident and there is no spherical aberration. Otherwise we must restrict ourselves to the paraxial region for sharp images. As with the case of a spherical refracting surface, we use an all-positive diagram and study the action of rays or of wavefronts.

3.8.1 Change in ray path

In Figure 3.20 the incident ray directed towards the axial object point B is reflected at the surface DAE so that it appears to be coming from the axial image point B'. The centre of curvature of the surface is C and so CD is a normal to the surface.

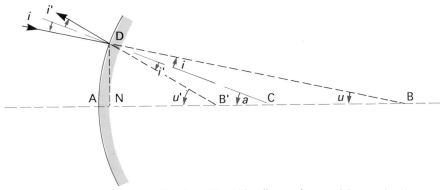

Figure 3.20 Reflection at a spherical surface. All-positive diagram for paraxial approximation

The law of reflection gives $i = i'$, which needs no approximation. In the diagram

$$\tan u = \frac{y}{AB}$$

but we approximate this to

$$u = \frac{y}{AB}$$

As AB is greater than NB and u is always smaller than $\tan u$, this is not a bad approximation.

In the same way,

$$u' = \frac{y}{AB'} \quad \text{and} \quad a = \frac{y}{AC}$$

In triangles DNC and DNB,

$$i = (90 - u) - (90 - a) = a - u$$

and in triangles DNC and DNB'

$$i' = (90 - a) - (90 - u') = u' - a$$

From the law of reflection,

$$a - u = u' - a$$

or

$$-u' = u - 2a$$

Substituting the above expression for u, u' and a, and dividing by y, we get

$$\frac{-1}{AB'} = \frac{1}{AB} - \frac{2}{AC}$$

As $AB' = l'$, $AB = l$ and $AC = r$, we have:

$$\frac{-1}{l'} = \frac{1}{l} - \frac{2}{r} \tag{3.23}$$

3.8.2 Change in vergence

Equation 3.23 can also be obtained using the ideas on vergence developed in Section 3.2. Again, we use an all-positive diagram (Figure 3.21). Light, converging towards a point B, meets a spherical mirror at A, and after reflection it is diverging from the virtual image B'. If we consider the incident wavefront GDEH, centred on B, the central part of this has already been reflected. The dotted curve DWE represents the location this wavefront would have reached but for the reflection. The reflected wavefront is DME, centred on B'. Because light travels at the same velocity after reflection as before, the distance MA is the same as AW. The line joining D and E cuts the axis at N. AN is the sag of the reflecting surface, while WN is the sag of the incident wavefront and MN is the sag of the reflected wavefront. Because AM equals AW, we have

$$MA = MN - AN = AN - WN = AW$$

Therefore,

$$MN = -WN + 2AN$$

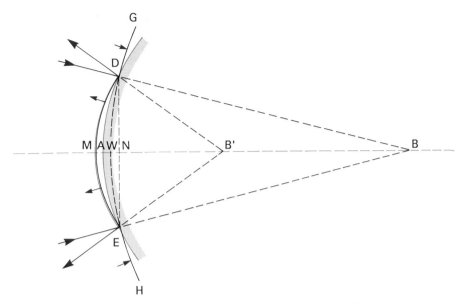

Figure 3.21 Reflection at a spherical surface. All-positive diagram for paraxial approximation

Using Equation 3.3 we can write

$$\frac{y^2}{2l'} = \frac{-y^2}{2l} + \frac{2y^2}{2r}$$

which simplifies to

$$\frac{1}{l'} = -\frac{1}{l} + \frac{2}{r} \tag{3.24}$$

where l' is the image distance, l the object distance and r the radius of curvature of the surface, and the equation is the same as Equation 3.23.

Because the reflected light is travelling in the opposite direction to the incident light, we can say that L', the image vergence, is given by:

$$L' = \frac{-1}{l'} \tag{3.25}$$

This means that the reflected light is divergent with negative vergence (L' has a negative value) when l' has a positive value; and it is convergent with positive vergence (L' has a positive value) when l' has a negative value.

Equations 3.23 and 3.24 may then be written:

$$L' = L - 2R \tag{3.26}$$

where R is the curvature of the spherical reflecting surface.

The term $2R$ is the amount by which L has to be changed to give L'. This power to change the vergence is called the dioptric power, F, of the mirror. We define this power as

$$F = -2R \tag{3.27}$$

Putting Equation 3.27 with Equation 3.26 gives

$$L' = L + F \qquad (3.28)$$

This is the same fundamental paraxial equation as was shown to apply to single refracting surfaces. It will be shown in the next chapter that it applies also to lenses and lens systems.

If we apply the sign convention for angles to the case of reflection, we see that the angles of incidence and reflection are always of opposite signs. The law of reflection can be considered a special case of the law of refraction with $n' = -n$ so that $n' \sin i' = n \sin i$ becomes $i' = -i$ where i' is the angle of reflection.

Using this, any equation for spherical refracting surfaces can be used for reflection if $-n$ is substituted for n'. Thus Equations 3.9 and 3.11 become:

$$\frac{-n}{l'} = \frac{n}{l} - \frac{2n}{r} \qquad (3.29)$$

or

$$\frac{-1}{l'} = \frac{1}{l} - \frac{2}{r}$$

in air, and this is the same as Equation 3.23.

In the same way, Equation 3.16 becomes

$$\frac{-1}{f'} = \frac{-1}{f} = F = \frac{-2}{r} = -2R \qquad (3.30)$$

When the mirror is operating in some medium other than air, we have Equation 3.29 as the applicable equation and Equation 3.30 becomes

$$\frac{-n}{f'} = \frac{-n}{f} = F = \frac{-2n}{r} = -2nR \qquad (3.31)$$

where n is the refractive index of the surrounding medium.

3.9 Focal properties of spherical reflecting surfaces

The same methods as were used for the single refracting surface can be used for the spherical mirror. Parallel light from a distant object reaching a concave spherical mirror (Figure 3.22) is caused to converge to a real image in front of the mirror. The position of this image is the **second principal focus**, F', of the mirror and its distance AF', from the mirror, is the second focal length, f'.

As the object is moved closer to the mirror, the image moves away. The object and image therefore move closer together until a point is reached where they coincide. This is at the centre of curvature of the mirror. The light rays from the object are now incident normally on the mirror surface and, for a point object, the point image is free of aberrations. This is the basis of an optical method for measuring curvature (Section 4.8).

When the object reaches the **first principal focus**, the image is at infinity and the reflected light is parallel to the axis. With spherical mirrors the position of the first principal focus coincides with the position of the second principal focus. As the object moves closer to the mirror than the principal focus, the reflected light is

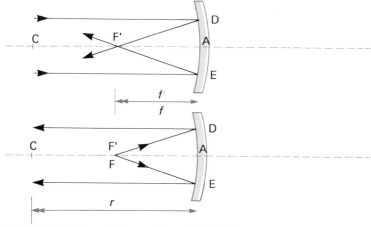

Figure 3.22 Positive reflecting surface: principal foci

divergent from a virtual image behind the mirror and at a greater distance from the mirror than the object is in front. When the object reaches the mirror surface the object and image again coincide.

With a convex mirror, the light is always made more divergent and so a virtual image is always formed with a real object (Figure 3.23). When the object is very distant, this virtual image is at the second principal focus, F', and its distance from the mirror, AF', is the second focal length, f'. When the incident light is converging to this same point, the reflected light is parallel. This point is, therefore, the first principal focus also. When the incident light is converging towards the centre of curvature of a convex mirror, the reflected light is diverging from the same point. Virtual image and virtual object therefore coincide at the centre of curvature of a convex mirror in the same way that a real object and real image coincide at the centre of curvature of a concave mirror. Object and image also coincide when they are at the mirror surface.

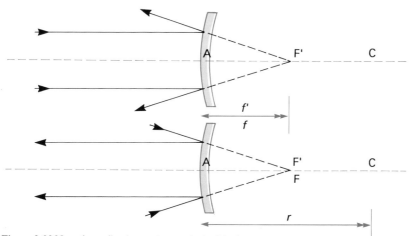

Figure 3.23 Negative reflecting surface: principal foci

When $l = \infty$, $l' = f'$ and when $l' = \infty$, $l = f$. Therefore, from Equation 3.27, we have

$$f' = \frac{-1}{F} = \frac{1}{2R} = \frac{r}{2} \tag{3.32}$$

which is the same as Equation 3.30. The principal focus of a spherical mirror is therefore located mid-way between the mirror surface and the centre of curvature.

Methods of measuring the focal length of spherical mirrors using an optical bench are described in Section 4.8. Instruments for this measurement are described in Chapter 6.

3.10 Imaging by spherical surface reflection

In the same way as described in Section 3.7, off-axis points are here considered to be within the paraxial region even though their off-axis heights are exaggerated in

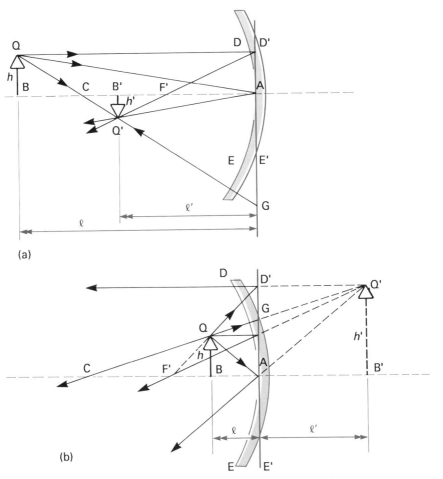

Figure 3.24 Image formation by positive reflecting surface (concave mirror)

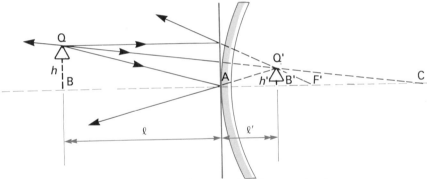

Figure 3.25 Image formation by negative reflecting surface (convex mirror)

the diagrams. In Figure 3.24, the object BQ is in front of the spherical mirror DAE. The position of the image B'Q' can be found using the fundamental paraxial equation. It is also possible to find this position by drawing constructional rays on a scale diagram if the position of the principal focus is known. The central paraxial part of the surface DAE, when exaggerated, becomes a shallow curve and we can approximate this to the straight line D'AE'.

The position of the image Q' of the field point Q can be found by drawing two rays. The principal ray through C is undeviated on reflection, as it is incident normally on the mirror even though the exaggerations of the diagram do not show this. A second ray drawn from Q parallel to the axis is reflected so that it passes through the principal focus, F; Q' is located where these two rays intersect. Two other rays could have been used or could be drawn as a check. The ray from Q through the principal focus F is reflected parallel to the axis. Lastly, the ray incident at A, the vertex of the mirror, is reflected at an equal angle on the opposite side of the mirror. This ray occurs only with mirrors and has been drawn, in preference to the third ray, in Figure 3.24(a). This shows the formation of an inverted real image outside the principal focus of a concave mirror.

If the object is moved to be within the focal length of the mirror, as shown in Figure 3.24(b), a virtual erect and enlarged image is formed behind the reflecting surface. This is the situation found when looking into a concave bathroom mirror with your face inside the principal focus.

Figure 3.25 shows a virtual, erect and diminished image of a real object in a convex mirror typically found on wing mirrors of motor vehicles. The diminished image allows a wider field of view and therefore more of the object is seen than would be the case with a plane mirror of the same size.

In Figures 3.24 and 3.25 the lateral magnification, m, is given by

$$m = \frac{h'}{h} = \frac{-AB'}{AB} = \frac{-l'}{l} = \frac{L}{L'} \tag{3.33}$$

This may also be expressed in terms of f':

$$m = \frac{L}{L'} = \frac{L' - F}{L'} = 1 - \frac{l'}{f'}$$

Other equations can be found which relate the object and image positions in terms of their distances from the centre of curvature or from the principal focus (Newton's relation).

Exercises

3.1   The radius of the circle containing the three fixed legs of a spherometer is 1⅛ in. The instrument is placed on a convex spherical surface of curvature 3 D. By how much will the central leg have to be raised above the outer legs to be just in contact with the surface?

3.2   An optician's lens measure has its outer points ¾ in apart, and a depression of the centre point of ½₂ in gives a reading of +9.25 D on its scale. For what refractive index was it scaled?

3.3   A lens measure applied to a brass tube of 9 in external diameter gave a reading of 4.75 D. For what refractive index was it adjusted?

3.5   A pencil of light is made to converge by means of a lens to a point 20 cm from the lens. Find the curvature of a wavefront (a) on leaving the lens, (b) at 1 cm from the lens, (c) at 19.9 cm from the lens, (d) at 21 cm from the lens.

3.5   A small source of light is placed 75 cm from a lens; what is the curvature of the wavefront at the lens? If the effect of the lens is to add 4 D convergence to the light, find the position at which the light will focus.

3.6   Where must a small source of light be placed for the wavefront of the divergent pencil to have a curvature of 7 D on reaching a lens? If the lens brings the light to a focus at a point 30 cm beyond the lens, what will be the change in curvature of the wavefront produced by the lens?

3.7   Show that when a pencil of convergent light is intercepted by a plane parallel plate of glass of thickness t and refractive index n, the position of the focus is moved outwards a distance $t - t/n$.

3.8   A plane wavefront is incident in air on a curved glass surface of +5 cm radius of curvature, refractive index of the glass 1.5, the line bisecting the wavefront at right angles passing through the centre of curvature of the surface. Find by Huygens' construction the form of the refracted wavefront. What do you notice about its form?

3.9   An object 1 cm high is placed 15 cm from a convex spherical refracting surface of 20 cm radius of curvature, which separates air from glass ($n = 1.52$). Find the position and size of the image. What are the positions of the principal foci of this surface?

3.10   A spherical refracting surface of radius +8 cm separates air on the left from glass of refractive index 1.5 on the right. A parallel pencil is incident on it, the chief ray of the pencil passing through the centre of curvature. Find by graphical construction the refracted rays corresponding to incident rays at 3, 4, 5, 6 and 7 cm above the chief ray. Note where they cut the chief ray.

3.11   What is the difference in the apparent thickness of a biconvex lens ($n = 1.5$) having radii of curvature of 8 and 20 cm and central thickness 2 cm, the lens being examined first from one side and then from the other?

3.12   A glass sphere has a diameter of 10 cm, and refractive index 1.53. There are two small bubbles in the glass. One appears to be exactly at the centre of the sphere and the other midway between the centre and the front surface. Find their actual positions.

3.13   A glass paperweight ($n = 1.53$) has a picture fastened to its base. Its top surface is spherical and has a radius of curvature of 2½ in, and its central thickness is 1½ in. Where will the picture appear to be when viewed through the top surface and what will be its magnification?

3.14   For the purpose of calculation the eye may be considered as a single refracting surface of +60 D power, having air on one side and a medium of refractive index ⁴⁄₃ on the other. What must be the distance of the retina – the receiving screen – from this surface, if the eye is correctly focused on a distant object? What will be the distance of the object that is focused by an eye in which the distance between refracting surface and retina is 24 mm?

3.15   Treating the cornea as a single spherical refracting surface of radius +7.8 mm separating air from the aqueous of refractive index ⁴⁄₃, find the position and size of the image of the eye's pupil, the latter being 3.6 mm behind the cornea and 4 mm in diameter.

What effects would be produced on the retinal image in an eye as the pupil is reduced in diameter?

3.16 A narrow beam of parallel light is incident normally on the flat surface of a glass hemisphere of 10 cm radius. Find where the rays of light are brought to a focus when the hemisphere is (a) in air, (b) in water. The refractive indices of glass and water may be taken to be ³⁄₂ and ⁴⁄₃ respectively.

3.17 A piece of glass 2 cm long and of refractive index 1.60 is plane at one end and has a convex spherical surface of radius of curvature 1.875 cm worked on the other end. It is set up with the curved surface facing a distant point source. Find the curvature of the wavefront emerging from the plane surface and the position of the point where the light comes to a focus.

3.18 Considering the eye as a single refracting surface of +5.56 mm radius of curvature, having air on one side and a medium of refractive index ⁴⁄₃ on the other, what will be the size of the image formed of a distant object that subtends an angle of 1° at the eye?

3.19 An object 3 in long is placed 15 in from the principal focus of a concave mirror, and a real image 1¼ in long is formed. Find the curvature of the mirror and the distance of object and image from the mirror.

3.20 When an object is 31 cm in front of a concave mirror the image is 19 cm in front of the mirror. Where will the image be if the object is a virtual one and 31 cm behind the mirror?

3.21 Find graphically and by calculation the positions and sizes of the images in the following cases, the object in each case being 3 cm long:
(a) Object 10 cm from convex mirror of 10 cm radius.
(b) Object 5 cm from concave mirror of 15 cm radius.

3.22 The radius of curvature of a convex mirror is 250 mm. Standing on its axis and 600 mm in front of the mirror is an object 5 mm high. Find the position and size of the image (a) graphically, (b) by calculation. Is the image real or virtual?

3.23 Find the position of the image of an object which is 9 in from a concave mirror whose radius of curvature is 12 in.

3.24 A concave mirror forms a real image of an object at 20 cm in front of the mirror; if the object is 133.3 cm from the mirror, find (a) the focal power, (b) the curvature, (c) the radius of curvature of the mirror.

3.25 Explain the formation of an image by reflection of light from a convex spherical mirror.
   A small object is placed on the axis of a convex spherical mirror at a distance of 200 mm from it, and the virtual image formed is observed to have half the linear dimensions of the object. Determine the focal length and the radius of curvature of the mirror.

3.26 A concave mirror has a radius of curvature of 2 ft; show graphically and by calculation where an observer must be situated to see a virtual image of his face magnified 2½ times.

3.27 A convex mirror having a radius of curvature of 25 cm forms, at 20 cm from the mirror, a virtual image of an object 2 cm in height. Calculate the magnification and give a diagram showing the path of the light.

3.28 The front surface of the cornea of the eye has a radius of curvature of 8 mm; what will be the size and position of the image formed by the reflection of an object 200 mm in front of the cornea and 60 mm long?

3.29 An object and its image are 1 m apart; if the object is four times the size of the image, find the radius of the concave mirror forming the image.

3.30 A concave and a convex mirror, each of 20 cm radius of curvature, are placed opposite to each other and 40 cm apart on the same axis. An object 3 cm high is placed midway between them. Find the position and size of the image formed when the light is reflected first at the convex and then at the concave mirror. Give a diagram showing the oath of the light from a point on the object to a corresponding point on the image.

3.31 A concave mirror is immersed in water. The image of an object 30 cm from the mirror is formed at 9 cm in front of the mirror. If both object and image are in the water, what is the radius of the mirror?

Chapter 4

Thin lenses

4.1 Forms of thin lenses

A lens is formed from two refracting surfaces. A **spherical lens** is formed from two spherical refracting surfaces. In a thin lens, the two surfaces are very close together. Examples of thin spherical lenses are:

1. Spectacle lenses
2. Trial case lenses
3. Contact lenses
4. Auxiliary lenses for cameras
5. Small magnifiers

Remember, however, that spectacle lenses are often toric (see Chapter 8) and the surface curves of magnifiers are often aspheric (see Chapter 8). All the above lenses are used with another lens system – an eye or a camera lens – and the trial case lenses (for testing sight) are often used in groups. A single thin spherical lens is rarely used on its own. We study it in this chapter because it forms the basis for lens systems and because the main equations developed in the last chapter will be shown to work for thin lenses and then for more complex conditions (Chapter 5).

The two surfaces of a spherical lens will each have a centre of curvature. The surface that refracts the incident light will be called the first surface, and the other surface which subsequently refracts the light will be called the second surface. The positions in the space through which the incident light is travelling will be said to be in front of the lens and, in the diagrams in this book, this will normally be to the left of the lens. Figure 4.1(a) shows two spheres forming a lens. The first surface has its centre at C_1 and the second surface centre is at C_2. The line C_1C_2 is a unique line and is called the **optical axis** of the lens. There is no other axis. The radii of the two spheres need not be equal or even on opposite sides of the lens, but the optical axis is always the line joining their centres of curvature. If one surface is plane, it may be thought of as a spherical surface of infinitely large radius. The optical axis is then the line through the centre of curvature of the other surface and perpendicular to the plane surface as shown in Figure 4.1(b).

The points A_1 and A_2 where the axis meets the surfaces are called the **front vertex** and **back vertex** of the lens respectively. The distance A_1A_2 is called the **centre thickness** of the lens and, when this distance is too small to be significant, the common vertex A is called the **optical centre** of the lens. Lenses are usually 'edged' so that when viewed along the axis they have a circular shape centred on the axis. If

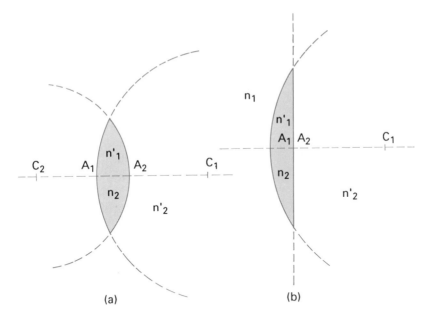

Figure 4.1 Optical axes of lenses

this is not the case, we say they are **decentred**. The individual lenses in a camera lens are usually edged so that the mechanical axis (of their overall circular shape) and the optical axis coincide to within a few microns. This is seldom the case with spectacle lenses, even when they are circular.

Each surface of a thin lens has a surface power and the total power of the lens is found by adding the two surface powers. In Chapter 5 it will be shown that the total power, F, of a lens is more accurately given by:

$$F = F_1 + F_2 - (d/n)F_1F_2$$

where F_1 and F_2 are the surface powers, d is the centre thickness and n the refractive index of the lens. If the $(d/n)F_1F_2$ part is ignored, because the thickness, d, is very small, we have:

$$F = F_1 + F_2 \tag{4.1}$$

It is clear that different combinations of F_1 and F_2 can be used to give the same value of F. For instance, a total lens power of $+6\,\text{D}$ can be obtained from combinations of $+3$ and $+3\,\text{D}$, $+4$ and $+2\,\text{D}$, $+6$ and $0\,\text{D}$, $+10$ and $-4\,\text{D}$, etc. This allows different **forms** or **bending** for a lens of any given power.

In Figure 4.2, the four lenses shown have the same total (positive) power. Lens A has surfaces of equal positive power. Lens B has surfaces of positive but unequal power. Lens C has one surface of zero power, while lens D has one negative surface which needs an excess of positive power on the other surface. All four lenses, assuming they are made of material of higher refractive index than their surroundings, will tend to make any incident wavefront more convergent. All of them can be called convex, convergent, positive lenses while their individual forms have the names given in Figure 4.2.

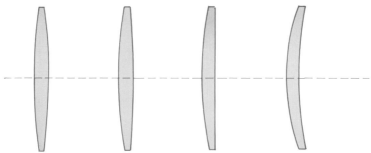

(a) Equi-convex (b) Bi-convex (c) Plano-convex (d) Meniscus

Figure 4.2 Convex, convergent positive lenses of equal power

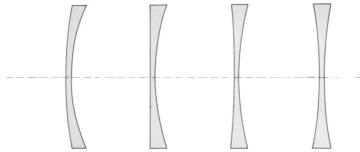

(a) Equi-concave (b) Bi-concave (c) Plano-concave (d) Meniscus

Figure 4.3 Concave, divergent negative lenses of equal power

Lenses that tend to diverge the light are called concave, divergent, negative lenses and these too can have various forms. For instance, a total power of $-8\,D$ can be obtained from combinations of -4 and $-4\,D$, -6 and $-2\,D$, -8 and $0\,D$, -12 and $+4\,D$. The lenses shown in Figure 4.3 have the same total negative power but obtained from different combinations of surface powers.

Note 1

While the centre thickness of negative lenses can be small for all powers, the positive lens needs to be thicker at the centre the more powerful it becomes. From the $(d/n)F_1F_2$ term shown above it is possible to calculate that, for an equi-convex lens, Equation 4.1 will give an error of less than 1% if the centre thickness is less than 1% of the total focal length times the refractive index. However, for a meniscus lens where F_1 is $3F$, the centre thickness must be less than 0.1% of the overall focal length times the refractive index. Thus, care must be taken with positive meniscus lenses when using them as *thin* lenses.

Note 2

The different forms or bendings of thin lenses give changes in the quality of the images they form. As might be expected from Figure 3.10, spherical aberration occurs at both surfaces of the lens. The residual aberration is found to be very dependent on the form of the lens and this is explained in Chapter 7. For this chapter we assume that all the object and image rays lie within the paraxial region.

4.2 Focal power

As with spherical refracting surfaces and with spherical mirrors, the effect of lenses is to change the vergence of the light arriving from an object so that it forms an image – real or virtual – in some different place. This change in vergence is due to the total power, usually called the **focal power** or **dioptric power** of the lens. It is calculated by using Equations 3.9 and 3.11 (the effect at a single surface) at each surface of the lens in turn. In Figure 4.4 the thickness, t, of the lens is assumed to be zero. The radii of curvature and the object and image distances can all be thought of as being measured from A, the optical centre of the lens.

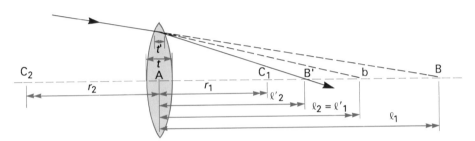

Figure 4.4 Refraction at a thin lens

At the first surface:

$$\frac{n_1'}{l_1'} = \frac{n_1}{l_1} + \frac{n_1' - n_1}{r_1} \tag{4.2}$$

where r_1 is the radius of curvature of the first surface; l_1 the object distance and l_1' the distance of the intermediate image b; n_1 the refractive index in front of the lens and n_1' the refractive index of the lens.

The intermediate image can now be taken as the object for the refraction at the second surface. This gives the equation:

$$\frac{n_2'}{l_2'} = \frac{n_2}{l_2} + \frac{n_2' - n_2}{r_2} \tag{4.3}$$

where r_2 is the radius of curvature of the second surface; l_2 the object distance and l_2' the final image distance; n_2 the refractive index of the lens and n_2' the refractive index behind the lens.

Because the lens is thin and t is assumed to be zero, the thickness t', where the ray passes through the lens, can also be assumed to be zero. Thus, the object distance l_2 to b is the same as the image distance, l_1'. The refractive index of the lens remains the same so n_1' is equal to n_2. Thus:

$$\frac{n_2}{l_2} = \frac{n_1'}{l_1'}$$

and Equations 4.3 and 4.4 may be put together to give:

$$\frac{n_2'}{l_2'} = \left(\frac{n_1}{l_1} + \frac{n_1' - n_1}{r_1}\right) + \frac{n_2' - n_2}{r_2}$$

or

$$\frac{n_2'}{l_2'} = \frac{n_1}{l_1} + \left(\frac{n_1' - n_1}{r_1} + \frac{n_2' - n_2}{r_2}\right) \tag{4.4}$$

which is the same as the fundamental paraxial equation with the two surface powers F_1 and F_2 combining to form the dioptric power, F, of the lens:

$$\frac{n_2'}{l_2'} = \frac{n_1}{l_1} + F_1 + F_2 = \frac{n_1}{l_1} + F \tag{4.5}$$

If the lens is surrounded by air, both n_1 and n_2' are equal to one. For simplicity the index of the lens will be called n and so $n_1' = n_2 = n$. We now have, from Equation 4.4:

$$\frac{1}{l_2'} = \frac{1}{l_1} + \frac{n + 1}{r_1} + \frac{1 - n}{r_2}$$

If the object distance, l_1, is now called l and the final image distance, l_2', is called l', we have:

$$\frac{1}{l'} = \frac{1}{l} + \frac{n - 1}{r_1} + \frac{1 - n}{r_2} = \frac{1}{l} + (n - 1)\left(\frac{1}{r_1} - \frac{1}{r_2}\right) \tag{4.6}$$

We define F, the dioptric power of the lens, as

$$F = (n - 1)\left(\frac{1}{r_1} - \frac{1}{r_2}\right)$$

$$= (n - 1)(R_1 - R_2) \tag{4.7}$$

where R_1 and R_2 are the curvatures of the first and second surfaces.

From Equations 4.6 and 4.5 we can write

$$L' = L + F \tag{4.8}$$

the fundamental paraxial equation, where L ($= 1/l$) is the object vergence for a thin lens in air and L' ($= 1/l'$) is the image vergence for a thin lens in air. Equation 4.8 may be stated as:

> For a thin lens in air the final vergence, L', of the light is equal to the incident vergence, L, plus the impressed vergence, F, the dioptric power of the lens.

or:

> What you get out equals what you put in plus what the lens does to it!

This equation should be remembered in the form given in Equation 4.8 as there are no negative signs in it to confuse the use of the sign convention when the actual values are being put in.

Note 3

The power of the lens as defined in Equation 4.7 is dependent on the refractive index of the material of the lens. As was noted in Section 1.10, the refractive index of most optical materials has a value which changes with the wavelength of light. For most optical glass types, this change in index from deep red to violet light amounts to between 1% and 4% of the refractive index. The effect was noted with refracting prisms and occurs with single refracting surfaces as well as with lenses. The term $(n - 1)$ in Equation 4.6 changes by 3% and 10% and so the lens power also changes by at least 3%. This means that light from a white point source forms images of different colours in slightly different places. This coloured blurring is called **chromatic aberration** and is described in Chapters 11 and 14. Because its effect is normally rather small, this aberration may be ignored completely in the paraxial region.

4.3 Focal properties of thin lenses

In the same way as with single spherical refracting surfaces and spherical mirrors, Equation 4.7 shows that, with thin lenses, for each object point B there is *one* corresponding image point B'. The distance of B' from the vertex of the lens can be calculated from Equation 4.7 by inserting the object vergence value and the dioptric power of the lens or from Equation 4.5 by inserting the object distance, the two radii of curvature and the index of the lens material. The distances must be given the correct signs according to the sign convention. As the path of the light is reversible, the positions of the object and image are reversible.

When the object is at an infinite distance from the lens, the object vergence is zero:

$$L = 0$$

and so

$$L' = F$$

Figure 4.5 shows this for a positive lens and Figure 4.6 for a negative lens. In *both cases* the image is said to be at the **second principal focus**, F', of the lens and its

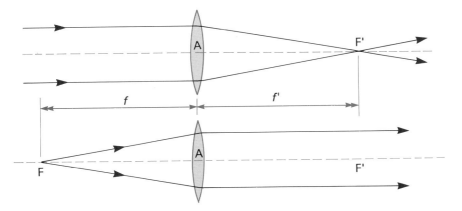

Figure 4.5 Positive lens: principal foci

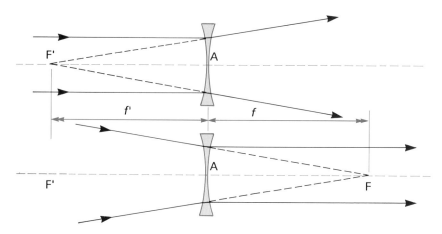

Figure 4.6 Negative lens: principal foci

distance *from* the lens is called the **second focal length**, f'. Because L is zero and l' has been redefined as f', we have from Equation 4.7

$$L' = \frac{1}{f'} = F \qquad (4.9)$$

With a lens of positive power, as shown in Figure 4.5, f' has a positive value and F' is the location of a *real* image. As the object moves towards the lens, the incident light becomes more and more divergent and so the emerging light becomes less convergent and the image moves away from the lens until for an object as shown in the second diagram of Figure 4.5, the emergent light is parallel, the image is at infinity and L' is zero. The object is now at the **first principal focus**, F, which is at a distance f from the lens.

With $L' = 0$, we have

$$0 = L + F = \frac{1}{f} + F \qquad (4.10)$$

Therefore,

$$L = \frac{1}{f} = -F \qquad (4.11)$$

If the object is brought still closer to the lens, the convergent power of the lens is not enough to overcome the divergence of the incident light and the emergent light diverges from a virtual image on the same side of the lens as the object but farther from the lens than the object.

With a negative lens, the effect is always to add divergence to the light so a virtual image is always formed of a real object and this image is always closer to the lens than the object. When the real object is moved out to infinity, the position of the virtual image is the second principal focus, F', and this is a negative distance from the lens as shown in Figure 4.6 and in Equation 4.9 where the negative value of F gives a negative value of f'.

The position of the object can be taken 'beyond infinity' if the incident light is made convergent. The point to which the incident light is converging may be thought of as a *virtual object*. A positive lens will always form a real image of a virtual object, but with a negative lens the incident light must have enough convergence to overcome the diverging action of the lens. When the convergence of the incident light is just equal (and opposite) to the power of the negative lens, Equation 4.10 applies and the emergent light is parallel, the image is at infinity and L' is zero. The virtual object is now at the first principal focus, F, of the lens which is at a distance, f, from the lens as shown in the second diagram of Figure 4.6. Equation 4.11 shows that f has a positive value when F is negative.

From Equations 4.9 and 4.11 it can be seen that for the thin lens in air, when the first and last media are of equal index, the two focal lengths are equal in distance but opposite in sign, the second focal length having the same sign as the power. Except for ophthalmic lenses, which are described by their focal power, most lenses are specified in terms of their second focal lengths. Equation 4.8 can also be written:

$$\frac{1}{l'} = \frac{1}{l} + \frac{1}{f'} \tag{4.12}$$

This equation may be used with any units of length – inches, millimetres, metres, etc. – as long as all the distances use the same units. A calculator with a reciprocal key is essential for this equation. If mixed equations such as 4.5 or 4.7 are being used, the distances should be measured in metres so that the reciprocals are automatically in dioptres.

4.4 Imaging with thin lenses

In a similar way to spherical refracting surfaces and spherical mirrors, the imaging properties of lenses for extended objects are worked out using point objects away from the axis. These field points will be assumed to lie within the paraxial region but in the diagrams the heights and angles are exaggerated for clarity. *All distances along the axis are drawn to scale.* The actual lens will be represented by a straight line with positive and negative lenses indicated by a dotted partial outline.

For either power of lens, the optical centre and principal foci may be used to find the directions of three rays of light which then define the position of the image of an off-axis object. These three rays are:

(a) The principal ray which passes undeviated through the optical centre of the lens. Because the lens is thin and its surfaces at this point are parallel (being both perpendicular to the axis) the ray effectively sees a thin parallel plate of glass. The optical centre is therefore the nodal point for a thin lens. A different situation occurs with thick lenses and lens systems.

(b) The F ray which *passes through* the first principal focus emerges from the lens as a ray parallel to the axis. The first principal focus is defined in this way and a ray passing through it is refracted in the same way as one starting from it.

(c) The F' ray which, if drawn parallel to the axis, is the same as a ray from a distant axial object and is, therefore, refracted through the second principal focus.

Figure 4.7 shows the case with a positive lens and a real object, BQ. Any two of the three rays drawn could be used to locate the image, but the third ray is useful as a check. Figure 4.8 shows the case with a negative lens and a real object. In both cases the principal ray (a) from Q passes through A, the optical centre of the lens. The ray (b) through the first principal focus is incident on the lens at E, having passed through F in the case of the positive lens. The refracted ray, now drawn through E and parallel to the axis, intersects the principal ray at Q′, the location of the image of Q. With the negative lens, the F ray is incident on the lens before it reaches F but the geometry is the same and the refracted F ray, parallel to the axis, appears to come from Q′. Similarly, the F′ ray is drawn from Q parallel to the axis to be incident on the lens at D. The refracted ray is then drawn from D to pass through, or appear to pass through, F′. This ray again defines the position of Q′ by its intersection with the principal ray.

Because these object and image points lie within the paraxial region, the image of an object BQ perpendicular to the axis is also perpendicular to the axis. The position of B′ is found by drawing a perpendicular line from Q′ to the axis. These object and image lines together with the axis and the principal ray form two

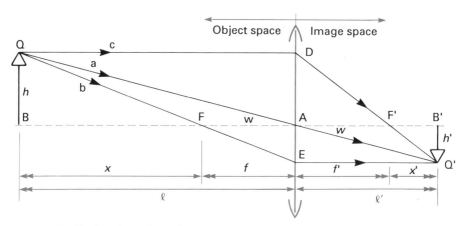

Figure 4.7 Positive lens: image formation

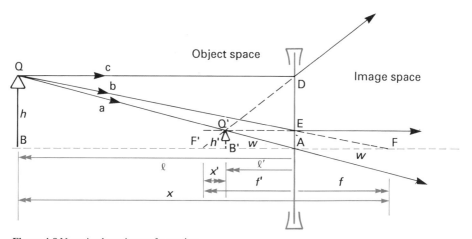

Figure 4.8 Negative lens: image formation

triangles ABQ and AB′Q′. In these triangles, the angle at A, between the principal ray and the axis, is common and the angles at B and B′ are right angles. Therefore, for both positive and negative lenses, the triangles ABQ and AB′Q′ are *similar*. By this we mean that their angles are the same and their sides are in the same ratio.

With continuous reference to Figures 4.7 and 4.8, we define the **lateral magnification**, m, of the image as the ratio h'/h. Because of the similar triangles, we have:

$$m = \frac{h'}{h} = \frac{B'Q'}{BQ} = \frac{AB'}{AB} = \frac{l'}{l} = \frac{L}{L'} \tag{4.13}$$

The magnification is positive when the image is erect with Q and Q′ on the same side of the lens as in Figure 4.8. The magnification is negative in Figure 4.7.

Other equations for the magnification can be found from Equations 4.13 and 4.8:

$$m = \frac{L}{L'} = \frac{L' - F}{L'} = 1 - \frac{F}{L'} = 1 - l'F = 1 - \frac{l'}{f'} \tag{4.14}$$

and

$$\frac{1}{m} = \frac{L'}{L} = \frac{L + F}{L} = 1 + \frac{F}{L} = 1 + lF = 1 + \frac{l}{f'} \tag{4.15}$$

In these equations l, l' and f' must be measured in metres when used with L, L' and F which are in dioptres.

If the object in Figure 4.7 is brought closer to the lens, the image moves further away and grows larger. At some point, it will be the same size as the object (but inverted). The magnification is then -1 where the negative sign shows the inversion. From Equation 4.13 we have:

$$m = -1 = \frac{l'}{l} \quad \text{or} \quad l' = -l$$

Putting this into Equation 4.12 gives

$$\frac{1}{l'} = \frac{-1}{l} = \frac{1}{l} + \frac{1}{f'}$$

and so

$$l = -2f'$$

Therefore

$$l' = 2f'$$

and the distance between object and image is

$$l' - l = 4f' \tag{4.16}$$

This is the closest possible distance between a real object and a real image using a positive lens. These positions are known as the **symmetrical planes**. With a negative lens the symmetrical planes are formed with a virtual object and a virtual image. At the symmetrical planes the magnification is -1 (the image is inverted).

The location of object and image when the magnification is $+1$ (the image is the same size and the same way up as the object) are known as the **principal planes** of a lens. With a single thin lens these planes are merely the case when the object and image coincide with each other at the lens itself. Principal planes become important with thick lenses and lens systems as described in Chapter 5.

In the same way as for a single refracting surface (Section 3.7), the locations of object and image may be found using their distances from the principal foci. Both Figure 4.7 and Figure 4.8 show x and x' and the directions in which they are measured. Because QD is parallel with the axis, AD is equal to h. Triangles F'AD and F'B'Q' are sinilar and so

$$\frac{h'}{h} = m = \frac{-f'}{x'} \tag{4.17}$$

Also, EQ' is parallel to the axis and so AE is equal to h'. Triangles QBF and EAF are similar, so

$$\frac{h'}{h} = m = \frac{-f}{x} \tag{4.18}$$

and

$$xx' = ff'$$

or

$$-xx' = f^2 = f'^2 \tag{4.19}$$

This is **Newton's equation** and is useful in measuring the focal lengths of lenses (see Sections 4.8 and 5.5).

If we consider an object B_1B_2 lying *along* the axis, the image will be $B_1'B_2'$ and each point can be found using Newton's relation:

$$-x_1x_1' = f^2 \qquad \text{and} \qquad -x_2x_2' = f^2$$

The **longitudinal magnification** of the image compared with the object is given by

$$\frac{x_2' - x_1'}{x_2 - x_1} = \frac{-\left(\dfrac{f^2}{x_1} - \dfrac{f^2}{x_2}\right)}{x_2 - x_1} = \frac{f}{x_1} \times \frac{f}{x_1} = m_1m_2 \tag{4.20}$$

where m_1 and m_2 are the lateral magnifications at B_1' and B_2'.

If the object is generally distant, so that x_1 and x_2 are much bigger than the difference between them, m_1 and m_2 are nearly equal and we can say that the

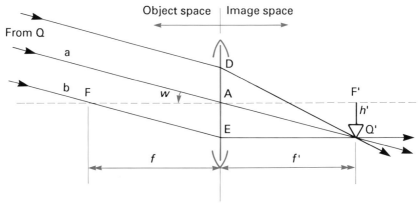

Figure 4.9 Positive thin lens. Image of a distant object

longitudinal magnification is equal to the square of the lateral magnification. If a camera image is, say, a hundred times smaller than the object being photographed, the 'depth' of the image is one ten-thousand times less than the depth of the object. This fact helps the depth of focus of fixed-lens miniature cameras.

For an infinitely distant object, the light rays reaching the lens are parallel to each other but at some angle to the axis. Figure 4.9 shows that only two construction rays, a and b, can be drawn. The ray through D has been added afterwards. As before, the principal ray passes undeviated through the optical centre of the lens and the ray through F is refracted parallel to the axis. Because f and f' are equal in size (but of opposite sign) the parallel ray meets the principal ray in the second focal plane.

If w is the angle subtended at the lens by a distant object standing on the axis (the apparent size of the object) the size of the image is:

$$\left. \begin{aligned} h' &= -f' \tan w = f \tan w \\ &= fw \quad \text{when } w \text{ is small} \end{aligned} \right\} \tag{4.21}$$

Often, the object is positioned symmetrically on both sides of the axis. Then:

$$\left. \begin{aligned} h' &= 2f \tan \frac{w}{2} \\ &= fw \quad \text{when } w \text{ is small and measured in radians} \end{aligned} \right\} \tag{4.22}$$

If the angle w is measured in prism dioptres, say P prism dioptres, the image size is then given by

$$h' = \frac{P \times f}{100} \quad \text{cm}$$

With a negative thin lens, the same rays, a and b, can be drawn (Figure 4.10). If the negative lens has the same magnitude of focal length as a positive lens, the virtual image formed of a symmetrical object is the same size as the real image with the positive lens as Equation 4.22 would predict. The real image is inverted.

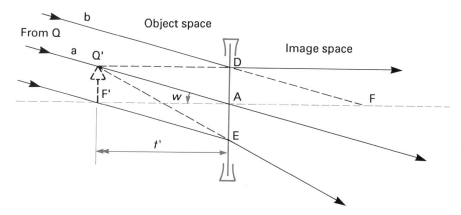

Figure 4.10 Negative thin lens. Image of a distant object

Note 4

With positive lenses, it is easy to associate the object and the object ray with the object space and the image and image rays with the image space as shown in Figures 4.7 and 4.9. With negative lenses (and all virtual images) it is more difficult. In Figures 4.8 and 4.10, the image space is to the right of the lens and the image rays only exist here. The image can only be seen in this space and does not exist in the object space to the left even though it appears to be there. It is of no importance that the object ray from Q through E, in Figure 4.10, appears to pass through the base of the image at F'. The image only 'exists' when viewed from image space.

4.5 Graphical methods

The use of graphical methods has already been shown in Figures 4.7 and 4.8 for a real object and Figures 4.9 and 4.10 for an object at an infinite distance (which cannot be specifically described as real or virtual because the parallel rays can be *going to* an infinitely distant virtual object just as easily as *coming from* an infinitely distant real object). If we start by assuming a distant object to the left of a positive lens and then move it in a left-to-right direction closer to the lens, the image which was at the second principal focus also moves in a left-to-right direction further away from the lens. When the object reaches the first principal focus the image is infinitely distant and further movement of the object makes the image reappear on the left as a virtual image but still moving from left to right. As the object moves to be in contact with the lens, the image also moves to be in contact with the lens. Imagine the object passing through the lens to become a virtual object on the right side. As the object moves further from the lens, the real image follows but at a slower rate so that it only reaches the second principal focus when the object gets to infinity; the same conditions as at the start.

All this can be shown in graphical form. In Figure 4.11, the line LL is meant to represent the track of a positive lens which is moving from top to bottom. A distant

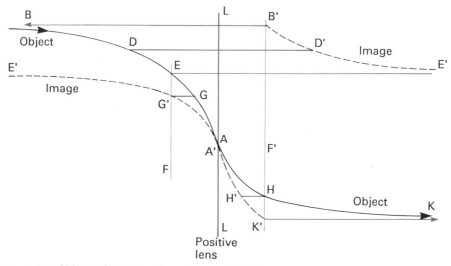

Figure 4.11 Object and image positions with a positive lens

Table 4.1

For a positive lens

Between B and E	Real object, real image (inverted)	Figure 4.7
Between E and A	Real object, virtual image (erect)	Figure 4.12
Between A and K	Virtual object, real image (erect)	Figure 4.13

For a negative lens

Between E' and A'	Real object, virtual image (erect)	Figure 4.8
Between A' and K'	Virtual object, real image (erect)	Figure 4.14
Between B' and E'	Virtual object, virtual image (inverted)	Figure 4.15

object at B forms an image at B'. As the lens descends, the object is brought nearer to it through D, E and G, meeting the lens at A and passing through it to H and K. The relevant locations of the image are shown by the dotted line. All objects to the right of the lens line are virtual, and all images to the left are virtual. An identical chart can be drawn for a negative lens if the object is brought along the dotted line; the image moves along the full line (F and F' are exchanged).

Between the infinity cases and the coincident case, three types of condition apply (see Table 4.1). In all the diagrams referred to in the table, distances along the axis are drawn to scale. Figures 4.12–4.15 show how the object and image positions can be exchanged using a lens of equal but opposite power. In each case, three rays are drawn, the principal ray (a), the ray through F (b), and the ray through F' (c),

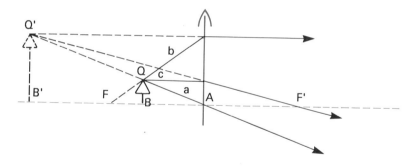

Figure 4.12 Positive lens, real object, virtual image (erect)

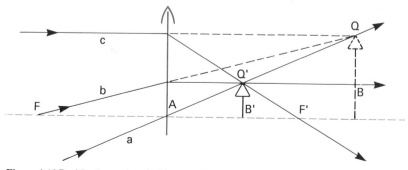

Figure 4.13 Positive lens, virtual object, real image (erect)

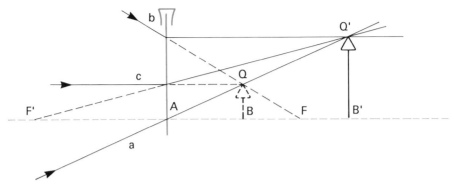

Figure 4.14 Negative lens, virtual object, real image (erect)

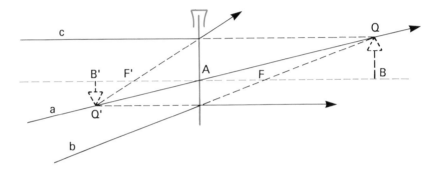

Figure 4.15 Negative lens, virtual object, virtual image (inverted)

following the rules described in Section 4.4. These rules can be extended to other constructional rays based on two further facts of refraction with thin lenses.

4. All rays that are parallel in the object space intersect, after refraction, in the second focal plane.
5. All rays from a point in the first focal plane are parallel after refraction.

This last rule is particularly useful for single ray refraction. As shown in Figure 4.16, the ray from B, an object on the axis, meets the lens at D, but it also cuts the first focal plane at E. Assume, for the moment, that a point object exists at E. A ray from E through the optical centre at A is undeviated. All other rays from E will be parallel to EA after refraction, and so the refracted ray DB′ can be drawn parallel to EA. This is the basis of the hanging Γ method developed for systems of lenses in Section 4.7. Figure 4.16 also shows that the deviation of the ray at D is the same angle as that constructed at E.

4.6 Prismatic effect and effective power

Because lenses are normally used in air and often together with other lenses, two effects can be described, within the paraxial region, which help with the study of lens systems. In the first case, a lens forms images by deviating the light. Figure

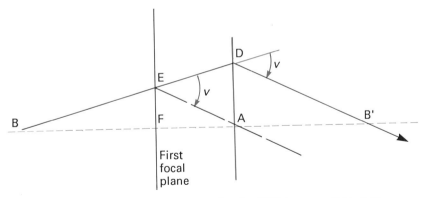

Figure 4.16 Basic construction for single ray refraction (DB′ drawn parallel to EA)

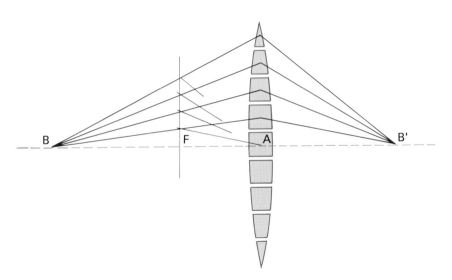

Figure 4.17 The thin lens as a series of thin prisms

4.16 shows how a ray incident on the lens at D is deviated by the angle v. This is seldom a large angle. For a focal length of 100 mm and AD equal to 20 mm, the angle DEA is about 11° rather than the 40° shown on our exaggerated diagram. Therefore, for most real lenses and certainly within the paraxial region, this deviation is about the same value as with ophthalmic prisms (Section 2.9).

With a thin lens, unlike a prism, the amount of this deviation is greater for points of incidence further from the axis. Figure 4.17 uses the construction for a number of rays and shows how a thin lens may be thought of as a collection of truncated thin prisms with apical angles which increase from zero on the axis to a maximum value at the edge of the lens. As Equation 2.20 shows, a thin prism deviates the light by an angle which depends only on the refractive index and the apical angle. Rays incident at different angles of incidence are deviated by about the same amount.

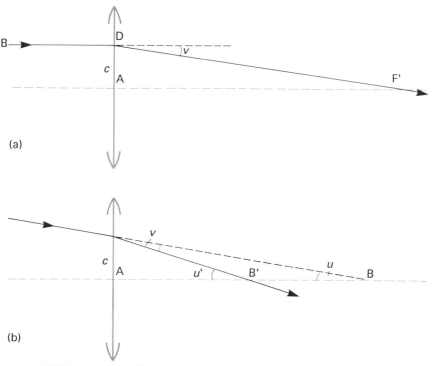

Figure 4.18 Prismatic effect of lenses

This **prismatic effect** of lenses is largely unaffected by changes in the angle of incidence, but it does depend on the distance of the point of incidence from the lens axis. Figure 4.18(a) shows a ray, parallel to the axis, incident at a height c on a lens of focal length f'. In triangle ADF' we have:

$$\tan v = \frac{c}{f'} = cF \qquad (4.23)$$

Where c is measured in metres.

Equation 2.19 defined $\tan v$ in terms of the prism power, P:

$$\tan v \times 100 = P \,(\text{prism dioptres}) \qquad (2.19)$$

Putting this into Equation 4.23 gives:

$$P\,(\text{prism dioptres}) = c\,(\text{in cm}) \times F\,(\text{in dioptres})$$

where c is known as the decentration of the ray.

In Figure 4.18(b), the incident ray is no longer parallel to the axis being directed towards the virtual object B. It is deviated by the lens towards B' and v, the angle of deviation, is:

$$v = u' - u$$

As the angles, within the paraxial region, are small,

$$u = \frac{c}{l} \quad \text{and} \quad u' = \frac{c}{l'}$$

and so

$$v = c\left(\frac{1}{l'} - \frac{1}{l}\right) = \frac{c}{f'} = cF \qquad (4.24)$$

which is the same as Equation 4.23 for small angles where tan $v = v$ (Section 3.4).

Therefore, for all paraxial rays meeting a lens at the same height, c, from the axis, the deviation produced by a thin lens of given power is constant and equal to c/f' rad (where c and f' are measured in the same units) or cF prism dioptres (c measured in centimetres and F in dioptres). The use of centimetres is against recommended practice but remains common.

The prismatic effect in lenses can be seen by looking through a lens at a fixed object. If now the lens is moved from side to side the image will also move sideways. Through a negative lens it moves in the same direction as the lens, that is, it moves *with* the lens. Through a positive lens it moves in the opposite direction to the lens, that is, it moves *against* the lens. With all thin lenses the image seen will be aligned with the object only when viewed through the optical centre of the lens where the two surfaces are effectively parallel and the prismatic effect is zero. No movement of the image from this aligned position when the lens or lens pair is moved is the basis of the power measurement called neutralization, which is described in Section 4.8.

As well as the deviation of light rays, the action of a lens is also explained by wavefronts and the change in vergence of these. When the image forming wavefront is just leaving a lens as at wavefront 1 in Figure 4.19, its vergence is L' where

$$L' = \frac{1}{l'}$$

As the wavefront moves towards the image its vergence is changing. At wavefront 2, a distance d from the lens, the vergence has changed to

$$L'_X = \frac{1}{l'_X} = \frac{1}{l' - d} = \frac{L'}{1 - dL'}$$

If the incident light is parallel, $L' = F$ and, using F_X for L'_X, we have:

$$F_X = \frac{F}{1 - dF} \tag{4.25}$$

where F is the actual power of the lens, d is the distance *in metres* to some point X, and F_X is the **effective power** of the lens at the point X.

In Figure 4.19 the effective power increases as the point X is moved nearer to the real image. For points beyond this image and for virtual images on the other side of

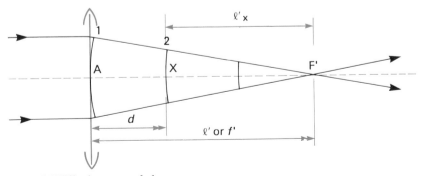

Figure 4.19 Effective power of a lens

the lens (as with negative lenses) the effective power decreases as the point X is moved further from the lens. The idea of effective power is very useful with lens systems (Chapter 5) and with spectacle lenses, which are placed a short distance in front of the eye they are meant to correct. A full treatment of this is given in H. Obstfeld, *Optics in Vision* (Butterworths, London).

4.7 More than one thin lens

Most optical instruments and 'lenses' such as camera and projector lenses have more than one lens and many of these constituent lenses are not 'thin'. The mathematics of 'thick' lenses and lens systems is described in Chapter 5 and the basics of optical instruments are explained in Chapter 6. Almost all lens systems and optical instruments have their lenses arranged on a single axis. This means that the size and position of the final image formed by a system or instrument can be found by using either the fundamental paraxial equation or the drawing methods *for each lens in turn*. The image formed by one lens becomes the object, real or virtual, for the next lens, and so on, stepping through each lens in turn.

For each lens we have

$$L_1' = L_1 + F_1$$

$$L_2' = L_2 + F_2$$

$$L_3' = L_3 + F_3 \quad \text{etc.}$$

We also have $l_2 = l_1'$ and $l_3 = l_2'$ so that $L_2 = L_1'$ and $L_3 = L_2'$ if the lenses are in contact.

Therefore we can say

$$\begin{aligned} L_3' &= L_2' + F_3 \\ &= L_2 + F_2 + F_3 \\ &= L_1' + F_2 + F_3 \\ &= L_1 + F_1 + F_2 + F_3 \end{aligned}$$

And so we can say

$$L_3' = L_1 + F \tag{4.26}$$

where F, the power of the complete system, is given by

$$F = F_1 + F_2 + F_3 \tag{4.27}$$

In fact, some space must exist between the optical centres as no lens has no thickness at all; Equation 4.27 gives approximate values.

If two thin lenses are on the same axis but separated by a distance, d, the vergence of the light leaving the first lens has changed before it reaches the second lens. We can no longer say that $L_1' = L_2$, and Equation 4.27 no longer applies.

We still have, for each lens,

$$\left.\begin{aligned} L_1' &= L_1 + F_1 \\ L_2' &= L_2 + F_2 \end{aligned}\right\} \tag{4.28}$$

but $l_2 = l_1' - d.$ $\tag{4.29}$

This does not allow a simple equation connecting L_1 and L_2' and the methods developed in Chapter 5 must be used to develop an analytical understanding of lens systems. However, we can work through the lenses to find the location and size of the final image using Equations 4.28 and 4.29 as many times as is necessary to deal with any number of lenses as long as they are all on the same axis.

Equation 4.28 uses vergences while Equation 4.29 uses distances. The extra work converting from one to the other can be reduced by changing Equation 4.29 into vergences as far as possible:

$$\frac{1}{L_2} = \frac{1}{L_1'} - d$$

which gives

$$L_2 = \frac{L_1'}{1 - dL_1'} \tag{4.30}$$

In using this equation the distance d must be in metres if, as usual, the vergences are in reciprocal metres.

Thus the step along method for three thin lenses would be:

$$
\left.
\begin{aligned}
L_1 &= \frac{1}{l_1} \\[1ex]
L_1' &= L_1 + F_1 \\[1ex]
L_2 &= \frac{L_1'}{1 - d_1 L_1'} \\[1ex]
L_2' &= L_2 + F_2 \\[1ex]
L_3 &= \frac{L_2'}{1 - d_2 L_2'} \\[1ex]
L_3' &= L_3 + F_3 \\[1ex]
l_3' &= \frac{1}{L_3'}
\end{aligned}
\right\} \tag{4.31}
$$

to give the final image distance from the last lens.

In Figure 4.20, ray drawing methods have been used on two separated positive lenses. For the first lens, the principal ray and the rays through F and F′ have been drawn (rays a, b and c as described in Section 4.4). For the second lens, the b ray

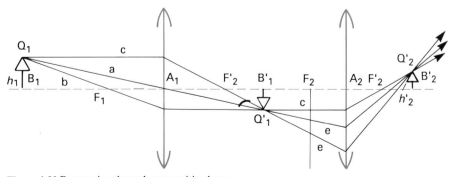

Figure 4.20 Ray tracing through two positive lenses

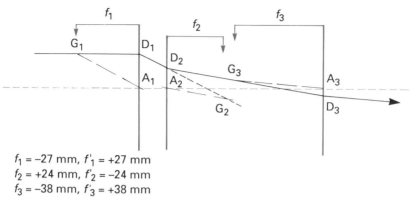

$f_1 = -27$ mm, $f'_1 = +27$ mm
$f_2 = +24$ mm, $f'_2 = -24$ mm
$f_3 = -38$ mm, $f'_3 = +38$ mm

Figure 4.21 The Γ diagram

becomes the c ray, because it is parallel to the axis before reaching the second lens. The other rays are continued through using the first focal plane method (ray e in Section 4.5).

This first focal plane method can be used on any number of lenses and a drawing technique which connects the first focal plane with lens position helps to avoid mistakes. Figure 4.21 shows this method used on three lenses. The central lens is negative so that its first focal plane is to the right of the lens. Each first focal plane is shown by drawing the first focal length from the top of the lens and ending each with a short vertical line. These focal lengths and the separation distances of the lenses are all drawn to scale. In the case shown it was convenient to make it full size.

The incident ray is drawn to meet the first lens at D_1 cutting the first focal plane at G_1. The refracted ray is simply drawn parallel to the line joining G_1 to A_1, where A_1 is the optical centre of the first lens. The same method is used for each lens in turn. The value of the drawn Γ shape (the greek letter gamma) is that it reduces the chances of mistakes caused by using the focal plane of a different lens.

To find the image positions from a finite object position, a second ray is needed drawn from the top of the object using the same method. For a full awareness of lenses and lens systems it is a very good exercise to draw a diagram of a series of lenses on a single axis with random focal lengths and separations. Two rays drawn as in Figure 4.21 will show up the final and intermediate image positions. Then give the diagram a scale and attempt to calculate the final and intermediate image positions using the step-along method. All the mathematics is very elementary, but practice is needed to avoid errors. An hour or two of practice will give the student a much better empathy with the optics of imaging systems.

If now we consider the sizes of the images rather than their positions, we find there is a lateral magnification between each image and its object, given by Equation 4.13 for each lens:

$$m = \frac{h'}{h}$$

Because the image height for the first lens is the height of the effective object for the second lens, we have the simple relationship that the overall magnification, m,

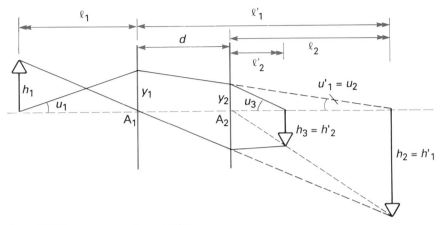

Figure 4.22 Convergence angle ray tracing

between the object and the final image for the two-lens system of Figure 4.22 is given by

$$m = \frac{h_3}{h_1} = \frac{h_1'}{h_1} \times \frac{h_2'}{h_2}$$

or

$$m = m_1 \times m_2$$

For more lenses, we have

$$m = m_1 \times m_2 \times m_3 \times m_4 \text{ etc.} \tag{4.32}$$

The fact that the intermediate magnifications, unlike the intermediate vergences, can be simply multiplied together allows the step-along mathematics to provide a value for an equivalent single lens which could replace the system. This is developed in Section 5.6.

In Figure 4.22 we have, from the pairs of similar triangles:

$$\frac{h_1}{h_1'} = \frac{l_1}{l_1'}$$

Therefore

$$\frac{h_1}{l_1} = \frac{h_1'}{l_1'} \tag{4.33}$$

and

$$\frac{h_2}{h_2'} = \frac{l_2}{l_2'}$$

therefore

$$\frac{h_2}{l_2} = \frac{h_2'}{l_2'} \tag{4.34}$$

Although $h'_1 = h_2$, we have seen that $l'_1 \neq l_2$ because of the distance between the lenses. Therefore, there is no obvious method to put Equations 4.33 and 4.34 together. If, however, we consider the axis crossing angles, or **convergence angles**, u_1, u_2 and u_3, of the *same* ray from the base of the object, we find that, within the paraxial approximation, we have

$$l_1 u_1 = A_1 D_1 = l'_1 u_2$$

and

$$l_2 u_2 = A_2 D_2 = l'_2 u_3$$

When these equations are combined with Equations 4.33 and 4.34 (by multiplying them together so that if $A_1 = A_2$ and $B_1 = B_2$ then $A_1 B_1 = A_2 B_2$) we find that

$$h_1 u_1 = h'_1 u_2 = h_2 u_2$$

and

$$h_2 u_2 = h'_2 u_3 = h_3 u_3$$

so that

$$h_1 u_1 = h_2 u_2 = h_3 u_3 \ldots \tag{4.35}$$

This equation can be extended to as many lenses as there are in the system. Each lens has a value of $h \times u$ which is invariant (unchanging) throughout the system. Equation 4.35 is the thin lens version of the **Lagrange invariant**, which was noted in Section 3.7 and will be developed further in Section 5.7. Its usefulness lies in the way it combines a ray from the axis point of the object with a ray (any ray) from the tip (or any other field point) of the object.

The convergence angles of a ray allow the fundamental paraxial equation to be expressed in terms of their values before and after refraction and the height of the ray from the lens axis at the lens. From Figure 4.22,

$$\frac{1}{l_1} = \frac{u_1}{y_1}$$

and

$$\frac{1}{l'_1} = \frac{u'_1}{y_1}$$

Therefore, from

$$\frac{1}{l'_1} = \frac{1}{l_1} + F_1$$

we have

$$u'_1 = u_1 + y_1 F_1 \tag{4.36}$$

and the transfer to the next lens is given by

$$y_2 = y_1 + d u'_1 \tag{4.37}$$

These equations can be used to trace paraxial rays through lens systems. Equation 4.36 is known as the **refraction equation** and Equation 4.37 as the **transfer**

equation. Using a computer, better accuracy is obtained if the thin lens approximation is dropped and the refraction equation applied at each surface. From

$$\frac{n'}{l'} = \frac{n}{l} + (n' - n)R$$

we obtain, by the same reasoning,

$$n'u' = nu + y\,(n' - n)R \tag{4.38}$$

The transfer equation remains the same but must be used inside each lens as well as between lenses.

4.8 Practical measurement of lens power (the optical bench)

The optical effect of a lens is described by its dioptric power. An unknown lens or newly made lens needs to have its power measured before it can be used with confidence. A number of methods are available and instruments may be bought which do the measurement automatically and provide a display or printed output showing the dioptric power.

The more simple (and troublesome) methods described here are important because the use of them teaches the fundamental optics of lenses. If the refractive index is known, the power of a lens can be found if the curvatures of its surfaces are measured. With a spherometer the values r_1 and r_2 (or R_1 and R_2) can be found and Equation 4.7 gives the power:

$$F = (n - 1)\left(\frac{1}{r_1} - \frac{1}{r_2}\right) = (n - 1)\,(R_1 - R_2) \tag{4.7}$$

where n is the known refractive index.

When spectacle lenses were all made of the same type of glass, an instrument could be made showing the power of each surface directly on its dial. This instrument is shown in Figure 3.4 but wrong values will be found if it is used on lenses made of materials other than spectacle crown glass.

Spectacles are prescribed for a patient after vision has been tested using trial lenses placed in a special empty spectacle frame worn by the patient. Trial lenses are available in extensive sets so that in combination any power from $-20\,\mathrm{D}$ to $+20\,\mathrm{D}$ in steps of $0.25\,\mathrm{D}$ or less can be obtained. With such a case of trial lenses, the power of an unknown lens can be found by **neutralization**. Because the prismatic effect (see Section 4.6) of even a weak lens or lens combination can be seen very easily, the unknown lens can be placed in contact with known lenses chosen from the trial case until the *combination* has zero prismatic effect and, therefore, zero power. The unknown lens is then known to have power equal and opposite to the lenses that neutralized it. With practice this method can be used for astigmatic lenses also.

All the other methods now to be described use the **optical bench**. This is a straight metal track or bar marked in centimetres and millimetres. Holders, which can carry lenses, screens, objects, etc., are designed to slide along this track. When the holders are properly adjusted, movement along the bench is the same as movement along the axis of the lens used. The optical bench, therefore, allows the ray diagrams in this book to be reconstructed with actual lenses and objects, etc., and the distances of these along the lens axis can be changed and measured.

The ray diagrams shown in this book are mostly describing the paraxial region and the distances above and below the axis have been exaggerated. On the optical bench, the lenses must be set in the holders with their axes parallel to the bench and, if more than one lens is used, their axes must coincide. This is easy when all the lenses are the same size, as in a trial set, and the holders are identical. If lenses of different sizes are used the holders must be adjustable and careful adjustments should be made with the holders placed close together.

The objects used must be well illuminated. A lamp placed behind a translucent screen with a cross on it or an opaque screen with a circle of holes in it makes a good object. Real images of these objects can be allowed to form on a translucent screen and these look brighter when the image is looked at *through* the screen.

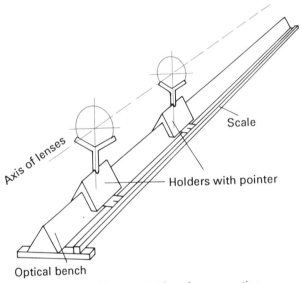

Figure 4.23 An optical bench with triangular cross-section

Optical benches as long as 3 m and as short as 250 mm are available in a number of shapes, but the most common are 1 or 2 m long with a triangular section as shown in Figure 4.23. In some designs the position scale is on the bench itself but sometimes it is a separate bar and the holders have a mark fixed to them which runs over the scale. When this mark coincides with the centre of the holder where the lens or screen is held, the distance from the thin lens to the screen is simply the difference between the two readings on the scale for the two holders. Sometimes the mark line is offset and this must be allowed for. With care, values correct to within a millimetre or two can be obtained. The mark line is sometimes replaced by a short vernier scale which allows the position of the holder to be read to 0.1 mm, but it is rare that the in-focus positions can be set to this accuracy.

4.8.1 Optical bench methods for positive lenses

Distant object

When a positive lens receives light from a very distant object, a real image is formed at the second principal focus. Light from a bright star can be used for this purpose by either tilting the optical bench so that it points towards the star or using a plane mirror to reflect the starlight along the horizontal bench. The focal length is then found from the difference between the lens holder position and the holder of the screen when it is receiving a sharp in-focus image. A street lamp 100 m away has a vergence at the lens of 0.01 D and the focal length measured in the same way will have an error of this amount. Thus a +4 D lens will appear to have a focal length of 250.62 mm. For a lamp at 10 m the apparent focal length will be 256.4 mm, a more serious error.

Collimation methods

Rather than build extra large laboratories, an object can be made to appear at infinity using a *collimator*. This instrument has a high-quality positive lens (usually a doublet lens corrected for aberrations) and an object fixed at its first principal focus. The lens and object are usually at each end of a metal tube and securely held at the correct distance apart. The object may be a narrow slit, a series of small holes or a crossline with a scale. A lamp is needed to illuminate the object and this appears to be at infinity to the lens under test as shown in Figure 4.24. When this is a positive lens, an in-focus real image can be found on a screen placed at the second principal focus. The distance between the lens and screen can be measured on the scale and is the second focal length of the lens. This method cannot be used for negative lenses.

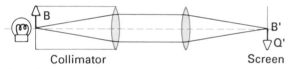

Collimator   Screen

Figure 4.24 Collimation method

If the object in the collimator is a scale of a known size, the image formed by the test lens has a size related to the focal length we are trying to measure. Using Equation 4.21 for the collimator (which is allowed because of the reversibility of light) we have:

$$h_o = f_o w$$

where h_o is the object size, f_o is the focal length of the collimator lens, and w is the angle subtended by the collimated image.

For the test lens, we can use the same equation:

$$h' = f'w$$

where h' is the measured size of the image, f' the second focal length of the test lens, and w the subtended angle as before. Putting these together, we obtain:

$$\frac{h_o}{f_o} = \frac{h'}{f'}$$

so that

$$f' = f_o \frac{h'}{h_o} \tag{4.39}$$

This equation applies only when the light between the lenses is collimated, but the lenses do not need to be close together. If the test lens is placed about the same distance from the collimator lens as the object, we find the size of the image to be measured does not change even if the screen is not exactly at the focus. This is the **telecentric principle** which is explained in Section 5.7.

4.8.2 Optical bench methods for positive and negative lenses

Telescope methods

The collimator method provides an object at infinity. The **telescope method** *looks* for an image at infinity. The various forms of telescopes are described in Chapter 6. The simplest form uses two positive lenses of unequal power arranged in an adjustable tube. A scale or crosswire is placed between the two lenses and this can be adjusted to be in sharp focus when viewed through the eyepiece. The whole instrument is used to view a distant object and adjusted so that the object and the crosswires appear in sharp focus together.

On the optical bench the telescope is set up looking through the lens to be measured. When an object is placed at the first principal focus of this lens as shown in Figure 4.25(a), its image is at infinity and this is seen in sharp focus with the crosswires in the telescope. The distance from object to lens is the first focal length. If the test lens is a negative lens, the first focal length is positive and a virtual object must be used. This can be provided by an extra positive lens. Figure 4.25(b) shows the arrangement when the object is in focus in the telescope. The position of the negative lens under test is noted. This lens is then removed and the position of the virtual object, now a real image, can be found by using a screen. The distance between these positions is the focal length.

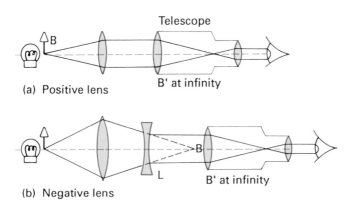

(a) Positive lens

(b) Negative lens

Figure 4.25 Telescope method

Autocollimation methods

If a real object is placed at the first principal focus of a positive lens, the image formed is at infinity and the light coming from the lens is parallel. A *plane* mirror will reflect this light so that it is returned to the lens and refocused *in the same place* as the object, as shown in Figure 4.26(a). If the object is moved closer to the lens, the image moves further away and if the object is moved further away, the image is

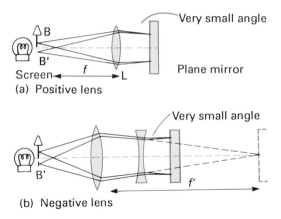

(a) Positive lens

(b) Negative lens

Figure 4.26 Autocollimation

formed closer to the lens. Coincidence means parallel light between the lens and the mirror, and this means that the object and image are both at the principal focus. The distance to the lens is the focal length.

With a negative test lens an extra positive lens is needed and, after the coincidence position has been found, the negative lens has to be taken away so that the initial object position can be found by receiving it on a screen. Figure 4.26(b) shows the arrangement.

The coincidence of object and image can be seen more easily if an object and screen are held in the same holder. With these methods the image is always formed with a magnification of −1. When the object is placed slightly to one side of the lens axis, the image is formed on the other side where the screen can be used to show the position at which the image is in sharp focus. If the object is a sharp bright pin, the image can be seen directly by an eye looking down the optical bench. The position of coincidence can be found when the image and object stay together when the head is moved from side to side, a condition of **no parallax**.

4.8.3 Optical bench methods for mirrors

Coincident methods

The same object and screen combination or the no parallax method can be used with spherical mirrors. In this case, there is no collimation and the coincidence is formed at the centre of curvature because at this position the light rays, from the axial point of the object, are incident normally on the mirror surface and are reflected back along the same path. With a concave mirror the practical arrangement is very simple, as shown in Figure 4.27(a).

For a negative convex mirror an extra positive lens is needed and this leads to a method which can be used on concave mirrors as well. The coincident image occurs when the light is striking the mirror normally, as shown in Figure 4.27(b), and the virtual image B is at the centre of curvature. However, if the mirror surface is moved back to B, the light is also reflected to give a coincident image. The distance moved is therefore the radius of curvature. This method, known as *Drysdale's method*, is most suitable for short-radius mirrors, either convex or concave, and is fully described in Chapter 6.

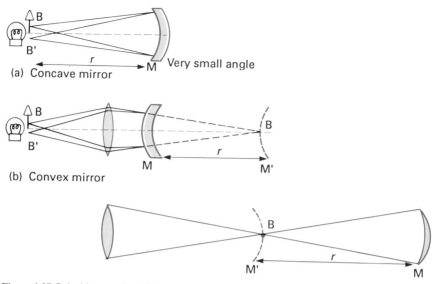

Figure 4.27 Coincident method. (b) is the basis of Drysdale's method (see Chapter 6)

4.8.4 General methods

Conjugate foci methods

The optical bench allows all the conjugate foci conditions described in Sections 4.4 and 4.5 to be set up with real lenses and the actual object and image distances measured. Real images are easily found using a small screen and virtual objects can be formed using the extra positive lens as described above. Virtual image positions are difficult to measure although an extra positive lens can be used for this. Generally these conjugate foci methods are not as accurate as the other methods described in this section because two measurements must be made, each of which is liable to error.

Exercises

4.1   The radius of curvature of one surface of a biconvex lens made of glass having a refractive index of 1.53 is six times that of the other surface. If the power of the lens is 5 D, what are its radii of curvature?

4.2   Explain what is meant by the focal power and the focal length of a thin lens, illustrating your answer by the case of both positive and negative lenses. A positive lens is to be made of glass having a refractive index of 1.55 and is required to have a focal length of 150 mm. What will be the radii of curvature of its surfaces if the lens is made (a) equi-convex, (b) plano-convex?

4.3   A thin plano-convex lens is to be made from glass of refractive index 1.62, so that its power will be +5 D. Find the radius of the tool with which the convex surface must be ground.

4.4   Draw, full size, a section through an equi-biconvex lens of 80 mm aperture and surface radii 50 mm. From a point on the axis 40 mm from the front surface draw a ray inclined to the axis at 30°. Trace its path through the lens and find the corresponding image point on the axis. Assume the refractive index to be 1.5.

4.5   Establish a formula that gives the focal lengths of a thin lens in terms of the positions of an axial object and its image.

A rod AB 15 cm in length lies along the axis of a thin converging lens of focal length 20 cm. If the nearer end of the rod is 25 cm from the lens, find the length of the image and show its position on a diagram.

4.6 Where must an object be placed in front of a convex lens of 8 cm focal length to give an image:
(a) 25 cm in front of the lens,
(b) 25 cm behind the lens.
State in each case whether the image is real or virtual.

4.7 An object is placed 333 mm from an equi-convex lens and a virtual image is formed 2 m from the lens. Calculate the radii of curvature of the lens if its refractive index is 1.523.

4.8 A small object is moved along the optical axis of a positive thin lens of focal length 25 cm. Find the position of the image and the vergence of the emergent pencil (or curvature of emergent wavefront) for each of the following object distances: (a) $-\infty$, (b) -200 cm, (c) -100 cm, (d) -50 cm, (e) -25 cm, (f) -10 cm.

4.9 An object 5 mm high is placed 600 mm in front of a negative lens of focal length 250 mm. Find (a) by calculation, (b) by graphical construction to scale, the position, size and attitude (erect or inverted) of the image.

4.10 A lamp and screen are 3 ft apart and a $+7$ D lens is mounted between them. Where must the lens be placed to give a sharp image, and what will be the magnification?

4.11 How does the transverse magnification produced by a lens of focal length f depend on the distance of the object from the lens? Calculate the focal length of the lens required to throw an image of a given object on a screen with a linear magnification of 3.5, the object being situated 20 cm from the lens.

4.12 What is meant by the magnification of an image? An object is placed 30 mm from a positive lens and a virtual image four times the size of the object is required. Find graphically and by calculation the necessary focal length of the lens.

4.13 In photographing a person 6 ft tall and standing 12 ft from the camera, it is required to obtain an image 4 in high. What must be the focal length of the lens?

4.14 An object 1 cm long is placed perpendicular to the optical axis of a positive lens, the lower end of the object being 5 cm above a point on the axis 20 cm from the lens. The lens has a focal length of 8 cm and an aperture of 3 cm. Find graphically the size and position of the image.

4.15 Apply carefully a graphical construction to find the position and size of the image of an object 2 mm high formed by a positive lens of focal length 25 cm. The object distance is (a) -50 cm, (b) -80 cm. Explain the reasons for the various steps in the construction.

4.16 Find the size and position of the image formed by a convex lens of 10 in focal length, of an arrow 1 ft 8 in long lying along the axis of the lens so that its middle point is 2 ft 1 in from the lens. What would be the size and position of the image if the arrow were turned through 90° about its middle point?

4.17 When an object is 10.5 in in front of a lens the virtual image formed is 4.3 in from the lens. What will be the position of the image when the object is 43 in in front of the lens?

4.18 A ray of light is incident in a direction inclined at 30° to the axis of a positive thin lens of 20 mm focal length, cutting the axis at a point 50 mm in front of the lens. A point moves along this ray from infinity on one side of the lens to infinity on the other. Find the conjugate ray graphically and trace the movement of the conjugate image point.

4.19 A lens of $+6.55$ D is 10 cm in diameter and is fixed horizontally 24 cm above a point source of light. Calculate the diameter of the circle of light projected on a ceiling situated 8.2 m above the lens.

4.20 A distant object subtends an angle of 5°; what focal length lens is necessary to form a real image 8 mm long?

4.21 A distant object subtends an angle of 10°; find the size of the image formed by a $+4$ D lens. Where must a -8 D lens be placed to form a real image four times the size of the original image?

4.22 A small object stands upright on the axis of a convergent lens A. Prove, by means of a diagram, that if another lens is placed in the anterior focal plane of the lens A the size of the image remains unaltered while its position is changed.

4.23 A converging lens forms an image of a distant object at a plane A and when a diverging lens is placed 1 in in front of A (between the converging lens and A) a real image is formed 5 in beyond A. Find the focal length and focal power of the diverging lens.

4.24 A thin converging lens of focal length f is used to form an image of a luminous object on a screen. Determine the least possible distance between object and screen.

4.25 A ray of light on passing through a $+5$ D lens is deviated through the same angle as it would have been when refracted by a prism of $10°$ refracting angle, $n = 1.62$. At what distance from the optical centre is the ray meeting the lens?

4.26 A thin biconvex lens having curvatures of 3 and 5 D and $n = 1.62$ is immersed in a large tank of water. An object is placed in the water 1 m from the lens; what will be the position of the image if that is also formed in the water?

4.27 In measuring the focal length of a lens by a telescope method, the telescope was focused on an object 8 ft away instead of on a distant object. With the telescope placed close to the lens a sharp image was obtained when the object was 6 in away from the lens. What was the focal length of the lens?

4.28 Describe carefully how you would determine the focal length of a weak positive lens, say of 0.5 D power. Illustrate the method by giving a typical set of bench readings, and show how the focal length could be found from these.

4.29 An illuminated object A is placed at the first principal focus of a 20 D positive lens L, on the other side of which, at some distance, is a telescope which is adjusted until the object is seen distinctly. When a lens of unknown power is placed at the second principal focus of L, between L and the telescope, it is found necessary to move the object 2 cm further away from L in order to see it distinctly in the telescope. Calculate the power of the unknown lens.

4.30 A thin convergent lens gives an inverted image the same size as the object when the latter is 20 cm from the lens. An image formed by reflection from the second surface of the lens is found to be coincident with the object at 6 cm from the lens. Find the radius of curvature of the second surface of the lens and, assuming $n = 1.5$, the radius of curvature of the first surface.

4.31 A thin converging lens and a convex spherical mirror are placed coaxially at a distance of 30 cm apart, and a luminous object, placed 40 cm in front of the lens, is found to give rise to an image coinciding with itself. If the focal length of the lens is 25 cm, determine the focal length of the mirror. Explain fully each step in your calculation.

4.32 A parallel beam of light falls normally on the plane surface of a thin plano-convex lens and is brought to a focus at a point 48 cm behind the lens. If the curved surface is silvered, the beam converges to a point 8 cm in front of the lens. What is the refractive index of the glass, and what is the radius of curvature of the convex surface?

4.33 A concave mirror is formed of a block of glass ($n = 1.62$) of central thickness 5 cm. The front surface is plane and the back surface, which is silvered, has a radius of curvature of 20 cm. Find the position of the principal focus and the position of the image of an object 20 cm in front of the plane surface.

Chapter 5

Thick lenses and systems of lenses

This chapter describes the mathematical analysis needed when the centre thickness of a lens is not taken to be zero. This gives more accurate results and we refer to such lenses as 'thick lenses' to show that we have used the more accurate mathematics rather than that their actual centre thickness is large. Almost the same mathematics applies to two thin lenses separated by a finite distance. In this case the 'thin lens' approximation still applies. Both cases are still restricted to the paraxial region and the approximations this entails.

However, well designed lens systems have definable paraxial parameters which are usable over a much wider region and these parameters can be measured accurately by the methods described in this chapter. Systems of lenses impose their own restrictions on the beams of light that can usefully pass through them. These limit the brightness and the extent of the images formed as described in the later sections of this chapter. The simple mathematics of stops and pupils is essential to the principles of optical instruments described in Chapter 6.

5.1 Focal properties of thick lenses

By 'thick lenses' we mean those lenses with a centre thickness that cannot be ignored as was done in Figure 4.4, where both t and t' were taken to be zero. With a thick lens we can no longer assume that the distance from the first surface to the image, B, formed by that surface is the same as the distance from the second surface to B as the effective object for the second surface. This means that the total power of a thick lens does *not* equal the sum of the surface powers:

$$F \neq F_1 + F_2 \quad \text{for thick lenses} \tag{5.1}$$

All lenses are thick lenses if we need to know the light paths and image positions accurately enough. All the 'thin lens' equations of Chapter 4 are approximations for real lenses and in this chapter we will study the thick lens as the more accurate calculation and see where it is essential to use it. One surprising case is the contact lens (Section 5.4) which, although thin in terms of actual thickness (about 0.1 mm), has to be considered as a thick lens to get accurate results. Of course, even these more accurate calculations are only valid within the paraxial region.

Because the lens in Figure 5.1 is a thick lens we accept that there are two distinct refractions, D_1 and D_2, for the ray coming from a distant object, before it crosses the axis at the point F' which we call the second principal focus in exactly the same

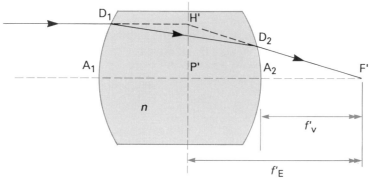

Figure 5.1 Thick lens: distant object

convention as in Chapter 4. By extending the ray paths as shown, it can be seen that these two refractions are equivalent to one refraction at H′. This suggests the idea of an **equivalent thin lens** which does not really exist but, as an imaginary idea, we can give it zero thickness so that the equations of Chapter 4 now apply accurately. This equivalent lens can be seen in the diagram to have an **equivalent focal length**, f'_E and, therefore, has an **equivalent power**, F_E, where

$$F_E = \frac{1}{f'_E} \tag{5.2}$$

This may seem an easy way to simplify the problem but, although the power of the equivalent thin lens is fixed by the curvatures, thickness and material of the thick lens, the *position* of the equivalent thin lens is found to be different for other object positions. In Figure 5.1, the distant object means that F′ is the **second principal focus**. We call H′P′ the **second principal plane** and P′ the **second principal point**.

The distance from the second vertex, A_2, of the thick lens to the second principal focus F′ is called the **second vertex focal length**, f'_v. It can be seen that in this case it is shorter than the equivalent focal length but it can be longer. We also define a **second vertex focal power**, F'_V which is given by:

$$F'_V = \frac{1}{f'_v} \tag{5.3}$$

The vertex and the principal focus are both physically real but the principal plane is a mathematical idea only.

In order to define the first vertex focal length we position the object at the first principal focus, F, so that the light finally emerges parallel to the axis, giving a distant image. Figure 5.2 shows the same lens as in Figure 5.1 but with the object at the first principal focus. Again, the light has two refractions, at the actual lens surfaces. However, when the ray paths are extended as before the equivalent single refraction is at the new position, H, as shown.

The new position, HP, of the equivalent thin lens is called the **first principal plane** and P is the **first principal point**. The equivalent focal length is found to be the same (but of opposite sign) as before and so the equivalent focal power is the same. The distance A_1F is the **first vertex focal length**, f_v, and this is usually different from the second vertex focal length. As before we define a **first vertex focal power**, F_V, so that:

$$F_V = \frac{-1}{f_v} \quad \text{(see Equation 4.11)} \tag{5.4}$$

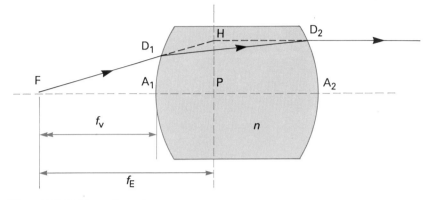

Figure 5.2 Thick lens: distant image

Once again, the vertex and the principal focus are both physically real but the principal plane is a mathematical idea only. Because of this, the equivalent focal length cannot be measured directly and the methods of Section 5.5 must be used.

For other objects and images it is found that the position of the equivalent lens moves. However, by choosing the way we measure object and image distances, it is possible to use only the principal plane positions and to do our calculations using the formula of Chapter 4. First of all we need to calculate the positions of these planes and the equivalent and vertex powers from the surface curvatures, thickness and refractive index of the thick lens itself.

In Figure 5.3, the incident light is from a distant object and so $L_1 = 0$. The power, F_1, of the first surface may be calculated as for a single spherical surface (Chapter 3) from

$$F_1 = (n - 1)R_1 \qquad (5.5)$$

where R_1 is the curvature of the first surface.

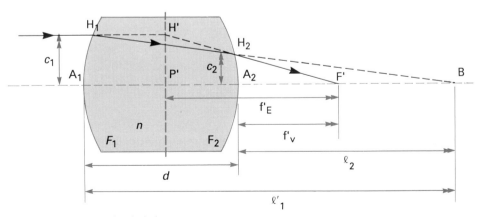

Figure 5.3 Focal length calculation

The image as formed by this surface is given by:

$$L_1' = \frac{n}{l_1'} = L_1 + F_1 = F_1$$

because $L_1 = 0$. Therefore,

$$l_1' = f_1' = \frac{n}{F_1} = A_1B$$

For the second surface, the object distance is now

$$A_2B = f_1' - d$$

$$= \frac{n}{F_1} - d$$

$$= \frac{n - dF_1}{F_1}$$

$$= l_2$$

The object vergence for this second surface is therefore

$$L_2 = \frac{n}{l_2}$$

$$= \frac{nF_1}{n - dF_1}$$

$$= \frac{F_1}{1 - (d/n)F_1} \tag{5.6}$$

and

$$l_2 = f_1' - d/n \tag{5.7}$$

The final image vergence is

$$L_2' = L_2 + F_2$$

$$= \frac{F_1}{1 - (d/n)F_1} + F_2$$

$$= \frac{F_1 + F_2 - (d/n)F_1F_2}{1 - (d/n)F_1}$$

$$= F_V' \tag{5.8}$$

because the original incident light was from a distant object. Also

$$f_V' = \frac{1}{F_V'} = \frac{1 - (d/n)F_1}{F_1 + F_2 - (d/n)F_1F_2}$$

$$= \frac{(f_1' - d/n)f_2'}{f_1' + f_2' - (d/n)} \tag{5.9}$$

using $F_1 = 1/f_1'$ and $F_2 = 1/f_2'$.

The first vertex focal power, F_V, and the first vertex focal length, f_V, can be found in the same way but with the object at the first principal focus. This is the same as transposing surface 1 for 2 and 2 for 1 so that Equations 5.8 and 5.9 give the expression for F_v and f_v if F_1 and F_2, f_1 and f_2 are interchanged. There is one difference – for the first focal length we have

$$f_v = -\frac{1}{F_V}$$

(see 5.4)

Therefore, from Equation 5.8,

$$F_V = \frac{F_1 + F_2 - (d/n)F_1 F_2}{1 - (d/n)F_2}$$

(5.10)

and

$$f_v = -\frac{f_1'\,(f_2' - d/n)}{f_1' + f_2' - d/n}$$

(5.11)

or

$$f_v = \frac{f_1'(d/n - f_2')}{f_1' + f_2' - d/n}$$

Note that in these equations we never use f_1 and f_2, the first focal lengths of the surfaces. The sign confusion possibilities are too great!

The equivalent focal power and equivalent focal lengths need different mathematics because we are dealing with an imaginary lens position. In Figure 5.3 we note there are two pairs of similar triangles. The first pair (H_1A_1B and H_2A_2B) and the second pair ($H'P'F'$ and H_2A_2F') have the height A_2H_2 common to both. The ray height at A_1 is c_1 and this reduces to c_2 at A_2. The equivalent ray height $P'H'$ is also c_1.

For the similar triangles H_1A_1B and H_2A_2B we have

$$\frac{c_1}{c_2} = \frac{f_1'}{l_2} = \frac{L_2}{F_1}$$

(5.12)

For the similar triangles $H'P'F'$ and H_2A_2F' we have

$$\frac{c_1}{c_2} = \frac{f_E'}{f_v'} = \frac{F_V'}{F_1}$$

(5.13)

Thus

$$\frac{f_E'}{f_v'} = \frac{f_1'}{l_2}$$

and so

$$f_E' = f_v' \frac{f_1'}{l_2}$$

$$= \frac{(f_1' - d/n)f_1' f_2'}{(f_1' + f_2' - d/n)\,(f_1' - d/n)}$$

which reduces to

$$f_E' = \frac{f_1' f_2'}{f_1' + f_2' - d/n}$$

(5.14)

From Equations 5.6, 5.8 and 5.13 or by transposing Equation 5.14, we find that the equivalent power is given by

$$F_E = F_1 + F_2 - (d/n)F_1F_2 \tag{5.15}$$

Both the vertex power equations (5.8 and 5.10) contain Equation 5.15 and so the three power equations can be written:

$$F_E = F_1 + F_2 - (d/n)F_1F_2 \tag{5.16}$$

$$F_V = \frac{F_E}{1 - (d/n)F_2} \tag{5.17}$$

$$F_V' = \frac{F_E}{1 - (d/n)F_1} \tag{5.18}$$

These equations contain the term d/n which is the apparent thickness of the glass between the spherical surfaces (see Equation 2.20).

When using actual values in these equations it is important to use complementary units for distance and power. *If, as is usual, the powers are in dioptres then d must be in metres.*

The three focal length equations can be written:

$$f_E' = \frac{f_1'f_2'}{f_1' + f_2' - d/n} \tag{5.19}$$

$$f_v = -f_E' \left(1 - \frac{d}{nf_2'}\right) \tag{5.20}$$

$$f_v' = f_E' \left(1 - \frac{d}{nf_1'}\right) \tag{5.21}$$

With these equations it is important that all the values are in the same units. If f_1' and f_2' are in metres then d must be in metres but if f_1' and f_2' are in millimetres then d must be in millimetres.

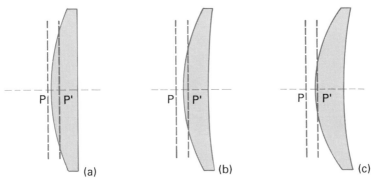

Figure 5.4 Worked example – constant back vertex focal length, principal plane positions change with bending

In the following example remember that, although the thick lens formulae give a more accurate calculation on every thickness of lens, they are still restricted to the paraxial region and have a limited accuracy in practice. The values here are given to three figures, which is somewhat more accurate than could be obtained in practice. Using values from this example, Figure 5.4 shows how the principal planes migrate as the lens becomes more meniscus. For an equi-convex thick lens they would be placed centrally.

Worked example   A spectacle lens made from glass of high index, 1.7, needs a second vertex power of $+10.0\,D$. If it has a centre thickness of $10\,mm$, find the second surface power; and the principal plane position if the first surface has a power (a) $+10.0\,D$, (b) $+12.0\,D$, (c) $+14.0\,D$.

The equation for second vertex power is

$$F'_V = \frac{F_E}{1 - (d/n)F_1}$$

$$= \frac{F_1 + F_2 - (d/n)F_1F_2}{1 - (d/n)F_1}$$

which gives

$$F'_V = \frac{F_1}{1 - (d/n)F_1} + F_2$$

or

$$F_2 = F'_V - \frac{F_1}{1 - (d/n)F_1}$$

Thus

$$F_2 = \quad -0.63\,D \qquad -2.91\,D \qquad -5.26\,D$$

for

$$F_1 = \quad +10.0\,D \qquad +12.0\,D \qquad +14.0\,D$$

Thus the equivalent focal power in each case is (from Equation 5.16)

$$F_E = +9.41\,D \qquad +9.30\,D \qquad +9.17\,D$$
$$f'_E = +106.3\,mm \qquad 107.6\,mm \qquad 109.1\,mm$$

Because F' is $100\,mm$ from A_2 ($F'_V = +10.0\,D$), the second principal plane position is known. The first vertex focal length, f_V, may be calculated in each case from

$$F_V = \frac{F_E}{1 - (d/n)F_2}$$

$$F_V = 9.38\,D \qquad 9.14\,D \qquad 8.89\,D$$
$$f_V = 106.7\,mm \qquad 109.4\,mm \qquad 112.4\,mm$$

These values are used in Figure 5.4, which shows how the principal planes migrate as the lens becomes more meniscus. For an equi-convex lens they would be placed centrally.

5.2 Focal properties of separated thin lens pairs

When two lenses are used together on a common axis, a ray of light through them is refracted four times. If the lenses are thin lenses, we may consider the ray to be refracted once at each lens – the usual thin lens approximation. The separated thin lens pair then has only two effective refractions and we find that the ideas and equations that described the thick lens in the last section also describe the thin lens pair. The only difference is that the space between them is air with a refractive index equal to one. These equations are restricted, of course, to the paraxial region and are therefore approximate. The equations when applied to a thick lens are more accurate than for a thin lens pair as the latter has the extra approximation of the 'thin' lens.

The equation for equivalent focal power, F, of a thin lens pair (from equation 5.15) is:

$$F_E = F_1 + F_2 - dF_1F_2 \tag{5.22}$$

where F_1 and F_2 are the focal powers of the two thin lenses and d is the distance between them, which must be in metres if F_1 and F_2 are in dioptres. Using the focal lengths f'_1 and f'_2 of the two thin lenses we have, from Equation 5.14, the equation for the equivalent focal length, f_E:

$$f'_E = \frac{f'_1 f'_2}{f'_1 + f'_2 - d} \tag{5.23}$$

These equations can, of course, be derived using the same methods as used in Section 5.1.

Figure 5.5 shows two thin lenses of the same powers and in the same positions as the surfaces of the thick lens in Figure 5.1. The lenses are represented by the single thin lines as was developed in Chapter 4. However, the faint outlines of the plano-convex lenses have been drawn with their curved surfaces on the lines because, as shown in Figure 5.4, this places their principal planes close to the representational line and so minimizes the thin lens error. However, when Figure 5.5 is compared with Figure 5.1, the second principal plane P' is *not* in the same place as in the thick lens case, nor has the equivalent focal length of the thin lens pair the same value as the thick lens. This is because the two surface powers of the

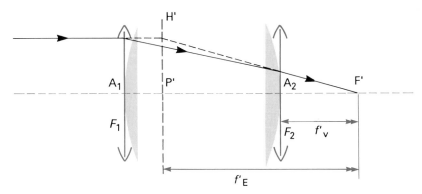

Figure 5.5 Separated thin lens pair

thick lens are separated by an optical distance equal to the 'apparent thickness' of the thick lens, d/n (see Section 2.5), while the thin lens pair is separated by d because we ignore the actual thicknesses of the lenses.

With two thin lenses of the same powers and in the same positions as the surfaces of the thick lens *and* a block of glass of the same index as the thick lens put between them, the thin lens pair is paraxially identical to the thick lens. The principal planes and principal foci will be in the same positions and Equations 5.15 and 5.14 both apply. Figure 5.6 shows this and is the same as Figure 5.1.

If the glass block has a smaller thickness, d_2, and spaces d_1 and d_3 remain between the lenses (Figure 5.7), Equation 5.22 becomes:

$$F_E = F_1 + F_2 - (d_1 + d_2/n + d_3)F_1F_2 \qquad (5.24)$$

The glass block must have plane faces which are normal to the axis of the lenses, but it can be anywhere between them. The expression $d_1 + d_2/n + d_3$ is the equivalent air separation distance, d.

When lenses are used together (with or without glass blocks between them) they form a **lens system** and when the optical axes of each lens form one single **system**

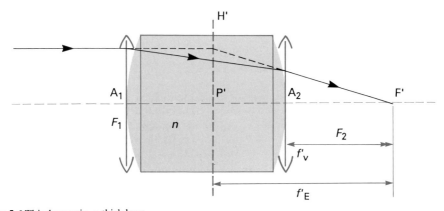

Figure 5.6 Thin lens pair → thick lens

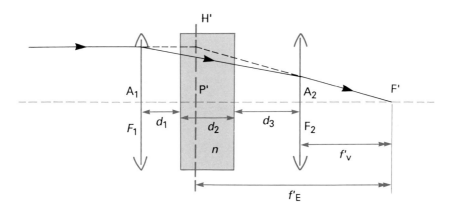

Figure 5.7 Thin lens pair with a glass block

axis, as in all the diagrams in this chapter, they are called **centred systems**. The general lens system is described in Section 5.6.

The positions of the principal points are found in the same way as for the thick lens, by using the vertex powers to locate the principal foci. Equations 5.17 and 5.18 apply in the case of thin lens pairs by merely putting the index equal to 1. Therefore for thin lens pairs we have:

$$F_V = \frac{F_E}{1 - dF_2} \tag{5.25}$$

$$F'_V = \frac{F_E}{1 - dF_1} \tag{5.26}$$

and

$$f_v = \frac{-f'_1(f'_2 - d)}{f'_1 + f'_2 - d} \tag{5.27}$$

$$f'_v = \frac{f'_2(f'_1 - d)}{f'_1 + f'_2 - d} \tag{5.28}$$

Using the values found for f_E, f'_E, f_v and f'_v we can locate P and P' by their distances from A_1 and A_2 respectively. Thus

$$A_1P = f_v - f_E = e \tag{5.29}$$

and

$$A_2P' = f'_v - f'_E = e' \tag{5.30}$$

Figure 5.8 shows the dimensions and their directions for a positive thin lens pair. Obviously Equations 5.29 and 5.30 can be used with Equations 5.27, 5.28 and 5.23 to provide direct expressions for e and e'. The fault with this approach is not only that there is more to remember but also that it encourages numbers to be blindly inserted into the equations. The author recommends that Equations 5.16–5.18 are committed to memory (plus possibly 5.14) with the knowledge that they reduce to the thin lens pair case by putting $n = 1$. Then F_E, F_V and F'_V can be calculated, f_E, f_v

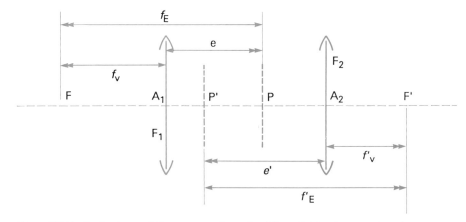

Figure 5.8 Cardinal points and distances, usual case (see Figure 5.9 for the general case)

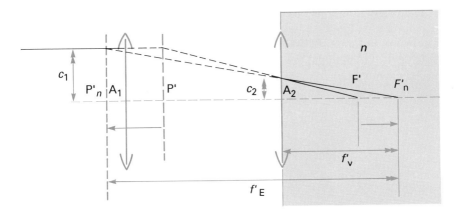

Figure 5.9 Thin lens pair: effect of unequal indices ($n = 1.5$)

and f'_v derived and a sketch diagram drawn very approximately to scale. Subsequent calculations on the imaging properties can be made and checked by drawing two of the construction rays as developed in Section 5.3.

On rare occasions the refractive index behind the lens system is different from that in front of it. In Figure 5.9, the same pair of thin plano-convex lenses is used. Again, the fiction that they are infinitely thin means that we make all the refractions take place at the solid line. If now we add material of higher index behind the second lens as a parallel sided block having zero power, the actual image moves to the right from F' to F'_n because the light ray is refracted so that $n \sin i = n'$ $\sin i'$. Because c_2 remains the same it is easy to see that, within the paraxial approximation, A_2F' is increased so that $A_2F'_n = n'(A_2F')$, or

$$f'_v = \frac{n'}{F'_V} \tag{5.31}$$

F'_V remains the same because we have not added any power.

Furthermore, when the new ray path is extended to find the location of the second principal plane we find that this too has moved. Because c_1 remains the same we have, again within the paraxial approximation, $P'_nF'_n = n'(P'F')$ or

$$f'_E = \frac{n'}{F'_E} \tag{5.32}$$

Again F'_E remains the same because we have not added any power.

These ideas are developed further in Section 5.3, but apart from the human eye there are very few situations where the image space index is different from that of the object space.

Returning to the simple case of two thin lenses in air, it is possible to draw a generalized graph showing the effect of changing the powers and the separation distance. If the first lens has a power F_1, the power of the second lens can be stated as a fraction or multiple of the first. If, for example, F_1 is $+10\,\mathrm{D}$ and F_2 is $+30\,\mathrm{D}$ then $F_2 = 3F_1$. Similarly, the distance between the lenses can be stated as a fraction or multiple of the focal length, f'_1, of the first lens. If d is 200 mm and f'_1 is 100 mm, then $d = 2f'_1$. Using F_1 and f'_1 as the 'units' we can draw the graphs shown in Figure 5.10.

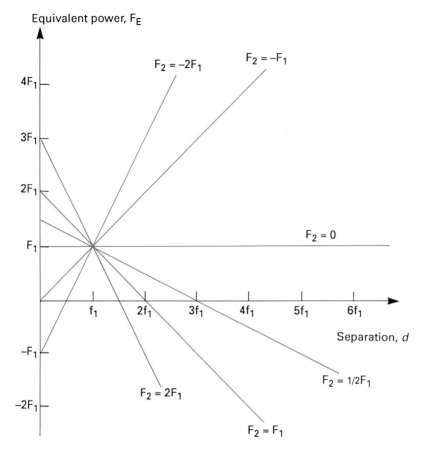

Figure 5.10 Equivalent power of a lens system in terms of the power and focal length of the first lens

When the value of F_2 is zero, the equivalent power is F_1 for any separation distance. When the value of F_2 is the same as F_1 then the equivalent power for $d = 0$ is that of two thin lenses in contact and $F = 2F_1$. As d is increased the equivalent power *decreases*. If F_2 is negative, then the equivalent power *increases* for larger d. For larger values of d the equivalent power may conceal the fact that the final image is erect and real because the second principal plane has been moved so far from the system as shown in Figure 5.51 (p. 166).

Indeed, it can be seen that for most combinations of lenses it is possible to select a separation distance for which the equivalent power is zero. The separation distance required is the sum of the focal lengths so that the second principal focus of the first lens is coincident with the first principal focus of the second lens. This makes the equivalent focal length infinite so that both the principal planes and principal foci are all located at infinity. Such systems are called **afocal systems** and have the special property that for all finite object and image conjugates the lateral magnification is constant. This special case will be excluded from the following sections although reference is made to it in Section 5.7. Such afocal systems are commonly used as telescopes and a full treatment is given in Section 6.5.

5.3 Imaging with thick lenses and thin lens pairs (the cardinal points)

The ray paths of light passing through a thick lens or separated thin lens pair can be drawn in the same way as a single thin lens if the positions of six important points are known. The six points are known as the **cardinal points**. For a thick lens or lens pair in air (or any medium which is the same in front and behind the lens) two pairs of points coincide and so, usually, only four positions can be seen.

The first pair of cardinal points are the first and second principal foci. Figure 5.11 shows a typical thick lens with F and F' marked and Figure 5.12 shows a less typical thin lens pair.

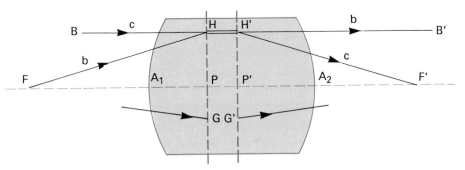

Figure 5.11 Thick lens: action of principal planes

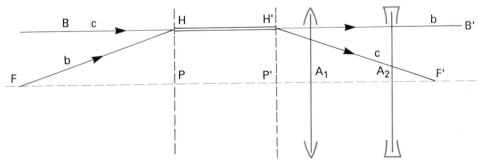

Figure 5.12 Thin lens pair: action of principal planes

The equivalent focal length can be calculated from the equations of Sections 5.1 and 5.2 and the principal points found where the equivalent lens is located for collimated incident light or collimated emergent light. These are shown at P and P' in the figures.

If a light ray c from a distant object is drawn with only the single imaginary refraction at the second principal plane P' and another ray b through the first principal focus is drawn with only the single imaginary refraction at the first principal plane P, it is seen in the figures that the rays coincide between the principal planes. If, now, we think of the rays BH and FH as being part of a beam of light forming an intermediate point object at H, the effect of the thick lens and the lens pair is to produce the rays H'B' and H'F' which could be part of a beam of light coming from a point image at H'.

This view of the ray paths means that H and H' are conjugate image points just the same as any other pair of conjugate points. PH can be thought of as an extended object and P'H' as the same-size image (because both rays between H and H' are parallel to the axis). Thus the magnification is +1 and the principal planes are sometimes called **unit planes** because of this. It therefore follows that *any* point, such as G, on the first principal plane has a corresponding point, G', on the second principal plane *at the same height* (PG = P'G') such that any incident ray going towards G will emerge as if coming from G'.

In Figures 5.11 and 5.12 the light rays are drawn as if the object space extended up to the first principal plane and the image space started at the second principal plane. In practice the actual ray paths are different within the largest region designated by P or P' or A_1 or A_2 but, allowing for the paraxial approximation, the theoretical ray paths are the same as the real rays outside this region. Thus, in Figure 5.11, theoretical and real rays coincide up to the first surface at A_1 and after the second surface at A_2. In Figure 5.12 theoretical and real rays coincide up to the first principal plane at P and after the last surface at A_2.

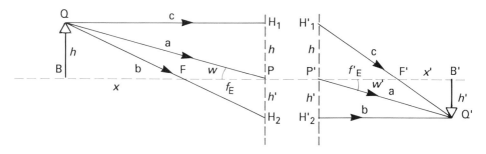

Figure 5.13 Image formation – positive thick lens or thin lens pair

Because the principal planes effectively take the place of the real lens or lenses we can use the constructional rays of Chapter 4. In Figure 5.13 only the principal planes and principal foci are drawn of a thin lens pair or thick lens. If BQ is a real object we can find the position of the image by drawing two of the three constructing rays for thin lenses, the ray through F (b) and the ray through F' (c) following the rules described in Section 4.4. It is as if the region between the principal planes did not exist. Figure 5.13 shows this effect and also an attempt to copy the principal ray, a, described in Section 4.4. Instead of directing this ray to the common vertex as with a thin lens and passing it through without deviation, a ray has been drawn from Q to P and another from P' to Q'. We know that if it is incident on the first principal plane at P it will emerge as if from P' because of the unit magnification law. We need to know if it is undeviated, that is, if angle w equals angle w'.

Because of the similar triangles QBF and FPH$_2$ we have

$$\frac{h'}{h} = \frac{-f_E}{x} \tag{5.33}$$

Where h and h' are the heights of the object and image and x is the distance from the first principal focus to the object.

Because of similar triangles $H_1'P'F'$ and $F'B'Q'$ we have

$$\frac{h'}{h} = \frac{-x'}{f_E'} \qquad (5.34)$$

where x' is the distance from the second principal focus to the image.

Thus $-f_E/x = -x'/f_E'$ and so Newton's equation (Section 4.4) holds for thick lenses and thin lens pairs.

$$xx' = f_E f_E' \qquad (5.35)$$

where f_E and f_E' are the first and second *equivalent* focal lengths.

In triangle QBP, $\tan w$ is given by

$$\tan w = \frac{h}{x + f_E}$$

In triangle Q'B'P', $\tan w'$ is given by

$$\tan w' = \frac{h'}{x' + f_E'} = \frac{-hf_E}{x}\left(\frac{1}{\dfrac{f_E f_E'}{x} + f_E'}\right)$$

$$= \frac{-hf_E}{xf_E'}\left(\frac{x}{f_E + x}\right)$$

$$= \frac{h}{f_E + x} \qquad (\text{because } f_E = -f_E')$$

Thus $\tan w = \tan w'$ and so $w = w'$; the ray is undeviated. The ray QP . . . P'Q' is therefore the ray a which was used in Section 3.7 and Section 4.4 and called the principal ray. We shall see later in this chapter that for lens systems and optical design work a different definition is used. We will therefore refer to ray, a, as the undeviated ray.

When the refractive index of the image space is different from that of the object space we find that the undeviated ray does not go through P and P'. In Figure 5.14, the refractive index in the image space is different from that in the object space. The object distance and equivalent focal power of the system are the same as in Figure 5.13. The change in image space index affects both P' and F', and $f_E = -f_E'$. By analogy with single refracting surfaces as shown in Figure 3.18 we suspect that

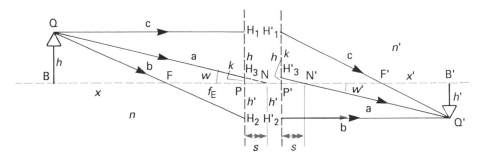

Figure 5.14 Image formation – positive thick lens or thin lens pair with different object and image space refractive index

the undeviated ray will not go through the principal points. The undeviated ray (a) in Figure 5.14 has therefore been drawn through two other points, N and N', and we now need to find where these are, knowing only that $w = w'$. In anticipation we call N and N' the **first and second nodal points**.

We do know that the intersections of this ray on the principal planes must be at the same height, k, and as $w = w'$ it therefore follows that $PN = P'N' = s$.

From the similar triangles QBN and Q'B'N', we have

$$\tan w = \frac{h}{x + f_E - s} \tag{5.36}$$

and

$$\tan w' = \frac{h'}{x' + f'_E - s} \tag{5.37}$$

We have *made* $w = w'$ so that, wherever N and N' are, we do know that

$$\frac{h}{x + f_E - s} = \tan w$$

$$= \tan w'$$

$$= \frac{h'}{x' + f'_E - s}$$

(In Figure 5.14, x and f_E are negative, while x', h' and s are positive.) Therefore

$$hx' + hf'_E - hs = h'x + h'f_E - h's$$

Using Equations 5.33 and 5.34, we have

$$-h'f'_E + hf'_E - hs = -hf_E + h'f_E - h's$$

which gives

$$(h - h')f'_E + (h - h')f_E = (h - h')s$$

so that

$$f'_E + f_E = s \tag{5.38}$$

From Figure 5.14 we can see that this means that the distance NF is equal to $-f'_E$ and the distance N'F' is equal to $-f_E$. Thus, these distances are the reverse of the equivalent focal lengths $PF = f_E$ and $P'F' = f'_E$, and the general case of Figure 5.14 is shown in Figure 5.15. When the refractive indices in front and behind the thick lens or thin lens pair are the same, then $f'_E = -f_E$; N coincides with P and N' coincides with P'. If the refractive indices remain different but the thick lens or thin lens pair reduces to a single surface power, we have the single spherical surface refraction. The principal points then coincide at the vertex of the surface and the nodal points coincide at the centre of curvature of the surface. In Figure 3.18 AF and AF' are the first and second focal lengths equal to f_E and f'_E respectively; CF and CF' are the nodal distances equal to $-f'_E$ and $-f_E$ respectively.

For the thick lens or thin lens pair case we can show the extent of the effect by extending Equation 5.38 using Equations 5.29–5.32, the definitions of the focal lengths in terms of the equivalent power. We have:

$$s = f'_E + f_E = \frac{n' - n}{F_E} \tag{5.38a}$$

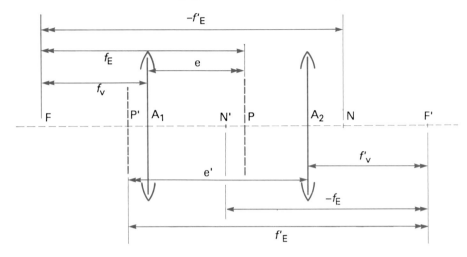

Figure 5.15 Cardinal points and distances, general case

This shows that s is zero when $n = n'$ and increases for a larger index difference.

Apart from the human eye, where the image on the retina is formed in the vitreous humour which fills the rear part of the eyeball, most systems are used with both the object and the image in air and so the nodal points N and N' coincide with the principal points P and P'. In the common case the ray, a, through P and P' is the undeviated ray. We find that not only Newton's equation but all the equations used for thin lenses can be applied to thick lenses and thin lens pairs if object distances are measured from the first principal plane and image distances from the second principal plane. If the ray paths in Figure 5.16 are compared to those in Figure 4.7 they are seen to be the same except that the thick lens has a gap between the principal planes. In fact the single thin lens is just a special case where this gap has become so small that we can take it to be zero and measure all our distances from a

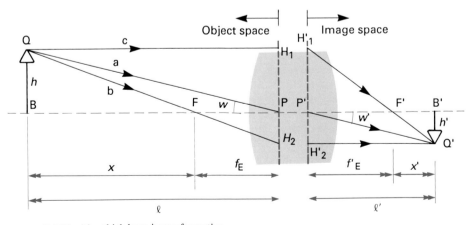

Figure 5.16 Positive thick lens: image formation

common principal plane at the centre of the thin lens. In Figure 5.16 it is seen that if the distance of the object from the first principal plane is l, then

$$l = x + f_E \qquad \text{and so} \qquad x = l - f_E$$

If the distance of the image from the second principal plane is l', then

$$l' = x' + f'_E \qquad \text{and so} \qquad x' = l' - f'_E$$

Using Newton's equation, $xx' = f_E f'_E$, we have

$$(l - f_E)(l' - f'_E) = f_E f'_E$$

$$ll' - f_E l' - f_E l + f_E f'_E = f_E f'_E$$

The $f_E f'_E$ terms cancel out and, remembering that $f'_E = -f_E$, we have

$$ll' + f'_E l' - f'_E l = 0$$

From this we obtain, by dividing each term by $ll' f_E$,

$$\frac{1}{f'_E} + \frac{1}{l} - \frac{1}{l'} = 0$$

This is the same as the fundamental paraxial equation

$$\frac{1}{l'} = \frac{1}{l} + \frac{1}{f_E} \tag{5.39}$$

and

$$L' = L + F_E \tag{5.40}$$

Thus the fundamental paraxial equation applies with thick lenses (and lens systems) if the object and image distances are measured from the appropriate principal planes.

It is not always obvious which principal plane relates to which ray. A thick lens with negative power is shown in Figure 5.17 for direct comparison with Figure 4.8. The points ADE in the earlier diagram are now replaced by PH_1H_2 and $P'H'_1H'_2$. Unlike in the positive thick lens example, the principal planes are now seen to overlap and the space between them is used both for the object space and for the image space. However, some thick lenses with negative power have a gap between the principal planes, and some thick lenses of positive power have an overlap. Often positive thin lens pairs have overlapping principal planes.

There is no value in learning any rule or equation regarding this. The best approach is careful working plus a sketch drawing.

In calculations of lenses like these only the vertex focal lengths are measured from the surfaces of the actual lens. All values to be used in the fundamental paraxial equation must be measured from one or other of the principal planes. The magnification formulae associated with thin lenses are valid for thick lenses and thin lens pairs if the object and image distances are measured in this way, Thus in Figure 5.16 the ray QPP'Q' is 'undeviated' so that QP and Q'P' are parallel. Thus triangles QBP and Q'B'P' are similar and

$$\frac{h}{l} = \frac{h'}{l'} \qquad \text{or} \qquad \frac{h'}{h} = m = \frac{l'}{l} \tag{5.41}$$

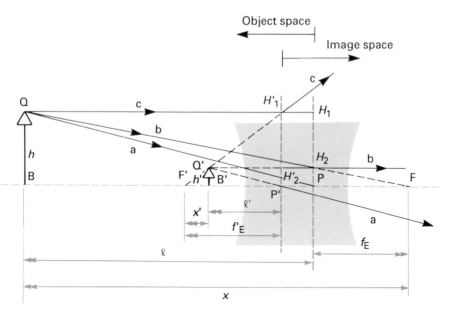

Figure 5.17 Negative thick lens: image formation

It can also be shown from Equations 5.41 and 5.39 that

$$m = 1 - \frac{l'}{f'_E} \quad \text{and} \quad \frac{1}{m} = 1 + \frac{1}{f'_E} \tag{5.42}$$

It has been shown that Newton's equation applies to the thick lens and lens pair case and we can therefore repeat the calculation of Section 4.4 to obtain the magnification along the axis. If two axial object points are defined by x_1 and x_2 we have:

$$x_1 x'_1 = f_E f'_E \quad \text{and} \quad x_2 x'_2 = f_E f'_E$$

Therefore:

$$\frac{x'_2 - x'_1}{x_2 - x_1} = \frac{\dfrac{f_E f'_E}{x_2} - \dfrac{f_E f'_E}{x_1}}{x_2 - x_1}$$

$$= \frac{f_E f'_E}{x_1 x_2}$$

$$= m_1 m_2 \tag{5.43}$$

Where m_1 and m_2 *are the lateral magnifications of* x_1 to x'_1 *and* x_2 to x'_2. Thus the longitudinal magnification is given approximately by the square of the lateral magnification as in the case of the thin lens alone.

These equations have been derived for the case of a conjugate image of an object at a finite distance. When the object is at infinity it has been previously shown that for spherical surfaces and thin lenses the image size h' is given by

$$h' = f \tan w$$

See Equations 3.21 and 4.17, where w is the angle subtended by the object and f is the first focal length.

This equation also applies in the case of the thick lens or the thin lens pair provided we use the first equivalent focal length, f_E. Thus

$$h' = f_E \tan w \tag{5.44}$$

Specifically, $f_E \tan w$ defines the height on the first principal plane of the ray through the first principal focus using f_E and w, both of which exist in object space. Because the system refracts this ray parallel to the axis, this height is also the image height. When the medium behind the thick lens or thin lens pair is different from that in front, Equation 5.44 still gives the correct answer because the image height does not change even though its location shifts.

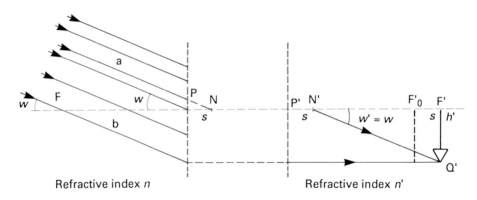

Figure 5.18 Action of nodal points

Figure 5.18 shows the general case. Of all the possible rays arriving from the extreme tip of the infinitely distant object, ray b through the first principal focus defines the image height while the undeviated ray, a, through the nodal points defines the image location by its intersection with b. If the index difference causing s were to disappear (without affecting any other values) the image would now be located at F_0' because N' would be coincident with P' without any change in the angle w. However, the height h' of the image would not change. Thus Equation 5.42 applies in all cases, including the human eye.

Worked example   A lens system comprises two thin lenses of powers $+10\,D$ and $-12\,D$ separated by 60 mm of which 30 mm is taken up by a glass block of index 1.5. Find the equivalent focal length and the location of the principal planes and principal foci. If the lens is used on a camera and focused for infinity, calculate the distance through which it must be moved to focus objects at 10 m and 2 m from the first lens. What is the magnification for each position?

The equivalent air separation distance is

$$30 + \frac{30}{1.5} = 50 \text{ mm}$$

The equivalent focal power (using Equation 5.7) is

$$F_E = 10 - 12 + 0.05 \times 10 \times 12 = 4 \text{ D}$$

$$\text{EFL} = f_E' = 250 \text{ mm}$$

$$F_V = 2.5 \text{ D} \qquad F_V' = 8 \qquad \text{(using Equations 5.25 and 5.26)}$$

$$f_v = -400 \text{ mm} \qquad f_v' = 125 \text{ mm}$$

The first principal plane is therefore 150 mm in front of the first lens and the second principal plane is 125 mm in front of the second lens. Figure 5.19 is a sketch of the layout. For objects at infinity, the film of the camera will be at F'. For nearer objects the lens must be refocused. If it is moved forward by a distance of g metres:

1. For the object 10 m from the first lens, the object distance, l, measured from the first principal plane, will be:

   $$l = -9.85 + g \quad \text{metres}$$

   and the image distance l' will be:

   $$l' = 0.250 + g \quad \text{metres}$$

   The equivalent focal length is 0.25 m and so the fundamental paraxial equation gives:

   $$\frac{1}{0.25 + g} = \frac{1}{-9.85 + g} + \frac{1}{0.25}$$

   From this, $g = 6.5$ mm if the g^2 term is ignored. The magnification is given by:

   $$m = \frac{l'}{l} = \frac{0.2565}{-9.8435} = -0.026 = \frac{-1}{38}$$

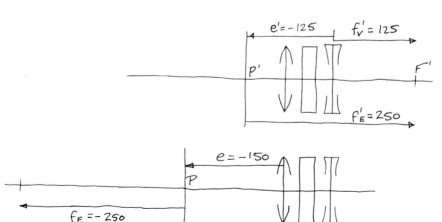

Figure 5.19

2. For the object 2 m from the first lens the same mathematics may be used but, as an alternative method, Newton's equation is employed. The extra-focal object distance measured from the first principal focus is

$$x = -1.6 + g$$

Where g is now the new distance through which the lens must be moved. The new extra-focal image distance is simply g and so we have from Newton's equation:

$$xx' = (-1.6 + g)g = -0.25 \times +0.25 = f_E f'_E$$

Ignoring g^2, we have

$$g = \frac{-0.25^2}{-1.6} = 39 \text{ mm}$$

If g^2 is not ignored, we have

$$g^2 - 1.6g + 0.0625 = 0$$

$$g = \frac{+1.6 \pm \sqrt{(2.46 - 0.25)}}{2} = 40 \text{ mm}$$

(The positive value of the square root puts the lens beyond the object.)

It is thus seen that, for objects as close as eight times the focal length, the simpler approach gives only a small error.

The magnification is given by

$$\frac{-x'}{f'_E} = \frac{-40}{250} = -0.16 = \frac{-1}{6.25}$$

5.4 The contact lens as a thick lens

Contact lenses are usually less than 0.2 mm thick and it may seem surprising that they need to be thought of as thick lenses. However, they are meniscus lenses with very high surface powers and the warning of note 1 on page 86 is appropriate to them. The concave rear surface of a contact lens has a typical value for its radius of curvature of 7.8 mm. For simplicity we assume a refractive index of 1.5 which, for the older hard lenses, is a good approximation. Thus, when the lens is in air, the power of this negative surface is -64 D! If the front surface had exactly the same radius of curvature then its surface power would be $+64$ D. On thin lens theory we would expect the power of the lens to be zero but when the thick lens formula (Equation 5.6) is used, with $d = 0.0002$ m, we find:

$$F_E = +64 - 64 + \frac{0.0002}{1.5} \times 64 \times 54$$

$$= +0.546$$

Thus the equivalent power is $+0.55$ D and the equivalent focal length is $f'_E = 1831$ mm.

The thickness, d, of the lens has an important influence. If d is reduced to 0.1 mm the equivalent power becomes $+0.27$ D. Contact lens powers are notoriously difficult to make accurately and this is one of the reasons.

The principal planes of the 0.2 mm thick lens are displaced 15 mm from the lens as is shown by using Equations 5.17 and 5.18:

$$F_V = \frac{0.546}{1 + \dfrac{0.0002}{1.5} \times 64} = 0.5415\,D$$

$$F'_{V'} = \frac{0.546}{1 - \dfrac{0.0002}{1.5} \times 64} = 0.5508\,D$$

Thus, the vertex focal lengths are $f_v = -1846.7$ mm and $f'_v = 1815.4$ mm which are 15.6 mm different from the equivalent focal length. Figure 5.20 shows the situation. Although the 15 mm displacement appears large it is a very small proportion of the focal length.

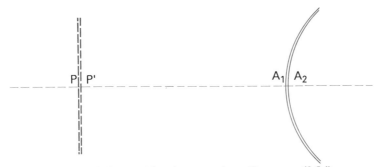

Figure 5.20 Principal plane positions for contact lens of low power (0.5 d)

These values are for the contact lens in air. When the lens is on the eye different conditions exist. However, contact lenses are usually measured in air on a **focimeter** (see Section 6.8) which gives a value based on the vertex power.

If we keep the back radius of curvature fixed at 7.8 mm but change the front radius of curvature then, for a lens thickness of 0.2 mm and a material of index 1.5, we can calculate the values as shown in Table 5.1. These values, accurate to two places of decimals, are calculated using (1, 2) Equation 3.15, (3) Equation 4.1, (4) Equation 5.16, (5) Equation 5.18.

Table 5.1

Back radius (mm)	(1) Back surface power (D)	Front radius (mm)	(2) Front surface power (D)	(3) Power of thin lens (D)	(4) Equiv. power of thick lens (D)	(5) Back vertex power (D)
7.8	−64.1	7.3	68.49	+4.39	+4.98	4.99
7.8	−64.1	7.5	66.67	+2.57	+3.14	+3.15
7.8	−64.1	7.7	64.94	+0.84	+1.40	+1.41
7.8	−64.1	7.9	63.29	−0.81	−0.27	−0.27
7.8	−64.1	8.1	61.73	−2.37	−1.84	−1.85
7.8	−64.1	8.3	60.24	−3.86	−3.35	−3.36

It can be seen that the back vertex power is nearly the same as the equivalent power but the equivalent power is very different from the 'thin lens power', which must not be used for contact lenses.

Worked example If the back radius of curvature of a contact lens is 7.6 mm on a material of index 1.48 and the required equivalent power and thickness are -5.5 D and 0.15 mm, what radius of curvature should be cut on the front?

Rear surface power:

$$F_2 = \frac{-1000}{7.6} \times 0.48 = -63.16 \text{ D}$$

(using $-1000/7.6$ to give the curvature in reciprocal metres when the radius is in millimetres).

Equivalent power:

$$F_E = -5.5 \text{ D}$$

From

$$F_E = F_1 + F_2 - (d/n)F_1F_2 \tag{5.16}$$

We have

$$F_1 = \frac{F_E - F_2}{1 - (d/n)F_2} \tag{5.45}$$

$$F_1 = \frac{+57.66}{1 + \dfrac{0.000\,15}{1.48} \times 63.16} = 57.29 \text{ D}$$

Front surface radius = 8.378 mm.

It is difficult to get any better accuracy than 8.4 mm which would give an equivalent power of -5.65 D.

5.5 Measurement of equivalent power and focal length

Although in the case of contact lenses the vertex power and equivalent power are very nearly the same, this is not true of most thick lenses and systems of lenses. The measurement of equivalent power or equivalent focal length is not straightforward. It is possible to find the position of the first principal focus and measure the distance to the first surface (or vertex) of the thick lens and so obtain the first vertex focal length, f_v. The same can be done on the second principal focus and last surface to give the second vertex focal length, f'_v, but the equivalent focal length, f_E (or EFL) must be measured from the principal plane, which is inside or somewhere outside the thick lens or lens system.

Generally, the vertex focal lengths and powers can be found using the methods described in Section 4.8 on the focimeter described in Section 6.8. For the measurement of equivalent focal length or power the following methods can be used.

5.5.1 Optical bench – Newton's method

On the optical bench (as described in Chapter 4) it is possible to make use of **Newton's equation** and obtain the equivalent focal length indirectly. In this method, the lens system is set up near the centre of the bench and the *positions* of the principal foci are found using a telescope or collimator (see Section 4.8). If an illuminated object is placed at a distance, x, outside the first principal focus, the location of the real image at a distance x' beyond the second principal focus can be found. The equivalent focal length can then be found because

$$f_E f'_E = xx' \qquad \text{Newton's equation (see Section 4.4)}$$

This method can only be used on positive systems and works most accurately when x and x' are nearly equal.

5.5.2 Optical bench – dual magnification method

Again on the optical bench it is possible to obtain the equivalent focal length directly by measuring lateral magnification at two positions. For positive systems a real object of known size is imaged by the system and the size of the image measured so that the magnification, m_1, is known. Then the object is moved along the bench and a new magnification, m_2, is measured. If the effective distance through which the object was moved is measured as d, $(l_2 - l_1)$ or the distance between the two image positions, d', $(l'_2 - l'_1)$ then the equivalent focal length can be found from:

$$f'_E = \frac{d'}{m_1 - m_2} \qquad (5.46)$$

or

$$f'_E = \frac{d}{\dfrac{1}{m_1} - \dfrac{1}{m_2}} \qquad (5.47)$$

The first of these equations is found by using Equation 4.14 twice:

$$1 - m_1 = \frac{l'_1}{f'_E} \qquad 1 - m_2 = \frac{l'_2}{f'_E}$$

Therefore

$$m_1 - m_2 = \frac{l'_1 - l'_2}{f'_E} = \frac{d'}{f'_E}$$

Equation 5.47 is found by the same method using Equation 4.15.

The same principle can be used with thick lenses and lens systems having negative equivalent power, if an auxiliary lens system is used to project a virtual object through the lens as shown in Figure 5.21. If a measuring instrument is built on this principle it can, of course, be used for positive systems as well.

5.5.3 Foco-collimator – single magnification method

If two distant point objects subtend a known angle, w, at a positive lens system, the images formed by that system will be separated by a distance h',

$$h' = 2f'_E \tan \frac{w}{2} \qquad \text{from Equation 4.22}$$

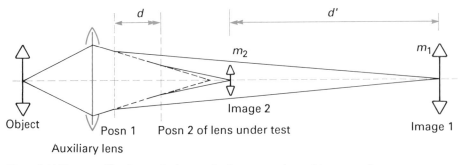

Figure 5.21 Two magnification method – negative lens system (or positive systems)

The two distant objects can be supplied by using two collimators at a fixed angle to each other or a single collimator with two point objects, provided the lens of the collimator has a good quality over the field used. The actual angle subtended may, in either case, be accurately measured using a theodolite (see Section 6.8). This angle must not be too large otherwise the field distortion of the lens being measured may give an incorrect result. If the angle is too small the magnification will be difficult to measure precisely because of the pupil aberrations of the lens being measured.

The principle is shown in Figure 5.22.

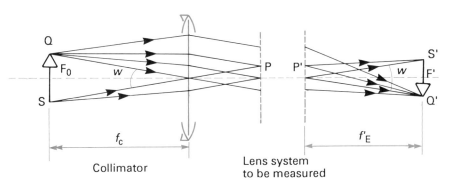

Figure 5.22 The foco-collimator

5.5.4 Rotation method – nodal points

The nodal points are points on the axis of a thick lens or lens system such that when an incident ray is aimed at the first nodal point it will emerge as if from the second nodal point at the same angle to the axis. For lens systems in air, the nodal points coincide with the principal points and so we can use them to locate the principal planes and so find the equivalent focal length.

The method uses a **nodal slide**, which is a short optical bench that can be rotated about a vertical axis which in turn is fixed to a holder that can be mounted on an optical bench. The lens to be measured is fitted to the short optical bench and

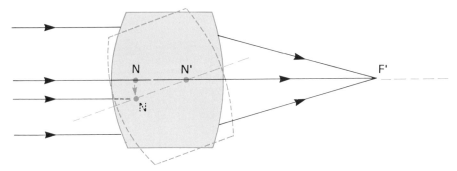

Figure 5.23 The nodal slide

moved along it during the measurement. At each position the lens is rotated from side to side while watching the image of a distant point object. Most times the image will also move from side to side, but when the lens is being rotated through its second nodal point, as shown in Figure 5.23, the image will be perfectly stationary. This is because the angle of the rays to the first nodal point N is not changing because they are parallel *and* the second nodal point N' is fixed on the optical bench because that is where the vertical axis is fixed. If the distance from this vertical axis to the in-focus image at F' is measured this gives the equivalent focal length directly.

Alternatively, if an object is located at the first principal focus and the lens system rotated about its first nodal point, the distant image viewed by a telescope will be stationary. When the stationary image is in focus and the telescope adjusted for infinity, the distance between object and rotation axis gives the equivalent focal length. This way is more accurate because of the magnification of the telescope.

5.6 The general lens system

The paraxial theory of thick lenses and separated thin lens pairs can be applied to complex lenses of any number of lens elements on the same system axis. The eye itself has a number of optical surfaces which have a common axis although the alignment is inexact. A typical camera lens may look like Figure 5.24, which contains six elemental lenses. Most lens systems are of positive equivalent power. They form the principal components in optical instruments and their complexity is largely due to the need to correct optical aberrations. Usually this involves the use of negative as well as positive lens elements.

There is very little agreement on terminology when describing these complex lens designs. A **lens system** is made up of **lens elements** but sometimes two or more adjacent lens elements are referred to as a lens group; a doublet or a triplet depending on their number. Figure 5.24 has two lens groups; a forward group and a rear group. It also contains two doublets; one a cemented doublet and the other a 'broken' doublet. Further aspects of lens system design are described in Chapter 14. These lenses have principal planes and equivalent focal lengths which can be measured using the methods of Section 5.5.

If the powers and separations are known of a group of thin lenses the step-along method of Section 4.7 may be used to calculate the position of the final image. It

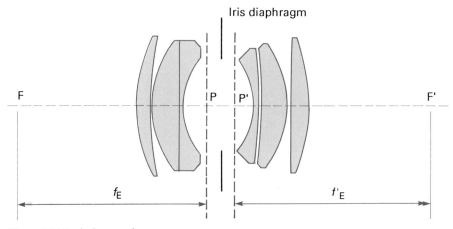

Figure 5.24 Typical camera lens

was also shown in that section that the intermediate magnifications could be multiplied to give the size of the final image. If this is done for an object at infinity, the size of the first intermediate image is given by:

$$h_1' = f_1 \tan w \tag{5.48}$$

where w is the angle subtended by the object and f_1 the first focal length of the first lens. From Equations 4.13 and 4.32 we can then see that the final image size, h', is:

$$h' = h_1' \left(\frac{L_2}{L_2'} \times \frac{L_3}{L_3'} \times \frac{L_4}{L_4'} \ldots \right) \tag{5.49}$$

However, this final image size is also given by Equation 5.44:

$$h' = f_E \tan w \tag{5.50}$$

where f_E is the first equivalent focal length of the system. Combining Equations 5.48–5.50 gives

$$f_E = f_1 \times \frac{L_2}{L_2'} \times \frac{L_3}{L_3'} \ldots \tag{5.51}$$

or

$$F_E = F_1 \times \frac{L_2'}{L_2} \times \frac{L_3'}{L_3} \ldots \tag{5.52}$$

and this appears to be a simple way of obtaining F_E except for the labour involved in the step-along method to get L_2, L_2', L_3, L_3', L_4, etc. For a two-lens system separated by a distance, d, Equation 5.52 becomes:

$$F_E = F_1 \times \frac{L_2 + F_2}{L_2}$$

$$= F_1 \left(1 + F_2 \left(f_1' - d \right) \right)$$

$$= F_1 + F_2 - dF_1F_2$$

previously given as Equation 5.15.

Using the same method for three thin lenses separated by d_1 and d_2 yields:

$$F_E = F_1 + F_2 + F_3 - d_1 F_1 F_2 - d_2 F_2 F_3 - (d_1 + d_2) F_1 F_3 + d_1 d_2 F_1 F_2 F_3 \qquad (5.53)$$

This is a simplified version of the equation for three surfaces which is developed below.

For more than three lenses, the equations become more and more cumbersome. Furthermore, the paraxial approximation and the thin lens assumption become more and more inaccurate. A computer can be programmed to treat every lens as a thick lens and to compare the paraxial values obtained with those found by ray tracing as described in Chapter 7. No matter how complex the system, it can still be considered as having an equivalent power and two principal planes. Often the positions of these are better defined than with simple systems because the complexity is there to correct the aberrations.

If Equation 5.53 is derived for three *surfaces*, as in a thick lens cemented doublet (see Section 14.3), the thin lens approximations do not apply and the equation is also useful in the analysis of the human eye (see Section 6.2). The paraxial approximations still apply, but reasonable accuracy is obtained on axis, where the eye manages a reasonable optical performance.

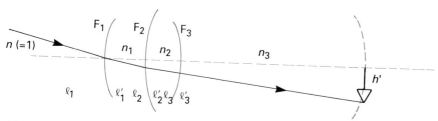

Figure 5.25 Basic optical parameters of the human eye (see also Section 6.2)

The basic equation remains Equation 5.52, but the reduced vergence values must be used for L_1, L'_1, etc. Figure 5.25 shows a basic sketch of the optics of the eye (see Section 6.2) in which the surface powers are shown and also the indices of the different media in which the vergences exist. Thus we have:

$$L_2 = \frac{n_1}{l_2} \qquad\qquad L'_2 = \frac{n_2}{l'_2}$$

$$L_3 = \frac{n_2}{l_3} \qquad\qquad L'_3 = \frac{n_3}{l'_3}$$

$$l_2 = l'_1 - d_1 = f'_1 - d_1 \qquad l_3 = l'_2 - d_2$$

and, of course,

$$L'_2 = L_2 + F_2 \quad \text{and} \quad L'_3 = L_3 + F_3 \qquad\qquad (5.54)$$

or

$$\frac{n_2}{l'_2} = \frac{n_1}{l_2} + F_2 \qquad \frac{n_3}{l'_3} = \frac{n_2}{l_3} + F_3$$

From Equation 5.52,

$$F_E = F_1 \left(\frac{l_2}{n_1} \frac{n_2}{l_2'} \times \frac{l_3}{n_2} \frac{n_3}{l_3'} \right)$$

$$= F_1 \left[\frac{l_2}{n_1} \frac{n_2}{l_2'} \times \frac{l_3}{n_2} \left\{ \frac{n_2}{l_3} + F_3 \right\} \right]$$

$$= F_1 \left[\frac{l_2}{n_1} \frac{n_2}{l_2'} \times \left\{ 1 + \frac{l_3 f_3}{n_2} \right\} \right]$$

$$= F_1 \left[\frac{l_2}{n_1} \frac{n_2}{l_2'} \left\{ 1 + \frac{l_2' F_3}{n_2} - \frac{d_2 F_3}{n_2} \right\} \right]$$

$$= F_1 \left[\frac{l_2}{n_1} \left\{ \frac{n_2}{l_2'} + F_3 - \frac{n_2}{l_2'} \frac{d_2}{n_2} F_3 \right\} \right]$$

$$= F_1 \left[\frac{l_2}{n_1} \left\{ \frac{n_2}{l_2} + F_2 + F_3 - \frac{n_2}{l_2} \frac{d_2}{n_2} F_3 - \frac{d_2}{n_2} F_2 F_3 \right\} \right]$$

$$= F_1 \left[1 + \frac{l_2}{n_1} F_2 + \frac{l_2}{n_1} F_3 - \frac{d_2}{n_2} F_3 - \frac{l_2}{n_1} \frac{d_2}{n_2} F_2 F_3 \right]$$

$$= F_1 \left[1 + \frac{f_1'}{n_1} F_2 - \frac{d_1}{n_1} F_2 + \frac{f_1'}{n_1} F_3 - \frac{d_1}{n_1} F_3 - \frac{d_2}{n_2} F_3 - \frac{f_1'}{n_1} \frac{d_2}{n_2} F_2 F_3 \right.$$
$$\left. + \frac{d_1}{n_1} \frac{d_2}{n_2} F_2 F_3 \right]$$

However, $F_1 = n_1/f_1'$; therefore

$$F_E = F_1 + F_2 + F_3 - \frac{d_1}{n_1} F_1 F_2 - \frac{d_2}{n_2} F_2 F_3 - \left(\frac{d_1}{n_1} + \frac{d_2}{n_2} \right) F_1 F_3 + \frac{d_1}{n_1} \frac{d_2}{n_2} F_1 F_2 F_3$$

$$(5.55)$$

In the eye, the final image is performed in a medium of refractive index n_3 so that the equivalent focal lengths are given by:

$$F_E = -\frac{1}{f_E} = \frac{n_3}{f_E'}$$

Equation 5.55 can be used to find image changes associated with small changes in the constituent parameters.

If the equivalent focal length of a given lens system is known and the positions of the principal foci can be easily located by experiment (as in the case for most positive systems), it is valid to use Newton's equation. This is useful when two complex systems are being used together. If their *equivalent* powers are F_1 and F_2, the equivalent power of the combination can be found by measuring the distance, g, from the second principal focus of the first system (or lens) to the first principal

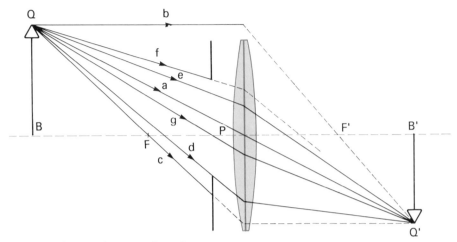

Figure 5.26 Construction rays and actual rays

focus of the second system (or lens). The distance, d, between the lenses (or between the principal planes of the systems) is then given by:

$$d = f_1' + f_2' + g \qquad (5.56)$$

When this is put into Equation 5.22 we have:

$$F_E = F_1 + F_2 - d(f_1' + f_2' + g)F_1F_2 = -gF_1F_2 \qquad (5.57)$$

where F_E is the equivalent power of the entire system. When using combinations of lens systems in the laboratory it is often the case that their equivalent powers (or focal lengths) are marked on them. Their principal foci positions can be found by experiment (locating the images of distant objects) and so Equation 5.57 allows a rapid calculation of the equivalent power of the complete experimental system. If one of the lens systems is a simple thin lens of known power, Equation 5.57 still applies provided g is measured between the relevant principal foci.

Having calculated the equivalent and vertex focal lengths and established the positions of the principal planes, the fundamental paraxial equation can be used to find image positions and magnifications. However, with systems of lenses the effects of their limited apertures is complicated. In our diagrams we usually assume unlimited apertures but this is never the real case. Even with a single lens as shown in Figure 5.26, the construction rays may not get through the lens if its usable diameter is not enough. The usable aperture and field of view of a single thin lens depend on what quality of image is acceptable. It is usually found that image quality is worse at large field angles and large apertures.

The beams and pencils passing through optical instruments and lens systems are limited by apertures, which may be the apertures of the lenses or prisms, etc., or may be additional holes known as **diaphragms** or **stops**. The iris diaphragm in a camera lens (see Figure 5.24) is an adjustable stop named after the limiting pupil in the human eye. The word 'stop' is used because its effect is to stop some rays such as ray f in Figure 5.26. This ray would be badly imaged by the real lens and actually reduce the quality of the image at Q' if it were allowed to get there. The bundle of rays from object point Q to image point Q' is now defined by rays d and e while the centre ray of the bundle is p, the principal ray.

The various stop and lens apertures in an optical system affect the amount of light passing through the system, the field of view and depth of focus. These aspects are described in the following sections of this chapter and further related to optical instrument principles in Chapter 6, while their effect on image quality is covered in Chapter 7 and their diffractive effects are considered in Chapter 13.

5.7 Aperture stop, entrance and exit pupils

The **aperture stop** is the main light limiting aperture in the lens system. If it is adjustable, as in most camera lenses, it is called the iris or iris diaphragm. The aperture stop is the smallest aperture or lens within the lens system that affects the light rays from an object on the lens axis. Sometimes this is a lens itself but usually it is an actual hole in a thin metal plate somewhere near the centre of the lens system. In Figure 5.27(a) the aperture stop is the first lens. In Figure 5.27(b) the aperture stop is the last lens. In Figure 5.27(c) the aperture stop is where the lens designer placed it so that it lets most light through the lens system without allowing the image to degrade too much. This type of design was used as a camera lens in the early years of the twentieth century. Now it is used mainly as a projection lens for slide projectors. The stop is usually close to the central negative lens; either just in front of it or just behind it.

In all these diagrams, B is shown as the long conjugate and B' as the short conjugate. In practice, however, B would normally be much further away than as shown. As a camera lens, B would be the scene to be photographed and B' the film receiving the image. As a projection lens B' would be the illuminated slide or film and B would be the screen. Because of the reversibility of light rays (Section 1.11) we can use the same mathematics for either case and will assume in Sections 5.7 and 5.8 that the light is travelling from left to right.

Figure 5.27(d) is the same lens design with the stop in a different position. The actual size of the stop A_2 is considerably different from that of stop A_1 but they allow equal amounts of light to get from the object at B to the image at B' (because the cone of rays from the object has the same size). Thus the actual size of the stop does not, on its own, determine the amount of light passing through the system. Instead we use the **entrance pupil**, which is the image of the aperture stop as seen from the object. This approach takes into account the effect of the lenses in front of the aperture stop. In Figure 5.28(a) and (b) the images of the stops (by the lens elements of 5.27(c) and (d) which are in front of them) have been found graphically and it is seen that they are the same apparent size as seen from B.

Only for the lens of Figure 5.27(a) is the pupil the same as the aperture stop, because there are no lenses in front of it. Usually the entrance pupil will be a different size from that of the aperture stop. The size of the entrance pupil is the controlling value which limits the rays from the object that eventually reach the image. For a given aperture stop, the entrance pupil does not significantly change with object position along the axis. Sometimes it is not possible to measure the actual aperture stop size, but the entrance pupil can be measured by using a traversing microscope with a long enough working distance to measure the distance between the sides of the entrance pupil. Although the microscope is focused on the actual aperture stop it will be looking at the entrance pupil because of the lenses between.

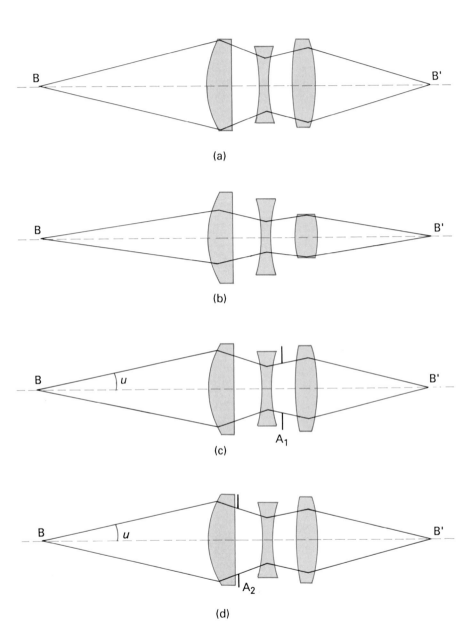

(a)

(b)

(c)

(d)

Figure 5.27 Aperture stops

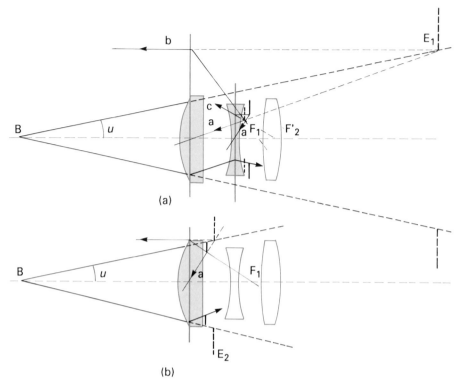

Figure 5.28 Entrance pupils

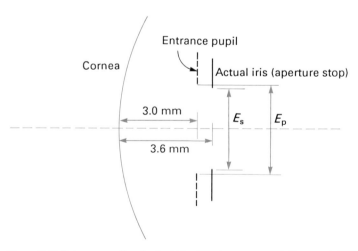

Figure 5.29 Entrance pupil and actual iris of the human eye (E_p diameter is 12% larger than E_s diameter)

Figure 5.29 shows the case for the eye where the actual iris is undoubtedly inaccessible. However, the controlling value is that of the entrance pupil and 'pupil size' in optometry and visual science means the size of the entrance pupil is measured by looking into the eye from the region of the object. The actual iris is smaller, as can be seen, but its size is never quoted.

The **exit pupil** has the same effect, but for the image rays. The exit pupil can also be used to define the amount of light reaching the image (see Section 5.9). Again, if the aperture stop is the last lens of the system, the exit pupil is the same as the aperture stop. More usually the aperture stop is inside the lens and the entrance and exit pupils are images of it as seen through the parts of the lens system in front of and behind the aperture stop. Figure 5.30 shows our standard lens with the aperture stop at A. This is imaged by the first lens to give a virtual image at E, which is the entrance pupil as seen from the object position at B. However, from the image position at B′, the aperture stop is seen imaged by the two rear lenses to

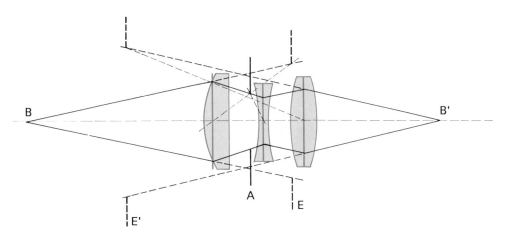

Figure 5.30 Entrance and exit pupils

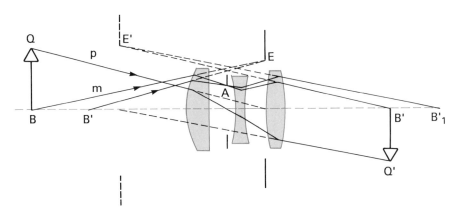

Figure 5.31 Marginal rays and principal ray

give a virtual image at E′ which is the exit pupil. Constructional rays through the centres of each lens element have been used to locate these images.

It follows from this that the whole lens may be considered to image the entrance pupil onto the exit pupil and vice versa. In Figure 5.31, the very common situation is shown where the entrance pupil is a virtual object for rays in object space while the exit pupil is a virtual image from which the rays in image space appear to be leaving. This is demonstrated by choosing two axial object positions B and B_1 and drawing rays from them which just graze the aperture stop. It can be seen that they could be going to E and emerging as if from E′.

Figure 5.31 also defines two important rays from the extended object BQ. For the total bundle or rays which leave B, the axial point of the object, the ray marked m is the most extreme. It marks the size of the accepted bundle from the axial object and is called the **marginal ray**. This is an important ray for lens design work (see Chapters 7 and 14). By the same reasoning, the ray p, from the extreme off-axis object point Q, which is directed towards the centre of the entrance pupil, is refracted to go through the centre of the aperture stop and leaves the lens as if from the centre of the exit pupil. As explained in Chapter 1, this ray defines the centre of the bundle of rays which pass through the lens from object Q to image Q′. This ray is called the **principal ray** and is also used in lens design work.

The principal ray and marginal ray are thus determined by the point objects chosen and the location of the aperture stop. They should not be confused with the a, b and c construction rays used in this and previous chapters to locate the image. In most lens systems, particularly camera lenses, the principal ray is close to ray c, which passes through the nodal points, but not invariably so. Sometimes, the aperture stop is placed outside the system and coincident with the first principal focus or the second principal focus. The former case is shown in Figure 5.32. This makes the principal ray the same as construction ray b. The value of this occurs in instruments that measure the size of the image, h'. With this arrangement any shift to S or S′ of the screen receiving the image will cause it to blur but will not immediately affect the size, h', of the image. Systems with principal rays like this are called **telecentric**. The case shown in Figure 5.32 is telecentric on the image side while an aperture stop at the second principal focus allows accurate height measurement of a near object and makes a system which is telecentric on the object side. The lens design shown in Figure 5.32 would not give good imagery with the stop so far away from the design position. A new design would be needed.

The afocal system briefly mentioned in Section 5.2 can be made telecentric for both object and image space by placing the aperture stop at the common principal focus between the two lenses. The principal ray is now parallel to the axis in object space and in image space so that the magnification of the system is constant for all

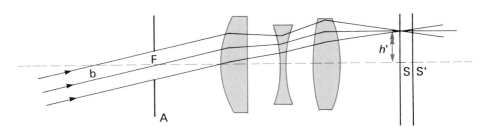

Figure 5.32 Telecentric stop

conjugates, but the range of useful conjugates is limited by the separation of the lenses. Afocal systems are normally used as telescopes and this additional telecentric aspect is discussed further in Section 6.5.

The definition of principal ray and marginal ray in a system allows values to be placed on the constituent parts of the Lagrange invariant which was introduced in Section 3.7 and defined for thin lens systems in Section 4.7. This used the axis crossing angles (Figure 4.22) of the ray we now call the marginal ray. This concept can be extended to give a paraxial ray tracing scheme for both the marginal and the principal rays.

In Section 4.7 two step-along equations were introduced which included the ray height, y, at each thin lens. For the first lens, the **refraction equation** is:

$$u_1' = u_1 + y_1 F_1 \quad (4.36) \tag{5.58}$$

and the **transfer equation** to the next lens is:

$$y_2 = y_1 + d_1 u_1' \quad (4.37) \tag{5.59}$$

where d_1 is the distance from lens 1 to lens 2.

Calculating this through a lens system gives at each lens the apertures required by the marginal ray. The aperture stop can be put in as a thin lens of zero power so that the y value can be made equal to the actual aperture. This is often done on a computer by trying various values of u_1 until the y value at the stop is correct. However, most computer programs use surface equations rather than thin lens equations as described in Sections 7.7 and 7.8.

The principal ray can be calculated using the same equations. Alternatively, the Lagrange invariant can be used with the marginal ray y values. We define the height of the principal ray at each lens using $\bar{y}$. In Figure 5.33 the m and p rays between lenses 1 and 2 are drawn. The values h_1' and u_1' are the components of the Lagrange invariant, H, for the lens system which has a single value:

$$H = hu = h_1 u_1 = h_1' u_1' = h_2 u_2 \ldots$$

(as drawn, l_1' and u_1' have negative values).

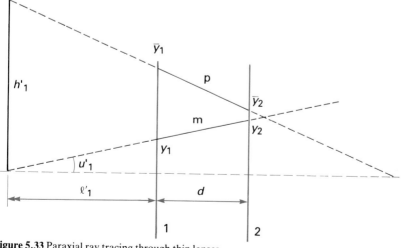

Figure 5.33 Paraxial ray tracing through thin lenses

From similar triangles on the marginal ray:

$$u'_1 = \frac{y_1 - y_2}{d} \quad \text{and} \quad l'_1 = \frac{y_1}{u'_1} \tag{5.60}$$

Similarly, on the principal ray:

$$\frac{h'_1 - \overline{y}_1}{-l'_1} = \frac{\overline{y}_1 - \overline{y}_2}{d} \tag{5.61}$$

Therefore

$$\frac{-(h'_1 - \overline{y}_1)u'_1}{y_1} = \frac{\overline{y}_1 - \overline{y}_2}{d}$$

and

$$-h'_1 du'_1 = \overline{y}_1 y_1 - \overline{y}_2 y_1 - \overline{y}_1 u'_1 d$$
$$= -\overline{y}_2 y_1 + \overline{y}_1 y_2$$

or

$$\frac{-dH}{y_1 y_2} = \frac{\overline{y}_1 y_2 - \overline{y}_2 y_1}{y_1 y_2} = \frac{\overline{y}_1}{y_1} - \frac{\overline{y}_2}{y_2} \tag{5.62}$$

The ratios $\overline{y}/y$ are given the term Q and express the eccentricity of the principal ray for lenses away from the stop position. At the stop, y is zero and so Q is zero. The value at other locations is then obtained sequentially (in either direction) by:

$$Q_2 - Q_1 = \frac{-dH}{y_1 y_2} \tag{5.63}$$

Q values are used in the definition of aberrations described in Chapter 7. The value of H for a system, depending as it does on the aperture stop size and field, gives a good indication of the design difficulty posed by the system.

5.8 Field stop, entrance and exit windows

Rays from an object on the axis of a lens system form a bundle the size of which is controlled by the aperture stop of the system. Ray bundles from the off-axis points of extended objects are controlled by the **field stop** working in conjunction with the

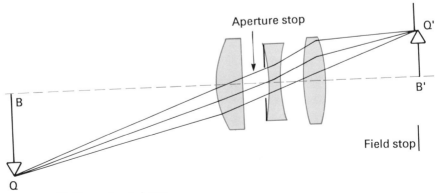

Figure 5.34 Field stop close to image

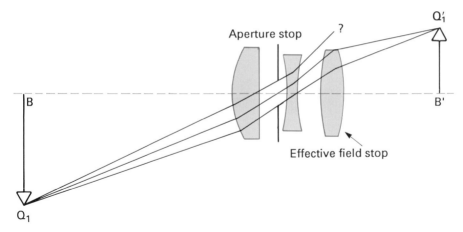

Figure 5.35 Field stop away from image

aperture stop. The field stop may be an actual aperture in a plate or one of the lenses in the system. In the simplest case, the field stop is located close to the image as in Figure 5.34. The ray bundles from object points further and further from the axis continue to fill the aperture stop until the image reaches the edge of the field stop beyond which no light is allowed to reach the image. The extent of the object which the lens actually images is called the **field of view** and is specified as the actual size for near objects or by the angle I between the extreme principal rays. Sometimes the semi-field is used, particularly in lens design descriptions.

When the field stop is in some other position the situation is more complicated. Instead of a sharp cut-off, the image brightness fades more gradually. This effect is called **vignetting** and is often used by optical designers to control the optical quality of the outer part of the image which will then remain sharp even if it is not as bright. In Figure 5.35 the field stop close to the image plane has been removed and the last lens now acts as the field stop. At the position Q'_1 shown, about half the light is reaching the image. The effect is better illustrated with a single lens stop as shown in Figure 5.36.

The off-axis bundle that contains the ray 0, which just grazes the front lens and the aperture stop, sets the limiting **field of view for full illumination** of the image. The off-axis bundle which contains the ray 1, which just grazes the front lens and passes through the centre of the aperture stop, defines (approximately – remember these apertures are usually circular) the field of view for half-illumination of the image, while the bundle containing the ray 2 grazing the lower edge of the first lens and the upper edge of the aperture stop is from the most off-axis object to get any light at all to the image. This defines the **total field of view**.

It will be realized that the edge of the field of view with this system is not sharp, depending on the closeness of the field stop to the object plane or image plane. In the system of Figure 5.35 the edge of the field of view is very diffuse because the last lens acting as the field stop is very close to the aperture stop. In most instruments a field stop is inserted close to the image or intermediate image. Its size is such that vignetting by other apertures and lens elements in the system does not reduce the brightness of the image by more than 50% before the sharp edge cuts it off entirely.

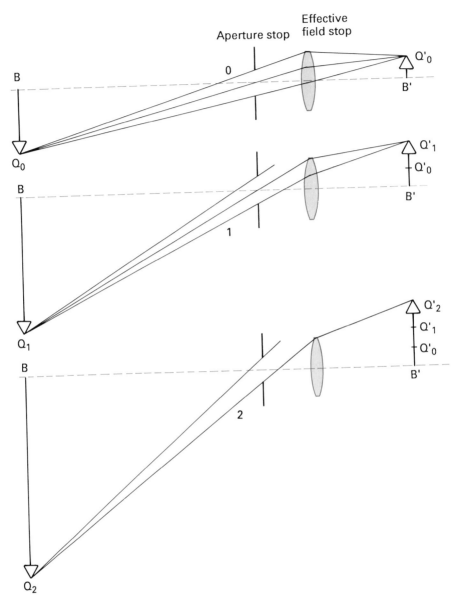

Figure 5.36 Vignetting action by effective field stop

Considering again our projection lens design, the field of view (assuming the image quality was satisfactory) could be increased by making the last lens larger, as in Figure 5.37. However, if the field size $B'Q_1'$ is sufficient, the designer would want to reduce the size of the first lens to save weight and cost. Indeed it is often difficult to see which lens is the effective field stop because the designer makes them all 'just big enough'! The central lens could be made smaller but the designer may leave it large so that it fits better into the body of the lens system and also acts as a light trap to remove stray light beams.

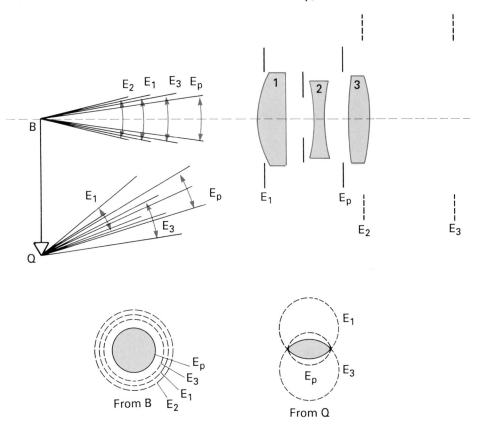

Figure 5.37 Entrance windows

The images of the limiting lens need to be viewed from the off-axis object (or image positions) to work out the vignetting action. When seen from the object space the effective field stop is known as the **entrance window** or **entrance port**.

When seen from the image space, the effective field stop is known as the **exit window** or **exit port**. In Figure 5.37 all the likely apertures (stop plus all the lenses) have been imaged into object space. The view from B shows that E_p is in fact the smallest and therefore the entrance pupil. From Q, the three smaller images are shown with their displacements due to Q being off and below the axis. It is seen that both E_1 and E_3 contribute to the size of the entrance pupil for the off-axis position. Thus both E_1 and E_3 are entrance windows. The same effect would be found if the apertures were imaged into image space and viewed from Q', the image of Q.

As stated above, it is likely that in a real system a field stop would be placed close to the image (or object) to provide a sharp cut-off at Q' (or Q) so that the illumination of the image has not fallen below 50% of the axial brightness as indicated by the relative area of the aperture viewed from Q compared with the aperture viewed from B.

Entrance and exit windows are therefore simple entities only in simple lens systems such as Figure 5.36, or when the field stop is placed close to the object or image.

Worked example   A lens system is assumed to be made from three thin lenses with the powers, diameters and separations as shown in Table 5.2, which also includes the stop position and diameter.

Table 5.2

Lens	Power (D)	Diameter (mm)	Separation (mm)
F_1	+15	40	
F_2	−30	36	20
Stop	−	24	5
F_3	+20	40	10

1. Draw a scale diagram of the system.
2. Calculate the EFL and cardinal point positions.
3. Calculate all the possible entrance and exit pupils.
4. Draw scale diagrams so that the fields of view for full and half-illumination can be found by drawing the limiting rays, assuming a distant object.

The equivalent power can be obtained by using Equation 5.53, remembering that $d_2 = (5 + 10)$ mm. This gives:

$$F_E = 15 - 30 + 20 + 0.02 \times 15 \times 30 + 0.015 \times 30 \times 20 - (0.02 + 0.015)$$
$$\times 15 \times 20 - 0.02 \times 0.015 \times 15 \times 30 \times 20$$

$$= 9.8 \, D$$

$$EFL = 102 \, mm$$

The alternative step-along method for a distant object gives:

$L_1 = 0.0 \, D$ $\qquad\qquad\qquad$ $F_1 = +15 \, D$
$L_1' = L_1 + F_1 = +15 \, D$
$l_1' = 0.067 \, m$ $\qquad\qquad\qquad$ $d_1 = 0.020 \, m$
$l_2 = l_1' - d_1 = 0.047 \, m$
$L_2 = 21.43 \, D$ $\qquad\qquad\qquad$ $F_2 = -30 \, D$
$L_2' = L_2 + F_2 = -8.57 \, D$
$l_2' = -0.117 \, m$ $\qquad\qquad\qquad$ $d_2 = 0.015 \, m$
$l_3 = l_2' - d_2 = -0.132 \, m$
$L_3 = -7.59 \, D$ $\qquad\qquad\qquad$ $F_3 = +20 \, D$
$L_3' = 12.40 \, D = F_v'$
$l_3' = f_v' = 0.0806 \, m = 80.6 \, mm$

From Equation 5.52,

$$F_E = 15 \times \frac{-8.57}{21.43} \times \frac{12.40}{-7.59}$$

$$= 9.8 \, D$$

confirming the value above.

The step-along method for a distance image gives:

$$L_3' = 0\,\text{D} \qquad\qquad\qquad F_3 = +20\,\text{D}$$
$$L_3 = L_3' - F_3 = -20\,\text{D}$$
$$l_3 = -0.050\,\text{m} \qquad\qquad\quad d_2 = 0.015\,\text{m}$$
$$l_2' = l_3 + d_2 = -0.035\,\text{m}$$
$$L_2' = -28.57\,\text{D} \qquad\qquad\quad F_2 = -30\,\text{D}$$
$$L_2 = L_2' - F_2 = +1.43\,\text{D}$$
$$l_2 = 0.699\,\text{m} \qquad\qquad\qquad d_1 = 0.020\,\text{m}$$
$$l_1' = l_2 + d_1 = 0.719\,\text{m}$$
$$L_1' = 1.39\,\text{D} \qquad\qquad\qquad F_1 = +15\,\text{D}$$
$$L_1 = L_1' - F_1 = -13.6\,\text{D} = F_v$$
$$l_1 = f_v = -0.0735\,\text{m} = -73.5\,\text{mm}$$

Again the equivalent power can be confirmed using:

$$F_E = F_3 \times \frac{L_2}{L_2'} \times \frac{L_1}{L_1'}$$

Thus

$$F_E = 20 \times \frac{1.43}{-28.57} \times \frac{-13.6}{1.39}$$

$$= 9.79\,\text{D}$$

A scale diagram showing the principal foci and principal points is given as Figure 5.38.

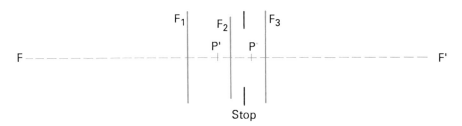

Figure 5.38 Scale diagram – worked example lens

Possible entrance pupils are calculated for each of the three lenses and the stop using F_1 as the zero position.

E_1 is coincident with F_1 and of the same diameter.

E_2 is the image of F_2 in F_1 or, more accurately, the virtual object (in object space) which is conjugate with F_2 after refraction at F_1. Thus,

$$l_1 = +0.02\,\text{m}$$
$$L_1' = 50\,\text{D} \qquad F_1 = 15\,\text{D}$$
$$L_1 = L_1' - F_1 = 35\,\text{D}$$
$$l_1 = +0.029\,\text{m, the position of }E_2\text{ with respect to }F_1$$

The size is given by

$$36\,\text{mm} \times \frac{50}{35} = 51.4\,\text{mm}$$

E_s is the image of the stop in F_2 and F_1 using the same method:

$$l_2' = 0.005 \text{ m}$$
$$L_2' = 200 \text{ D} \qquad\qquad F_2 = -30 \text{ D}$$
$$L_2 = L_2' - F = 230 \text{ D}$$
$$l_2 = 0.004 \text{ m} \qquad\qquad d_1 = 0.020 \text{ m}$$
$$l_1' = l_2 + d_1 = 0.024 \text{ m}$$
$$L_1' = 41.07 \text{ D} \qquad\qquad F_1 = +15 \text{ D}$$
$$L_1 = L_1' - F_1 = 26.07 \text{ D}$$
$$l_1 = 0.038 \text{ m, the position of } E_s \text{ with respect to } F_1$$

Its size is given by

$$24 \text{ mm} \times \frac{200}{230} \times \frac{41.07}{26.07} = 32.9 \text{ mm}$$

E_3 is the image of F_3 in F_2 and F_1 using the same method.

$$l_2' = 0.015 \text{ m}$$
$$L_2' = 66.67 \text{ D} \qquad\qquad F_2 = -30 \text{ D}$$
$$L_2 = L_2' - F_2 = 96.67 \text{ D}$$
$$l_2 = 0.0104 \text{ m} \qquad\qquad d_1 = 0.020 \text{ m}$$
$$l_1' = l_2 + d_1 = 0.0304 \text{ m}$$
$$L_1' = 32.95 \text{ D} \qquad\qquad F_1 = +15 \text{ D}$$
$$L_1 = L_1' - F_1 = 17.95 \text{ D}$$
$$l_1 = 0.056 \text{ m, the position of } E_3 \text{ with respect to } F_1$$

Its size is given by

$$40 \times \frac{66.67}{96.67} \times \frac{32.95}{17.95} = 50.6 \text{ mm}$$

Possible exit pupils are calculated for each of the three lenses and the stop using F_3 as the zero position. These are more straightforward because the exit pupils are images formed in image space with the light going through the lens in the normal left-to-right direction.

E_1' is the image of F_1 formed by F_2 and F_3. Thus,

$$l_2 = -0.020 \text{ m}$$
$$L_2 = -50 \text{ D} \qquad\qquad F_2 = -30 \text{ D}$$
$$L_2' = -80 \text{ D}$$
$$l_2' = 0.0125 \text{ m} \qquad\qquad d_2 = 0.015 \text{ m}$$
$$l_3 = l_2' - d = -0.0275 \text{ m}$$
$$L_3 = -36.36 \text{ D} \qquad\qquad F_3 = +20 \text{ D}$$
$$L_3' = -16.36 \text{ D}$$
$$l_3' = -0.061 \text{ m, the position of } E_1' \text{ with respect to } F_3$$

Its size is given by

$$40 \text{ mm} \times \frac{-50}{-80} \times \frac{-36.36}{-16.36} = 55.6 \text{ mm}$$

E_2' is the image of F_2 formed by F_3

$$l_3 = -0.015 \text{ m}$$
$$L_3 = -66.67 \text{ D} \qquad\qquad F_3 = +20 \text{ D}$$
$$L_3' = L_3 + F_3 = -46.67 \text{ D}$$
$$l_3' = -0.0214 \text{ m, the position of } E_2' \text{ with respect to } F_3$$

Its size is given by

$$36 \text{ mm} \times \frac{-66.67}{-46.67} = 51.4 \text{ mm}$$

E'_s is the image of the stop formed by F_3

$l_3 = -0.010 \text{ m}$
$L_3 = -100 \text{ D}$ $F_3 = +20 \text{ D}$
$L'_3 = L_3 + F_3 = -80 \text{ D}$
$l'_3 = -0.0125 \text{ m}$, the position of E'_s with respect to F_3

Its size is given by

$$24 \text{ mm} \times \frac{-100}{-80} = 30 \text{ mm}$$

E'_3 is coincident with F_3 and of the same diameter.

The results of these calculations are shown on the scale diagrams in Figures 5.39 and 5.40. For a distant object, light rays (1 and 2) parallel to the axis can be drawn, on Figure 5.39, through E_s without any other pupil restricting them. E_s is, therefore, the entrance pupil and can be renamed E_p. For rays at other angles, the interaction at the lowest angle occurs for the ray (3) drawn from E_p to E_1, showing that for the greater angles the entrance pupil is partially restricted by E_1. This is the limiting angle for the field of full illumination and so E_1 is the entrance window.

By convention, we define the field of half illumination as the angle at which the limiting ray (5) from E_1 goes through the centre of the entrance pupil E_p. In the diagram it can be seen that, at this angle, the other side of the entrance pupil is beginning to be restricted by E_3 (ray 6).

Applying the same principles to Figure 5.40, we must now draw converging rays to positions of the image plane through F'. A ray (3) just touching E'_1 and E'_p (E'_s) reaches the image plane at Q'_0. The other limiting ray (4) to the top of E_p is not limited by any other pupil and so Q'_0 is the actual image size for full illumination. The angle this makes in object space can be found by connecting Q'_0 with P', the second principal point. This is measured to be the same as in Figure 5.39. E'_1 is the exit window.

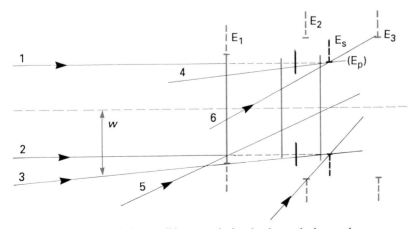

Figure 5.39 Entrance window candidates – scale drawing for worked example

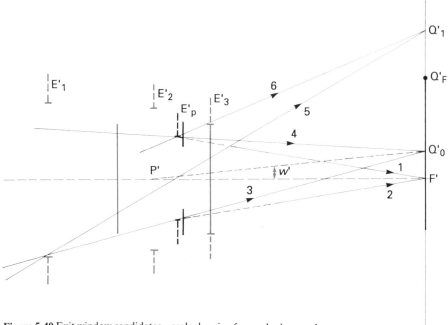

Figure 5.40 Exit window candidates – scale drawing for worked example

The field for half-illumination can be found by the ray (5) connecting the edge of E_1' with the centre of the exit pupil. This ray reaches the image plane at Q_1'. The ray (6) drawn from Q_1' to the top of E_p just clears E_3'. In a real system it is likely that a real field stop would be inserted close to the image at about Q_F', where the illumination is still well above the 50% value.

Considering Figures 5.39 and 5.40, each pair of numbered rays is in fact the same ray shown before and after travelling through the lens. Up to the field of half-illumination the first lens forms the effective field stop.

The lens system shown is a typical projection lens. We have considered it as an imaging (camera) lens which was its original design purpose. The principle of reversibility allows us to use the same mathematics provided the long conjugate remains on the left (object) side and the short conjugate on the right (image) side. As a projection lens this means that the slide or film would be placed in the F' plane of Figure 5.38 and illuminated from the right so that the image would be formed in the distant screen. Although not at infinity, there would be only a minor difference in the calculations as can be seen in the worked example in Section 5.3. See Section 6.7 for projection lens and condenser lens design.

5.9 Numerical aperture and (geometric) depth of focus

In the previous sections it has been said that the aperture stop limits the 'number of rays' and therefore the illumination of the image. Light rays as such do not exist and cannot be numbered in this way. However, for a point object that is emitting light uniformly, the amount of light reaching the image point does depend on the size of the solid cone in space from the point object to the, usually circular, entrance pupil of the lens. Therefore, for a given object distance, the greater the area of the entrance pupil the more light reaches the image.

For a distant object the same rule holds, but it will be seen in Chapter 10 that the image brightness or illumination per unit area is also dependent on the size of the image and for a distant object this depends on the equivalent focal length of the lens. For two lenses with the same size of entrance pupil the one with the shorter focal length has the smaller image to receive the same light. An image of, say, half the size has an area reduced by four times so that the image brightness is increased by four times.

Therefore, the image brightness is proportional to the square of the entrance pupil diameter and inversely proportional to the square of the equivalent focal length. In the early years of photography the lens makers unfortunately chose to describe their lenses by the ratio of equivalent focal length to the entrance pupil diameter and then to write this as an inverse fraction! This must rank as one of the more silly decisions in the history of optics. The inverse fraction is called the **f-number** of the lens and is written $f/8$, $f/16$, etc., where this shows that the equivalent focal length is eight times or 16 times the entrance pupil diameter.

However, the image brightness is given by the square of this fraction and so photographic lenses are generally arranged with adjustable aperture stops or iris diaphragms which have a series of settings giving f-numbers as in Table 5.3. Because the image brightness varies with the square of the f-number the series is arranged so that each setting is the previous value reduced by $\sqrt{2}$ and so the image brightness is halved. In photographic terms, this means the exposure time is doubled and representative values are given in the table.

Table 5.3

f-number	Exposure time
$f/1.4$	1
$f/2$	2
$f/2.8$	4
$f/4$	8
$f/5.6$	16
$f/8$	32
$f/11$	64
$f/16$	128

When the object is not at infinity an adjustment must be made depending on the focal length of the lens. This is because the new image distance has increased and the aperture must also be increased if the effective f-number is to be maintained. This is known as **reciprocity failure** because the exposure time for near objects must be increased over that shown on an independent light meter. In the case where the image is the same size as the object the image distance has increased to $2f'$ so that the intrinsic image area has increased by four times and the image brightness reduced by four times. Cameras that measure the image brightness directly (through the lens metering) give the correct exposure for all image distances.

A better scientific definition for the calculation of image brightness is **numerical aperture**. This uses the actual cone of light reaching the image and was developed

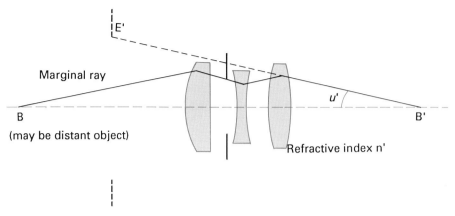

Figure 5.41 Definition of numerical aperture

by Abbe to define the maximum possible resolution of an optical system when limited by diffraction effects as described in Chapter 13. These depend on the refractive index of the image space and the equation includes this and the sine of the angle which the marginal ray (see Section 5.7) makes with the axis in image space. Figure 5.41 shows the marginal (m) ray just clearing the aperture stop (and exit pupil) and making an angle, u', with the axis. This is the semi-angle of the cone of light forming the point image.

The numerical aperture (NA) is defined as:

$$NA = n' \sin u' \tag{5.64}$$

where n' is the refractive index in which u' is measured. Any changes in image conjugate are taken into account by this method although when the numerical aperture of a lens, such as a microscope objective, is engraved on the lens housing, the value is that for the normal working conjugates of the lens.

For a lens with a distant object we have an approximate equation linking f-number with numerical aperture:

$$f\text{-number} = \frac{1}{2NA} \tag{5.65}$$

This is valid up to about $f/2$ at which the NA is 0.25. At larger numerical apertures, $\sin u'$ does not allow a simple relationship. The good thing about numerical aperture is that the numbers increase for lenses with bigger apertures and brighter images:

NA	0.01	0.03	0.1	0.25
f-number	$f/50$	$f/16$	$f/5$	$f/2$

Field effects with camera lenses are described in Chapter 6. It is well known that a camera lens with a large aperture will take pictures with shorter exposure times but suffers because all parts of a general scene may not be in focus. This is because they are at different distances from the camera and the term **depth of field** is used to describe the problem when considered in the object space and **depth of focus** when looked at in the image space. Depth of focus depends on the idea that a small amount of blur is allowable. Thus in Figure 5.42 with a lens of NA 0.25 ($f/2$) the allowed blur gives a range of image screen positions as shown.

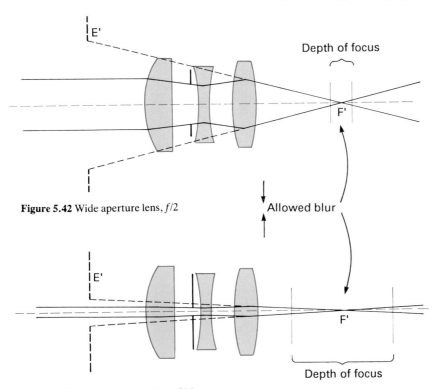

Figure 5.42 Wide aperture lens, $f/2$

Figure 5.43 Narrow aperture lens, $f/10$

A marginal ray has been drawn above and below the axis and the positions marked on the axis either side of the second principal focus where the marginal rays match the allowed blur. In Figure 5.43 where the lens is NA 0.05 ($f/10$) the depth of focus for the same allowed blur is greater, in general terms 5× greater than in the $f/2$ case. This is not necessarily the case when diffraction effects are affecting the image and the names **geometrical blur** and **geometrical depth of focus** are often used to show that these values have been obtained from ray drawings and calculations. Diffraction effects are considered in Chapter 13. The main point here is that, in the usual case, a smaller pupil gives a narrower cone of light in the image space and a receiving screen (or detector) has a wider tolerance of position for a given blur value.

When depth of field is considered, this can be regarded simply as the conjugate of the depth of focus. However, the question here is not by how much the screen can be moved but what range of object distances can be adequately received for a screen in a fixed position. In Figure 5.44, two objects have been defined on the axis but at different distances. The greatest possible distance is at infinity and the marginal rays, parallel in object space, have been drawn with solid lines that focus at F'. A closer object yields two marginal rays which cross the previous rays at the stop edges and are imaged at B'. These are drawn with broken lines.

If the allowed blur is as shown, the maximum depth of field is obtained when the screen is placed midway between the two image positions *and* experiences the maximum allowed blur. Other images between F' and B' will give less blur and the

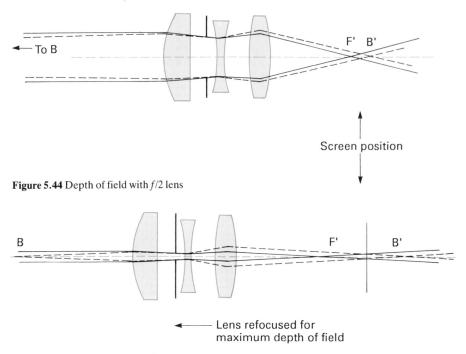

Figure 5.44 Depth of field with $f/2$ lens

Figure 5.45 Depth of field with $f/10$ lens

depth of field is the range of conjugate objects between infinity and B. Note that the screen position is not now at F'.

For the lens of reduced NA shown in Figure 5.45, the depth of field is greater and it is clear from these diagrams that the maximum depth of field is obtained by focusing the lens so that the maximum allowed blur occurs for infinitely distant objects and then objects nearer than this give sharper images then poorer images until the maximum allowed blur occurs again. The actual object distance which is in focus is called the **hyperfocal distance**. If x' is the depth of focus given by the allowable blur, then the best focus is at $x'/2$. By Newton's equation, the conjugate distance is

$$x = \frac{2ff'}{x'} = \text{Hyperfocal distance} \tag{5.66}$$

Objects from infinity to half this distance will then be in satisfactory focus.

5.10 Complex optical systems

Optical systems become complex for two main reasons. Even if the purpose of the system is relatively simple, such as an objective or camera lens providing a real image of a distant object, the necessary control of aberrations can demand a large number of surfaces. On the other hand, if the purpose itself is complicated, many lenses will be needed. The first case is considered in Chapters 7 and 14. The second is more fundamental and is considered here. The commonest complex purpose is

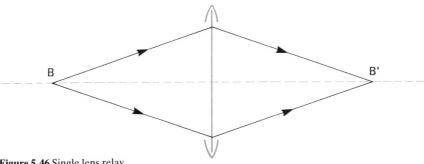

Figure 5.46 Single lens relay

that of transferring an optical image from one end of an instrument to the other. Examples of this include submarine periscopes and other military viewing systems. Often laboratory systems for experiments on vision need this sort of complexity. Such systems are generally known as **relay systems**.

Obviously, the conjugate foci of a simple thin lens constitute the basis of a relay system. In Figure 5.46 an image is being relayed over a distance BB′. The complex system may want to make this as short as possible or as long as possible. In the first case (Figure 5.46) BB′ is equal to 4f, the symmetrical planes condition of Section 4.4 where the magnification is −1. At this same magnification, Figure 5.47 shows the single thin lens replaced by two equal lenses with a separation equal to the focal length of either. BB′ is now equal to 3f and so the powers of these two lenses are less than that of the single lens in Figure 5.46. They are also smaller in diameter. Surprisingly, the equivalent power of the system is also less.

When $d = f_1 = f_2 = f$ (the focal length of either lens),

$$f'_E = \frac{f^2}{f + f - f} = f \qquad \text{(using Equation 5.19)}$$

Figure 5.47 shows how the principal planes coincide with the lenses but also overlap so that a distance equal to f is saved. It is also possible to make the system partially telecentric and symmetrical by placing an aperture stop midway between the lenses. The symmetry means that the light rays from off-axis parts of the object pass through different parts of each lens but as the ray height increases for one lens it reduces for the other. Obviously if the lenses are not big enough some vignetting will occur but in Figure 5.48 the lenses are still weaker and smaller than the single lens in Figure 5.46 and therefore their aberrations will tend to be less. The image brightness is the same because the entrance and exit pupils of Figure 5.48 accept the same cones of rays as the aperture of the lens in Figure 5.46.

When relaying an image over a large distance a single lens or even a two-lens pair cannot accept a reasonable cone of rays without needing very large diameters. The alternative is to cover the distance in a number of conjugate images. In Figure 5.49 light from the object at B is successively imaged at B′, B″, by the lenses at A_1, A_2. However, when light from Q, an off-axis part of the object, is considered very little of it reaches the second lens within a reasonable aperture. The usual solution to this problem is a positive lens, A_3, at the position of the intermediate image, which images the first lens onto the second lens. This extra lens increases the field of full illumination of the system and is therefore known as the **field lens** even though its purpose is to image the aperture stop at the first lens onto the second lens. In Figures 5.49 and 5.50 the powers and apertures of each lens are all identical.

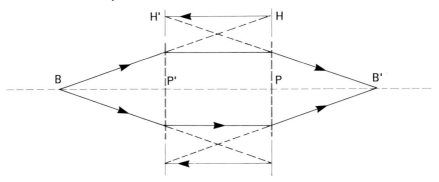

Figure 5.47 Principle of the lens pair relay

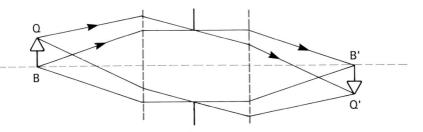

Figure 5.48 The lens pair relay in practice

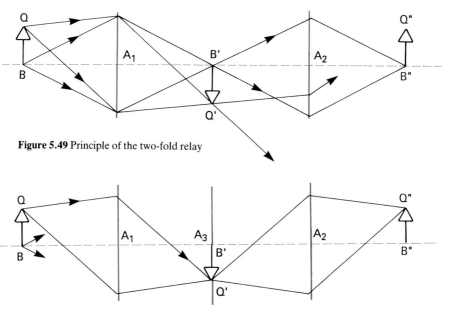

Figure 5.49 Principle of the two-fold relay

Figure 5.50 The two-fold relay in practice

In the two-lens case the separation distance is equal to $4f$ and the equivalent power is given by Equation 5.22. With $F_1 = F_2 = F$ and $d = 4/F$ we have:

$$F_E = 2F - \frac{-4F^2}{F}$$

$$= -2F$$

In the three-lens case both the separation distances are equal to $2f$ and so the equivalent power is given by Equation 5.48. With $F_1 = F_2 = F_3 = F$ and $d_1 = d_2 = 2/F$ we have:

$$F_E = 3F - 2F - 2F - 4F + 4F$$

$$= -F$$

It may seem surprising that these systems both giving real images should have different and negative equivalent powers. The situation is clarified in Figures 5.51 and 5.52, where the locations of the second principal planes and second principal foci are found by tracing parallel rays to find the equivalent lens position. The P and F are then located symmetrically because both systems are the same from either direction. Because the final image in both cases is the same size *and orientation* as the object it has a magnification of $+1$. In lens system analysis terms this means that the object and final image lie on the principal (unit) planes of each system and the different equivalent power has no effect.

The useful number of relay stages in a given system depends crucially on the aberration of the individual lenses. This is considered more extensively in Chapter 7. The use of afocal systems as potentially telecentric relays is discussed under telescope principles in Section 6.5.

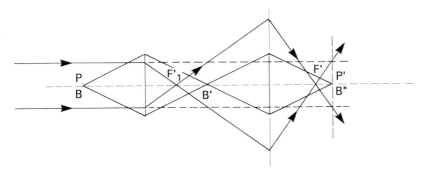

Figure 5.51 Two-fold relay, analysis of two-lens system

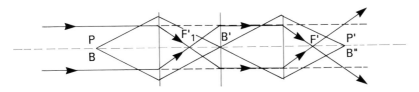

Figure 5.52 Two-fold relay, analysis of three-lens system

Exercises

5.1   A positive meniscus lens of central thickness 7.60 mm is made of glass of refractive index 1.520, the back surface power being $-6 D$. What power must be given to the front surface in order that the back vertex power of the lens shall be $+9.7 D$?

5.2   Find the equivalent power, back vertex power and position of the second principal focus of a meniscus lens (in air) with the following constants:

Radius of first surface $+50 \, mm$
Radius of second surface $+250 \, mm$
Central thickness   15 mm
Refractive index   1.6

5.3   A bi-convex lens of glass ($n = 1.52$) has curvatures of 5 in and 8 in radius respectively and a central thickness of 1½ in. Find the size and position of the image of an object 2 in long placed 15 in from the first surface of the lens.

5.4   A slab of glass ($n = 1.5$) 3 in long has its end faces ground convex so that each has a radius of curvature of ½ in. What are the optical properties of the thick lens so made?

5.5   A spherical electric bulb 3 in in diameter is immersed in water; what will be the equivalent focal length of the optical system thus produced?

5.6   Two positive thin lenses of focal length 80 mm and 60 mm respectively are coaxial and are separated by 35 mm. Find by graphical construction the equivalent focal length of the combination and the positions of its principal and focal points. Mark these positions clearly on your drawing. Check the results by calculation.

5.7   Find the equivalent focal length of a lens system consisting of a $+4 D$ and a $-4 D$ lens separated by 15 cm. What will be the position and size of the image formed by this system of an object placed 1 m in front of the first lens?

5.8   A spectacle lens of power $+10 D$ is mounted coaxially with and 5 cm in front of a second lens of power $-8 D$. Find the power of the combination and the position of the principal points. Of what well known optical system is this a model?

5.9   What are meant by the equivalent focal power and principal points of a lens system? What will be the size of the image of a distant object which subtends an angle of $8°$ at a lens system consisting of two lenses of $+10 \, in$ and $-5 \, in$ focal length, respectively, separated by 7 in?

5.10   A lens system consists of a $+4 D$ lens and a $+3 D$ lens separated by 10 cm. Find the equivalent focal length and the positions of the principal points. An object 3 cm long is situated 50 cm in front of the first lens. What will be the position and size of the image produced by the combination?

5.11   Find the equivalent focal power and the position of the principal points of a system consisting of a $+6 D$ and a $-8 D$ lens separated by 8 cm. Use the values obtained to find the position and size of the image of an object 3 cm long placed 50 cm from the first lens.

5.12   The positions of object and image such that the magnification is -1 are known as the symmetrical points. Show that the distances of these from the principal points are respectively $2f$ and $2f'$.

5.13   A Huygens eyepiece consists of two thin lenses having focal lengths of 2.5 and 1 in respectively, separated by 1.75 in. Find the equivalent focal length and the positions of the principal points. Show in a diagram, drawn to scale, the positions of these points and the focal length.

5.14   Find the positions of the focal and principal points of a Ramsden eyepiece consisting of two lenses of 2 in focal length separated by ⅔ in.

5.15   A contact lens has a front surface radius of 8 mm and a back surface radius of 7.5 mm. If it is made from a material of refractive index 1.435, calculate the back vertex power for a lens of (a) centre thickness 0.05 mm, and (b) centre thickness 0.08 mm.

5.16   In measuring the focal length of a lens system by the foco-collimator, the collimator used had a lens of 10 in focal length and a graticule in the focal plane with two marks

0.35 in apart. The size of the focused image formed by the lens being tested was 0.25 in. Find the focal length of the lens.

5.17 An equi-convex lens having curves of 10 cm radius, a central thickness of 2 cm and refractive index 1.61 closes one end of a long tube filled with water. An object 5 cm long is placed 60 cm in front of the lens: find the position and size of the image.

5.18 A +10 D equi-convex lens of glass of $n = 1.62$ is mounted in the centre of a tank 20 cm long and having thin plane glass ends; the tank is filled with water ($n = 1.33$). Find the power of the system and the position of the focus when parallel light is incident on one end of the tank.

5.19 What are the properties of the principal and nodal points of a lens?

A double convex lens of glass of refractive index 1.53 having curves of 5 and 8 D respectively and thickness of 4 cm has its second surface immersed in water ($n = 1.33$). Find the focal lengths and positions of the principal and nodal points.

5.20 Calculate the equivalent focal length of a lens system consisting of three thin positive lenses of 4, 5 and 6 in focal length respectively, each separation being 1 in.

5.21 Find the equivalent focal length and the positions of the principal points of the Rapid Landscape photographic lens having the following constants:

$r_1 = -120.8$ mm		$n_1 = 1$
$r_2 = -34.6$ mm	$d_2 = 6$ mm	$n_2 = 1.521$
$r_3 = -96.2$ mm	$d_3 = 2$ mm	$n_3 = 1.581$
$r_4 = -51.2$ mm	$d_4 = 3$ mm	$n_4 = 1.514$
		$n_5 = 1.000$

5.22 Find from the following data the equivalent focal length and the positions of the principal points of the Cooke Series IV photographic objective:

$r_1 = +19.44$ mm		$n_1 = 1.000$
$r_2 = -128.3$ mm	$d_2 = 4.29$ mm	$n_2 = 1.6110$
$r_3 = -57.85$ mm	$d_3 = 1.63$ mm	$n_3 = 1.000$
$r_4 = +18.19$ mm	$d_4 = 0.73$ mm	$n_4 = 1.5754$
$r_5 = +311.3$ mm	$d_5 = 12.9$ mm	$n_5 = 1.000$
$r_6 = -66.4$ mm	$d_6 = 3.03$ mm	$n_6 = 1.6110$
		$n_7 = 1.000$

5.23 A system consists of two thin lenses having focal length +10 and +15 cm respectively, separated by 5 cm; an aperture of 2 cm diameter is placed coaxially between the lenses and 3 cm from the shorter focus lens, which is turned towards the object. Find the positions and sizes of the entrance and exit pupils and ports.

5.24 If the lenses of the system in Exercise 5.3 have diameters of 3 cm and the object is situated 50 cm from the first lens, find the field of view of full illumination and the total field.

5.25 A +6 D lens, having an aperture euqal to one-twelfth the focal length, is correctly focused on a very distant object. Find the position of the nearest object that would have its image apparently well defined in the same plane, assuming the image to be well defined when images of points do not exceed 0.1 mm in diameter.

5.26 A camera fitted with a lens of 6 in focal length and aperture $f/8$ is correctly focused on an object 10 ft away. What will be the range of apparent good focus on either side of the object position if the image is considered sharp when blur circles do not exceed 0.005 in diameter?

5.27 With reference to the photographic camera, explain (a) the purpose of the adjustable aperture or stop, and (b) the meanings and significance of depth of focus.

If the correct exposure with an aperture of $f/8$ is $^1/_{30}$ s, what would be the exposure with the $f/16$ aperture?

If the depth of focus with the $f/8$ aperture is 0.016 in, what would be the depth of focus with the $f/16$ aperture?

Chapter 6

Principles of optical instruments

This chapter looks at the general arrangements of lenses that form optical instruments in common use. Almost all the instruments discussed are visual instruments designed for use by the eye or eyes. However, the same basic principles are involved even when the eye is replaced by a photo-detector.

The lenses are assumed to be free of aberrations so that the optical principles can be explained in terms of their size and power. Later in this book (Chapter 14) it will be possible to build on these principles by incorporating aberration theory (Chapter 7), optical material limitations (Chapter 11) and diffraction effects (Chapter 13) as a basic introduction to optical instrument design.

6.1 The camera

The camera is an optical instrument which uses a positive lens system to form a real image in a flat plane containing photographic film or, in the case of a video camera, an electronic detector such as a vidicon image tube. The most common

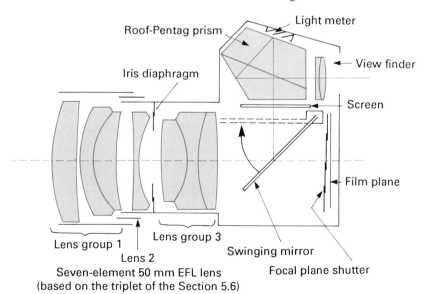

Seven-element 50 mm EFL lens
(based on the triplet of the Section 5.6)

Figure 6.1 35 mm SLR camera

photographic camera uses film 35 mm wide (Figure 6.1) while movie cameras and video cameras (for amateur use) have images about 16 mm wide. The quality of image provided by still cameras is far higher than that produced by movie or video systems. The fundamentals of camera lens design will be dealt with in Chapter 14.

The size of the film or image tube used is called the **format** of the camera (Figure 6.2) and, because this is usually a fixed size and shape, the field of view of the camera is defined by the focal length of the lens which, in a zoom lens, is adjustable. Each exposure of the film is controlled by a shutter which is commonly placed close to the film. This is called a **focal plane shutter** and allows the light from the lens to reach the film for a short period of time, usually a small fraction of a second. The camera lens contains an adjustable aperture stop, the **iris diaphragm**, which controls both the amount of light passing through the lens and the depth of field (Section 5.9). Separating the shutter (which controls the exposure time) from the iris (which controls the exposure intensity) allows the lens to be changed for different field sizes without exposing the film in the camera. A typical arrangement is shown in Figure 6.1.

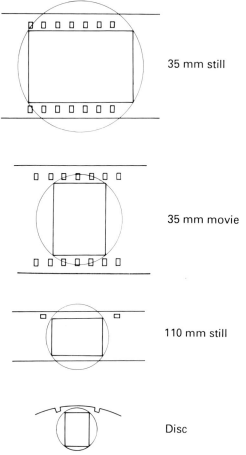

35 mm still

35 mm movie

110 mm still

Disc

Figure 6.2 Photographic formats

The photographer needs to view the size and content of the scene and the best method is called **single lens reflex (SLR)** which gives the view through the actual camera lens. Better quality cameras now have **through the lens (TTL) metering**, which measures the brightness of the scene through the actual camera lens. Both these features need a mirror behind the lens,which rapidly moves out of the way when the film is being exposed. The space needed for this mirror means that the shorter focal length camera lenses may have a back vertex focal length which is longer than the equivalent focal length.

The aperture stop and individual lenses (lens elements) of a camera lens are almost invariably circular. The format of photographic film and video systems is almost invariably rectangular. The '35 mm camera' uses film that is 35 mm wide and rolled onto a spool. Each individual photograph is 24 mm wide to allow space for a series of holes along each edge, which engage on sprockets in the camera and control the winding-on of the film from one exposure to the next. The length of each photograph along the film is 36 mm, which gives a 1.5:1 ratio between width and height as shown in Figure 6.2. The circular field of view of the camera lens has to be sufficient to reach into the corners. Thus the designed field of view of the lens is about the same as the diagonal field of the format.

In general terms,

$$\text{FOV} = 2 \tan^{-1} \frac{D/2}{\text{EFL}} \tag{6.1}$$

where D is the size of the format diagonal.

In the 35 mm case the diagonal of the format is 43.27 mm. Thus a camera lens of 50 mm equivalent focal length will need a usable field of view given (Figure 6.3) by:

$$\text{FOV} = 2 \tan^{-1} \frac{43.27}{2 \times 50} = 46.8°$$

The film format will allow photography of a distant scene 40° by 27°. If the lens is focused on a nearer scene the field of view will decrease because the new lens–film distance is longer. Thus, when the object distance is 2 m from the camera, the image distance from the second nodal point to the film for best focus is 51.25 mm.

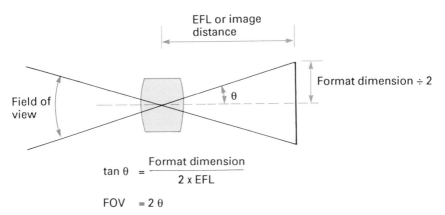

Figure 6.3 Field of view, format and EFL

Standard movie cameras also use 35 mm film but the format is across the film with a dimension along with film roll one-half that for still cameras. This gives the dimensions 24 mm × 18 mm, a ratio of 1.33:1. The same ratio is used in TV and video cameras and cine cameras using 16 mm film. More recent developments in cheap lightweight cameras, 110 format and disc, use smaller formats of 17 mm × 13 mm and 10.5 mm × 8 mm which retain the 1.33:1 ratio. The smallest cine film in common use is 8 mm wide with a format size approximately 4.8 mm × 3.7 mm, which also maintains the 1.33:1 ratio (Figure 6.2).

With each of the format sizes above, a lens of a given focal length will provide a different field of view. To maintain a consistent field of view of 40° × 30°, the 110 camera needs a lens of 23 mm EFL and the disc camera a lens of 14.5 mm EFL.

$$\text{EFL} = \frac{D/2}{\tan{(\text{FOV}/2)}} \tag{6.2}$$

However, with a given format such as that of the common 35 mm camera, the field of view can be changed only by changing the equivalent focal length of the lens. For detailed photography of distant objects a long focal length lens is needed. This will make the field of view small and therefore fill the available format with a magnified view of the scene. For example, a 250 mm EFL lens will give a field of view of 8° × 5.5° and a standing person at 20 m will just fill the short dimension. A lens of this focal length must project a long way in front of the camera, making it very unwieldy. An optical solution to this problem is the **telephoto lens**.

From the equation for the equivalent power of a thin lens pair it is possible to get a low power (long EFL) by combining a negative lens with a positive lens. In Figure 6.4, lens 1 has a power of 10 D and lens 2 a power of −12 D. With a separation of 50 mm (0.05 m) the equivalent power is +4 D. However, F'_v the back vertex power has a value of +8 D. The net result of this is a lens whose equivalent focal length and field of view for the camera format is that of a 250 mm lens but the lens itself only projects a distance of 175 mm (125 mm + 50 mm) from the camera. A more extreme design could incorporate two lenses of powers +20 D and −40 D with a separation of 30 mm. The equivalent power remains +4 D but F'_v is now 10 D. The lens now projects a distance of 130 mm (100 mm + 30 mm) from the camera.

The ratio of F/F'_v is called the **telephoto power** and values up to about 3 are possible, but the lens design becomes more difficult so more elements are needed and the weight may increase even though the lens is shorter.

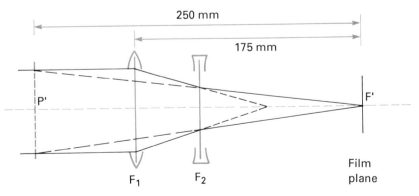

Figure 6.4 Telephoto lens

For wide angle lenses a short focal length is required. The difficulty now is that the mirror needed for the single lens reflex viewfinder must cover the field but still have room to move out of the way. This means that the back vertex focal length of the lens must not be less than about 34 mm. With all the correcting lenses in place, this is not too difficult down to an equivalent focal length of 40 mm. For lenses of shorter EFL than this an **inverted telephoto** or **retrofocus lens** is needed.

As the first name implies, the arrangement is the opposite to that of the telephoto lens. Here, the negative lens is placed first and, in Figure 6.5, a design using $F_1 = -20$ D, F_2 of +40 D with a separation of 30 mm gives an EFL of 23 mm but a back vertex focal length of 36 mm. Such lens powers are high and the optical design of such a lens tends to conceal the simple thin lens pair on which it is based. The details of a short focus 35 mm camera lens will reveal negative power to the front and positive to the rear.

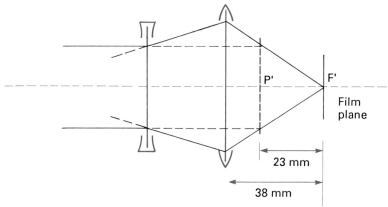

Figure 6.5 Inverted telephoto lens

With camera lenses where the design uses the principle of a positive lens and a negative lens separated by a significant distance, a variation in power may be effected by changing the separation and the locations of lenses in a continuous manner. This is the basis of the **zoom lens** which in practice requires at least two moving elements. The control of aberrations to provide sufficient image quality over the range of focal lengths is the major cause of its complex design. The ability of film to record detail provides an upper value for the allowed blur of the image. This not only tells the designer how good to make the lens but also controls the depth of field for a given aperture (f-number) as described in Section 5.9. The allowed blur size is much the same for all film and therefore lenses for use with the smaller format cameras will have the same depth of focus (for a given aperture) as lenses for larger formats, but because they have, in general, shorter focal length their depth of field as calculated by Equation 5.66 will be much greater because their hyperfocal distance is less. With movie and video cameras, the allowed blur is also greater and these cameras often seem to have unlimited depth of field.

6.2 The eye and its refractive correction

This section deals only with basic principles of the eye. A description of the eye as an optical instrument is given in Chapter 15. A full treatment of the eye from a

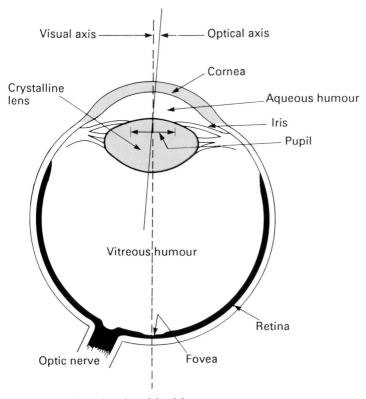

Figure 6.6 Horizontal section of the right eye

clinical standpoint can be found in A. B. Bennett and R. B. Rabbetts, *Clinical Visual Optics* (Butterworths, London).

Like the camera, the eye uses a positive optical system to image the outside world onto a photosensitive surface, the **retina**, which is on the inside surface of an approximately spherical eyeball (Figure 6.6). The retina comprises an array of detectors sensitive to light and colour. The optical system comprises the **cornea** and the **crystalline lens** and has a focal length of about 17 mm, +60 D. The cornea provides between 40 and 45 D and the lens makes up the rest. Unlike the camera, the image is not formed in air but in the **vitreous humour**, a transparent gel of index 1.336. Between the lens and the cornea is the **aqueous humour** of similar index and just in front of the lens is the aperture stop of the system, the **iris**, the size of which is controlled by the **pupillary response**. For light approaching the eye, the iris becomes the entrance pupil which has a diameter variable between 2 mm in bright light and 7–8 mm in darkness. The *f*-number therefore varies between *f*/8 and *f*/2 and the numerical aperture between 0.06 and 0.25.

The field of view of the eye is very wide because the retina covers most of the inside of the eyeball. However, the peripheral vision quality is very poor because the detector array is coarse in the retinal periphery. To overcome this the eye rotates to **fixate** any object of interest and this causes the image of that object to fall on a small central part of the retina called the **fovea**, where the retinal structure has the detectors packed close together. This allows high-quality vision and relieves the optical system of the need to form a good image over a field of view of more than a few degrees. In discussing the paraxial properties of the eye attention is restricted to this narrow field of view providing **foveal vision**. This ignores aberrations of both the cornea and the lens (see Chapter 15) and the fact that the axis of the cornea–lens system, the **optical axis**, is not quite aligned with the axis through the centre of the lens to the fovea, the **visual axis**, as shown in Figure 6.6.

The surface curvatures of the cornea and crystalline lens, their refractive indices and distances between them are all involved in providing a real image of the outside world which is in focus on the retina. An eye that accurately focuses distant objects onto its retina is said to be **emmetropic**. While the refractive indices are the same for all human eyes, considerable variations are found in the curvatures and axial distances of eyes even when they are emmetropic. How the growing eye correlates these values to achieve emmetropia is not known. In many people it fails to do so and the resultant eye either has too much power for its overall length and only close objects are focused on the retina, a condition called **myopia**; or too little power for the overall length, a condition known as **hypermetropia** (or **hyperopia**). The common terms for these conditions are short-sightedness and long-sightedness respectively. Some eyes also suffer from axial astigmatism, which is briefly considered in Chapter 8.

Up to middle age the eye has the ability to increase the curvatures of the crystalline lens, a process called **accommodation**, and thereby increase its overall optical power. This may relieve the visual problems of mild hypermetropia but cannot help myopia. For both conditions, the usual corrective measure is to wear spectacles, the lenses of which have just the right power so that the whole system of spectacle lens plus non-accommodating or **relaxed** eye has its second principal focus coincident with the retina. Thus distant objects in the outside world are sharply focused on the retina. The accommodation then allows objects at closer distances to be clearly seen. The closest possible object for comfortable and sustained clear vision is called the **near point**. The relaxed eye is focused on its **far point**, which is at infinity if the eye is emmetropic or properly corrected. The word 'relaxed' may not be accurate because it is found that in darkness many eyes settle to a focus condition nearer than infinity.

The optical system of spectacle lens plus eye is not often a centred system because the spectacle lens remains stationary while the eye rotates to look at different objects. An alternative means of correction is the contact lens. This lens rests on the eye and moves with the eye as it rotates, although in practice it may not rest in exactly the correct place to give a perfectly centred system at all times (see Section 5.4).

Because real eyes show considerable variations in surface curvatures, as well as having non-spherical surfaces and inhomogeneous crystalline lenses, it is normal practice for optical calculations to use a **schematic eye** or **model eye**. Much thought has gone into these over the last 140 years to the extent that it is often forgotten that they are merely meant to be representative of an optical organ which shows wide variations in its detailed values. Schematic eyes are dealt with more completely in Chapter 15.

The commonest form of schematic eye has three optical surfaces. The average curvature and location values of the crystalline lens are retained but the cornea is reduced to a single surface bounding the aqueous humour. Figure 6.7 shows the Emsley schematic eye as a relaxed emmetropic eye and the calculations for the paraxial optics of this are given in full as a worked example so that the student can

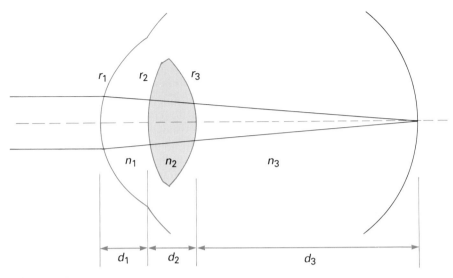

Figure 6.7 Emsley schematic eye

r_1 = 7.8 mm	d_1 = 3.6 mm	n_1 = 1.3333
r_2 = 10.0 mm	d_2 = 3.6 mm	n_2 = 1.4160
r_3 = −6.0 mm	d_3 = 16.7 mm	n_3 = 1.3333

work through them following the procedures of Section 5.6. The location of the principal planes, PP′, are shown on Figure 6.8, which is to scale. The first and second focal lengths have different values because the image is in a medium of different refractive index from that of the object. This places the nodal points, NN′, in a different location and much closer to C, the centre of rotation of the eye (see Section 5.3).

The centre of rotation is important when considering the spectacle lens plus eye combination. Of course, an emmetropic eye does not need a spectacle lens and so a

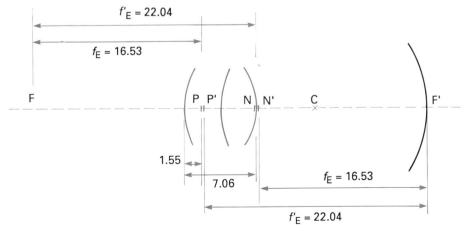

Figure 6.8 Schematic eye – principal points

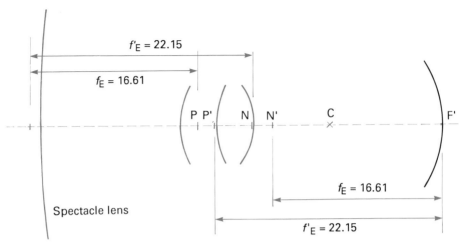

Figure 6.9 -5 D prescription eye – principal points (see worked example)

schematic version of a myopic eye (or a myopic version of a schematic eye!) is needed when considering spectacle lenses of negative power and of a hypermetropic eye for positive spectacle lenses. Such representative eyes may be called **prescription eyes** and may be adapted from the schematic eye by changing the overall axial length, which is most easily done by altering d_3. In real eyes this parameter has the most uniform influence on refractive error but the radii of curvature, particularly that of the cornea, and the separation distances also vary. Axial astigmatism (see Section 8.1) may also occur. Figure 6.9 shows a -5.00 D myopic prescription eye by virtue of d_3 being increased. The correcting lens is assumed to be 14 mm in front of the cornea. The worked example includes the calculation for this eye showing the value of d_3 to be 18.5 mm. Although total correction may be achieved in this way when the eye is looking along the axis of the spectacle lens, the usefulness of prescription eyes lies in the results obtained when the eye is looking obliquely through the lens.

In middle age the crystalline lens of the eye becomes too stiff for the accommodation muscles to increase its power. The usual result is that the emmetropic eye (or corrected eye) can see distant objects clearly but cannot focus near objects, especially print. This condition is called **presbyopia**. Spectacles with low positive power form a distant virtual image of a near object at their first principal focus and the presbyopic eye can see clearly to read the print. The effect is similar to that of a magnifier (Section 6.3) although no magnification is involved when the final apparent image size is compared with that at the near point. Unfortunately distant objects are then imaged in front of the retina and are therefore not seen distinctly. Spectacles with lenses having a power in their lower part which is 2 or 3 D more positive allow the wearer to select the part used for different object distances. Such **bifocal** spectacles are explained more fully in Chapter 8.

Worked example   Calculate the increase in axial length of a prescription eye which needs a spectacle lens of -5.00 D back vertex power, 14 mm in front of the cornea for a distant object to be imaged onto the retina. Calculate the equivalent power and principal points of the whole system. Assume the spectacle lens to be thin and use the Emsley schematic eye as the starting point. Draw a scale diagram to illustrate your results.

The image formed by the spectacle lens of a distant object lies 200 mm in front of its rear surface which is 14 mm in front of the cornea. The object distance for the cornea is -0.214 m.

In this calculation the emmetropic eye without a spectacle lens is also calculated with the results shown in brackets. In this case the object distance from the cornea is infinity. These values are used for Figure 6.8.

The step-along calculation gives:

$l_1 = 0.214$ m   (∞)
$L_1 = -4.673$ D   (0.0 D)
    $F_1 = 0.3333 \div 0.0078 = +42.731$ D
$L_1' = L_1 + F_1 = 38.058$ D   (+42.731 D)
$l_1' = n_1/L_1' = +0.035\,03$ m)   (+0.031 20 m)
$l_2 = l_1' - d_1 = +0.031\,43$ m   (+0.027 60 m)
$L_2 = n_1/l_2 = +42.417$ D   (+48.304 D)
    $F_2 = 0.0827 \div 0.010 = 8.27$ D
$L_2' = L_2 + F_2 = +50.687$ D   (+56.574 D)
$l_2' = n_2/L_2' = +0.027\,94$ m   (+0.025 03 m)
$l_3 = l_2' - d_2 = +0.024\,34$ m   (+0.021 43 m)
$L_3 = n_2/l_3 = +58.184$ D   (+66.078 D)
    $F_3 = -0.0827 \div (-0.006) = +13.783$ D
$L_3' = L_3 + F_3 = +71.968$ D   (+79.861 D)
$l_3' = n_3/L_3' = 0.018\,53$ m $= d_3$ (+0.016 69 m)

(d_3 is equivalent to f_v').

Thus, for the emmetropic eye, $d_3 = 16.69$ mm, while for this (-5.0 D) prescription eye, $d_3 = 18.53$ mm. The increase in axial length is therefore 1.84 mm.

Including the power of the spectacle lens as F_s, we have, by Equation 5.52:

$$F_E = F_s \times \frac{L_1'}{L_1} \times \frac{L_2'}{L_2} \times \frac{L_3'}{L_3} \qquad\qquad \left(F_E = F_1 \times \frac{L_2'}{L_2} \times \frac{L_3'}{L_3} \right)$$

$$= -5.00 \times \frac{38.058}{-4.673} \times \frac{50.687}{42.417} \times \frac{71.968}{58.184}$$

$$= +60.19 \text{ D} \qquad\qquad\qquad (+60.49 \text{ D})$$

Thus the corrected myopic eye has a slightly lower equivalent power than the emmetropic eye and will therefore receive an image which is 0.5% larger. If the comparison is made between the corrected myopic eye and the uncorrected myopic eye, it is the experience of spectacle wearers that the corrected image is smaller. This is because the uncorrected eye is seeing a blurred image roughly on a parallel ray through the original location of N', close to N.

Knowing F_E for both types of eye, we can calculate f_E and f_E' to be:

$$f_E = \frac{-1}{F_E} = +16.61 \text{ mm} \qquad\qquad (+16.53 \text{ mm})$$

$$f_E' = \frac{n_3}{F_E} = +22.15 \text{ mm} \qquad\qquad (+22.04 \text{ mm})$$

Knowing the position of F′ we can locate P′ and N′ but we need to calculate the position of F to locate P and N. For this calculation we assume collimated light in the vitreous humour and calculate forwards through the lens in the opposite direction to the normal direction of the light. The two calculations are identical until the spectacle lens is included:

$$L_3' = 0.00\,\text{D}$$
$$F_3 = +13.783\,\text{D}$$
$$L_3 = L_3' - F_3 = -13.783\,\text{D}$$
$$l_3 = n_2/L_3 = -0.010\,273\,\text{m}$$
$$l_2' = l_3 + d_2 = -0.099\,13\,\text{m}$$
$$L_2' = n_2/l_2' = -14.28\,\text{D}$$
$$F_2 = +8.27\,\text{D}$$
$$L_2 = L_2' - F_2 = -22.55\,\text{D}$$
$$l_2 = n_1/l_2' = -0.059\,12\,\text{m}$$
$$l_1' = l_2 + d_1 = -0.055\,52$$
$$L_1' = -24.016\,\text{D}$$
$$F_1 = +42.731\,\text{D}$$
$$L_1 = L_1' - F_1 = -66.747\,\text{D}$$
$$l_1 = -0.014\,98\,\text{m} \qquad (f_v = -0.014\,98\,\text{m})$$
$$l_s' = l_1 + d_s = -0.000\,98$$
$$L_s' = -1020.4\,\text{D}$$
$$F_s = -5\,\text{D}$$
$$L_s = L_s' - F_s = -1025.4\,\text{D}$$
$$l_s = 0.000\,97\,\text{m} = 0.97\,\text{mm}$$

This gives the position of the first principal focus, F. P and N can be found knowing the values of f_E and f_E'. The locations of P and N in Figure 6.9 are very little different from the emmetropic eye of Figure 6.8 mainly because l_s' and l_s differ by so little in the calculation above. However, the location of P′ and N′ has moved because F′ has moved by 1.84 mm and so the separation distances between PP′ and NN′ have been increased.

It should be noted that with real eyes an increase in the vitreous depth (d_3) by 1.84 mm would not necessarily produce −5 D of myopia because changes in the values of other parameters are quite likely, which tend to compensate.

6.3 The simple magnifier (loupe) and biocular magnifier

Magnifiers are designed to work in conjunction with the eye. In this section the eye is assumed to be emmetropic (or corrected) and to be able to accommodate. The extent of its accommodation is conventionally assumed to be 4 D, that is a near point at 250 mm. Most eyes up to the age of 40 years can accommodate more than this but 250 mm represents a comfortable viewing distance.

For the finest possible detail of an object to be seen, the retinal image should be made as large as possible. The retinal image size is greatest when the object is brought close to the eye, but this is of no value to see detail if the retinal image is blurred because of defocus. The effect of a **magnifier** or **loupe** is to bring the retinal image of a close object into focus without appreciably changing its size.

Figure 6.10 shows an object close to the eye giving rise to a large but blurred image on the retina. Figure 6.11 shows a strong positive lens inserted between the

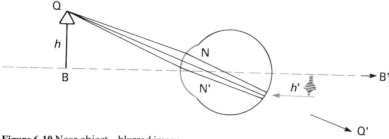

Figure 6.10 Near object – blurred image

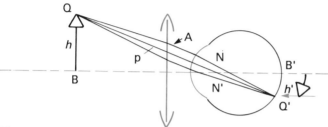

Figure 6.11 Near object – action of loupe (magnifier)

eye and the object so that the retinal image is now in sharp focus while still being approximately the same size as before.

The intermediate ray bundle at A has been collimated or partially collimated by the lens and, although the principal ray, p, would be deviated by the lens to provide a slightly larger retinal image than before, this is a minor bonus. Another approximation is that the principal ray through the centre of the pupil does not, in the normal eye, pass through the nodal points and, furthermore, at the image size shown in Figure 6.11, the eye would rotate about its centre of rotation to fixate Q'. All definitions and calculations concerning magnification are therefore approximate.

In Figure 6.12 a single thin positive lens is again acting as a magnifier. The object to be magnified has been placed at the first principal focus and the emmetropic eye with accommodation relaxed sees an erect image at infinity. The retinal image size, h', is determined by the angle, w'_1, of the ray from the tip of the object, which, after refraction, passes through the nodal point of the eye. The image at infinity means that all rays from the tip are parallel and have the same angle to the axis. Thus the retinal image size does not vary with eye position, d, or rotation. However, because the lens acts as a field stop, the largest field of view is when the eye is placed close to the magnifier.

The **magnifying power**, M (or **apparent magnification**) of the magnifier may be defined as the ratio of retinal image size when using the magnifier compared to the largest focused retinal image obtained by the unaided eye without the magnifier. The actual retinal image sizes are dependent on the focal length of the observer's eye and the angles subtended by the magnified and unmagnified object at the first nodal point of the eye. The focal length and other details of the eye are not known

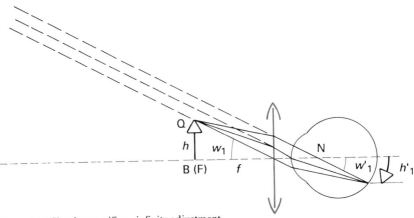

Figure 6.12 Simple magnifier – infinity adjustment

but they are common to both parts of the ratio so the best definition of magnifying power is:

$$M = \frac{\text{Angle subtended by the magnified image}}{\text{Angle subtended by the object viewed directly}}$$

$$= \frac{w'}{w} \qquad (6.3)$$

The angle, w, which any object or image subtends at the eye is often called the **visual angle**. To give a standard basis of comparison the directly viewed object is assumed to be at the comfortable near point defined above and shown in Figure 6.13 as q, the object distance $-250\,mm$ from the eye. In the USA, q is often taken as $-10\,in$, which is then sometimes expressed as a conversion value of $-254\,mm$. This will give a 2% higher value to the magnifying power of any system and is perhaps justified when it is recognized that an object requiring $4\,D$ of accommodation is $-250\,mm$ from the cornea of the eye and therefore about $-257\,mm$ from the nodal point of the eye. However, the convention is to use the value $-250\,mm$ *as if it were from the first nodal point of the eye.*

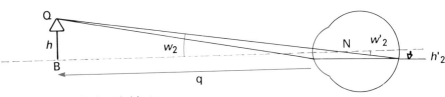

Figure 6.13 Directly viewed object

For the magnified image located at infinity (Figure 6.12), the angle subtended everywhere in the eye space is given by:

$$\tan w_1' = \frac{h}{-f} = \frac{h}{f'} = w_1' \qquad \text{for small angles}$$

$(= w'$ in Equation 6.3).

By convention the angle subtended by the object at the distance q (in Figure 6.13) is:

$$\tan w_2 = \frac{h}{-q} = \frac{h}{+250} = w_2 \qquad \text{for small angles}$$

$(= w$ in Equation 6.3).

Thus the magnifying power, M, is given by:

$$M = \frac{w_I'}{w_2} = \frac{h}{f'} \times \frac{-q}{h} = \frac{-q}{f'} = \frac{250}{f'} = \frac{F}{4} \tag{6.4}$$

Where F is the focal power of the magnifier. From this it can be seen that f' must be shorter than 250 mm if any useful magnification is to be obtained.

If the magnified image is formed at the distance q, as shown in Figure 6.14, a slightly increased magnifying power is achieved. Assuming that the eye is close to the lens, an approximate expression for the new magnification, M', is obtained by setting the image distance l' equal to q (-250 mm). Thus

$$\frac{1}{l'} = \frac{1}{-250} = \frac{1}{l} + \frac{1}{f'} \qquad \text{or} \qquad \frac{1}{-l} = \frac{1}{250} + \frac{1}{f'}$$

and

$$\tan w_3 = w_3 = \frac{h}{-l}$$

We assume this is equal to w_3' because the lens is close to the eye. Therefore,

$$M' = \frac{w_3'}{w_2} = \frac{h}{-l} \times \frac{-q}{h}$$

$$= \left(\frac{1}{250} + \frac{1}{f'}\right) 250$$

$$= 1 + \frac{F}{4} = 1 + M \tag{6.5}$$

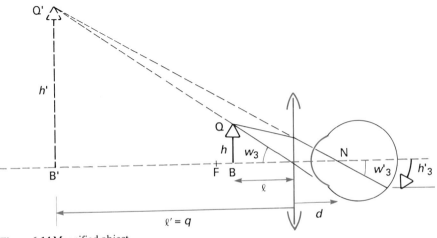

Figure 6.14 Magnified object

This is the maximum magnification obtainable if the viewed image is not to be closer than q.

Typically, magnifiers are used with the image between infinity and $-250\,\text{mm}$. The image location, expressed as dioptres vergence at the eye, is known as the **dioptric setting** of the magnifier. A common setting is $-2\,\text{D}$ at which the magnifying power is $M + \frac{1}{2}$.

When the image is at distances closer than infinity the magnification obtained is dependent on eye position. The maximum field of view is still obtained when the eye is placed close to the magnifier, although the useful field may be smaller than that given by the diameter of the magnifier because of aberrations. More comfortable viewing is sometimes obtained when the magnifier is held at a distance from the eye, as with a reading glass of the Sherlock Holmes type. If the image is at infinity there is no change in magnification. For nearer images the magnification is found by calculating, from the size and position of the image, the angular subtense at the eye and comparing this with the angular subtense of the object without the magnifier.

Referring again to Figure 6.14; for a given object size, h, the image size is $h' = hl'/l$ where l and l' are the object and image distances from the *lens*.

If the lens is at a distance d (positive) from the eye the object–eye distance is $d - l$ and the image–eye distance is $d - l'$. Therefore

$$
\text{Magnification} = \frac{w'}{w} = \frac{\dfrac{hl'/l}{d - l'}}{\dfrac{h}{d - l}}
$$

$$
= \frac{(d - l)l'}{(d - l')l}
$$

$$
= \frac{dL - 1}{dL' - 1} \tag{6.6}
$$

where L and L' are the object and image vergences *at the lens*. This gives the magnification in comparison with the unaided view of the object *in the same place*.

If the comparison is with the object at the distance of distinct vision, q, the expression becomes:

$$
\text{Magnification} = \frac{qL}{dl' - 1} \tag{6.7}
$$

Either of these may be used but it should be made clear which.

A particular application of large magnifier optics is to view electro-optical displays in military systems. Collimating optics are used in high-speed aircraft to give the pilot a view of instrument information which is viewed via a partially reflecting mirror so that it appears to be superimposed on the outside world. These systems are called **head-up displays**. In military vehicles such as tanks, non-collimating but high-power magnifiers are used to view night-vision displays.

These larger devices are called **biocular magnifiers** to distinguish them from binoculars where separate optical systems are used for each eye. The design of biocular magnifiers of magnifications greater than $3\times$ is difficult because neither eye is on the lens axis and the aberrations of the lens system must be related to the requirements of binocular as well as monocular vision.

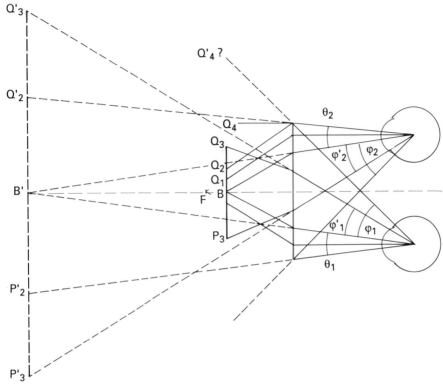

Figure 6.15 Biocular magnifiers

In Figure 6.15 the magnifier aperture is large and both eyes can be used to view the magnified image. Even though the central part only is seen by both eyes, the viewing is normally comfortable and can be maintained indefinitely provided the edges defining the central **binocular field of view** are not in sharp focus while viewing the image. When the viewed image is not at infinity it is difficult to define the extent of the field of view when two eyes are involved. In Figure 6.15 the full **monocular field of view** may be found by calculating the angle, φ_1, from the centre of the image to the extreme edge viewed by the right eye and adding this to the similar angle, φ_2, for the left eye. Because these angles are large the directions are taken from the centre of rotation of the eyes. Even if the head is moved somewhat off-axis so that these field angles are no longer equal, the value $\varphi_1 + \varphi_2$ remains a good description of the monocular field of view. In practical instruments the monocular field of view is often restricted to $\varphi'_1 + \varphi'_2$ by the extent of the display P_3Q_3. The full binocular field of view is given by $\theta_1 + \theta_2$ on the same reasoning.

6.4 The compound microscope

Simple magnifiers have limited magnifying power and field of view. The compound microscope is an instrument with two lenses or systems of lens which magnify in turn and therefore compound (multiply) together to give very large magnifying powers. Figure 6.16 shows the basic principle where the first lens is close to the object, and therefore called the **objective**, provides a magnified image which is in turn magnified by a second lens which looks like a simple magnifier but is, in fact, different and is therefore called the **eyepiece**, being close to the eye.

The action of the objective is that of a positive lens providing a real inverted magnified image, I'. This means that l'_o must be larger than l_o and both are larger than f'_o, the focal length of the objective. In the normal instrument, the objective and eyepiece are held at opposite ends of a metal tube and the intermediate image I' is caused to be coincident with the first principal focus of the eyepiece by moving the whole system nearer to or further from the object. This allowed the distance between the lenses to be standardized and x'_o, the distance between the foci, became known as the **optical tube length**, g, with a common value of 160 mm (or 170 mm). The magnification of the objective (by Equation 4.15) is therefore given by g/f'_o and higher values are obtained by using objectives of shorter focal length. It can be seen in Figure 6.16 that with objectives of shorter focal length the object being studied is closer to the instrument.

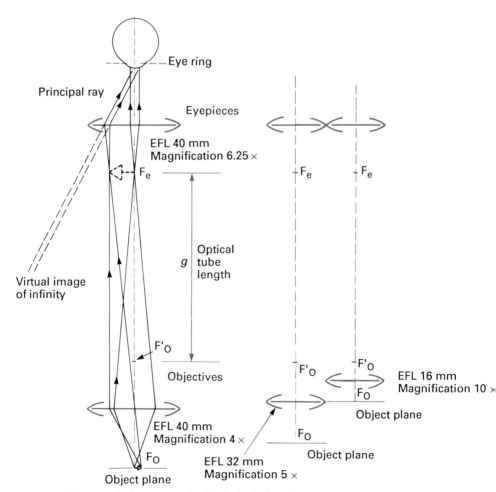

Figure 6.16 Compound microscopes – fixed optical tube length

To achieve this the metal tube had to be a standard length with the body of the objective lens so designed that when screwed fully into the tube its second principal focus is 160 mm from the first principal focus of the eyepiece when this is pushed fully home at the other end of the tube. For most objectives and eyepieces this **mechanical tube length** was also standardized at 160 or 170 mm. Modern microscopes, however, comprise a main housing which can be racked near to or further away from the object and a draw tube into which the eyepiece is pushed. The draw tube is usually calibrated or, at least, marked at the position giving the mechanical tube length but is normally adjustable to be longer or shorter than this, as shown in Figure 6.17.

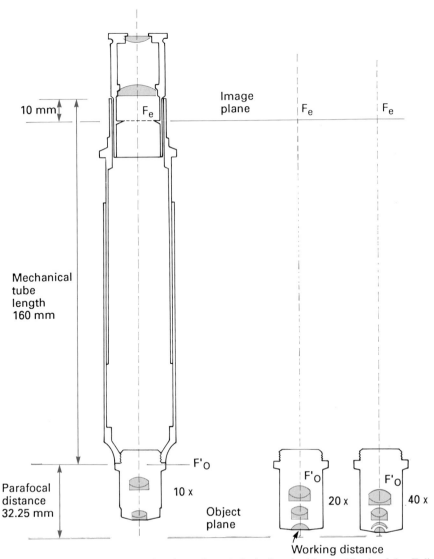

Figure 6.17 Microscope mechanics: change in optical tube length with magnification (after Ealing Beck)

A difficulty arises when objectives of different focal lengths are mounted on a rotating turret so that they can be rapidly swung into position by the user, who then obtains a range of magnifications. Ease of use would be lost if the instrument had to be refocused each time. Objectives are now designed to place the image I' at a fixed distance from the object. This distance comprises the parfocal distance from the object to the objective end of the tube (normally set at 32.25 mm) plus 150 mm of the mechanical tube length allowing for the first principal focus of the eyepiece to be 10 mm into the draw tube. This principle of **parfocality** is not compatible with a fixed optical tube length so that with modern objectives their effective optical tube lengths vary with their magnification as shown in Figure 6.17.

Most objectives are engraved with their magnification, their numerical aperture and sometimes their focal length. The magnification is that obtained when they are being used at the conjugates and, therefore, the optical tube length for which they were designed. Microscope objectives are generally powerful lenses and their optical performance falls off rapidly when they are used at incorrect conjugates. When the magnification and the focal length are both marked it is a simple matter to calculate conjugates and, therefore, the optical tube length as shown below.

The eyepiece lenses magnify the intermediate image I' to give an easily viewed final image I'' at infinity or some comfortable nearer distance. This is obtained either by sliding the eyepiece nearer to the objective, if the draw tube adjustment allows, or by moving the whole microscope slightly nearer to the object with the penalty due to the objective now being used at incorrect conjugates. The difference between an eyepiece lens and a magnifier arises because the light forming the image I' has very restricted ray bundles due to the limited aperture of the objective. Although the position and size of the final image can be found by using the usual constructional rays, the actual rays are limited by the apertures of the lenses as described in Section 5.7. This difference affects the detail design of the eyepiece and in particular allows some of its lens element to be positioned before the intermediate image as described in Section 6.6. Some eyepieces, therefore, cannot be used as magnifiers because their first principal focus is *inside* the lens system. However, the magnification formula for eyepieces is the same as that developed for magnifiers in Section 6.3.

The simplest eyepiece is a single lens and so the basic microscope comprises two separated lenses of short focal length. The object to be studied must be strongly illuminated. Often the object is a very thin slice of material which is therefore transparent and can be illuminated from its other side. The optics of illumination systems are considered in Section 6.8. The objective, if properly filled with light by the illumination system, is the aperture stop of the instrument and its entrance pupil. The narrow bundles passing through the eyepiece all pass through the exit pupil which is the image of the objective/aperture stop formed by the eyepiece. In normal adjustment the exit pupil is between 10 and 20 mm behind the eyepiece lens. Only if the eye is positioned so that its pupil is coincident with the exit pupil can the observer see the whole field of view. On visual instruments such as microscopes and telescopes, the exit pupil is called the **eye ring** because it shows where the eye should be placed.

With the eye in the correct position, the field of view is defined by the size of the eyepiece lens. This has a vignetting effect, as described in Section 5.8, which can be distracting. The usual practice is to fit a field stop coincident with the intermediate image, I'. This gives a sharp in-focus edge to the field of view all of which has equal brightness provided the eye stays within the eye ring. If the pupil of the eye is smaller than the eye ring (in which case the observer's iris is the aperture stop of whole microscope + eye system) some movement of the head is allowed before any illumination or field of view is lost. Surprisingly, the same thing is the case if the eye pupil is larger than the eye ring. However, it will be seen in Chapter 15 that the quality of vision is usually best when the eye pupil is between 2 and 3 mm. A bright image to give this pupil size and an eye ring of 7 mm or more diameter would give very comfortable viewing conditions. Although this may occur with telescopes and binoculars (Section 6.5) it is rarely the case with microscopes.

With an eyepiece of given focal length, the eye ring size of a compound microscope is dependent on the diameter of the objective because the conjugates of the objective and eye ring are limited by the optical tube length. The objective is a short focus lens and rarely exceeds 10 mm in aperture diameter. The examples below will show that microscope eye rings are usually only 1 or 2 mm in diameter.

The magnification of the compound microscope occurs in two stages. With an objective of equivalent focal length f_o', the intermediate image will be laterally magnified by:

$$M_o = \frac{-g}{f_o'} \tag{6.8}$$

where g is the optical tube length and the negative sign indicates an inverted image. This value of M_o is usually engraved (without the negative sign) on the objective body as, for example, $20\times$ or $20{:}1$.

The eyepiece has the same formula as for magnifiers,

$$M_e = \frac{q}{f_e'} \tag{6.9}$$

where q is the standard distance of distinct vision. This value is also engraved on the eyepiece.

Thus the magnifying power of a compound microscope when used at the designed conjugates and in infinity adjustment (final image at infinity) is given by:

$$MP_{microscope} = M_o \times M_e = (-)\frac{g}{f_o'} \times \frac{q}{f_e'} \tag{6.10}$$

$$= \frac{-160}{f_o'} \times \frac{-250}{-f_e'} = \frac{F_o F_e}{24}$$

when F_o and F_e are in dioptres.

When M_o and f_o' are marked on the objective, the designed optical tube length g is found by $g = M_o \times f_o'$ and the distance of the image from the second principal plane is $l_o' = g + f_o'$. Because the objective is a complex lens system the principal planes are generally inside and for high power lenses the distance from the lower lens to the object, known as the **working distance**, is very small. Also, because of the complexity of the design, the field of view is usually small. All the design effort goes into obtaining the best quality image with the largest numerical aperture consistent with an acceptable price for the item. A large NA minimizes the effects of diffraction (Chapter 13) and gets as much light as possible to the eye.

Worked example   The objective of a microscope is marked:

$20\times$ NA 0.5   EFL 9 mm

and the eyepiece is marked $12.5\times$. Calculate the design optical tube length and the distance between the lenses, assumed thin, when the microscope is in infinity adjustment.

To place the final image at the distance of distinct vision calculate how much nearer to the object the instrument must be moved or by what distance towards the objective the eyepiece must be moved. Draw a scale diagram of the latter case and calculate the final magnification in each case.

The eyepiece magnification is calculated from

$$M_e = \frac{q}{f'_e}$$

where q is $-250\,mm$, the standard distance of distinct vision (measured from the eyepiece lens). From this, f'_e is found to be 20 mm.

The objective has its equivalent focal length marked as 9 mm and so its design optical tube length is calculated by $M_o = g/f'_o$ from which:

$g = 180\,mm$

Assuming that both lenses are thin lenses (most unlikely) the distance between them for infinity adjustment is then $9 + 180 + 20 = 209\,mm$. Note that the magnification of the instrument is now $20 \times 12.5 = 250\times$.

In order for the final image to be 250 mm from the eye lens, the intermediate image distance from this lens is given by:

$$\frac{1}{-250} = \frac{1}{l_e} + \frac{1}{20}$$

from which

$l_e = -18.5\,mm$

The eyepiece must be moved nearer the objective by 1.5 mm and the magnification of the eyepiece is now $M_e + 1$ so that the magnification of the whole instrument is now $20 \times 13.5 = 270\times$.

Alternatively, to move the intermediate image to be 190.5 mm from the objective, the new object distance is given by:

$$\frac{1}{+190.5} = \frac{1}{l_0} + \frac{1}{9}$$

from which

$l_o = -9.4463\,mm$

The working distance for infinity adjustment is given by:

$$\frac{1}{189} = \frac{1}{l} + \frac{1}{9}$$

from which

$l = -9.45\,mm$

Thus the instrument must be moved only 0.004 mm nearer to the image and the new magnification is now:

$$\frac{190.5}{9.4463} \times 13.5 = 27/2\times$$

– a very marginal increase over the eyepiece shift.

It is interesting to note that the depth of focus associated with the normal (4 D) accommodating eye is only 4 μm. Even if the allowed blur is equivalent to a defocus to the eye of 0.25 D at each end of this range, the geometrical depth of focus becomes 4.5 μm. Thus the mechanical parts of a microscope must be accurately adjustable and very stable.

6.5 Telescopes and binoculars

Telescopes and binoculars are generally used to give a magnified view of a distant object but they can also be used to define a line of sight, with an accurate direction and position, for surveying purposes. Then they are mounted on a rotatable support to form a **theodolite**. A telescopic system has an objective and an eyepiece so that it is a compound instrument similar to a microscope, but the objective has much less power and can therefore be of larger diameter. Telescopes may use reflecting elements, particularly when the objective is of large size as with modern astronomical systems. In the main these mirrors are non-spherical and reference is made to them in Chapter 8. In this section, only refracting systems are considered.

In normal use, light from a distant object arrives as a collimated or near-collimated beam and is focused by the objective at its second principal focus, F_o' as shown in Figure 6.18. If this is the same place as the first principal focus of the eyepiece, the light is recollimated through the eye ring and a relaxed emmetropic eye placed there can see a clear magnified image at infinity. The telescope is said to be in **infinity adjustment**.

The exit pupil of the telescope–eye system is either the pupil of the eye or the eye ring, whichever is the smaller. In the latter case the objective acts as the aperture stop. The field stop can only be the eyepiece and because this is close to the eye a very wide field can be obtained, but the optical quality at the edges will be poor due to aberrations on the eyepiece design. Normally a sharp edge to the field of view is

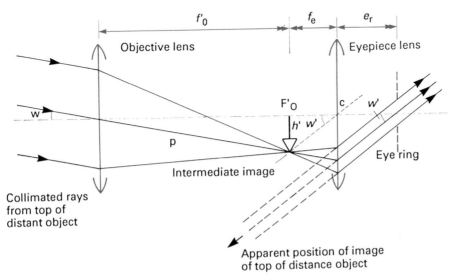

Figure 6.18 The astronomical telescope

formed by inserting a field stop at the location of the intermediate image. Unfortunately, the image viewed through the system of Figure 6.18 is inverted. This type of instrument is commonly called an **astronomical telescope**, and in this application the upside down image does not matter. The inverted image may be made erect by:

1. A system of prisms.
2. A further telescope system of low magnification.
3. An eyepiece of negative power.

Method 1 is the most commonly used and is described more fully later in this section. Method 2, called an **erecting telescope**, makes for the very long instrument used by sea captains in the seventeenth and eighteenth centuries. The basic principles are discussed at the end of this section.

Method 3 is called a **Galilean telescope** and gives a short instrument as shown in Figure 6.19. The negative eyepiece recollimates the light *before* it forms an image and the eye ring is then seen to lie *inside* the instrument. This means that the eye cannot be placed at the same position. While the pupil of the eye remains the aperture stop of the telescope–eye system, this internal eye ring now becomes the exit window, making the objective the field stop. This severely limits the field of view to the eye even if a very large diameter objective is used. However, for low magnifications such as ×2 or ×3, this design gives a cheap instrument with a moderate field of view. When arranged as a pair for binocular viewing they are commonly called opera glasses and are often available in older theatres for viewing the stage from the more distant seats.

Returning to method 1, the use of reflecting prisms to invert an image was described in principle in Section 2.10. By far the most common are the Porro systems, and Figure 6.20 shows the general arrangement and light paths. The light is reflected first at prism 1 and from the top of the object it is reversed along its own path by the roof edge action of this prism. Prism 2 reflects as if it was a single mirror

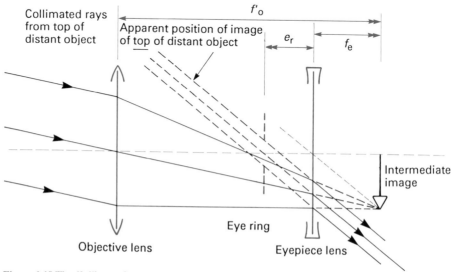

Figure 6.19 The Galilean telescope

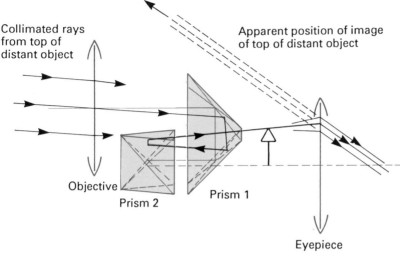

Figure 6.20 Prism erecting system for telescope

because its roof lies in the same plane as the light ray. Hence the light goes upwards towards the intermediate image and is then refracted downwards by the eyepiece lens.

The complexity of this can be simplified by using a **tunnel diagram** which lays out the optical path. This conceals the inverting action but it can be seen in Figure 6.21 that the vergence change to the light beam on entering the prisms actually makes the light path longer even though the intermediate image size is unchanged. However, the folding of the axis by the prisms makes the resulting instrument shorter overall.

When incorporated into a binocular system the Porro prisms displace the axes and allow the objectives to be set further apart than the eyepieces and this enhances stereopsis. Neater but more expensive instruments incorporate prisms that have a central roof edge and give no displacement of the light beam. Again, a tunnel diagram can be drawn and the basic calculations given below apply to any form of the telescope or binocular containing an objective and a positive eyepiece. The size and location of the inverting prisms have no first-order effect.

When the simple two-lens telescope is in infinity adjustment the second principal focus of the first lens is coincident with the first principal focus of the second lens.

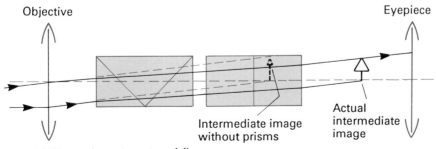

Figure 6.21 Porro prism system – tunnel diagram

The separation distance, d, is given by:

$$d = f_1' - f_2$$
$$= f_1' + f_2'$$
$$= \frac{1}{F_1} + \frac{1}{F_2} \tag{6.11}$$

When this value is put into the thin lens pair equation for equivalent power, we have:

$$F_E = F_1 + F_2 - dF_1F_2$$
$$= F_1 + F_2 - \left(\frac{1}{F_1} + \frac{1}{F_2}\right)F_1F_2$$
$$= 0 \tag{6.12}$$

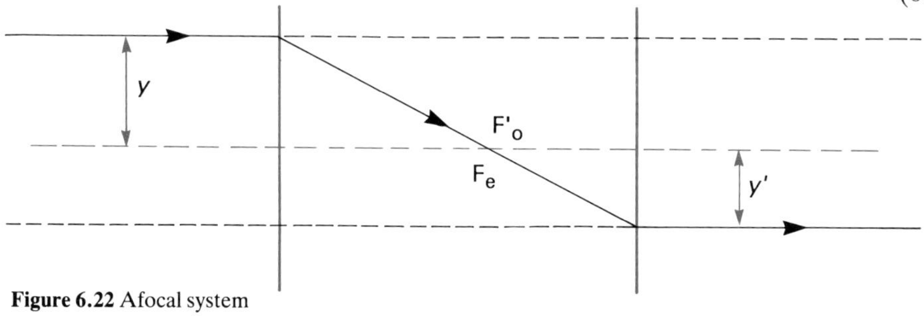

Figure 6.22 A focal system

As noted in Section 5.2, such systems are called **afocal systems** and, because they have an infinite focal length, the principal planes and principal foci are all located at infinity. An important feature of such systems is that the lateral magnification is constant for all conjugates. In Figure 6.22, for example, a light ray parallel to the axis in object space is refracted through the common focus and then emerges parallel to the axis in image space. From the similar triangles between the lenses it is clear that the distance y of the object ray from the axis is linked to y', the distance of the image ray from the axis, by:

$$\frac{y'}{y} = \frac{f_e}{f_o'} \tag{6.13}$$

This means that an object placed anywhere on the object ray will be imaged somewhere on the image ray. Thus an extended object of height $h = y$ will have an image of height $h' = y'$. The lateral magnification therefore has a constant value, m, given by:

$$m = \frac{h'}{h} = \frac{f_e}{f_o'} \tag{6.14}$$

One possible location for the object is in the plane of the first principal focus of lens 1. After refraction at the first lens the light rays will be collimated as shown in Figure 6.23 before refraction at lens 2 and forming an image in the focal plane of the second lens. If a stop is placed at the common focal plane as shown, the bundles of rays leaving the object and forming the image will be telecentric and no change

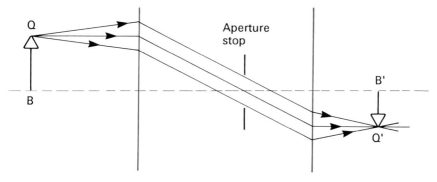

Figure 6.23 Telecentric afocal system

in size will occur when either is defocused. Even when the light between the lenses is not collimated, this telecentric action will be maintained by the aperture stop. If, however, the object is moved further from the first lens by more than $x = f_o^2/f_e$, the image will be formed before the second lens and the final image will be virtual, although still of the same magnification. In spite of this restriction on object position the afocal or near-afocal system forms a useful relay system (Section 5.10).

The longitudinal magnification of this relay system cannot be straightforwardly found from the equivalent focal length, which is infinite, or from the application of Newton's equation to the principal foci of the system, because these are located at infinity. However, it is a simple matter to apply Newton's equation to each lens in turn, because F_1' is coincident with F_2 and so $x_1' = x_2$. We have:

$$x_1 x_1' = f_o f_o'$$

and

$$x_2 x_2' = f_e f_e'$$

Therefore

$$\frac{x_2'}{x_1} = \frac{f_e f_e'}{f_o f_o'} = m^2 \tag{6.15}$$

Defining a new axial object point at $\bar{x}_1$, its image will be at $\bar{x}_2'$ so that:

$$\frac{\bar{x}_2'}{\bar{x}_1} = m^2$$

and therefore:

$$\frac{x_2' - \bar{x}_2'}{x_1 - \bar{x}_1} = \frac{x_1 m^2 - \bar{x}_1 m^2}{x_1 - \bar{x}_1} = m^2 \tag{6.16}$$

and is a constant throughout all object space and image space. As with most optical systems, the image formed is still a distorted version of the object when all three dimensions are considered. However, if the lenses of this telescopic relay system are given identical focal lengths:

$$m = -1 \quad \text{and} \quad m^2 = 1$$

so that the image is an undistorted replica of the object. If the lenses in Figure 6.23 have focal lengths both equal to f, the image B' is formed so that BB' is equal to $4f$. Because the longitudinal magnification is 1 every other axial object will form an axial image at a distance $4f$ from the object no matter whether the image is real or virtual. Off-axis object positions will be inverted, but the relay is a true unit magnification system in every other respect, and as far as is allowed by the aberrations of the lenses.

When the afocal system has unequal lenses and is used as a telescopic system as in Figure 6.18 the apparent magnification is obtained because the eye sees the image off-axis by a greater angle than if it were looking at the object directly. The magnifying power of the telescope, M, is defined as the ratio:

$$M = \frac{\text{Angle subtended by the image seen by the eye}}{\text{Angle subtended by the object seen without the telescope}} \tag{6.17}$$

Because of the large distances involved the angle subtended by the object at the objective of the telescope is effectively the same as that subtended at the unaided eye. In Figure 6.18, these angles are shown as w and w'. Because the intermediate image is in the first focal plane of the eyepiece lens the light ray emerges parallel to the construction ray c, which therefore subtends w' to the axis. Using the triangles within the lenses we have, from Equation 6.17,

$$M = \frac{w'}{w} \simeq \frac{\tan w'}{\tan w} = \frac{h'}{f_e} \times \frac{f_o}{h'} = \frac{f_o}{f_e} \tag{6.18}$$

Thus the magnifying power is greater than one when the focal length of the objective is larger than that of the eyepiece lens. This does not seem to fit with Equation 6.14, which showed that the lateral magnification, m, is greater than unity when f_e is larger than f_o; in fact it would appear that

$$M = \frac{1}{m} \tag{6.19}$$

The mystery is solved by remembering that real-world objects stretch from the objective of the telescope out towards infinity. The image viewed by the eye is virtual and also stretches out towards infinity. For any finite object distance, however, the value of the longitudinal magnification is m^2 and so the virtual image is much closer than the object. The actual viewed image really is m times smaller than the actual object, but it is also m^2 times nearer.

The full expression for Equation 6.19 therefore gives the magnifying power or apparent magnification, M, as:

$$M = \frac{m}{m^2} = \frac{1}{m} \tag{6.20}$$

If, for example, a telescope has a magnifying power of $10\times$ using an objective of focal length 200 mm and an eyepiece lens of 20 mm, the relative object and image distances from F_O and F'_e respectively (using Newton's equations as in Equation 6.15) are given in Figure 6.24.

It can be seen that changes in image distances are always 10^{-2} of the object distance changes. It may seem that the image has a lot of catching up to do in order to meet the object at infinity! The answer is that infinity even when divided by 100 is still equal to infinity. It is because the axial object and image distances are

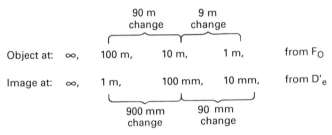

Figure 6.24

measured from the instrument itself that the low value of the longitudinal magnification makes the depth of focus of a telescope very small so that not even the accommodating eye can increase it greatly (see Section 6.8).

The constant lateral magnification of the telescope allows a simple measurement of its magnifying power. The eye ring of this instrument can easily be seen on a screen held near to the eyepiece lens if the objective is well illuminated. Using Figure 6.23, the object BQ is now made coincident with the objective and the image B'Q' moves down the axis a short distance but the ratio of the two diameters is still the lateral magnification of the instrument. If the measured diameter of the objective is D_o and of the eye ring is D_e, then

$$\frac{D_e}{D_o} = \frac{f_e}{f_o} = m \quad \text{and} \quad \frac{D_o}{D_e} = \frac{f'_o}{f_e} = M \tag{6.21}$$

Telescopes and binoculars are commonly described by their magnifying power and their objective diameter in millimetres. For example a pair of 8 × 50 binoculars has a magnification of 8× and objectives that are 50 mm in diameter. From Equation 6.21 it can be seen that the eye rings are 6.25 mm in diameter and so form a good match with the pupils of the eye up to quite low levels of illumination. On the other hand, binoculars designated 10 × 30 have eye rings only 3 mm in diameter and would give an apparently dim image under lower levels of illumination when the natural eye pupil is large.

These calculations also apply when the eyepiece is a lens system rather than the single lens as shown in the diagrams so far in this section. Equation 6.18 was derived from Figure 6.18 and the same letters are used in Figure 6.25 where the

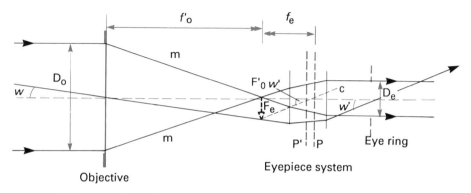

Figure 6.25 Telescope with eyepiece system

eyepiece is a two-lens system having an *equivalent* focal length f'_e. As usual the marginal ray (through either side of the aperture) defines the exit pupil. The principal ray passes through the centre of the aperture stop (the objective lens) and out through the centre of the exit pupil. The angular magnification is determined by w' and w, and these give the same equation as Equation 6.18 as long as the equivalent focal length of the eyepiece system is used.

Not all eyepiece systems can be used as magnifiers but Equation 6.4, derived for a magnified image at infinity, is valid when the equivalent focal power and equivalent focal length are used. The basic principle of eyepieces, developed in this and the next section, apply equally to telescopes (including binoculars) and microscopes although the latter cover less aperture and field of view and are therefore less complicated systems.

The starting point for all eyepiece systems is an extra lens coincident with the intermediate image. This is called the **field lens** because in this position it acts as a field stop. In practice, as described in Section 6.6, it is displaced from this position, but for the moment we will assume it is at the intermediate image where its principal action is to bend the light, without affecting the intermediate image, so that the original lens, now called the **eye lens**, can be of smaller diameter. The reduction in diameter of the eye lens allows this to be of shorter focal length for the same level of aberration control and so the objective can also be made more powerful and, for a given magnification, the instrument becomes shorter.

The action of the field lens also has an effect on the eye ring, which is moved closer to the eye lens as shown in Figure 6.26. The distance from eye lens to eye ring is called the **eye relief distance**. Comfortable viewing occurs when this is between 10 mm and 20 mm. When the field lens is coincident with the intermediate image it is also coincident with the first principal focus of the eye lens for a system in infinity adjustment. Figure 5.10 shows that the equivalent power remains equal to that of the single eye lens. The magnification of the telescope is therefore unchanged.

Although the eye ring *position* has been changed by the field lens, the diameter of the eye ring is unchanged because Equation 6.21 is unchanged. As a system, therefore, the eyepiece has the same first principal focus and first principal plane as the single eye lens but the second principal focus and second principal plane have been moved even though the equivalent focal lengths remain the same as for the

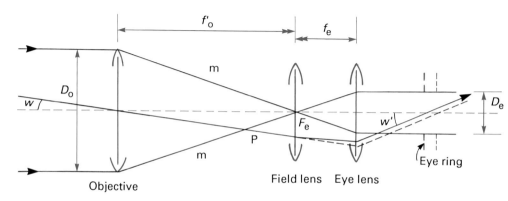

Figure 6.26 Telescope with field lens

single lens. Figure 6.25 shows the location of the second principal plane to the left of the first principal plane.

Obviously, the power of the field lens must not be made too high or the eye ring will be formed in front of the eye lens. However, it is clear that the field lens can be used to control the eye relief distance without affecting the other parameters of the telescope. One disadvantage of the field lens coincident with the intermediate image is that any dust on its surface will be in focus with the final image. In measuring systems a graticule or crosswire is needed (see Section 6.7) which must be coincident with the intermediate image and therefore the field lens is normally moved either forwards or backwards from this point. This leads to the two main types of eyepiece, the basic principles of which are described in Section 6.6.

The need for field lenses is well demonstrated in the **erecting telescope** briefly introduced at the beginning of this section. Figure 6.26 shows the basic principle of an erecting telescope where a ×4 inverting telescope is followed by a ×2 inverting telescope, thereby giving an erect image. One principal ray and two marginal rays have been drawn to indicate the sizes of the lenses, each lens needing to allow the two marginal rays even when offset about the principal ray. The first lens of the erecting system, F_{e2}, has been placed at the original eye ring and both F_{e2} and F_{e3} are relatively small. However, the instrument is more than 50% longer than the original ×4 inverting telescope. It might be thought possible to shorten this length by moving F_{e2} close to F_{e1} but, as shown in Figure 6.27(b), the location of the final eye ring is unchanged and the diameters of the ×2 system lenses have increased considerably.

This effect is important when designing complex instruments or setting up a chain of lenses for a laboratory experiment. As well as the required image, the system must image the aperture stop in an economical way. Lenses F_{e1} and F_{e2} can be considered as a 1:1 relay. In Figure 6.27(b) they relay the image but in Figure 6.27(a) they relay the aperture stop as well. If field lenses are introduced into the Figure 6.27(b) design then, as shown in Figure 6.28, they can be chosen to relay the aperture almost independently of the other lenses. F_{f1} is given a power which causes the eye ring of the first telescope to fall close to the eye lens, that is $F_{f1} \simeq F_{e1}$ while F_{f2} is chosen to give a comfortable eye relief distance.

6.6 Eyepieces

The basis of two-lens eyepiece design is a field lens placed coincident with the intermediate image and an eye lens placed with its first principal focus coincident with the intermediate image (for infinity adjustment). The optical action of this arrangement is described in Section 6.5. However, this position for the field lens means that its surfaces and any dust or marks on those surfaces will be in focus with the final viewed image. If the field lens and eye lens are connected together, as is usual, adjustment of the eyepiece to suit the observer's eye may leave the field lens on one or the other side of the intermediate image and so change the apparent size and quality of the image. To avoid this, the field lens is deliberately moved away from this position. Eyepiece designs therefore fall into one of two kinds depending on which way this lens is moved. In its new position this lens is not really a true field lens. The following discussion continues to refer to the two lenses as field lens and eye lens, but in the diagrams and equations the designations L_1 and L_2, etc., will be used.

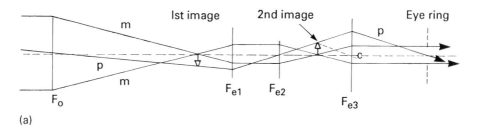

(a)

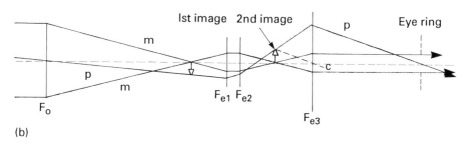

(b)

Figure 6.27 Erecting telescope: (a) F_{e2} at original eye ring, (b) F_{e2} near F_{e1}

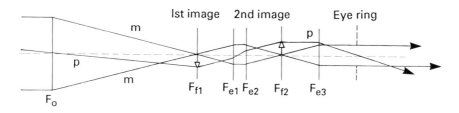

Figure 6.28 Erecting telescope: effect of field lenses

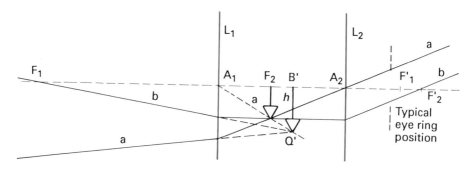

Figure 6.29 Huygens eyepiece – reverse ray tracing

6.6.1 The Huygens eyepiece

In the *Huygens eyepiece* the field lens is moved towards the objective to give the overall system shown in Figure 6.29. The focal length of the field lens is usually between two and three times the focal length of the eye lens. The distance between the lenses is made equal to the average of the focal lengths because this corrects the aberration lateral colour (see Section 14.3). It is this separation that causes the first principal focus to lie inside the system and moves the eye ring rather too close to the eye lens for a comfortable eye relief distance.

In Figure 6.29 the lens powers and separations are based on

$$f_1 = 2.5f_2 \quad \text{and} \quad d = \frac{f_1 + f_2}{2} \tag{6.22}$$

The rays are drawn for the eyepiece in infinity adjustment and, because no details of the objective are given, the emerging parallel rays to the right of the eye lens are the first to be drawn. The basic construction ray, a, through A_2, the centre of the eye lens, is drawn at a reasonably steep angle because eyepieces usually provide a wide apparent field of view. We know that the image must be at the first principal focus, F_2, of the eye lens, but this is not the position of the intermediate image formed by the objective. A useful construction ray is b, which is drawn through F_2', parallel to the first ray. Between the lenses this ray is parallel to the axis and must have come from the objective to the field lens via F_1.

These rays confirm the image at F_2. The further construction ray, a, through A_1, the centre of the field lens, is shown dotted. The position of the intermediate image can now be found by projecting this ray and the ray through F_1. This defines Q', the position at which the image *as formed by the objective* must be located for the eyepiece to be in infinity adjustment. The field stop of the eyepiece, however, is positioned at F_2.

The internal focus of this eyepiece means that it cannot be used as a simple magnifier. It is, however, commonly used in microscopes, and the effective magnification of the compound microscope given by Equation 6.9 includes the term $M_1 = q/f_e'$, which is the magnifying power of the eyepiece part of the microscope. This was developed in Section 6.3 assuming a single positive lens with an external focus, although not practical. The same reasoning applies in the case of the Huygens eyepiece. If the image at $B'Q'$ was a real object of height h, the apparent angular size when viewed unaided at the distance of distinct vision q would be h/q; while the apparent angular size when viewed as a virtual object via the *whole* eyepiece is w', where

$$\tan w' = \frac{-h}{f_e} = \frac{h}{f_e'}$$

and w' is the angle subtended at the first principal plane of the eyepiece. Thus the magnifying power of the eyepiece is given by:

$$M_E = \frac{q}{f_E'} \tag{6.23}$$

where f_E' is the equivalent focal length of the eyepiece.

The position of the cardinal points of a Huygens eyepiece are given in Figure 6.30.

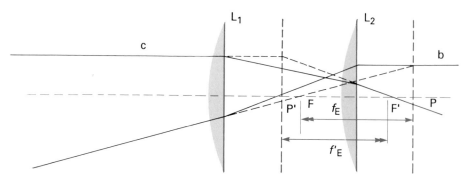

Figure 6.30 Huygens eyepiece – cardinal points (identical lenses to those in Figure 6.29 and the worked example)

This is the same lens system as in Figure 6.30 and in the worked example. It can be seen that the principal planes are so far overlapping that they include the principal foci between them. The second principal focus is located so close to the eye lens that the eye ring may also be formed too close for a comfortable eye relief distance. If, for example, f'_E were 20 mm it can be seen in Figure 6.30 that only about 7 mm (f_v) would be clear of the eye lens.

With

$$f_1 = 2.5f_2 \quad \text{and} \quad d = \frac{f_1 + f_2}{2}$$

the equivalent focal length is found from

$$f'_E = \frac{f'_1 f'_2}{f'_1 + f'_2 - d}$$

(Equation 5.23) to be

$$f'_E = \frac{2.5f'_2 \times f'_2}{2.5f'_2 + f'_2 - \frac{3.5}{2} f_2}$$

$$= 1.43f_2 \tag{6.24}$$

(for a standard '2.5' design Huygens eyepiece).

This shows that the eye lens must be 43% more powerful than the required eyepiece power.

Without knowing the aperture and focal length of the objective (or aperture and optical tube length in the case of a microscope) it is not possible to establish the size and position of the eye ring. The following worked example gives some typical values for a 10× eyepiece constructed with a 2.5:1 ratio of focal length, the prescribed separation distance and an objective located 200 mm from the intermediate image. A comparison with 2:1 and 3:1 designs is also made. The two lenses are normally made plano-convex with the convex surfaces of both facing the objective.

Worked example – Huygens eyepiece   A 10× eyepiece has an equivalent focal length of 25 mm. Calculate the individual lens powers for a Huygens-type eyepiece with

$$f_1' = 2.5f_2' \quad \text{and} \quad d = \frac{f_1' + f_2'}{2}$$

Calculate the eye relief distance for an objective 200 mm in front of the intermediate image. Assume infinity adjustment.

We have

$$f_1' = 2.5f_2'$$

$$d = \frac{f_1' + f_2'}{2}$$

$$f_E' = \frac{f_1'f_2'}{f_1' + f_2' - d}$$

Therefore

$$\text{EFL} = f_E'$$

$$= \frac{5f_2'}{3.5} = +25 \text{ mm} \qquad \text{(for 10× power)}$$

Therefore

$$f_2' = \frac{3.5 \times 25}{5} = +17.5 \text{ mm}$$

$$f_1' = 2.5f_2' = +43.75 \text{ mm}$$

$$d = 30.625 \text{ mm}$$

$$f_v' = \frac{f_2'(f_1' - d)}{f_1' + f_2' - d}$$

$$= \frac{17.5\,(13.125)}{30.625} = 7.5 \text{ mm}$$

By Newton's equation, $xx' = ff'$; with the objective distance $x = 200$ mm we have:

$$x' = \frac{25 \times 25}{200} = 3.125 \text{ mm}$$

Therefore,

Eye relief distance $= f_v' + x'$

$$= 10.625 \text{ mm}$$

This position is shown to scale on Figure 6.28 and it suggests that the construction rays a and b approximate to the actual rays from the top and bottom of the objective assumed in this example.

For the same 10× eyepiece but with different power ratios we have, for $f_1' = 2f_2'$:

$$\text{EFL} = 25 \text{ mm} = \frac{4f_2'}{3}$$

Therefore

$$f_2' = 18.75 \text{ mm} \qquad f_1' = 37.5 \text{ mm} \qquad d = 28.125 \text{ mm}$$

$$f_v' = 6.25 \text{ mm}$$

Therefore eye relief $= 6.25 + 3.125 = 9.375$ mm.

For $f_1' = 3f_2'$,

$$\text{EFL} = 25 \text{ mm} = \frac{3f_2'}{2}$$

Therefore

$$f_2' = 26.67 \text{ mm} \qquad f_1' = 37.5 \text{ mm} \qquad d = 28.125 \text{ mm}$$

$$f_v' = 8.34 \text{ mm}$$

Therefore eye relief $= 8.34 + 3.125 = 11.46$ mm.

Thus, departures from the optimum power ratio for aberration control do not greatly help the eye relief problem. However, the Huygens eyepiece, sometimes called a negative eyepiece for no good reason, is a robust economical design.

For the standard $f_1' = 2.5f_2'$, $d = (f_1' + f_2')/2$ Huygens design it is possible to formulate the separation (d), the component powers $(f_1'$ and $f_2')$ and the second vertex focal length, (f_v') in terms of the equivalent focal length (f_E') as follows:

$$\left. \begin{array}{l} d = 1.224f_E' \\ f_1' = 1.75f_E' \\ f_2' = 0.7f_E' \\ f_v' = 0.3f_E' \end{array} \right\} \quad \text{for a standard design Huygens eyepiece} \qquad (6.25)$$

6.6.2 The Ramsden eyepiece

In the *Ramsden eyepiece* the two lenses have equal or nearly equal focal lengths. The separation is reduced to be the sum of the focal lengths divided by 3. This places the first principal focus outside the system and this eyepiece can be used as a magnifier. The Ramsden eyepiece is also better for measuring instruments (Section 6.8) because a graticule placed at the first principal focus is magnified by both eyepiece lenses in an identical way to the intermediate image. Various types of graticule are described later in this section.

Figure 6.31 shows the overall system with the objective assumed to be off the page to the left. For infinity adjustment the emergent rays can be drawn in reverse with ray a passing undeviated through A_2 towards the image at F_2. Ray b is drawn through F_2' parallel to ray a and is refracted between the lenses parallel to the axis. This ray must have come from the objective to the first lens via F_1. A ray, a, from the F_2 image through A_1, the centre of the first lens, is used to define the intermediate imge required by this eyepiece when in infinity adjustment.

The cardinal points are simply calculated when $f_1' = f_2' = f'$ and

$$d = \frac{f_1' + f_2'}{3} = \frac{2f'}{3}$$

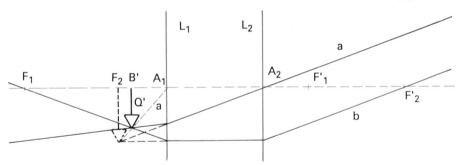

Figure 6.31 Ramsden eyepiece – reverse ray tracing

The equivalent focal length, f'_E, is

$$f'_E = \frac{f'^2}{2f' - 2f'/3} = \frac{f'^2}{4f'/3} = \frac{3f'}{4}$$

(6.26)

(for a standard design Ramsden eyepiece) so that each lens is less powerful than the overall magnifier. Because both lenses are equal the cardinal points are symmetrically arranged (Figure 6.32). The equation for f_v is the same (sign reversed) as that for f'_v:

$$f'_v = \frac{f'_2(f'_1 - d)}{f'_1 - f'_2 - d'}$$

$$= \frac{f'(f' - d)}{2f - d}$$

$$= \frac{f'f'/3}{4f'/3}$$

$$= \frac{f'}{4}$$

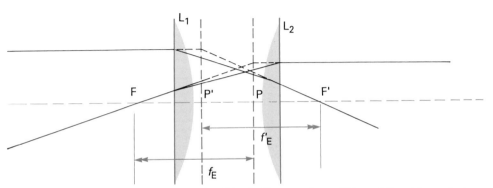

Figure 6.32 Ramsden eyepiece – cardinal points (identical lenses to those of Figure 6.31 and the worked example)

Alternatively, knowing the equivalent focal length, f_E' from the designated magnifying power of a Ramsden eyepiece, it is possible to define the separation, d, the component lens power, f', and the vertex focal length, f_v', as:

$$\left.\begin{aligned} d &= \frac{8}{9} f_E' \\[2mm] f' &= \frac{4}{3} f_E' \\[2mm] f_v' &= \frac{1}{3} f_E' \end{aligned}\right\} \quad \text{for a standard design Ramsden eyepiece} \qquad (6.27)$$

Without knowing the aperture and focal length of the objective it is not possible to establish the size and position of the eye ring. The following worked example gives some typical values for a $10\times$ eyepiece working with an objective of 200 mm focal length. The results may be compared with those for the Huygens eyepiece above. In the Ramsden eyepiece the two lenses are normally made plano-convex with the convex surfaces facing inwards.

Worked example – Ramsden eyepiece   A $10\times$ eyepiece has an equivalent focal length of 25 mm. Calculate the individual lens powers for a Ramsden type eyepiece with

$$f_1 = f_2 \quad \text{and} \quad d = \frac{f_1 + f_2}{3}$$

Calculate the eye relief distance for an objective 200 mm in front of the intermediate image. Assume infinity adjustment.

From $f_E' = 25$ mm and Equations 6.26, we have

$$d = \frac{8}{9} f_E' = 22.22 \text{ mm}$$

$$f' = \frac{4}{3} f_E' = 33.33 \text{ mm}$$

$$f_v' = \frac{1}{3} f_E' = 8.33 \text{ mm}$$

From Newton's equation, $xx' = ff'$ with the objective 200 mm in front of the intermediate image (in infinity adjustment), $x = -200$ mm. Therefore

$$x' = \frac{25 \times 25}{200} = 3.125 \text{ mm}$$

The eye ring is therefore $f_v' + x'$ from the eye lens, an eye relief distance of 11.45 mm, which is little different from the Huygens eyepiece in standard form.

6.6.3 Comparison of the Huygens and Ramsden eyepieces

Bearing in mind that the x' distance for the eye ring is dependent only on the equivalent focal length of the eyepiece and the objective distance and not on its internal construction, we can see from the final expression in Equations 6.14 and 6.27, for f'_v, that there is little to choose between the designs for eye relief and that, for high power units of, say, $25\times$, $f'_E = 10\,mm$ gives eye relief distances of $3.5\,mm$ and $3.8\,mm$ in each case respectively. Both are therefore limited at higher powers.

In practice, the Huygens has better basic aberration control; the Ramsden is often improved by using a doublet for the second lens. Sometimes the Huygens eyepiece is called a negative eyepiece and the Ramsden a positive eyepiece. This is bad terminology as both eyepieces have positive power. The main difference lies in the intermediate image location. The internal image of the Huygens means that it cannot be used as a magnifier while the Ramsden can be a magnifier and can more accurately be used with a graticule.

6.6.4 Graticules

The graticules commonly used with eyepieces may be of the simple cross-hair type which requires no glass support. When using these to align on vertical lines in the image it is better to set the cross-hairs at $45°$, as shown in Figure 6.33, so that the eye can judge the best position of symmetry. For setting exclusively on lines, very high visual accuracy can be obtained with the graticule in Figure 6.34 when the two lines are placed symmetrically each side of the image line. The interaction of the eye with the graticule is important. A telescope can be focused more consistently with a fixed graticule, which helps the eye to reduce involuntary changes in accommodation. Sometimes the graticule is fixed at the first principal focus of the eyepiece and moves with the eyepiece as the instrument is focused, as in Figure 6.17. More often, in measuring microscopes, the graticule is fixed in the draw tube so that the eyepiece can be focused as desired on the graticule and then the eyepiece/graticule combination can be focused onto the image from the objective.

Some graticules are used for direct measurement and have a millimetre scale, full size or reduced, by which the eye can measure the apparent size of features in the image. In most simple instruments a change in focus makes a difference to the size of the image. Graticules can be used with Huygens eyepieces, in less demanding applications, as well as with Ramsden eyepieces. When a scale is used with either eyepiece it must be calibrated by placing a known scale in the working plane of the microscope. If an aperture stop can be fitted at the first principal focus of a telescope objective the light beams forming the image become telecentric as

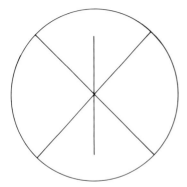

Figure 6.33 Traditional crosswires: 45° setting

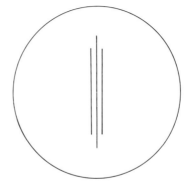

Figure 6.34 'Line pair' – symmetry setting graticule

described in Section 5.7. Any slight defocus of the eyepiece does not change the size of the image against the graticule. This aperture stop appears to be at infinity as far as the eyepiece is concerned and so the eye ring is now located at the second principal focus of the eyepiece. The principles of measuring instruments are described more fully in Section 6.8.

6.7 Projectors and condenser lenses

The most common type of projector is used with 35 mm slides made with photographic film which has been exposed in a 35 mm camera. When developed, slide film carries a record of the image in the form of transparent dyes. When viewed against a bright background the slide transmits the colours of the original scene. The format size, 24 × 36 mm, needs some magnification and small hand-held viewers are available which contain a simple magnifier (Section 6.3), usually with an aspheric surface.

For general display it is necessary to project a magnified image onto a screen, usually a white diffusing surface, and for this the slide must be made to appear very bright in front of a positive lens which forms a real magnified image at some convenient conjugate distance. Because the size of the transparent part of the slide is fixed, the distance of the screen from the projector and the size of the image on it combine to define the focal length of the projection lens. The distance from projector to screen is often called the **throw** of the projector and because the magnification is usually between ×30 and ×100 the throw can be considered equal to the image distance l' or the extra-foveal distance x' with little loss of accuracy. From Newton's equation (Equation 4.19), the magnification, m, is given by:

$$m = \frac{x'}{-f'}$$

and so

$$f' = \frac{x'}{-m} \tag{6.28}$$

where the negative sign indicates an inversion of the image. This is usually corrected by turning the slide upside down. From Equation 6.28 it is seen that for a given focal length of projection lens the throw increases with increasing magnification. This may seem convenient because a larger magnification giving a larger final image size suggests a larger audience which in turn needs a larger room. A simple table (Table 6.1) shows the relationship for three lenses with approximate image sizes in metres.

Table 6.1 Image size for 35 mm slides

Projection lens EFL (mm)	Throw		
	5 m	10 m	20 m
50	×100 (2.4 × 3.6)	×200 (5 × 7.5)	×400 (10 × 15)
85	×60 (1.4 × 2.1)	×120 (3 × 4)	×240 (6 × 9)
135	×37 (0.9 × 1.3)	×74 (2 × 3)	×150 (4 × 6)

Partly because of the size-constancy effect in human vision, more distant projected images do not need a proportional increase in size. Consequently, the 50 mm, 85 mm and 135 mm lenses would be considered optimum for throw values of 5 m, 10 m and 20 m respectively, and projection lenses which can zoom from 70 mm to 140 mm are now on the market. Taking the 85 mm EFL lens as representative of most 'fixed focus' projection lenses and noting also that most have an aperture of $f/3$ we can refer to Figure 6.35, which shows a lens of 28 mm aperture positioned 86 mm ($f + x$ for a magnification of ×85) from a 24 mm × 36 mm slide, 44 mm maximum (diagonal) dimension. The imaging rays from the top and centre have been drawn in, showing that the lens must have a field angle of 14.3°. In practice the projection lens would be a lens system similar to a camera lens for the same focal length and f-number. The first principal plane, GPH, would be close to, but not necessarily coincident with, the entrance pupil as shown in Figure 6.35.

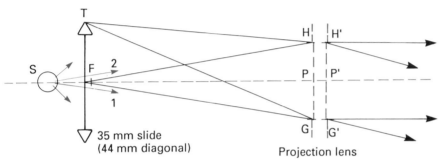

Figure 6.35 Slide projector requirements

The transparent slide, T, needs light to pass through it *and* through the projection lens. A bright source placed at S would firstly melt the slide and, secondly, illuminate only a small part of the slide with light which then passes through the lens; that is, light between ray 1 and ray 2 which are incident at G and H respectively. Even if a large source was placed behind the slide a large amount of the light emitted by it would fail to pass through the lens aperture.

The action of a **condenser** is to create an apparently large source which is uniform and directed mainly at the projection lens. The condenser is a positive lens or lens system of short focal length. Projectors commonly use quartz–halogen lamps which are described in Section 9.5 and, typically, have a luminous area of approximately 3 mm × 6 mm. Figure 6.36 shows a condenser lens system forming a magnified image of the source, S, in the entrance pupil, P, of the projection lens, EFL 85 mm. The use of two plano-convex lenses as shown is usual and the curved surfaces are normally non-spherical as described in Chapter 8. The simplified ray tracing in Figure 6.36 assumes one principal plane for the condenser lenses because they are normally very close.

The short focal length of the condenser system is determined by the size of the light source and the aperture of the projection lens. One of these is going to be the effective entrance pupil of the whole system, while the condenser system and slide surround form the field lens and stop. The system has the same optical principles as a two-lens eyepiece: the lamp has taken the place of the objective, while the

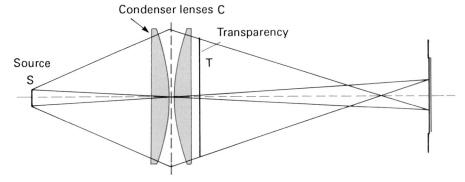

(a) Condenser too weak

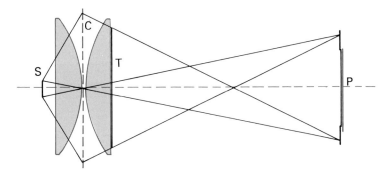

(b) Condenser too powerful

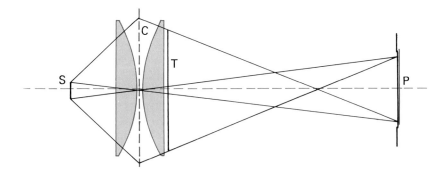

(c) Optimum

Figure 6.36 Condensers for slide projectors

condenser and projection lens replace the field lens and eye lens respectively. Like the Huygens eyepiece the slide takes the place of the intermediate image and is located between the condenser and projection lenses.

Figure 6.36(a) shows a condenser of long focal length where the lamp is at too great a distance and light from it is wasted at the same time as leaving unused parts of the projection lens aperture. Figure 6.36(b) shows a condenser of short focal length but parts of the source are unused because the image of the source is too large for the projection lens. Figure 6.36(c) shows the optimum arrangement where the light source image is just smaller than the projection lens aperture (to allow for focusing movements). For the lens and lamp quoted the paraxial power of the condenser is found from $m = -4$ and $l' = 100\,$mm (allowing 14 mm for lens thickness and slide plus 86 mm to the lens):

$$f'_c = \frac{l'}{1 - m} = \frac{100}{5} = 20\,\text{mm} \tag{6.29}$$

Thus the condenser needs to have a very short focal length and yet cover an aperture of 44 mm. The action of the slide projector condenser described above utilizes the concept of the **Maxwellian view**; when viewed from the position of the projector lens the condenser lens and the slide in front of it appear to be fully and evenly filled with light. The condenser is said to be completely flashed. The evenness of the illumination is very important as it is very distracting to have extra variations in brightness in the viewed image on the screen.

If the screen brightness is insufficient either a brighter source or a larger source must be used. There are practical limits to brightness with incandescent sources because of the large amount of heat they radiate (see Section 9.4). With a larger source there is no improvement unless the projector lens is made larger. A more powerful condenser as in Figure 6.36(b) only makes matters worse. However, with a larger source it is possible to make the condenser less powerful and therefore cheaper.

The condensers shown in Figure 6.36 have been drawn assuming an unobtainable refractive index of about 3.5! Ordinary glass would need much greater curvatures than those indicated by the outlines. Normally condensers are made with aspheric curves (Section 8.4) but these cannot fully cope with the problem. Another way to reduce the requirement for such powerful lenses is to increase the apparent size of the source by using a spherical mirror with the lamp located to one side of its centre of curvature. Figure 6.37 shows how this creates a lamp image which can be made adjacent to the real source to give an effective source of twice the size. The real source is below the axis and light from the rear of this, previously wasted, is now reflected by the mirror to form an apparent source immediately above the axis.

Where the size of the source and the size of the illuminated area are more similar it is possible to make the condenser image of the source coincident with the object. In Figure 6.38 the source, S_0S_1, has been made larger and the illuminated area of the transparency, T, smaller. The condenser, C, in Figure 6.38(a), the Maxwellian view system, is seen to accept only a narrow cone of rays from the source through to the projection lens, P. As the condenser and source are moved away the projection lens remains the aperture stop of the system but the source fills it in a different way so that in Figure 6.38(c) the total illumination projected onto the screen is the cone angle, β, integrated across the source. In Figure 6.38(a) the total projected illumination is α integrated across the source. There is little difference between the two cases but, with the **direct illumination** of Figure 6.38(c), the condenser lens has

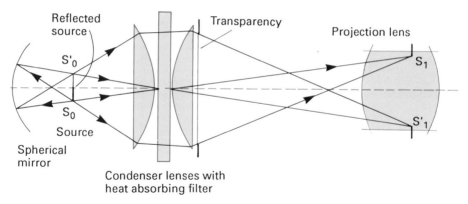

Figure 6.37 Slide projection system

much more space available to it. It should also be noted that with the Maxwellian view there is no advantage in increasing the aperture of the projection lens without changing the magnification source of the condenser and this makes it more difficult to fit all the components in. On the other hand an increase in the apertures of both the condenser and projection lens with the direct illumination will give a brighter projected image.

Direct illumination is used in movie projectors where there is less concern with temperature because the film is moving. Large incandescent arc sources are used. This type of illumination is also used with microscopes when looking at thin samples which are semi-transparent. These substage condensers must somehow provide a light beam that fills the numerical aperture of the microscope objective. Figure 6.38(c) would suggest that the numerical aperture of the substage condenser must be even greater than that of the microscope objective. Fortunately, for very high-powered objectives, the illuminated area is very small and the two apertures become equal. Substage condensers are therefore very similar in design to microscope objectives.

6.8 Measuring instruments

This section is concerned with visual measuring instruments that use the eye to measure along the axis – by selecting the best focus; or to measure across the axis – by obtaining some coincidence between two images or one image and a reference mark such as a crosswire or graticule scale.

Other visual instruments exist that allow a comparison in terms of colour or contrast, either in a side-by-side situation or alternating with time (Chapter 10). Many new non-visual optical instruments are now in use which use optical fibre technology or interferometry (Chapter 12) followed by computer analysis of the results. Still others replace the eye by a television camera followed by electronic picture analysis. None of these can avoid the optical principles developed earlier in this book and the following examples of visual measuring instruments demonstrate the use of mirrors, prism, lenses and lens systems previously described as well as the abilities and limitations of the eye.

When the eye is used to judge best focus it is limited by the depth of focus associated with its pupil size (or the eye ring size of the instrument) as well as its

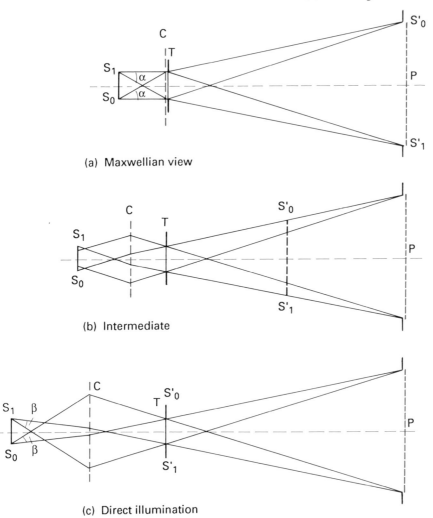

(a) Maxwellian view

(b) Intermediate

(c) Direct illumination

Figure 6.38 Condenser principles

own accommodation. Young observers, in particular, when using one eye only can exercise 5–6 D of accommodation without realizing it. When this is applied to a simple monocular telescope in infinity adjustment such as that shown in Figure 6.39 this eye is not able to distinguish the difference between an apparent image at infinity (solid lines) and one at 200 mm (dotted lines). By Newton's equation, the nearer intermediate image at I_a is closer to the eye lens by a distance x_e given by:

$$x_e = \frac{f_e f'_e}{200 \text{ mm}}$$

$$= \frac{f_e f'_e}{1000/D_e}$$

$$= x'_o \tag{6.30}$$

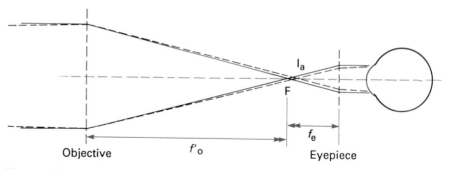

Figure 6.39 Focus inaccuracy with telescopes

where D_e is the dioptres of accommodation and the eye is assumed to be very near the second principal focus of the eye lens. This new position is x_o' from the second principal focus of the objective and so the object position x_o which is conjugate with this is given again by Newton's equation:

$$x_o = \frac{f_o f_o'}{x_o'}$$

$$= \frac{f_o f_o'}{f_e f_e'} \frac{1000}{D_e}$$

$$= \frac{1000}{D_o} \tag{6.31}$$

Where D_o is the dioptres value for this object position. Thus, the dioptre error at the objective is reduced by the square of the telescope magnification, M, as might be expected from the calculation of longitudinal magnification in Section 6.5.

$$D_o = D_e \frac{(f_e f_e')}{(f_o f_o')} = \frac{D_e}{M^2} \tag{6.32}$$

In a telescope of ×7 power, the 5 D error at the eye is only 0.1 D error in the object space. However, the situation is worse in a camera viewfinder where the lateral magnification may be only ×2 (see below), and in the case of the microscope where the small diameter of the eye ring can increase the depth of focus in addition to the accommodative error indicated.

When a graticule is introduced at the intermediate image plane, the accommodative error is reduced if the observer tries to set the image and the graticule marks in focus at the same time. However, untrained observers can unconsciously change their accommodation when changing their attention from one to the other. Binocular viewing also helps to reduce the accommodative fluctuations because the accommodative response of the eye is linked to the convergence between the two eyes needed to fuse the two images. Again this can be negated with untrained observers, who often suppress the vision in one eye when looking into eyepieces.

A diffusing screen at the intermediate image position can be beneficial because out-of-focus blur can no longer be compensated by accommodation and the scattering action makes the effective eye ring larger, thereby reducing the depth of focus. Unfortunately this is accompanied by considerable loss of image brightness and the structure of the screen makes it difficult to see the fine detail in the image by which best focus is judged.

6.8.1 The focimeter

The **focimeter** is an instrument that measures the power of lenses and relies on an 'in-focus' judgement by the eye. This instrument is based on the telescope method for measuring negative lenses which is described in Section 4.8 and shown in Figure 4.25(b). The principle uses an auxiliary lens and is equally applicable to positive lenses and to negative lenses. The basic system is refined so that the lens to be tested is always located at the second principal focus of the auxiliary lens by having a circular support against which the lens under test can be held, as shown in Figure 6.40. The auxiliary lens, A, is made more powerful than the strongest lens to be tested. This means that the apparent object for the test lens will always be a virtual object but its position may be changed by moving the actual object (target) along the axis, and its distance from the test lens can be calculated using Newton's equation, which defines image distances from F'_A, the position of the test lens.

In the telescope focimeter, the viewing telescope is fixed in infinity adjustment so that a clear image of the target is seen by the eye only when the light entering the telescope is collimated. For this to happen the apparent object for the test lens must be at its first principal focus. Because the test lens is at the second principal focus of the auxiliary lens we have

$$l_T = f_T = x' = \frac{1}{F_T} \tag{6.33}$$

where f_T and F_T are the focal length and focal power of the lens under test.

If this occurred when the target was at the first principal focus of the auxiliary lens, the light before and after the test lens would be collimated so that F_T would have to be zero and this point on the scale could be marked 0.0 D.

If the target must be moved a distance x to give a sharp image in the telescope for a non-zero test lens, then

$$x' = \frac{1}{F_T} = \frac{f_a f'_a}{x}$$

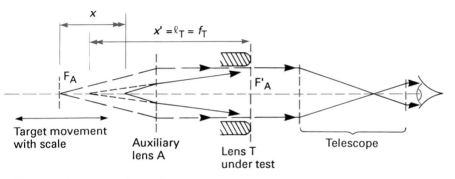

Figure 6.40 Focimeter (telescope)

or

$$F_T = \frac{x}{f_a f_a'} \tag{6.34}$$

where f_a and f_a' are the focal lengths of the auxiliary lens. Thus the distance the target moves is *directly proportional* to the power of the test lens. If the test lens is negative the target moves away from the auxiliary lens and, if positive, the target moves towards the auxiliary lens. Thus, the scale may be marked in dioptre divisions with negative and positive values either side of the zero position.

In the projection focimeter the telescope is replaced by a translucent screen through which the image formed by the auxiliary lens and test lens combination can be viewed. The target must now be well illuminated with a small condenser system. The screen is usually placed about 1 m from the test lens and some recalibration of the scale is needed to allow for this (see the worked example). In principle the projection system should be more accurate than the telescope because the eye must focus on the screen. However, both systems are more often limited by the aberrations of the lens under test and the accuracy with which it can be positioned. Figure 6.41 shows a general system with some of the mirrors that are used to fold the light path to give a compact instrument.

Worked example If a telescope focimeter is modified to be a projection focimeter with a screen located 1 m from the test lens, how should the reading on the original dioptre scale be modified to obtain the correct value, assuming that the telescope had been set in infinity adjustment?

In the original case (o), the vergences at the test lens of power F are:

$$L_o' = 0 = L_o + F$$

In the modified case (m), the vergences at the test lens are:

$$L_m' = 1 = L_m + F$$

Therefore

$$L_m = 1 - F = 1 + L_o$$

Thus the focimeter will read L_o as though it had a lens 1 D weaker ($L_m - 1$), and all readings must be changed by $+1$ D.

Alternatively, imagine a test lens that consists of two thin lenses in contact; $+3$ D and $+1$ D. When correctly in focus on the modified focimeter, the $+1$ D will be focusing collimated light onto the screen. The original part of the focimeter will read as though it were causing collimated light to emerge from a 3 D lens. Therefore the 4 D test system will be read as a 3 D lens. Perhaps it would be better to change the scale markings!

6.8.2 The radiuscope

Another measuring instrument, which also relies on focus setting, is the **radiuscope**. This is used for measuring the surface curvatures of short radius optical devices such as contact lenses, and is based on the coincidence principle described in Section 4.8 and shown in Figure 4.27. Known as the **Drysdale method**, the system uses a microscope with an objective of about 20 mm focal length. An illuminated

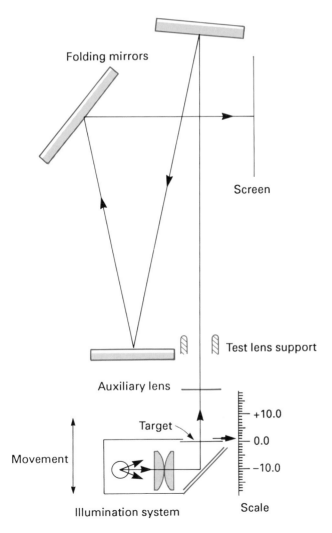

Figure 6.41 Projection focimeter

graticule, G, is effectively located at the same position as the intermediate image plane by reflection in a semi-transparent mirror, M, as shown in Figure 6.42. This graticule is then focused by the objective into the normal object plane conjugate with the intermediate image. The actual object is supported on a stage with a calibrated vertical movement. If a reflecting surface is placed on the stage and moved to coincide with the normal object plane the light is reflected back to the objective and then to the eye via the intermediate image plane. Thus the normal object plane can also be called the 'expected image plane' for the microscope.

Even with short-radius curved reflectors receiving the image, the reflected image 'is the same distance behind the mirror as the object is in front' when the distances involved are very small. Thus in Figure 6.43 the expected image plane is above the surface and so the actual image is $2d$ below this until the surface is moved up by the

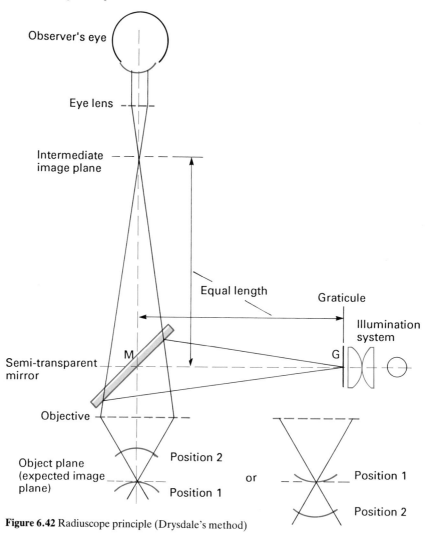

Figure 6.42 Radiuscope principle (Drysdale's method)

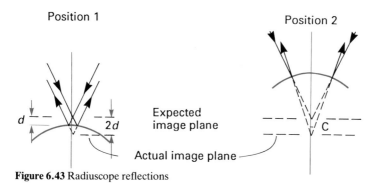

Figure 6.43 Radiuscope reflections

distance d in order to coincide with the expected image plane. The same effect occurs in position 2 where the centre of curvature, C, must coincide with the expected image plane for the actual image to be there and therefore seen in focus in the eyepiece. Because the actual image moves by twice the error in the surface position there is a doubling of the accuracy of the focus judgement by the eye. Nevertheless care and experience are necessary to obtain the 0.05–0.025 mm accuracy required for contact lenses. It should be noted that the system works on spherical surfaces and can be made to work on toric surfaces, but on aspheric surfaces and badly made surfaces, where there is no single centre of curvature, the image quality for position 2 will be poor and prevent an accurate reading. Because the radius of curvature is found by subtracting one reading from the other there is a final error of about 1.4 times the mean single reading error.

6.8.3 Range finders and view finders

Although the in-focus judgement of the eye is always suspect, the use of the eye to judge the coincidence of two lines is very accurate because the eye–brain system possesses a remarkable **vernier acuity** which, surprisingly, is better than its normal acuity. Many cameras therefore have a focusing system comprising two prisms in the viewfinder system, which converts the out-of-focus image into a vernier displacement. In the same way an optical range finder uses the compensation for different viewpoints to measure the distance of an observed target. In both these cases the vernier acuity used is that shown in Figure 6.44(a). The displacement of the upper line with respect to the lower line may subtend an angle as low as 5″ of arc and yet the eye will detect it and correctly judge the direction. This is particularly important when the camera lens is used at a large aperture as the depth of focus of the lens is much smaller than that of the eye.

In the camera view finder and the military range finder the viewed image is divided so that the lower part appears to be displaced with respect to the upper part (Figure 6.44(b)). The observer needs to select a vertical edge in the scene which straddles the dividing line and adjust for zero displacement (Figure 6.44(c)). The camera view finder is a simple prism telescope and the field lens of the eyepiece is in the form of a positive Fresnel lens (see Chapter 8) at the centre of which are two thin prisms, one of which has its base to the left and the other to the right. Figure 6.45(a) shows the case when the intermediate image coincides with the field lens; the prisms merely deflect the light a little differently into the eye ring. Figure 6.45(b) shows, much exaggerated, the case when the device is out of focus and the image is beyond the field lens. Then the prisms deflect the light *and* displace the image in opposite directions. In each case the dotted lines show the ray paths in the absence of the prisms. Obviously there is the difficulty that the prisms must be

(a) High accuracy   (b) Displaced   (c) Aligned   (d) Edge aligned

Figure 6.44 Vernier acuity → range finding

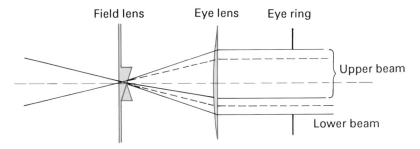

(a) Seen from above – in-focus image

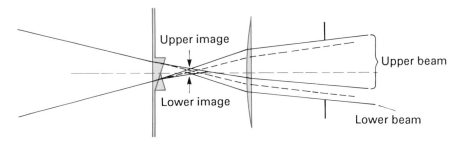

(b) Seen from above – out-of-focus image

Figure 6.45 Camera focusing by split-field viewfinder

powerful enough to show up significant defocus but not so powerful that they deflect the light out of the eye ring. Most camera users experience the need to locate the eye accurately in order to see both images. The effect can be seen very clearly if the camera view finder is observed at arm's length.

In the **optical range finder,** two objectives are used to collect light from the scene as in Figure 6.46. The distance between them is called the **base line**. The larger the base line the higher the accuracy over longer ranges. Before the development of radar and lasers, naval and shore battery range finders used base lines of up to 30 m, but at these extreme sizes mechanical flexing can limit the accuracy actually achieved. In the infinity setting, mirrors M_1 and M_2 are parallel to each other and mirrors M_3 and M_4 are also parallel to each other. Mirror M_2 reflects the lower part of the image and M_4 reflects the upper part. Light from an infinitely distant point object not only arrives at each lens L_1 and L_2 in collimated beams but the two beams are parallel to one another. They remain so through the two mirror systems $M_1 + M_2$ and $M_3 + M_4$ and so the upper and lower parts of the scene appear coincident to the observer. For a nearer object the two arriving beams are no longer parallel and the situation shown by the dotted lines in Figure 6.46 applies. The parts of the split image are displaced in opposite directions as in Figure 6.44(b), but this can be compensated by rotating one or both of mirrors M_1 and M_3. If M_3 is rotated, the dotted ray 2 becomes parallel ray 1 to give Figure 6.44(c). The total amount of rotation needed to realign the images is a function of the object distance (range) and distance, B, between the axes (the base line). Remembering that reflected light is deviated by twice the angle of the rotated mirror the measured mirror rotation angle, φ, is equal to θ/2 where θ is the angle subtended by the base line, B, at the object. Then we have the range R, where

$$R = \frac{B}{\theta} \qquad \text{for small values of } \theta \qquad (6.35)$$

When the range is already known, the actual width of the light-house at the level of the dividing line can be calculated from the difference, say δθ, between the mirror rotation value in Figure 6.44(c) where the two images are aligned and in Figure 6.44(d) where opposite sides of the light-house are aligned; that is, the image is sheered across the dividing line. This indicates that the width, W, of the light-house subtends the angle δθ at the distance R, so that

$$W = R.\delta\theta \qquad (6.36)$$

The feature is useful for measuring the size of inaccessible objects such as the image formed by reflection in the cornea in the keratometer discussed later in this section. With large base-line range finders the higher accuracy will not be achieved if the instrument flexes in use and inadvertent mirror rotation occurs. Much of this can be eliminated if the reflections at M_1 and M_2 are given a fixed 90° deviation by using pentag prisms (Section 2.10). The controlled deviation of the converging light beams to give image displacement to obtain coincidence may then be by a pair of contra-rotating prisms or a single prism moving along the base line.

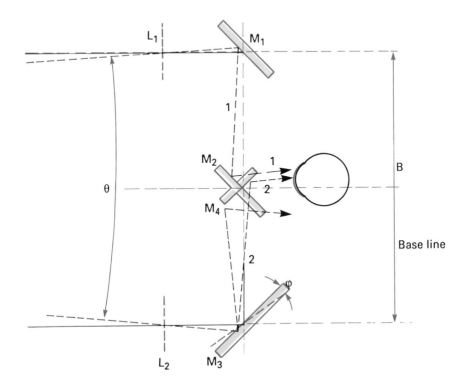

Figure 6.46 The range finder principle

6.8.4 Measurement of small objects

Considering now the lateral measurement of smaller objects, the starting point is a microscope or telescope depending on the object distance. For small near objects the methods start with a simple scale in the microscope eyepiece as shown in Figure 6.47(a). When the image of any object is in focus with the scale the dimension can be found from two readings against the scale. Sometimes two scales can be used for orthogonal measurements, or the eyepiece may be rotated. Inaccuracies can occur because the scale is not orientated exactly in the direction of the required dimension, but these are usually very small because the difference in length is given by $1 - \cos \theta$ where θ is the misalignment. Even when θ is $5°$, which should be obvious in most circumstances, the error is less than 0.4%.

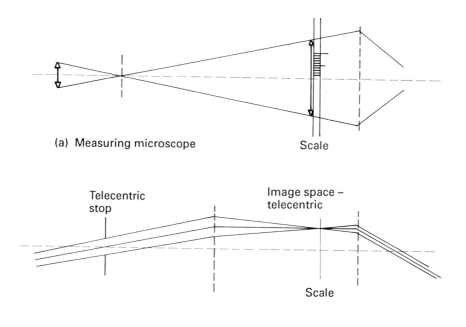

(a) Measuring microscope

(b) Telecentric telescope

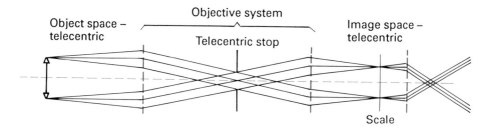

(c) Telecentric cathetometer

Figure 6.47 Lateral size measurement

A more serious error is that in the normal microscope the apparent size of the image varies with focus. If the dimensions of the instrument and its application allow, the telecentric principle described in Sections 5.7 and 6.6 may be used. In the case of a relatively distant object where the measuring instrument is a telescope, the telecentric principle is met with a stop at the first principal focus of the objective (Figure 6.47(b)). Any small defocus of the eyepiece with respect to the intermediate image will not give rise to error. However, it is not possible to have a single lens telecentric on both object and image sides and, if the object is relatively near, any defocus caused by an incorrect object-to-objective distance will introduce error. For short-focus telescopes (or long-focus microscopes, sometimes called **cathetometers**) it is possible to use a lens system as the objective, comprising two lenses separated by the sum of their focal lengths with the stop at the common focus. This condition is shown in Figure 6.47(c) and is used in machine shop instruments known as **contour projectors**.

A better but slower way of making lateral measurements is with a **travelling microscope**. This instrument takes the scale outside the optical system and arranges for the whole microscope body to move in one (or two orthogonal) directions (Figure 6.48). The eyepiece scale now reduces to either a crosswire as in Figure 6.33 or a pair of parallel lines as in Figure 6.34. The crosswire has a more general use, but when setting onto line images it should be used diagonally across the line as shown in Figure 6.33 and not aligned with the line because the masking of object by the crosswire makes the visual judgement more difficult. As well as the vernier acuity mentioned above, the eye–brain system possesses a remarkable acuity for

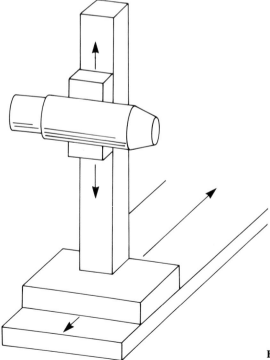

Figure 6.48 Travelling microscope

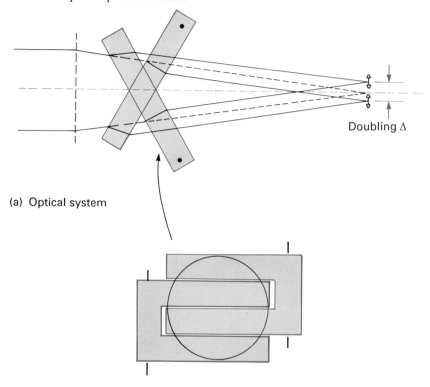

(a) Optical system

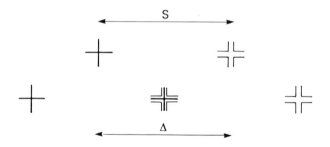

(b) Interleaved micrometer

(c) Appearance when doubling, Δ equals object size, S

Figure 6.49 Size measurement by doubling

symmetry. When setting two lines about a third line as shown in Figure 6.34 very small imbalances can be detected and accurate settings achieved.

If an object is designed on which two lines are formed at one end and a single line at the other, its overall size can be measured using doubling techniques. These are similar to the image shear method for size measurement with the optical range finder, but now the whole image is doubled. One such technique uses the parallel plate micrometer principle described in Section 2.7. In a convergent beam this micrometer displaces the image by an amount dependent on the tangent of the incidence angle. A pair of interleaved parallel plate micrometers is placed close to the objective of a measuring instrument (Figure 6.49). The whole image is doubled and the extent of this can be altered by rotating the plates. For the object shown in Figure 6.49(c), the size S can be measured by doubling the image so that one end of the object is apparently sitting inside the other end. The rotation of the plates can be calibrated either by calculation or by measuring known object sizes.

6.8.5 The keratometer

This principle may be used to measure the radius of curvature of the cornea with an instrument called a **keratometer**. The special object is attached to a long-focus microscope so that its 'size' is across the objective. The object is reflected by the cornea and the reflection viewed by the microscope containing a doubling system which allows the size of the image to be measured and, because the actual object size is known, the magnification may be calculated. In Figure 6.50, the magnification, m, is related to the focal length, f, and therefore the radius, r, of the reflecting surface by Newton's equation:

$$m = \frac{h'}{h} = \frac{f'}{-x} = \frac{r}{-2x} \tag{6.37}$$

where x is the distance of the original object from the first principal focus of the reflecting surface. Therefore

$$r = -2xm \tag{6.38}$$

Obviously x cannot be measured directly into the eye but it is possible to fix the focus setting for the microscope so that (within the error of the eye setting) the distance, d, between the object and its in-focus image is known.

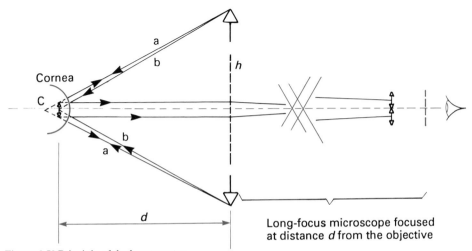

Figure 6.50 Principle of the keratometer

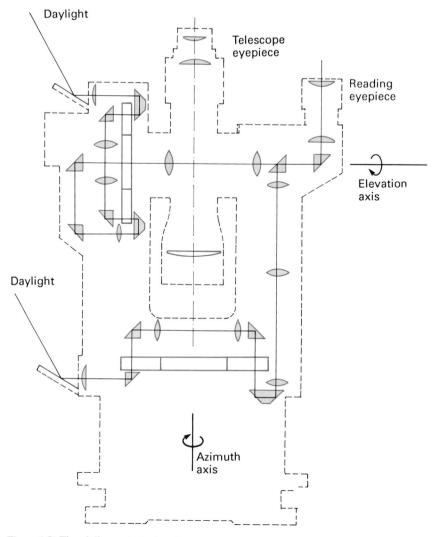

Figure 6.51 Theodolite – principal optics

With spherical reflecting surfaces the first principal focus is coincident with the second principal focus. The extra focal distances in Newton's equation are therefore measured from the same points to give:

$$d = -(x - x')$$

where x is normally negative but d is measured positive, and

$$x' = \frac{-ff'}{x} = xm^2$$

Therefore

$$d = -x(1 - m^2)$$

and Equation 6.38 becomes

$$r = \frac{2md}{1 - m^2} \qquad (6.39)$$

Because the magnification is usually less than 0.05, m^2 is very small and the equation

$$r = 2md \qquad (6.40)$$

is sufficient.

6.8.6 Measurement of large distances

When the size measurement required is of large distances, as for mapping and surveying, the problem can be converted into that of angle measurement. This is carried out using a telescope, which can be rotated and the angle of rotation carefully measured. Such instruments need suitable crosswires, which are aligned on target poles by the observer looking through the instrument. Usually the telescope can be rotated vertically (elevation reading) as well as horizontally (azimuth reading) and are called **theodolites**. The magnifying power of the telescope of a theodolite may be ×20 or more. This allows an experienced observer to align it with target poles to an accuracy of about 1″ of arc.

Just as important as the telescope, therefore, is the method of measuring angles to this accuracy. Older instruments used extra optics to view both sides of a finely divided circular scale. This reduced errors due to centring inaccuracies. Figure 6.51 shows that the complexity of the telescope is less than that of the optics needed to allow the operator to read the scales accurately! In these days a theodolite is used only to measure the angles. Distances are measured using a laser range finder, which utilizes the known speed of light.

In recent years visual measuring instruments have benefitted from photo-electronic systems, which use moiré fringes and shaft-encoders to provide the scale-reading information, which can then be printed out or stored directly in computers. As well as making the operator's job easier, these methods considerably reduce the number of errors, which are more frequently associated with scale-reading and recording rather than the actual setting of an instrument. This amount of electronics is relatively cheap, and so electronically assisted visual measuring instruments form a useful half-way house between the older designs and fully automatic systems with their attendant high costs and limitations.

Exercises

Note: In the following exercises the lenses – objectives, eyepieces, etc. – are to be considered as thin.

6.1   A camera is fitted with a lens of 50 mm focal length. Three of the marked apertures have effective diameters of 12.5, 8.8 and 4.54 mm. Express these as f numbers (see Chapter 5). Calculate the exposure needed for the largest and smallest aperture if that for the 8.8 mm aperture is $\frac{1}{60}$ of a second. How must the effective apertures change if the markings are to remain correct when the lens is a zoom lens and adjusted to have an equivalent focal length equal to 75 mm?

6.2   What is the significance of the f number of a photographic lens? Describe the system of numbering which is commonly adopted and explain the reasons for its choice.

6.3   Explain briefly the following terms, with diagrams:
(a) Myopic eye.
(b) Hypermetropic eye.
(c) Crystalline lens.
(d) Far point.
Why is a negative lens used to correct a myopic eye and a positive lens a hypermetropic eye for distance? Give diagrams.

6.4   What must be the position of an object in order that it may be seen distinctly through a +10 D lens placed 5 cm in front of an eye, the eye being accommodated for a distance of 40 cm? If the lens has a diameter of 1 cm, what length of object can be seen through the lens?

6.5   State what is meant by (a) the lateral magnification of an image, (b) the apparent magnification of an image as seen by an eye.

6.6   Show fully how a positive lens produces an apparent magnification of an object seen through it. What will be the magnification produced by a +15 D lens 50 mm from an emmetropic eye when the eye is (a) unaccommodated, (b) accommodated 4 D?

6.7   State exactly what is meant by the magnification of a microscope. What will be the magnification of the microscope having an objective of ⅔ in focal length, an eyepiece of 1¼ in focal length and a tube length of 6 in?

6.8   A compound microscope has an objective of 15 mm focal length and an eyepiece of 30 mm focal length, the lenses being 180 mm apart. What will be the position of the object and the magnification of the microscope when focused for an emmetropic person with accommodation relaxed?
    Find the equivalent focal length of the microscope as a complete system and hence the magnification considering the system as a magnifier.

6.9   Give a careful diagram showing the path of light through an astronomical telescope. A telescope 12 in long is to have a magnification of eight times. Find the focal length of objective and eyepiece when the telescope is (a) astronomical, (b) Galilean.

6.10   Show by means of a diagram how the magnification is produced in the case of a Galilean telescope. A Galilean field glass magnifying five times has an objective of 180 mm focal length; find the focal length of the eyepiece and its distance from the objective when the field glass is focused for a myope of 4 D.

6.11   An astronomical telescope has an objective 150 mm in diameter and a focal length of 850 mm. If the eyepiece has a focal length of 25 mm and a diameter of 15 mm calculate the angular magnification, the diameter of the exit pupil, the eye relief distance and the apparent field of view (from the centre of the exit pupil).

6.12   A Huygens eyepiece consists of two lenses having focal length of 2½ and 1 in, separated by 1¾ in; find the equivalent focal length and the positions of the principal foci. What will be the magnification of a telescope having an objective of 20 in focal length with the above eyepiece?

6.13   Find the locations of the cardinal points of an eyepiece comprising two equal lenses of focal length 40 mm separated by 30 mm. If this system is used with an objective of focal length 600 mm and diameter 60 mm, find the size and location of the exit pupil. Assume infinity adjustment.

6.14   Describe all the elements in a slide projector from the lamp to the screen. What effect will (a) transparency size and (b) room size have on the choice of condenser lens and projection lens?

6.15   A horizontal telescope contains a pair of horizontal crosswires 2.5 mm apart. The telescope is focused on a vertical staff 10 m away from the objective which has a focal length of 100 mm. Find the length of the staff apparently intercepted between the wires.

6.16   Describe the construction and action of a simple telescope, and show how, by adding (a) lenses, (b) prisms, an erect image may be obtained instead of an inverted one.

6.17   The objective of a reading telescope has a focal length of 250 mm and the eyepiece 50 mm. The telescope is focused on an object 1 m away, and the image is formed at the distance of most distinct vision (250 mm). Draw a diagram showing the path of rays through the telescope. Determine the magnification and the length of the telescope.

6.18   A Galilean telescope has an objective of 120 mm focal length and magnifies five times when used by an emmetrope to view a distant object. What adjustment must be made when the telescope is used by (a) a myope of 10 D, (b) a hypermetrope of 5 D? Find the magnification in each case.

6.19   The finder on a camera consists of a 6 D concave lens having a rectangular aperture of 25 × 20 mm and a peep-hole 75 mm behind it; what extent of object at 10 m distance can be seen in this finder?

6.20   The microscope of Exercise 6.8 is used to form a real image on a photographic plate 0.5 m from the eyepiece (photomicrography). How far and in which direction must the microscope be moved with respect to the object in Exercise 6.8 and what will be the lateral magnification of the image?

6.21   Give a diagram showing the path of light through a compound microscope and deduce the formula giving the magnification of the instrument.

6.22   Two thin convex lenses when placed 250 mm apart form a compound microscope whose apparent magnification is 20. If the focal length of the lens representing the eyepiece is 40 mm, what is the focal length of the other?

Chapter 7

Aberrations and ray tracing

7.1 Out of the paraxial

The approximations and assumptions of the paraxial region conceal the fact that an optical system does not form a point image of a point object except under very rare conditions. Normally the image is a blurred region because rays from the point object which pass through different parts of the lens are not refracted to a single image point. The discrepancies between other rays and the principal ray (through the centre of the exit pupil) are called **aberrations**. An exact value for the aberration of a particular ray can be found by calculating the ray path through the lens or system and comparing this with the path of the principal ray. This is called **ray tracing** and is best done on a computer as described in Section 7.6.2.

Although ray tracing gives accurate results it does not provide any explanation of the aberrations. In order to understand their causes and, hopefully, reduce their damaging effects on image quality, a theory of aberrations is needed. This turns out to be yet another approximation! However, we are now calculating approximate values of the defect, which is the departure from the perfect case. Because the main aim of optical design is to reduce these defects (aberrations) to negligible values it does not matter if the equations we use to calculate them are not quite accurate.

This chapter concentrates on the aberrations of thin lenses which, as was shown in Chapter 5, are approximations to real lenses. However, most optical systems and instruments can be regarded as arrangements of thin lenses and this aberration theory will give an overall view of the optical quality to be expected.

Another limitation of this chapter is its restriction to monochromatic aberrations. Because the refractive index of most optical materials shows a significant change with the colour of the light, the aberrations of a given lens for blue light will be different from those for red light. However, it will be shown in Chapter 14 that any lens may have its principal colour aberrations corrected by making it a doublet; that is, two lenses in contact, using different optical materials.

Thus the theory of aberrations developed in this chapter gives a basic understanding of the likely image defects of all lenses and lens systems from spectacle lenses to photographic objectives. The full theory of optical design is beyond the scope of this book, but Chapter 14 attempts the basic principles of optical design and describes some of the concepts in current use.

7.2 Third-order theory

It has been shown that refraction at a lens surface follows Snell's law: $n \sin i = n' \sin i'$. In Sections 2.4 and 3.4, it was seen that $\sin i$ is a non-linear function of i, and so Snell's law does not give a simple relationship between i and i'. However, $\sin i$ can be represented by a series of terms:

$$\sin i = i - \frac{i^3}{6} + \frac{i^5}{120} - \frac{i^7}{5040} + \ldots$$

where i is measured in radians. As long as i is much less than 1 (1 rad is approximately 57°) each extra term is much smaller than the preceding term. The assumption, $\sin i = i$, gives the paraxial approximation, which has been used so far in this book. It is called first-order theory. The next better approximation is $\sin i = i - i^3/6$. When this is used it is possible to derive five independent ways in which an image may be an imperfect rendering of a point object. These were studied by Ludwig von Seidel, who published the mathematical expressions in 1855–56. Thus these five aberrations are often known as the **Seidel aberrations** and sometimes as the **primary aberrations**. This theory, discussed in this chapter, is called **third-order theory** because the extra term in the approximation contains i^3.

An even closer approximation is

$$\sin i = i - \frac{i^3}{6} + \frac{i^5}{120}$$

which gives **fifth-order theory**. This yields a great many aberration types, most of which remain small. Fifth-order theory with its **secondary aberrations** is not widely used in lens design, as the equations are complex, and third-order theory is usually enough for the initial design work, which can then be followed by exact ray tracing (Section 7.6).

The five Seidel aberrations are:

S_I Spherical aberration
S_{II} Coma
S_{III} Astigmatism
S_{IV} Curvature of field
S_V Distortion

Each of these will be dealt with in this chapter. They all occur together in differing amounts, and the purpose of lens design is to reduce the sum of their effects to an acceptable level. It is possible to derive equations of varying complexity for each of these aberrations for refraction (and reflection) at spherical surfaces, aspheric surfaces and thin lenses. This chapter will concentrate on thin lenses.

A converging wavefront must be perfectly spherical to form a point image. An aberrated image is formed by a non-spherical wavefront. One way of expressing the aberration is in terms of the differences, usually very small, between the aberrated wavefront and a perfectly spherical reference wavefront. This is called the **wavefront aberration** and the mathematical basis for this gives an overall view, as follows.

The accurate formula for the sag of a spherical curve is:

$$x = r - (r^2 - \rho^2)^{1/2} = r\left[1 - \left(1 - \frac{\rho^2}{r^2}\right)^{1/2}\right] \tag{7.1}$$

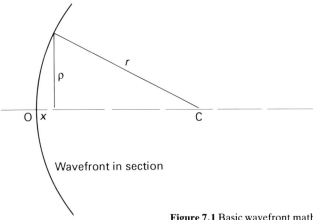

Wavefront in section

Figure 7.1 Basic wavefront mathematics

where the symbols are as defined in Figure 7.1 and ρ has been used rather than y because it turns out to be more complex.

By the binomial expression, Equation 7.1 can be written:

$$x = \frac{\rho^2}{2r} + \frac{\rho^4}{8r^3} + \frac{\rho^6}{16r^5} + \frac{\rho^8}{128r^7} + \text{higher terms} \qquad (7.2)$$

if spherical, and

$$+ A\rho^4 + B\rho^6 + C\rho^8 + \text{higher terms} \qquad (7.3)$$

if not spherical.

These extra 'correction' terms have been added deliberately and indicate what is needed to obtain a completely general expression for a wavefront surface which has been deformed from a spherical shape. These added terms give rise to third-order theory, fifth-order theory and seventh-order theory respectively. We will consider only $A\rho^4$, third-order theory.

The value of ρ in real systems is slightly complicated by the use of stops. In Figure 7.2 the off-axis image uses different parts of each lens depending on how far the surface is from the stop. The 'corrections' to the wavefront due to a particular surface are a function of ρ^4 (and ρ^6 and ρ^8, if we wish to include them), but only a part of this is effective as shown in Figure 7.3. We can express ρ in terms of a

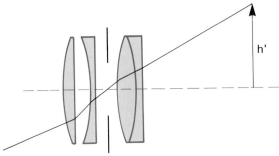

h'

Figure 7.2 Usual lens system conditions

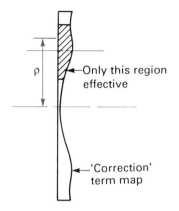

Figure 7.3 Selected region for aberration effect

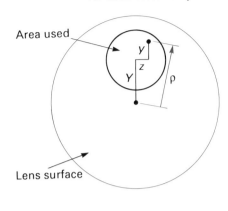

Figure 7.4 Effective point location mathematics

combination of position within the stop (y,z) and the displacement (Y) of the centre of the stop on the particular surface. From Figure 7.4, we have

$$\rho^2 = [(Y + y)^2 + z^2]$$

and so the third-order 'correction' term is:

$$A\rho^4 = A[(Y + y)^2 + z^2]^2 \tag{7.4}$$

This is a perfectly general description (to this level of accuracy) of the way in which, via Equation 7.3, changes in Y, y and z can modify x, the axial location of any point on a wavefront. It is called a correction term because, with it, we hope to obtain a correct mathematical description of the aberrated wavefront. We are not yet at the point of correcting the aberration! In fact the correction term equals the wavefront aberration because without it we have a perfect spherical (unaberrated) wavefront.

Thus we can expand Equation 7.4 to give the wavefront aberration, W:

$$W = A(y^2 + z^2)^2 + AY[4y(y^2 + z^2)] + AY^2(6y^2 + 2z^2) + AY^3(4y) + AY^4 \tag{7.5}$$

where the terms have been grouped in increasing power of Y. From Figure 7.2, Y is proportional to h' and because we need a completely general equation we allow each coefficient to be different from the others, to give:

$$W = \tfrac{1}{8}A_\mathrm{I}(y^2 + z^2)^2 + \tfrac{1}{2}(A_\mathrm{II}h'[y(y^2 + z^2)] + \tfrac{1}{4}A_\mathrm{III}h'^2(3y^2 + z^2) + \tfrac{1}{2}A_\mathrm{V}h'^3(y)$$

Spherical aberration (circular function)	Coma (displaced circular function)	Astigmatism (elliptical function)	Distortion (displacement function)

$$\tag{7.6}$$

The last term in Equation 7.5 is omitted because it does not change the wavefront shape. Dividing through by 8 brings the terms into a better correspondence with the spherical reference wavefront.

W is merely the distance (in the x-direction) between the aberrated wavefront and the spherical reference sphere for each point on its surface as defined by values of y, z and h'. Remember that y and z are the axes across the pupil and h' is proportional to the displacement, Y, of the pupil from the symmetric axis of the

234

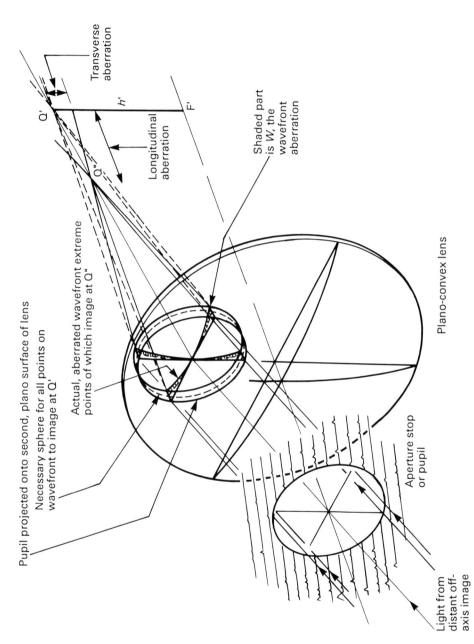

Transverse aberration

Longitudinal aberration

Q'

Q"

h'

F'

Pupil projected onto second, plano surface of lens

Necessary sphere for all points on wavefront to image to Q'

Actual, aberrated wavefront extreme points of which image at Q"

Shaded part is W, the wavefront aberration

Plano-convex lens

Aperture stop or pupil

Light from distant off-axis image

Figure 7.5 The relationship between longitudinal, transverse and wavefront aberration of a thin lens

lens. If the lens were perfect the image would be at the paraxial point Q', as shown in Figure 7.5. Q' is therefore the centre of the spherical (non-aberrated) reference sphere and W is the difference (for all the different values of y, z and h') between the reference sphere and the actual wavefront.

Often a better image can be found a little nearer to the lens. This type of adjustment is called **defocus** and is needed because the paraxial image position is rarely the best when aberrations are present. A defocus means that the reference sphere has a change of curvature but otherwise stays spherical. Our description of the aberrated wavefront must therefore be changed by an equal and opposite spherical amount. Strictly a sphere needs all the terms of the binomial expansion. However, for third-order theory we can use simply $A(y^2 + z^2)/2$, which is the well known sag formula $s = y^2/2r$ (Equation 3.3b), except that A is the curvature rather than the radius of curvature, r.

In particular, when defocus is needed for an off-axis point we use

$$A_{\mathrm{IV}} \frac{h'^2}{2} \frac{(y^2 + z^2)}{2}$$

where the $h'^2/2$ term is due to the actual paraxial image being curved as explained in Section 7.4. This term is therefore added to Equation 7.6 and fits with the astigmatism term, A_{III}. It will be shown in Section 7.4 that, if sufficient defocus is given,

$$\frac{A_{\mathrm{IV}}}{4} h'^2(y^2 + z^2)$$

is subtracted from

$$\frac{A_{\mathrm{III}}}{4} h'^2(3y^2 + z^2)$$

and, depending on the relative values of A_{III} and A_{IV}, the result could be

$$\frac{A_{\mathrm{III}}}{4} h'^2(2y^2) \qquad \text{or} \qquad \frac{-A_{\mathrm{III}}}{4} h'^2(2z^2)$$

which gives straight line images because there is no aberration on the other axis.

In Equation 7.6 the coefficients have the dimensions $(\text{length})^{-3}$ because they are substituting for $1/r^3$ in the binomial expansion, *but* it has become the usual practice to calculate the value of each coefficient for the extreme ray allowed by the stop and at the extreme angle of field. These are then called S_{I}, S_{II}, etc., the Seidel coefficients. This means that the maximum values of y, z and h' are used to calculate S_{I}, S_{II}, S_{III}, etc. Then if we should wish to calculate the wavefront aberration, W, for a particular ray or a different size of pupil we can use S_{I}, S_{II}, S_{III}, etc., for the coefficients and insert the *fractional* values of y, z and h' for that ray or pupil edge with respect to the maximum values.

This wavefront aberration remains the actual distance between the aberrated wavefront and a reference sphere centred on the paraxial image as calculated in previous chapters (flat object and image planes). Therefore, when we include the curvature of field modification, Equation 7.6 now becomes:

$$W_{(yzh')} = \tfrac{1}{8}S_{\mathrm{I}}(y^2 + z^2)^2 + \tfrac{1}{2}S_{\mathrm{II}}h'[y(y^2 + z^2)] + \tfrac{1}{4}(3S_{\mathrm{III}} + S_{\mathrm{IV}})h'^2y^2$$
$$+ \tfrac{1}{4}(S_{\mathrm{III}} + S_{\mathrm{IV}})h'^2z^2 + \tfrac{1}{2}S_{\mathrm{V}}h'^3y \qquad (7.7)$$

where y, z and h' are all less than one as a proportion of their maximum values. This also means that they are dimensionless and so the coefficients S_I, S_{II}, S_{III}, etc., have the dimension of length. These are usually very small and often given in terms of the wavelength of light. In optical design usage they are more often calculated for single surfaces and given by W with suffixes which show the powers of yz and h' so that S_I is given by W_{40} and S_{II} by W_{31}. This method allows higher aberration coefficients to be included.

Because a full development of the theory is beyond the scope of this book and the emphasis in the following section is on thin lenses rather than single surfaces, the older usage, S_I, S_{II}, etc., will be retained. At this level it is very difficult to present the subject as much better than a hotch-potch of ill-connected relationships. Equation 7.7 serves as the main guiding beacon indicating the way the constituent aberrations affect the wavefront.

In full system design work there is the added advantage that the path differences (between the aberrated wavefront and reference sphere) caused by one surface can be added directly to those due to the next surface and so on throughout the system. The total wavefront aberration can then be used to find the residual transverse or longitudinal aberration of the system. Although this can be done with thin lenses it would inevitably contain the inaccuracies of the thin lens approximation. Nevertheless, for the study at this level, the thin lens approach offers the quickest route into an appreciation of aberrations and their effects.

From a value of the wavefront aberration it is possible to calculate the transverse aberration, TA, from the equation:

$$\text{TA} = l' \frac{\partial W}{\partial y} \tag{7.8}$$

where l' is the final image distance (or radius of the reference sphere). Equation 7.8 gives the transverse 'miss-distance' in the **meridional plane**, which is the plane containing the axis and the off-axis image point. The transverse aberration due to distortion is largely that given as the miss-distance of the blurred image from the paraxially calculated image point Q'. On the other hand, the transverse aberration due to the coma in the lens would be better shown as the miss-distance of particular rays from the principal ray.

The longitudinal aberration, LA, can be calculated from the relevant TA by the equation:

$$\text{LA} = \frac{l'\text{TA}}{y} \tag{7.9}$$

All these equations are approximations, and particularly so when the aberrations are large. However, they are sufficiently correct for third-order theory. Both longitudinal and transverse aberrations will be used in the following sections where appropriate. The wavefront aberration will be used to show the relationship between the different types of aberration.

Plate ● shows the beauty and the complexity of aberrations. The effects of chromatic aberration are clearly seen and this is described in Chapter 14. The lens used to create the aberrations of Plate ● was a single lens at about $F/3$. This allows the monochromatic aberrations to dominate the chromatic aberration in these images.

7.3 Spherical aberration

This aberration is unique among the Seidel aberrations because it is the only one affecting axial images. This is because the value of all the other aberrations, whose terms in Equation 7.7 contain h', is zero for on-axis points where $h' = 0$. However, spherical aberration should not be thought of as an 'axial aberration' because it affects off-axis image points as well. It is *independent* of h' and so its value does not change with field. The oblique aberrations of Section 7.4 therefore affect off-axis points *in addition* to spherical aberration.

The general nature of spherical aberration has been briefly discussed in Section 3.3. In Section 3.4 the paraxial approximation was described but, in Section 3.5, the paraxial equation for a spherical surface was found using *two* approximations. We firstly assumed that Snell's law was $ni = n'i'$ (Equation 3.8) and, further, that:

$$u = \frac{y}{AB} \qquad u' = \frac{y}{AB'} \qquad \text{and} \qquad a = \frac{y}{AC}$$

in Figure 3.13.

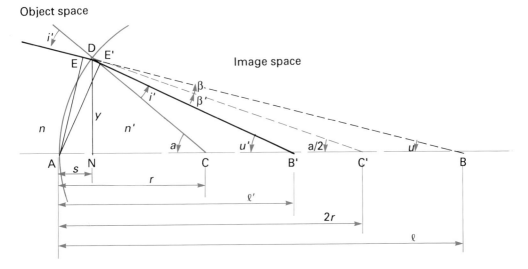

Figure 7.6 Refraction at a spherical surface – all-positive diagram

Working now from Figure 7.6, with a marginal ray incident on a single spherical surface at non-paraxial values of i and i', we know $n \sin i = n' \sin i'$. In triangle DBC we can apply the sine rule to give:

$$\frac{\sin i}{l - r} = \frac{\sin u}{r} \tag{7.10}$$

or

$$\sin i = l \sin u \left(\frac{1}{r} - \frac{1}{l} \right) \tag{7.11}$$

The same equations can be obtained for the refracted ray using triangle DB'C.

$$\sin i' = l' \sin u' \left(\frac{1}{r} - \frac{1}{l'} \right) \tag{7.12}$$

Joining 7.11 and 7.12 using Snell's law, we get:

$$n \left(\frac{1}{r} - \frac{1}{l} \right) l \sin u = n' \left(\frac{1}{r} - \frac{1}{l'} \right) l' \sin u' \tag{7.13}$$

In Figure 7.6, $l \sin u = EA$, the perpendicular drawn from the incident ray to A making ABE a right-angled triangle. In the same way, $l' \sin u' = E'A$ in the triangle AB'E'. Equation 7.13 is the same as the paraxial equation for a single refracting surface if $l \sin u = l' \sin u'$ and so the difference in length between EA and E'A is the key to spherical aberration.

Only when l is infinite or equal to r does EA equal y, the paraxial assumption, and the same applies to E'A. At all other times there is a small discrepancy. We can see from Figure 7.6 that:

$$l \sin u = l \left(\frac{Y}{DB} \right)$$

$$= l \left\{ \frac{y}{[y^2 + (l - s)^2]^{1/2}} \right\}$$

$$= y \left\{ 1 - \frac{2s}{l} + \frac{s^2 + y^2}{l^2} \right\}^{-1/2} \tag{7.14}$$

Using the binomial expansion and letting O indicate higher terms we have:

$$s = r - r \left(1 - \frac{y^2}{r^2} \right)^{1/2} = \frac{y^2}{2r} \left[1 + \frac{y^2}{4r^2} + O \left(\frac{y}{r} \right)^4 \right]$$

and

$$s^2 = \frac{y^4}{4r^2} \left[1 + \frac{y^2}{2r^2} + O \left(\frac{y}{r} \right)^4 \right]$$

Therefore

$$l \sin u = y \left\{ 1 + \frac{y^2}{rl} \left[1 + \frac{y^2}{4r^2} + O \left(\frac{y}{r} \right)^4 \right] + \frac{y^2}{l^2} \left[1 + \frac{y^2}{4r^2} + O \left(\frac{y}{r} \right)^4 \right] \right\}^{-1/2}$$

and again by the binomial expression

$$l \sin u = y + \frac{y^3}{2rl} - \frac{y^3}{2l^2} + \frac{y^5}{8r^3l} + \frac{y^5}{4r^2l^2} - \frac{3y^5}{4rl^3} + \frac{3y^5}{8l^4} + O(y)^7 \tag{7.15}$$

A similar expression can be obtained from $l' \sin u'$ and, putting these into Equation 7.13, we have, as far as the third order:

$$n \left(\frac{1}{r} - \frac{1}{l} \right) + \frac{ny^2}{2l} \left(\frac{1}{r} - \frac{1}{l} \right)^2 = n' \left(\frac{1}{r} - \frac{1}{l'} \right) + \frac{n'y^2}{2l'} \left(\frac{1}{r} - \frac{1}{l'} \right)^2 \tag{7.16}$$

This shows that the aberration in l' (the longitudinal aberration) varies as y^2. This agrees with Section 7.2 when Equations 7.6 and 7.7 are applied to the wavefront aberration, S_I, which was shown to vary with y^4. Equation 7.16 also agrees with the paraxial case if $y^2/2l$ is considered too small to be included, and so the second term on both sides is removed. This outline of the derivation is included here to indicate the general approach, which is very much that of successive approximations carefully handled. No further derivations will be included here and the interested reader is directed to the books referenced at the end of this chapter.

Going back to Equation 7.13 and Figure 7.6 it can be seen that equality for all values of n and n' occur when $l = l' = 0$ and when $l = l' = r$. Thus there is no spherical aberration at these points, which are cases where $EA = E'A = y$. There is a further condition for equality when $EA = E'A$. This is when they are inclined at equal and opposite angles to DA. This gives (Figure 7.6)

$$\beta = \beta' = \frac{a}{2} - u = u' - \frac{a}{2}$$

This is equivalent to

$$\frac{1}{l'} = -\frac{1}{l} + \frac{1}{r}$$

which may be used with Equation 3.9 to show that at this position

$$\frac{l}{r} = \frac{n + n'}{n} \tag{7.17}$$

and

$$\frac{l'}{r} = \frac{n + n'}{n'} \tag{7.18}$$

Figure 7.7 shows these **aplanatic points**. The absence of spherical aberration at the last of these points can be shown graphically using Young's construction (Section 3.3). Aplanation means freedom from spherical aberration and coma. In none of these cases it is possible to have a real image of a real object, but a combination of the second and third condition is often used in the design of large aperture lenses to be free from spherical aberration and coma. Thus the lens shown in Figure 7.8 is aplanatic for the particular object point shown because the first surface of the lens is centred on the object and the second surface satisfies Equations 7.17 and 7.18. The light rays *from this particular object* B are therefore imaged at B' without spherical aberration. The use of these lenses in corrected optical systems is described in Section 14.3.3.

When Equation 7.16 is applied to the thin lens case, considerable interaction is found between l and l', the object and final image distances, and between r_1 and r_2, the radii of curvature of the first and second surface. Two new factors may be defined, which substantially reduce the complexity of the aberration equations. These factors are defined as

$$X = \frac{r_2 + r_1}{r_2 - r_1} = \frac{R_1 + R_2}{R_1 - R_2} \tag{7.19}$$

which is dimensionless and known as the **bending factor**, and

$$Y = \frac{l' + l}{l' - l} = \frac{L + L'}{L - L'} \tag{7.20}$$

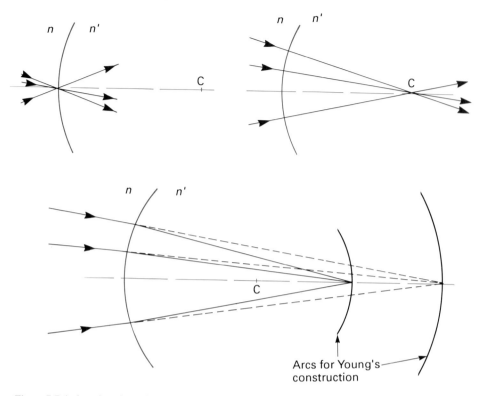

Figure 7.7 Aplanatic points of a spherical refracting surface

which is also dimensionless and known as the **conjugates factor**. (Some writers give different definitions, but these give all positive signs in Equations 7.21 and 7.51.)

Using these it is found that the difference between the paraxial and marginal foci is given by

$$l' - l'_m = \tfrac{1}{2}y^2l'^2F^3(\alpha X^2 + \beta XY + \gamma Y^2 + \delta) \tag{7.21}$$

where

$$
\left.
\begin{aligned}
\alpha &= \frac{n + 2}{4n(n - 1)^2} \\[2mm]
\beta &= \frac{n + 1}{n(n - 1)} \\[2mm]
\gamma &= \frac{3n + 2}{4n} \\[2mm]
\delta &= \frac{n^2}{4(n - 1)^2}
\end{aligned}
\right\} \tag{7.22}
$$

and l'_m is the axial crossing distance of the marginal ray which is incident on the lens at height y from the axis.

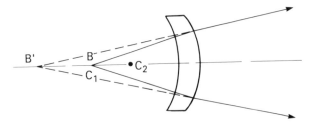

Figure 7.8 An aplanatic lens

Table 7.1

n	α	β	γ	δ	ϵ	ξ
1.4	3.79	4.28	1.11	3.06	2.14	1.36
1.5	2.33	3.33	1.08	2.25	1.67	1.33
1.6	1.56	2.71	1.06	1.78	1.35	1.31
1.7	1.11	2.27	1.04	1.48	1.13	1.29

The values of these coefficients for four values of refractive index are given in Table 7.1 (for ϵ and ξ see Section 7.4.2). As they all reduce with increasing index, a lens made out of high-index glass will have less spherical aberration than one of equal power made from low-index glass. For a plano-convex lens ($n = 1.5$) of 80 mm EFL, the spherical aberration is shown to scale, using Equation 7.21, in Figures 7.9 and 7.10. The differences between the two drawings amply demonstrate the effect of the bending factor, X. The length LA is longitudinal third-order spherical aberration. The distance TA is the transverse third-order spherical aberration. Clearly TA = LA tan u', which compares with Equation 7.9. It is also clear that the best image obtainable is a good deal better than the value shown for TA. The smallest diameter blur occurs three-quarters of the way to the marginal

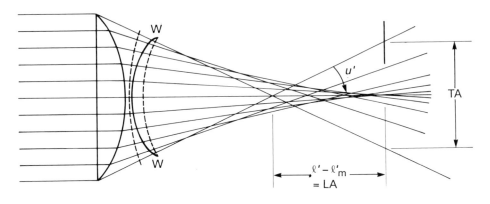

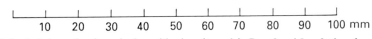

Figure 7.9 Spherical aberration for a single positive lens (to scale). $R_1 = 0$ and $L = 0$, therefore $X = -1$ and $Y = -1$

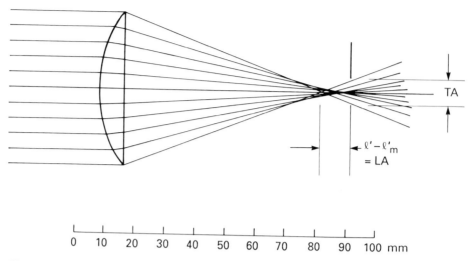

Figure 7.10 Spherical aberration for a single positive lens (to scale). $R_2 = 0$ and $L = 0$, therefore $X = +1$ and $Y = -1$

focus and is necessarily TA/4. The shift of the utilized image plane is known as defocusing even though it improves the image.

This shift of focus can be examined from the point of view of the wavefront aberration. The wavefront W–W is drawn in Figure 7.9 with each ray normal to it. It is apparent that it is not a sphere and the optical path differences between this wavefront and a reference sphere constitute the wavefront aberration. Different reference spheres result from defocusing and give different values for the aberration. The thin lens contribution to the maximum wavefront aberration due to spherical aberration with respect to the paraxial focus is

$$W_{SA} = \tfrac{1}{8}(S_I) = \tfrac{1}{8}y^4 F^3 (\alpha X^2 + \beta XY + \gamma Y^2 + \delta) \tag{7.23}$$

which reduces to Equation 7.21 when Equations 7.8 and 7.9 are applied. Note that Equation 7.23 does not contain l' directly (only as a component of Y) and straightforward summation over a number of lenses is possible. The wavefront aberration varies with y^4, the transverse with y^3 and the longitudinal with y^2.

The shape of the wavefront aberration can be shown in pictorial form as in Figure 7.11, where the hatched areas show the wavefront difference from a paraxial reference sphere. An alternative presentation of the aberration is by means of a **spot diagram** in which rays, equally spaced in the aperture of the lens, are traced by computer and points of intersection with the chosen image surface are plotted. The regions of greatest density of dots show the brighter parts of the image. The actual intensity profile of the image may also be computed and forms the **point spread function**. Often the intensity across the image of a line object is calculated, giving the **line spread function**, which can be utilized with the MTF concept discussed in Sections 14.7 and 14.8.

Yet again, calculations may be made to obtain the locations of the **caustic surfaces**. These are three-dimensional surfaces in space which define the intersections of adjacent rays in image space. In general, only adjacent rays along two directions across a lens intersect, thus giving two surfaces. In the case of spherical aberration these caustic 'sheets' comprise a trumpet-shaped surface and a

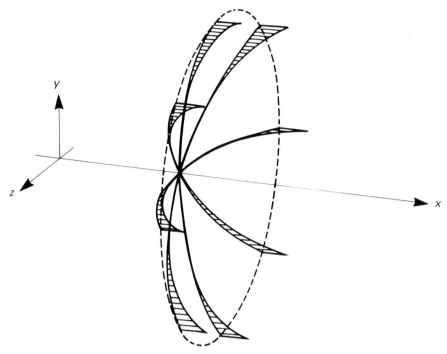

Figure 7.11 Representation of spherical aberration. The broken lines represent a spherical reference wavefront advancing along the axis. The solid lines represent the aberrated wavefront and the hatching the aberration

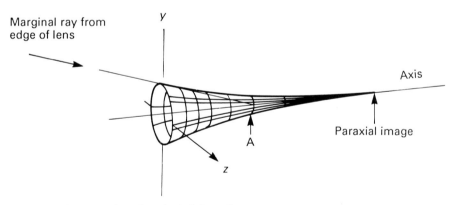

Figure 7.12 Caustic surfaces for spherical aberration

a straight line on the axis (as a degenerate surface) shown in Figure 7.12. An image plane cutting through these sheets will show a bright spot and ring. A ring only is seen when nearer to the lens than A. This approach can be useful when the image plane is indeterminate. By far the most common representation is a simple graph showing where a ray from a given height on the lens intersects the axis, as used in Section 14.4. Optical designers commonly use transverse aberration values plotted

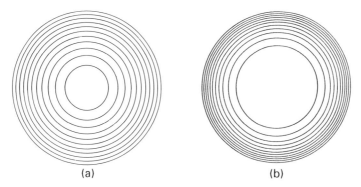

Figure 7.13 Contour maps of wavefronts: (a) spherical wavefront – no aberration, (b) aberrated wavefront – spherical aberration

with y as the horizontal axis! Most of the representations used in this book are for explanatory purposes. A further method which is used more in Chapter 8 for surface shapes can be ·used here for the wavefront shape. Figure 7.13 shows a contour map of a spherical wavefront compared with an aspherical wavefront similar to that shown in Figure 7.10. The contours are drawn for equally spaced distances along the optical axis. In the spherical case the diameters of the contours increase with a simple square root dependence. In the aberrated case (for spherical aberration) they have a fourth root dependence, emphasizing how a small increase in lens aperture can allow a large increase in aberration.

The bending factor, X, can be changed without altering the power of the lens. Thus, for any given power and conjugates the spherical aberration may be varied throughout a range by changing X. Since Equation 7.23 is quadratic in X, the result is a parabolic variation with a minimum value but with no intersection with the x-axis. Thus, the spherical aberration of a single positive lens is never zero and always in the same direction. This may be termed positive aberration or negative aberration according to different conventions. Ambiguity can be avoided by referring to it as undercorrect spherical aberration.

The bending for minimum spherical aberration is found by differentiating Equation 7.23 with respect to X, regarding Y as constant:

$$X_{\min} = \frac{-2(n^2 - 1)}{n + 2} Y \tag{7.24}$$

and the coefficient at this value is given by

$$S_{\text{I(min)}} = y^4 F^3 \left[\frac{n^2}{4(n - 1)^2} - \frac{n}{4(n + 2)} Y^2 \right] \tag{7.25}$$

Because of the F^3 dependence, a negative lens will always have negative spherical aberration. When a positive and negative lens are designed as an achromatic doublet (Section 14.3) only their respective powers are defined, and their curvatures can be changed to bend the lenses so that their spherical aberrations become equal and opposite and so cancel out (Section 14.3.2). Combinations of achromatic doublets and aplanatic single lenses can be used to make aberration-corrected systems of large aperture (Section 14.3.3).

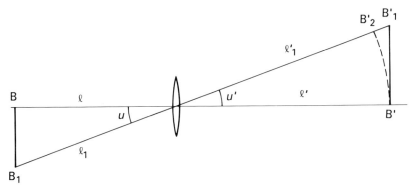

Figure 7.14 Image curvature

7.4 Oblique aberrations

7.4.1 Curvature of field and oblique* astigmatism

In the preceding section the paraxial approximation was shown to be insufficient to describe the image of an axial object when this is formed by a lens of appreciable aperture. Different problems arise when the lens aperture is small but the object is considerably off-axis. Figure 7.14 shows the familiar example of a lens forming an image $B'B_1'$ of the object BB_1. However, it is clear that the length l_1 is longer than l. If, therefore,

$$\frac{1}{l'} = \frac{1}{l} + \frac{1}{f}$$

and

$$\frac{1}{l_1'} = \frac{1}{l_1} + \frac{1}{f}$$

are both to be true then l_1' must be shorter than l'. The image must be curved to be somehow like $B'B_2'$. Strictly, it has a paraboloidal shape and is known as the **Petzval surface**.

Calculations for the single surface case (see Figure 3.17) show that this curvature may be expressed approximately as a sphere of radius r_i' given by

$$\frac{1}{n'r_i'} - \frac{1}{nr_i} = \frac{n - n'}{nn'r} \tag{7.26}$$

where r is the actual radius of curvature of the surface and r_i the radius of curvature of the object (which may not be plano for later surfaces in a system). The right-hand side of Equation 7.26 is the *contribution* of each surface to the curvature of the final image. If the contributions of two surfaces are added together we easily obtain the thin lens contribution to field curvature as

$$C = \frac{-F}{n} \tag{7.27}$$

---

* Axial astigmatism due to toroidal surfaces is covered in Chapter 8.

where n is the index of the lens material. If a lens system is to provide a flat image field from a flat object, the sum of these contributions must be zero:

$$\Sigma_k \frac{F_k}{n_k} = 0 \qquad (7.28)$$

which is known as the Petzval condition. Because Equation 7.27 is not affected by lens shape or image conjugates, only by combining lenses of different refractive index can this condition be satisfied and field curvature eliminated.

The curvature referred to here is *not* that of the wavefront. The maximum wavefront aberration is given by

$$W_{\text{Field Curve}} = \tfrac{1}{4}S_{IV} = \tfrac{1}{4}H^2\frac{F}{n} \qquad (7.29)$$

where H is the Lagrange invariant, $n'h'u'$, which was developed in Section 3.7. For a thin lens is air, this reduces to $h'y/l'$. If Equations 7.8 and 7.9 are applied to Equation 7.29, we find that the longitudinal aberration, that is, the distance from the paraxial image plane to the relevant point on the Petzval surface, is $\tfrac{1}{2}h'^2F/n$. This is also obtained if the sag formula is applied to Equation 7.27.

A further consideration of the off-axis image is the difference between the section of the wavefront lying in the plane of the paper (T) and that containing the ray B_1B_2' (Figure 7.15) but perpendicular to the paper (S). The optical path

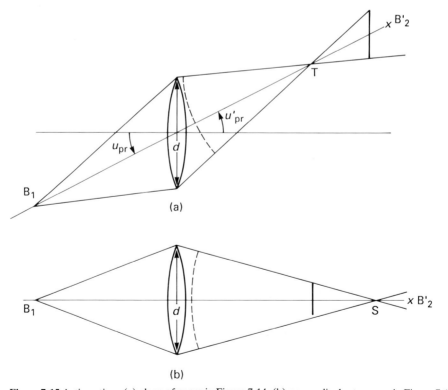

(a)

(b)

Figure 7.15 Astigmatism: (a) plane of paper in Figure 7.14, (b) perpendicular to paper in Figure 7.14

difference due to the lens is zero at its edge if we make its thickness there zero. At the centre, the optical path difference is the same for each section of the wavefront as it is that experienced by the ray B_1B_2', which is called the **principal ray**. However, in the case of the S section this occurs across the whole diameter, d, of the lens, while for the T case the diameter is foreshortened by the angle u_{pr} to be $d\cos u_{pr}$ (or $d\cos u_{pr}'$). The radius of curvature of the sections of the emerging wavefront can be obtained from the sag formula and, as the sags are equal, we have

$$\frac{y^2}{2r_S} = \frac{y^2\cos^2 u_{pr}'}{2r_T} \tag{7.30}$$

If the incident beam is from a distant object, these radii become focal lengths. We then have

$$f_T' = f_S'(1 - \sin^2 u_{pr}')$$

and, when the difference is a small fraction of the paraxial focal length,

$$f_S' - f_T' = f' \sin^2 u_{pr}' \tag{7.31}$$

This is called the astigmatic difference or **Sturm's interval**. This longitudinal aberration is *not* reduced by stopping down the lens as it is independent of y.

A pictorial representation of the wavefront aberration is given in Figure 7.16. The actual wavefront has a toroidal shape similar to the surfaces described in Section 8.1. In this case, however, the astigmatism is occurring with a spherical lens receiving the light obliquely. In ophthalmic texts this astigmatism is often referred to as oblique astigmatism to distinguish it from that occurring (on-axis) with toric lenses. Oblique astigmatism arises at all spherical surfaces used off-axis, the wavefront aberration giving rise to the fan of rays, as in Figure 7.16, which passes

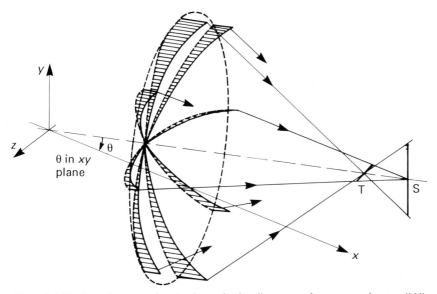

Figure 7.16 Astigmatism – wavefront and rays: broken line arcs, reference wavefront; solid line arcs, actual wavefront; hatched areas, wavefront aberration

through the short line at T and a second perpendicular line at S. There is, therefore, a general similarity in the form of the beam in the two cases.

The length of each of these lines depends on the astigmatic difference and the lens aperture: the transverse aberration depending on y as predicted by Equation 7.9. Between the two lines, roughly midway, there is, if the aperture is circular, a point where the blur is circular. This is the circle of least confusion, which is also proportional to y.

Because the longitudinal aberration is independent of y it is possible to develop an imitation of paraxial mathematics in a form that applies along narrow *oblique* pencils of light. In Figure 7.17 AEDGH is a spherical surface between two media. C is the centre of curvature and CA a radius in the plane of the paper. Q is an object point in the plane of the paper and this plane is called the meridional plane or tangent plane. The principal ray, QA, the central ray of the bundle is in this plane and remains in it after refraction to Q'. The plane perpendicular to the meridional plane and containing the principal ray is the equatorial or sagittal plane. This plane is tilted on refraction. The points Q, A, G, H, T, S, N, P lie in the sagittal planes (approximately).

Consider four limiting rays of the pencil, which is supposed to be very narrow, but is shown with its width exaggerated. Rays QD and QE are in the plane of the paper while rays QG and QH (not drawn) are above and below the plane of the paper. QD and QE remain in the plane of the paper and meet the principal ray at T. The ray QG is refracted back towards the paper and the ray QH forwards towards the paper and they meet the principal ray at S. Other rays are refracted in such a way that they pass through a short line at T and a second perpendicular line at S. These focal lines are really small caustic surfaces; that at T is called the **tangential line focus** and is a short arc of a circle centred on the axis of the surface, while that at S is called the **radial** or **sagittal line focus** and is a narrow figure of eight. Note that the *actual line* of T lies in the sagittal plane and vice versa!

The reason for the shapes and positions of these lines is clear when their relationship with the basic caustic surfaces of spherical aberration is recognized. In Figure 7.17, the axis for which the object and image would be regarded as axial has been shown as a dashed line. All the oblique narrow pencil does is to select a small non-axial aperture on the surface and the relevant parts of the caustics.

When the aperture is very small, the marginal ray paths with respect to the principal ray show very small changes in the angles of incidence and refraction. It is possible to work out a sort of first-order mathematics which uses the principal ray as if it were an axis even though it has a bend in the middle. As usual this is an approximation, but it works surprisingly well.

In Figure 7.17, line QAQ' is the principal ray (axis). Lines QDT and QET are marginal rays in the meridional plane and lines QGS and QHS are marginal rays in the sagittal plane. Snell's law may be differentiated to give

$$n \, \delta i \cos i = n' \, \delta i' \cos i' \tag{7.32}$$

where i and i' will be taken to refer to the principal ray, and δi and $\delta i'$ will be the difference angles between the marginal rays and the principal ray.

In Figure 7.17 we can use the triangles KDC and KAQ to obtain

$$di = i_{\mathrm{m}} - i = \angle C'DQ - \angle KAQ = \angle DCA - \angle AQD$$

remembering that $\angle$ AQD as drawn is a negative angle. The last two angles can be approximated to give

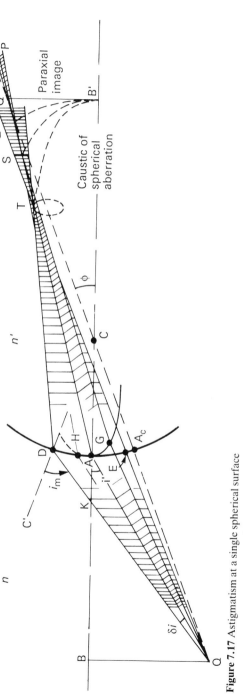

Figure 7.17 Astigmatism at a single spherical surface

$$di = \frac{AD}{r} - \frac{AD \cos i}{t}$$

where r is the radius AC and t is the object distance along the principal ray in the tangential plane. By the same method we find

$$di' = \frac{AD}{r} - \frac{AD \cos i'}{t'}$$

where t' is the image distance in the tangential plane. Putting these expressions into Equation 7.32 gives

$$\frac{n' \cos^2 i'}{t'} - \frac{n \cos^2 i}{t} = \frac{n' \cos i' - n \cos i}{r} \tag{7.33}$$

where i and i' are the incident and refracted angles of the principal ray.

In the other plane we know that QAS and QGS intersect at S on the line QCS as this is the axis of symmetry. In the triangle AQC, by the sine rule,

$$\frac{QC}{\sin i} = \frac{s}{\sin \angle ACQ}$$

and similarly

$$\frac{SC}{\sin i'} = \frac{s'}{\sin \angle ACS}$$

where s and s' are equal to the object and image distances along the principal ray in the sagittal plane.

As $\angle ACS + \angle ACQ = 180°$,

$$\frac{SC}{s' \sin i'} = \frac{QC}{s \sin i}$$

which together with Snell's law gives

$$\frac{n' SC}{s'} = \frac{n QC}{s}$$

If we project SC and QC onto the line C'AC by multiplying both sides by $\cos \varphi$ we have

$$QC \cos \varphi = s \cos i - r$$

$$SC \cos \varphi = s' \cos i' - r$$

so that

$$\frac{n'}{s'} - \frac{n}{s} = \frac{n' \cos i' - n \cos i}{r} \tag{7.34}$$

Equations 7.33 and 7.34 are commonly known as the **Coddington equations** and may be used in a method known as differential ray tracing to obtain the location of astigmatic images of oblique pencils through a lens system by taking distances along the principal ray and not the axis. The value

$$\frac{n' \cos i' - n \cos i}{r}$$

is called the oblique power, K, of the surface. For a thin lens, $s_1' = s_2$ and $t_1' = t_2$ in the usual way so the oblique power becomes

$$K = (n' \cos i' - n \cos i) \left(\frac{1}{r_1} - \frac{1}{r_2} \right)$$

$$= \left(\frac{n' \cos i' - n \cos i}{n - 1} \right) F \qquad (7.35)$$

where n is the refractive index of the lens material and F the paraxial power.
    We then have the equations

$$S' = S + \left(\frac{n \cos i' - \cos i}{n - 1} \right) F = S + F_s \qquad (7.36)$$

and

$$T' = T + \left(\frac{n \cos i' - \cos i}{(n - 1) \cos^2 i} \right) F = T + F_t \qquad (7.37)$$

where S, T, S' and T' are the object and image vergences along the principal ray in the two planes of regard.

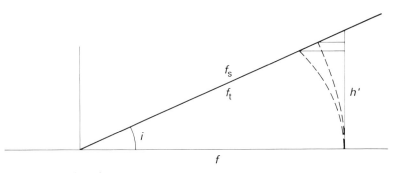

Figure 7.18 Astigmatism curvatures

    In Figure 7.18 the sags for the curved images at height h' on the paraxial image plane are

$$\left(\frac{f}{\cos i} - f_s \right) \cos i \qquad \text{and} \qquad \left(\frac{f}{\cos i} - f_t \right) \cos i$$

or, approximately,

$$\frac{F_s - F \cos i}{F^2} \qquad \text{and} \qquad \frac{F_t - F \cos i}{F^2} \qquad (7.38)$$

By the binomial expansion,

$$\cos i = 1 - \frac{h'^2}{2f^2}$$

$$\cos i' = 1 - \frac{h'^2}{2n^2 f^2}$$

and

$$\cos^2 i = 1 - \frac{h'^2}{f^2}$$

all to third order. The respective sags may be found, using Equations 7.36–7.38, to be

$$\left(1 + \frac{1}{n}\right) \frac{h'^2 F}{2} \quad \text{and} \quad \left(3 + \frac{1}{n}\right) \frac{h'^2 F}{2} \tag{7.39}$$

The general shape of these curves is given in Figure 7.19, the so-called cup and saucer diagram. If ABCD is a plane object, a point P upon it is imaged as a short arc on the tangential image surface and a short radial line on the sagittal image surface. If P is already part of a line, the image of the line may appear to be in focus on either of the surfaces if the object line is orientated in the same direction as the image of the point. Thus the circle on ABCD will appear to be in focus at I_t and the spokes at I_s, if the aperture is small. Object lines at other orientations do not form a good image.

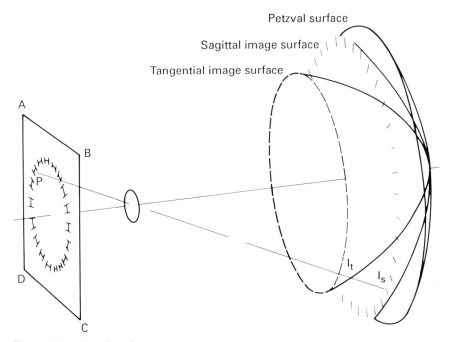

Figure 7.19 Astigmatic surfaces

The diagram shows a third surface, the Petzval surface. This is not an image surface at all in the presence of astigmatism. If astigmatism is corrected all images of objects on ABCD will be formed on this surface (assuming narrow pencils). Third-order astigmatism is the difference between the image surfaces and the Petzval surface. The tangential image is always three times further from the Petzval surface even when the astigmatism is negative. Astigmatism in a thin lens cannot be controlled except by having a stop some distance from the lens (see Section 7.5). The maximum wavefront aberrations for a thin lens are given by

$$W_{\mathrm{Astig\,S}} = \tfrac{1}{4}(S_{\mathrm{III}} + S_{\mathrm{IV}}) \qquad \text{for the sagittal image} \tag{7.40}$$

and

$$W_{\mathrm{Astig\,T}} = \tfrac{1}{4}(3S_{\mathrm{III}} + S_{\mathrm{IV}}) \qquad \text{for the tangential image} \tag{7.41}$$

where $S_{\mathrm{III}} = H^2 F$; and $S_{\mathrm{IV}} = H^2 F/n$, as given in Equation 7.29, is the effect of field curvature. Equations 7.40 and 7.41 assume a circular lens. They constitute the third and fourth terms of the general equation (7.7) and include the effect of field curvature. H is the Lagrange invariant, $n'h'u' = h'y/l'$ in the thin lens case. If Equations 7.8 and 7.9 are applied to Equations 7.40 and 7.41, we obtain the longitudinal astigmatic aberration which is the sag value obtained as in Equation 7.39. If we apply the sag formula to Equation 7.39, we can obtain approximate curvatures for the image surfaces of Figure 7.19, which are the very simple formulae

$$\text{Sagittal curvature} = -\left(1 + \frac{1}{n}\right) F \tag{7.42}$$

and

$$\text{Tangential curvature} = -\left(3 + \frac{1}{n}\right) F \tag{7.43}$$

These oblique aberrations with narrow pencils do not depend on the lens bending even though Figure 7.17 suggests that astigmatism could be regarded as the spherical aberration of the principal ray. This is because the 'y' distance for the point A away from QCQ' as an axis depends on r in such a way that the effects for the first and second surfaces of the lens have a cancelling effect if $1/r_1 - 1/r_2$ is a constant, which is the case when bending a thin lens of constant power.

The difference between the two expressions of Equation 7.39 may be shown to be approximately equal to Equation 7.31, where $f'_s - f'_t$ was obtained by much simpler reasoning, if u_{pr} of the latter is set equal to i, and $\sin i = h'/f \cos i$ is used.

7.4.2 Coma

Although coma is usually categorized as the second monochromatic aberration it is dealt with here after astigmatism and curvature of field because it requires both an off-axis object *and* an appreciable pupil size. Its form gives it an unsymmetrical appearance somewhat like the tail of a comet – hence its name. It will be shown later that it depends on the first power of the off-axis angle, making it the first oblique aberration to appear as the field is increased. In early optical instruments it was therefore a very troublesome aberration, particularly when accurate measurements were required. Fortunately, it is not very difficult to correct in a single thin lens, which may have zero coma.

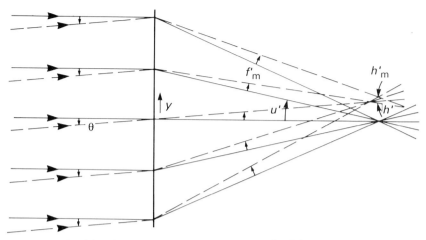

Figure 7.20 Straight principal plane – coma occurs at the off-axis image

For a single refracting surface it is possible to regard coma as part of the spherical aberration. However, this is not very rewarding as the coma effect for moderate field angles is only a small part of the spherical aberration (and we cannot, as in the case of astigmatism, reduce the spherical aberration by considering a small aperture). In order, therefore, to demonstrate coma we consider a case where the spherical aberration is zero. In Figure 7.20, parallel light from a distant axial object is assumed to be refracted at the second principal plane of a thick lens in air. Furthermore, the image formed is assumed to be free from spherical aberration.

If now the object is moved to cause a small change, say θ, in the angle of the incident rays (shown with broken lines) it is possible to construct the new refracted rays by shifting them through the same angle. This is valid for small angles up to about 5°. When these new rays are produced towards their intersection they are found to be aberrated. This is coma.

In Figure 7.21 the same procedure has been applied but the second principal plane is assumed to be curved! When the centre of its curvature is coincident with

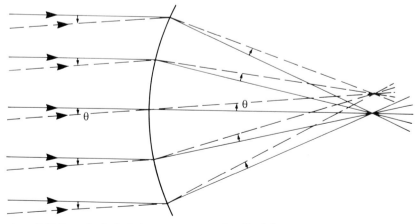

Figure 7.21 Curved principal plane – no coma at the off-axis image

the image we find the off-axis image to be unaberrated. Before discussing the importance of this we will consider the form of the aberration illustrated in Figure 7.20. It can be seen from this diagram that the image size for the outer zones of the lens is greater than for the central zone. Whereas spherical aberration is a difference in axial location of the image for different zones of the lens, coma is a difference in magnification.

Figure 7.20 shows only the tangential coma. When various parts of the lens are designated as in Figure 7.22, the related parts of the image are as in Figure 7.23. The region A at the top of the lens *and* the region A at the bottom combine to produce the image at A. The line CcPcC is in the sagittal plane of the lens and so the distance PC on the image is called the sagittal coma. As with astigmatism, the tangential coma, PA, is three times the sagittal coma, PC. Because the light from the annulus ABCD is spread over a considerable area, the actual intensity of the image is mainly in the triangle Pdb. Thus the sagittal coma value is a much better indication of the image blur. With such a complicated ray pattern the wavefront representation, Figure 7.24, gives a clearer description of the aberration.

In Section 3.7, it was shown that for objects and images close to the axis the magnification expression gave rise to the Lagrange invariant,

$$nhu = n'h'u' = H \qquad (3.20)$$

within the paraxial region. We may extend this by remembering that the ray through the centre of curvature of a spherical surface is undeviated. Therefore, correct magnifications may be obtained from it for the *sagittal* images of Figure 7.17 which lie along it. In Figure 7.25 this ray has been drawn together with refracted rays to the axial image. The object is at B and the image at B'. This is an all-positive diagram.

By similar triangles,

$$\frac{h}{l - r} = \frac{h'}{l' - r} \qquad (7.44)$$

We also have, from Equation 7.13,

$$n(l - r) \sin u = n'(l' - r) \sin u'$$

Putting this with Equation 7.44, we obtain

$$nh \sin u = n'h' \sin u' \qquad (7.45)$$

This is known as the **sine condition**, discovered independently by Abbe and Helmholtz in 1873. Although applying only to sagittal rays, this famous relationship is of great value in optics as it links together the axial and off-axis images. It does not matter that the ray through the centre of curvature does not exist. The important feature is the invariant nature of the condition allowing it to be applied successively to a number of surfaces. Thus, it certainly applies to thin lenses and also to lens systems.

In applying it to the distant object case of Figure 7.20, we have $\sin u = y/l$ for each sagittal ray, even though u is very small. Thus

$$h' = \frac{h}{l} \left(\frac{n}{n'} \right) \frac{y}{\sin u'}$$

and, for h' to be constant over different values of y, we must have $y/\sin u'$ equal to a constant, as u' is the only other variable. This term is equal to the slant refracted

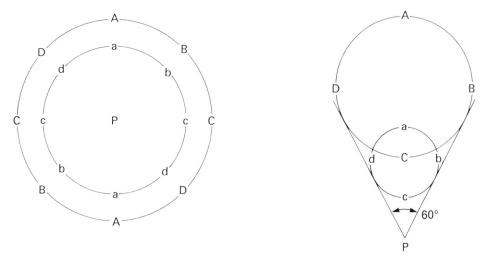

Figure 7.22 Coma designation of the lens surface

Figure 7.23 Coma – form of image

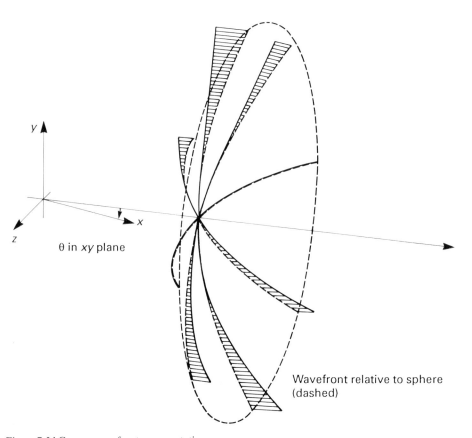

y

z

x

θ in *xy* plane

Wavefront relative to sphere (dashed)

Figure 7.24 Coma – wavefront representation

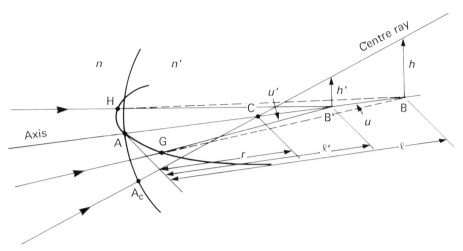

Figure 7.25 Construction for the sine rule

ray lengths, and so it is a constant only when the equivalent surface is a sphere. In Figure 7.20, where coma is present, we can see that $y/\sin u'$ is not a constant because the slant lengths vary.

For a distant object we may call the extreme slant length the marginal focal length, f'_m. If we define the extent of the mismatch as $(h' - h'_m)/h'$, we have

$$\frac{h' - h'_m}{h'} = 1 - \frac{h'_m}{h'}$$

$$= 1 - \frac{y_m \sin u'}{y \sin u'_m}$$

$$= 1 - \frac{f'_m}{f'} \tag{7.46}$$

This expression, equal to zero for zero coma, is called the **offence against the sine condition** or OSC. The transverse sagittal coma is then given by

$$\text{Transverse coma}_s = h'(\text{OSC}) \tag{7.47}$$

In the tangential plane, as shown in the figure, $\cos i$ enters in a similar manner to astigmatism and, to a third-order approximation,

$$\text{Transverse coma}_t = 3h'(\text{OSC}) \tag{7.48}$$

For a thin lens the wavefront aberration is given by

$$W_{\text{Coma}} = \tfrac{1}{2}(S_{\text{II}}) = \tfrac{1}{2}y^2HF^2(\epsilon X + \xi Y) \tag{7.49}$$

where H is the Langrange invariant ($h'y/l'$ for a thin lens) and X and Y are defined as in Equations 7.19 and 7.20 (giving a positive sign inside the brackets). The coefficients ϵ and ξ are given by

$$\epsilon = \frac{n + 1}{2n(n - 1)} \quad \text{and} \quad \xi = \frac{2n + 1}{2n} \tag{7.50}$$

These are tabulated, for four refractive indices, on page 241. When Equation 7.49 is treated by Equation 7.8, we obtain the equation

$$\text{Transverse coma}_t = \tfrac{3}{2}y^2 h' F^2 (\epsilon X + \xi Y) \tag{7.51}$$

and the transverse coma$_s$ is one-third of this (but this simplified treatment does not readily give this).

This coma is often referred to as **central coma**, being that exhibited when the stop is adjacent to the lens. The simple expression in the brackets of Equation 7.51 allows bending of the lens, X, to give zero coma for a given conjugates factor, Y. The correction of coma in a doublet lens is briefly described in Section 14.3.2.

7.5 Aberrations and stop position – distortion

The previous sections have dealt with monochromatic aberrations in an order most suited to learning and have assumed that in the thin lens case the controlling aperture was at the lens. As a thin lens under these conditions exhibits no distortion ($S_V = 0$) this aberration has been left until now.

The order of the aberrations given by Seidel and indicated in Section 7.1 is very important when considering the effect of relocating the aperture stop. Coma, astigmatism and distortion are all affected by stop position if one of the preceding aberrations is present. Curvature of field, as suggested by the development of Equation 7.29, is not affected by stop position. Spherical aberration also is not affected by stop position (except that the amount of the lens used may alter, with finite conjugates), but coma is affected provided there is some spherical aberration present. Likewise astigmatism is affected if there is spherical aberration or coma. Distortion changes with stop position if there is spherical aberration, coma, astigmatism or curvature of field.

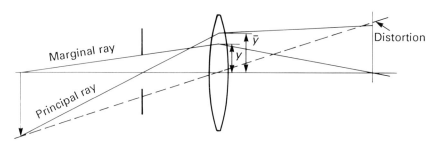

Figure 7.26 Definition of eccentricity, $\bar{y}/y$

In Figure 7.26 the principal ray, by definition, passes through the centre of the entrance pupil, stop and exit pupil. When the stop is at the lens, $\bar{y}$ is zero. For other stop locations the eccentricity of the principal ray may be defined as

$$Q = \frac{\bar{y}}{y} \tag{7.52}$$

Following the notation of H. H. Hopkins (*Wave Theory of Aberrations*, Oxford University Press, Oxford, 1950) we can define new Seidel coefficients for a displaced stop as follows:

$$S_I^* = S_I$$
$$S_{II}^* = S_{II} + QS_I$$
$$S_{III}^* = S_{III} + 2QS_{II} + Q^2S_I$$
$$S_{IV}^* = S_{IV}$$
$$S_V^* = S_V + Q(3S_{III} + S_{IV}) + 3Q^2S_{II} + Q^3S_I \qquad (7.53)$$

where the new coefficients are indicated by the asterisk and the 'old' coefficients for the stop at the lens are given by Equations 7.23, 7.49, 7.40, 7.41 and 7.29, and $S_V = 0$, for a thin lens. The new values S_{II}^*, S_{III}^* and S_V^* can be used in these equations in place of the old values to give the wavefront aberration under the new conditions.

The relatively simple expressions in Equations 7.53 arise because of the use of maximum values for h', y (and z) given by the two rays in Figure 7.26. The aberrations at other field points and other parts of the pupil are obtained by using the fractional values of h', y (and z) in Equation 7.7. The simple expression for the coma coefficient in Equation 7.53 is apparent when the similarity between Figure 7.20 and Figure 7.10 is recognized. Provided the coma is not as large as that shown, it is possible, in the presence of spherical aberration, to restrict the area of the lens used so as to centralize the caustic sheets about the image by choosing a new principal ray. This is what is achieved by relocating the stop and the extent of this is measured by Q.

Distortion is evident in Figure 7.26, but only if there is aberration between the ray that would have passed through the centre of the lens and the actual principal ray. Because the principal ray crosses the axis at the centre of the stop and emerges as if from the centre of the exit pupil, it is sometimes convenient to regard distortion as spherical aberration of the principal ray. Figure 7.27 shows undercorrect spherical aberration of the stop image, the exit pupil, which gives rise to a reduced magnification at the outer parts of the image. If the object in this case were a square, the image would appear as in Figure 7.29, having a characteristic barrel shape. For a stop on the other side of the lens the image shape of Figure 7.28 would occur.

The terms **barrel distortion** and **pincushion distortion** are commonly used to describe these two types. For centred systems these are the only distortions that can occur although, for some corrected lenses, the central parts of the field may suffer from one type and the outer parts from the other.

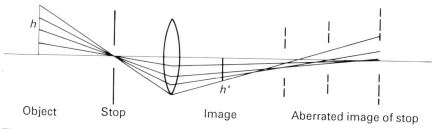

Figure 7.27 Distortion with a thin lens

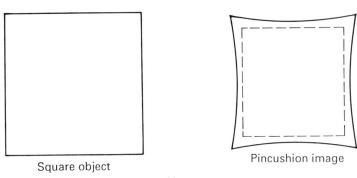

Square object

Pincushion image

Figure 7.28 Distortion types – pincushion

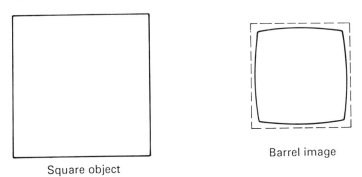

Square object

Barrel image

Figure 7.29 Distortion types – barrel

Obviously the term for S_V^* is rather complex, but a few general points can be obtained directly from Equation 7.7, from which

$$W_{\mathrm{DIST}} = \frac{1}{2}S_V^* h'^3 y \qquad (7.54)$$

The direct dependence on y shows that this is a simple tilt of the wavefront. When this expression is reduced to the transverse aberration, by Equation 7.8 we have

$$\mathrm{Distortion} = \frac{1}{2}\left(\frac{S_V^*}{y}\right) l'h'^3 = h_d' - h' \qquad (7.55)$$

where h' is the expected image height ($h \times$ magnification) and h_d' the aberrated height of the principal ray. (The division by y occurs here because the S^* term contains y. That in Equation 7.54 is the *fractional* value of y.) Thus the actual distortion varies as the cube of the image height. A common measure of distortion is as a percentage error,

$$\mathrm{Percent\ distortion} = \frac{h_d' - h'}{h'} \times 100\%$$

$$= \frac{1}{2}\left(\frac{S_V^*}{y}\right) l'h'^2 \times 100\% \qquad (7.56)$$

This is quoted for the maximum field angle, and values for intermediate field points can be obtained by noting that this measure of distortion varies with the square of the image height.

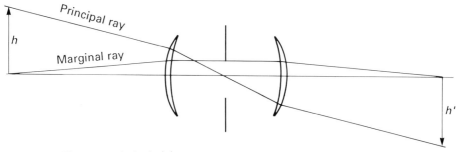

Figure 7.30 The symmetrical principle

When looking through a single lens, the eye acts as a distant stop so that distortion is seen. This means that all spectacle lenses suffer from distortion as S_I, S_{II}, S_{III} and S_{IV} cannot be zero for a single lens. When the lens is placed very close to the eye the distortion is very much less.

The correction of distortion arises mainly from its change of sign as the stop is moved from one side to the other side of the lens. In the case in which a lens system has to work at a magnification of -1, the use of two identical lenses symmetrically placed each side of the stop offers some advantages. In Figure 7.30 the eccentricity factors for lenses A and B are clearly subject to:

$$Q_A = - Q_B$$

Similarly,

$$X_A = - X_B \qquad \text{Bending factors}$$
$$Y_A = - Y_B \qquad \text{Conjugates factors}$$

From these,

$$S_I^A = S_I^B$$
$$S_{II}^A = - S_{II}^B$$
$$S_{III}^A = S_{III}^B$$
$$S_{IV}^A = S_{IV}^B$$

When these values are used in Equation 7.53, we find that

$$S_V^{*A} = - S_V^{*B}$$

so that the residual distortion is zero. We also obtain zero coma, but spherical aberration, astigmatism and field curvature remain uncorrected. The use of doublets in this symmetrical layout is discussed in Section 14.4.

With a single lens, the spherical aberration never goes to zero. The coma is therefore always susceptible to correction by the use of a relocated stop. Furthermore, the equation for astigmatism may be examined for the effect of stop position. S_{III}^* has a maximum or minimum value when

$$\frac{\partial S_{III}^*}{\partial Q} = 0$$

From Equation 7.53,

$$\frac{\partial S_{III}^*}{\partial Q} = 2S_{II} + 2QS_I = 2S_{II}^*$$

Thus the correction of coma occurs at the same time as astigmatism is a maximum or minimum. With undercorrect spherical aberration, the coma-free stop position gives the most backward curving astigmatism. In the design of single lens eyepieces and magnifiers it is common practice to choose a bending which gives zero coma at the eye position as this also gives minimum astigmatism.

7.6 Ray tracing

7.6.1 Meridional rays

Previously we have seen how Snell's law governs the refraction of a ray of light at a surface:

$$n' \sin i' = n \sin i$$

The formula also applies to reflection in air or inside an optical medium if we set n' equal to $-n$. Thus, this equation is exact and applicable at all the interfaces likely to be encountered in an optical system. The more straightforward application of it is to rays from axial objects and rays from off-axis objects when the ray remains in a single plane. Thus, in Figure 7.17, rays QDT and QET are **meridional** rays. Rays QGS and QHS do not remain in a single plane and are called skew rays (see Section 7.6.2) by optical designers.

In Section 3.3 it was seen that rays could be traced through a surface using a graphical construction to an accuracy limited only by the accuracy of the drawing. The same process may be done mathematically by taking sines and arcsines in sequence to evaluate Equation 7.13. Ray tracing in itself does not give an analytical description of a lens system, but it is possible to develop a sequence of ray traces that assess the performance of a given lens system, and then by changing the system parameters slightly to do the sequence of ray traces again to find out if the second system is better than the first. If it is, the changes can be made again in the same direction and the whole process repeated until the best system is found.

This process of iteration is called an **optimization program** and considerable work has gone into the design of such programs so that they reach the best design in the shortest amount of computing time. As a lens of five elements will have up to ten surfaces, nine thicknesses and spacings, and five refractive indices, rather complex mathematics is required to compute which changes are most likely to improve the performance. This is a technology in its own right and well beyond the scope of this book.

A considerable amount of information can be obtained about a simple lens by ray tracing in two dimensions. Almost every ray diagram in this book has used meridional rays, which lie in the plane of the page before and after refraction. Thus, the normal to the surface at the point where the ray strikes it must also lie in the plane of the page. In optometry, with toroidal surfaces abounding and the eye not at the centre of the lens, most rays lie in different planes and a full three-dimensional skew ray trace is needed. The basic procedure is the same for both (but skew rays need a much larger number of computing steps), and may be set out as follows:

1. Set up the incident ray.
2. (a) Define the surface.
   (b) Calculate the intersection point with the surface.
   (c) Calculate the normal to the surface at the intersection.
   (d) Calculate the angle of incidence and, hence, refraction.
   (e) Calculate the details of the refracted ray ready for the next surface.
3. Repeat 2 for the following surfaces.
4. Repeat 1, 2 and 3 for other rays.
5. Use information on all rays to calculate image position and extent of aberrations.

In the two-dimensional meridional case, parts of stage 2 in the procedure above may be condensed in the mathematics so that it fits into a small calculator, but overall the procedure is followed.

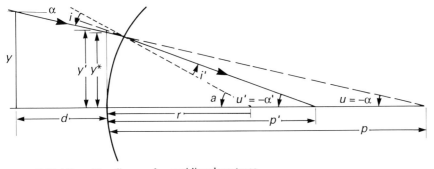

Figure 7.31 All-positive diagram for meridional ray trace

In Figure 7.31, the familiar all-positive diagram is used with the object ray coming from an off-axis object. The first stage of the procedure is to define the incident ray. For simplicity we assume that the object of height y is at $x = 0$. The ray leaves it at gradient angle α to the axis, travelling in a medium of refractive index n. Note that the value of u is positive for the all-positive diagram when the gradient angle α of the ray is negative. Although slopes as defined in Section 3.4 assist the derivation of formulae, in ray tracing gradients (and direction cosines) are positive when x, y and z are all increasing.

The values of y, α and n define the incident ray (stage 1). The definition of the surface is contained in the parameters, r, d and n', which are the radius of curvature, the axial distance from the object (or previous surface) and the subsequent refractive index. These will need changing as the calculation progresses through the lens system.

The simplest refraction mathematics comes from

$$i + u = a = i' + u' \tag{7.57}$$

and

$$n \sin i = n' \sin i' \tag{7.58}$$

for every refraction, and

$$\frac{p - r}{\sin i} = \frac{r}{\sin u} = \frac{-r}{\sin \alpha} \tag{7.59}$$

for every ray (from Equation 7.10).
    In Figure 7.31,

$$y^* = y + d \tan \alpha \tag{7.60}$$

$$p = \frac{y^*}{- \tan \alpha} \tag{7.61}$$

From Equations 7.59–7.61,

$$\sin i = (y + d \tan \alpha + r \tan \alpha) \frac{\cos \alpha}{r} \tag{7.62}$$

Also,

$$\sin i' = \frac{n}{n'}(y + d \tan \alpha + r \tan \alpha) \frac{\cos \alpha}{r}$$

$$= (y' + r \tan \alpha') \frac{\cos \alpha'}{r} \tag{7.63}$$

because $d' = 0$.
    The first expression gives i' so that α' is known from Equation 7.57. The second expression then gives

$$y' = r \left(\frac{\sin i' - \sin \alpha'}{\cos \alpha'} \right) \tag{7.64}$$

This approach includes stages 2(b) and 2(c) only by implication, but we have calculated the angles of incidence and refraction (2(d)) and if we put the values y', α' and n' into the same computer locations that had $y\alpha$ and n (2(e)) then we may proceed to the next surface (3) and repeat the process. If a plano surface is encountered, it is a simple matter to insert a very large radius of curvature, but then inaccuracies arise. Most computer programs use curvatures rather than radii, and angles with intercept heights rather than lengths, as this avoids large numbers. The main inaccuracy in the scheme above is that, when r becomes large, $\sin i' - \sin \alpha'$ becomes very small.
    If the above scheme is programmed into a personal computer, off-axis images can be found by ray tracing two rays. The location (x_i, y_i), where they intersect, can be found by calculating with respect to the pole of the last surface using:

$$x_i = - \left(\frac{y_1' - y_2'}{\tan \alpha_1' - \tan \alpha_2'} \right) \tag{7.65}$$

and y_i from $y_i = x_i \tan \alpha_1' + y_1'$.
    Such calculations will yield values for tangential coma, tangential astigmatism and distortion.

7.6.2 Skew rays

As with the previous section, the mathematics described here is not necessarily that used in computer ray tracing. The aim here is to provide an introduction to the procedures required, together with equations that best represent the mathematics involved. These can be programmed into a home computer, where the purpose is that of producing actual figures for educational use rather than speed and efficiency.

The procedure given in the previous section must be followed, but we start this exposition with the actual refraction. Snell's law applies as usual:

$$n \sin i = n' \sin i' \tag{7.66}$$

It is necessary to find i and i' in terms of the incident ray, the normal to the surface and the refracted ray. This is obtained from the very simple equations

$$\cos i = \alpha_1 \alpha_0 + \beta_1 \beta_0 + \gamma_1 \gamma_0 \tag{7.67}$$

where $\alpha_1 \beta_1 \gamma_1$ and $\alpha_0 \beta_0 \gamma_0$ are the direction cosines of the incident ray and the normal, and similarly

$$\cos i' = \alpha_1' \alpha_0 + \beta_1' \beta_0 + \gamma_1' \gamma_0 \tag{7.68}$$

where $\alpha_1' \beta_1' \gamma_1'$ are the direction cosines of the refracted ray.

We require the last three values independently, so that Equation 7.68 is no real use as it stands. Taking Equations 7.66–7.68 together, plus the property that for any refraction the incident ray, the normal to the surface at the point of incidence and the refracted ray will all lie in one plane (although this plane will change for different rays), it can be shown that:

$$n(\alpha_1 - \alpha_0 \cos i) = n'(\alpha_1' - \alpha_0 \cos i')$$
$$n(\beta_1 - \beta_0 \cos i) = n'(\beta_1' - \beta_0 \cos i')$$
$$n(\gamma_1 - \gamma_0 \cos i) = n'(\gamma_1' - \gamma_0 \cos i') \tag{7.69}$$

Alternatively, we can set these out as

$$n'\alpha_1' = n\alpha_1 + (n \cos i - n' \cos i')\alpha_0$$
$$n'\beta_1' = n\beta_1 + (n \cos i - n' \cos i')\beta_0$$
$$n'\gamma_1' = n\gamma_1 + (n \cos i - n' \cos i')\gamma_0 \tag{7.70}$$

Thus the refracted ray can be calculated if we know the direction cosines of the incident ray, the point of interception of the incident ray with the surface, and the direction cosines of the normal to the surface at that point.

The direction cosines of the incident ray may be defined by a knowledge of two points on that ray – usually the object itself and some point in the entrance pupil. Defining these (Figure 7.32) as $x_1 y_1 z_1$ and $x_2 y_2 z_2$, we have

$$(x_2 - x_1)^2 + (y_2 - y_1)^2 + (z_2 - z_1)^2 = T^2 \tag{7.71}$$

where T is the length from $x_1 y_1 z_1$ to $x_2 y_2 z_2$, and then

$$\alpha_1 = \frac{x_2 - x_1}{T} \quad \text{or} \quad \left(\frac{x_0 - x_1}{T_0}\right)$$

$$\beta_1 = \frac{y_2 - y_1}{T} \quad \text{or} \quad \left(\frac{y_0 - y_1}{T_0}\right)$$

$$\gamma_1 = \frac{z_2 - z_1}{T} \quad \text{or} \quad \left(\frac{z_0 - z_1}{T_0}\right) \tag{7.72}$$

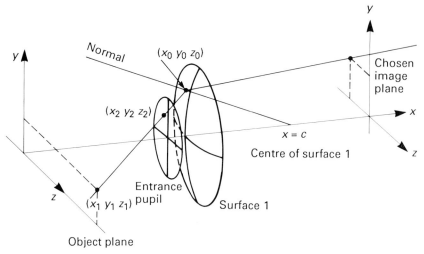

Figure 7.32 Skew ray tracing

However, we need the intercept point $(x_0 y_0 z_0)$ with the surface. If this is a spherical surface, radius r, we have

$$x_0^2 + y_0^2 + z_0^2 = r^2 \tag{7.73}$$

if the centre of the surface is at the origin of the coordinate system; or

$$(x_0 - r)^2 + y_0^2 + z_0^2 = r^2 \tag{7.74}$$

if the pole of the surface is at the origin; or

$$(x_0 - c)^2 + y_0^2 + z_0^2 = r^2 \tag{7.75}$$

if the centre of the sphere is at $x = c$, which may be calculated from the actual thicknesses, separations and radii of the system. From Equations 7.72 and 7.75, we find that the distance T_0 from $x_1 y_1 z_1$ to the point of interception is given by

$$T_0^2 + 2[(x_1 - c)\alpha_1 + y_1\beta_1 + z_1\gamma_1] T_0 + [x_1^2 + y_1^2 + z_1^2 - 2x_1c + c^2 - r^2] = 0 \tag{7.76}$$

from which T_0 can be calculated and $x_0 y_0 z_0$ found by using T_0 in Equation 7.72 (α and T are positive for rays left to right).

From here it is a simple matter, with a sphere, to obtain $\alpha_0 \beta_0 \gamma_0$ as

$$\alpha_0 = \frac{x_0 - c}{r} \qquad \beta_0 = \frac{y_0}{r} \qquad \gamma_0 = \frac{z_0}{r} \tag{7.77}$$

These are all the equations needed to trace a skew ray through a spherical surface centred on the x-axis. If Equations 7.77 are put straight into Equation 7.70, the last two terms become

$$\frac{n \cos i - n' \cos i'}{r}$$

which is the oblique power, K, developed in Equations 7.33 and 7.34 for the Coddington formulae.

To recapitulate, the procedure laid down in Section 7.6.1 is followed (1) by letting the ray find its own direction cosines from a given object point and one of an array of points in the entrance pupil. Then, with the first surface defined (2(a)), the interception point is found via Equation 7.76 (2(b)). Stages 2(c), (d) and (e) follow using Equations 7.77, 7.67, 7.70 and 7.68. Rather than the sine version of Equation 7.66, $\sin i = \sqrt{(1 - \cos^2 i)}$ can be used.

When a number of rays have been traced a spot diagram can be built up. Retaining the direction cosines of the final rays allows this to be converted to longitudinal aberration.

For further reading on skew rays see:

1. Welford, W. T. (1986) *Aberrations of Optical Systems*, Adam Hilger, Bristol, UK
2. Fry, G. A. (1970) *Ray Tracing Procedures*, College of Optometry, Ohio State University
3. Kingslake, R. (1978) *Lens Design Fundamentals*, Academic Press, New York
4. Smith, W. J. (1966) *Modern Optical Engineering*, McGraw-Hill, New York

Exercises

7.1   Explain what is meant by the spherical aberration of a lens. An object is viewed over the edge of a card placed near to the eye. When the card is moved up and down, the object appears to move, the motion being *with* the card when the object is distant, and *against* the card when the object is near. What optical defects of the eye does this indicate? Illustrate your answer with a diagram.

7.2   Prove that the distances (from the vertex) of the aplanatic points of a spherical refracting surface are given by

$$r\left(1 + \frac{n'}{n}\right) \quad \text{and} \quad r\left(1 + \frac{n}{n'}\right)$$

What is the magnification at these points?

7.3   Explain from first principles (using the concepts of wave curvature, or otherwise) why a thin positive lens traversed obliquely by a pencil of light from a star does not yield a point image. Describe the appearances which could be found on a focusing screen held at different distances in the convergent beam if the lens were a +5 D 'sphere' and the light incident at 45° to the axis.

7.4   Apply Young's construction to determine the spherical aberration of the cornea. Assume the latter to be spherical and of radius 7.8 mm and the refractive index of aqueous humour to be 1.33. Take a ray parallel to the visual axis at an incident height of 4 mm.

7.5   Give a general description, with the aid of a diagram, of the defect of lenses called coma. How would you show this experimentally?

7.6   A narrow pencil of light from an infinitely distant object is incident on a spherical refracting surface of +5 cm radius separating air from glass of $n = 1.5$. The angle of incidence is 50°. Calculate the distance from the point of incidence to the primary and secondary focal lines.

7.7   A spherical surface of radius +8 in separates two media of refractive indices 1.36 and 1.70 respectively. Light from the left is converging in the first medium, towards an axial point 18 in to the right of the surface vertex. Find the position of the image point and show that there will be no spherical aberration.

7.8   A narrow pencil is incident centrally on a +5 D spherical lens in a direction making 30° with the optical axis of the lens. Explain briefly the nature of the refracted pencil.

7.9   A narrow pencil of parallel rays in air is refracted at a spherical surface of radius r into a medium of refractive index $\sqrt{3}$. If the angle of incidence is $60°$, show that the distance of the primary and secondary image points from the point of incidence are given by $\frac{3}{4}r\sqrt{3}$ and $r\sqrt{3}$ respectively.

7.10   Explain with the help of a diagram the following terms associated with the refraction by a lens of an oblique pencil: meridian plane, sagittal plane, astigmatic difference, place of least confusion.

7.11   What is meant by radial astigmatism in lenses? Give examples of the effect produced by it in obtaining an image of a star-covered field.

7.12   A spectacle lens of power $+5\,D$ is made to form the image of a distant axial source of light on a ground glass screen; the lens is then slightly tilted about one diameter. Explain with numerical details the changes thus produced in the image and the further changes observed as the screen is moved nearer the lens. Note: if F is the axial power, and F_S and F_T are respectively the sagittal and tangential powers, while θ is the angle of obliquity, then

$$F_S = F\left(1 + \frac{\sin^2 \theta}{3}\right) \text{ approximately}$$

$$F_T = F_S \sec^2 \theta \text{ approximately}$$

7.13   Write a short account of the aberrations in the image produced of a flat object by an uncorrected optical system. Give diagrams.

   Explain, giving reasons, which aberrations are of most importance in the case of (a) a telescope objective, and (b) a spectacle lens.

7.14   Explain the terms barrel and pincushion distortion and show how they are produced by a single convex lens and stop.

7.15   Draw the all-positive diagram for refraction at a spherical surface and from it deduce the five simple equations used in the trigonometrical computation of rays through such a surface.

   Write down also the corresponding equations for computing a paraxial ray and from these latter equations derive the fundamental paraxial equation:

$$\frac{n'}{l'} = \frac{n}{l} = \frac{n' - n}{r}$$

7.16   Write down the equations used in computing the path of a ray of light after refraction at a spherical surface. Apply them to compute the refracted ray in the following case:

   $r_1 = +377.85\,\text{mm}$ $n = 1.0$ $n' = 1.5119$

   The axial object point is situated $487\,\text{cm}$ in front of the surface; the incidence height of the ray is $12.7\,\text{cm}$. Show how to arrange the work in tabular form and calculate the first six quantities.

7.17   Trace by trigonometrical computation the course through the lens, the constants of which are given below, of a ray that crosses the axis at a point B $30\,\text{cm}$ in front of the first surface and is incident on the lens at an incident height of $2\,\text{cm}$. Find also the point where a paraxial ray incident through B crosses the axis after refraction.

   $r_1 = +10\,\text{cm}$ $r_2 = -20\,\text{cm}$ $d = 3\,\text{cm}$ $n = 1.624$

7.18   In general, the image formed by a simple spherical lens suffers from certain defects or aberrations. Enumerate these aberrations.

   If the object consists of a point source of white light situated on the optical axis of the lens, with which aberration or aberrations will the image be affected when the aperture of the lens is (a) small, (b) large?

   If the point source is moved away from the axis, what aberration or aberrations will be manifest when the aperture of the lens is small?

   Give clear diagrams.

7.19 List the aberrations from which lens systems may suffer, and explain how each occurs, using diagrams where applicable.

Which of the aberrations you describe are of importance in:

(a) A slide projector.

(b) A high power microscope?

7.20 List and describe the aberrations from which optical systems may suffer. State which of these are of importance in:

(a) Telescopes.

(b) Film projectors.

7.21 Explain briefly what is meant by:

(a) Disc of least confusion.

(b) Astigmatism.

(c) Field curvature.

(d) Distortion.

Explain qualitatively how the primary field curvature of a simple camera lens can be compensated by the introduction of overcorrected astigmatism to obtain a flat image field.

7.22 Trace by trigonometrical computation the path through a spherical surface of a ray that crosses the axis at a point B 500 cm in front of the surface and is incident on the surface at a height of 15 cm.

$$r_1 = +400 \, \text{mm} \qquad n = 1.000 \qquad n' = 1.512$$

7.23 A thin $+10 \, \text{D}$ lens 40 mm in diameter exhibits 0.50 D of positive spherical aberration for marginal rays.

(a) For a distant object, what is the amount of lateral spherical aberration (in millimetres) for this lens?

(b) If the lens is stopped down to only a 20 mm aperture how much longitudinal spherical aberration (in millimetres) does it then have for distant objects?

7.24 A parallel beam of light strikes the plane surface of a plano-convex lens normally. The radius of curvature of the convex surface of the lens is 20 cm and its refractive index is 1.5. Calculate the position of the second focal point for the paraxial rays, and, by using the laws of refraction, the longitudinal spherical aberration for a ray which strikes the lens 5 cm from the axis.

7.25 An equi-concave lens in air has radii of curvature equal to $+$ and $-$ 120 mm. If light from a distant axial object is incident on the first surface at a height of 25 mm from the axis find:

(a) the (paraxial) focal length,

(b) the bending factor,

(c) the conjugates factor,

(d) the longitudinal spherical aberration,

if the refractive index of the lens material is 1.85 and the lens is considered to be thin.

7.26 A thin plano-convex lens has a *second* surface of radius $-100 \, \text{mm}$. If the lens material is ophthalmic crown glass of refractive index 1.523, calculate

(a) The paraxial power.

(b) The bending factor.

(c) The conjugates factor.

(d) Calculte the longitudinal spherical aberration for a light ray incident 20 mm above and parallel to the axis.

7.27 A thin glass lens has a paraxial power of 8 D. It is designed to be free of coma for an object placed 150 mm in front of the lens. Assuming a refractive index of 1.523, calculate:

(a) The location of the image.

(b) The conjugates factor.

(c) The bending factor (for no coma).

(d) The radii of the first and second surfaces.

Chapter 8

Non-spherical and segmented optical surfaces

All the optical surfaces discussed in Chapters 1–7 have been spherical. Even flat surfaces are spherical surfaces with an infinitely long radius of curvature! Spherical surfaces are very common in optics, as they are easy to make because of the self-correcting action that occurs when a convex surface is ground and polished with a concave tool and vice versa.

Non-spherical surfaces that are accurate to some other shape are generally difficult to make. Surfaces that are controlled departures from the spherical form need extra precision and control to make them optically satisfactory. In recent times the development of high-precision movements and bearings has enabled computer-controlled lathes to be built which achieve accuracies well within the wavelength of light.

Non-spherical surfaces can take many forms, and this chapter describes some of them.

8.1 Toroidal and cylindrical surfaces

If a thin meniscus lens is squeezed as shown in Figure 8.1, the surface distorts from its spherical form. There is an increase in curvature along the section YAY', but along the section ZAZ' the curvature flattens as the lens flexes to take the strain. Every part of the surface now has two curvatures which are, in fact, the maximum and minimum values of a general curvature which changes with the orientation of the section. Thus, to a first-order approximation, the sag on the chord YY' is s and the sag on the chord ZZ' is also S, even though these chords are of unequal length. As chords at different orientations are chosen, such as VV', their length changes from a minimum at YY' to a maximum at ZZ' and back to YY' again twice in each rotation.

This concept has been used to make contact lenses that correct the axial astigmatism of the eye. The method is known as **crimping** and the lens material is squeezed while it is cut and polished as a spherical lens. When the crimp is removed it springs *back* to a two-curvature shape such as in Figure 8.1. On rigid materials such as glass a complicated rocking motion is needed to grind and polish the shape directly. This type of surface is called a **toroidal** surface and its shape is maintained even down to a very small elemental area such as T, where the curvature shows maximum and minimum values in sections which are at right angles. This applies at every point on the surface. An elemental area such as T will also have maximum

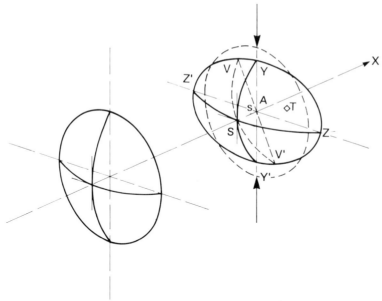

Figure 8.1 Distortion of a spherical surface into a toroidal surface

and minimum curvature values. Although these may not be orientated at the same angles as at S, they will still be at right angles to each other.

This toroidal surface with orthogonal maximum and minimum curvatures is a fundamental mathematical surface. A spherical surface is simply a special case of a toroidal surface where the maximum and minimum curvatures happen to be equal! A **cylindrical** surface is one in which the minimum curvature is zero, and a flat surface is one in which they are both zero. Figure 8.2 shows cylindrical surfaces and Figure 8.3 some practical toroidal surfaces. For all these surfaces, the maximum and minimum curvatures may be sections which are arcs of circles on the axis only. The other sections at oblique angles on the axis and all other places only approximate to circles, but over short lengths we assume that these are circles and have a simple curvature value.

All these surfaces therefore (except the flat plate) have curvatures that change between a maximum and a minimum value depending on the orientation of the section or **meridian** chosen. The orientations that contain the maximum curvatures are called the **principal meridians**. In the case of the cylindrical lens the axis of the cylinder, part of which forms the surface, is parallel to the principal meridian, which has zero curvature. This meridian is known as the **axis meridian** or the **cylinder axis**.

In Figure 8.4 it can be seen that for a cylindrical lens with a vertical cylinder axis the refraction of the light from a point object on the optical axis occurs only in the horizontal plane, which in this case is the principal meridian containing the curvature and, therefore, the power of the lens. This is known as the **power meridian**. Each light ray reaching the lens is refracted only in the power meridian. The result of this is a line image which is parallel with the cylinder axis; that is, the zero power meridian of the lens. This is because the line is due to the *absence* of power in, in this case, the vertical plane.

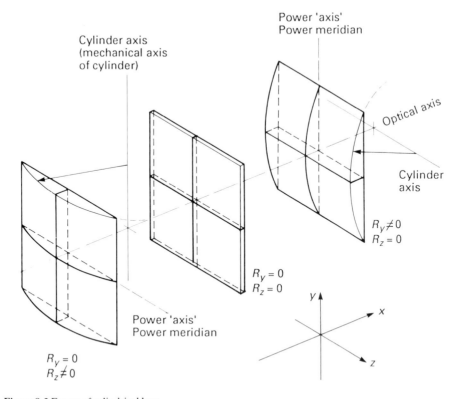

Figure 8.2 Forms of cylindrical lens

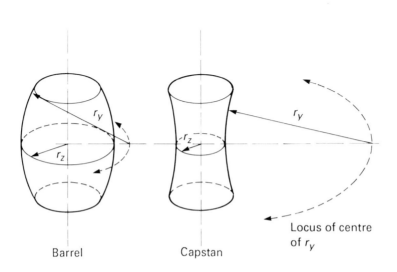

Figure 8.3 Practical toroidal surfaces

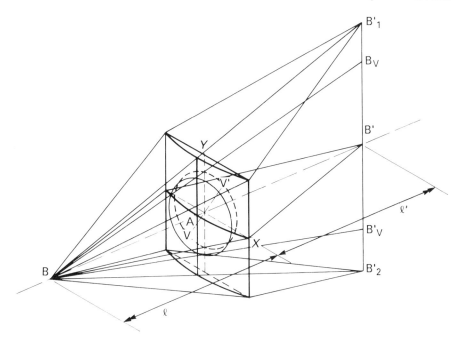

Figure 8.4 Action of a cylindrical lens

When the lens aperture is rectangular the line image of a point source has little change in intensity along its length. For a cylindrical lens of circular aperture the intensity reduces at the same rate as the effective width of the lens, becoming zero at points B'_1 and B'_2. Within the plane containing the power meridian there is nothing to distinguish the lens from a spherical lens. The lens power is given by

$$F_c = R_c(n - 1) = \frac{n - 1}{r_c} \tag{8.1}$$

where the suffix c indicates a cylindrical surface and n is the refractive index.

For a cylindrical lens where the other surface is zero the fundamental paraxial equations apply:

$$L' = L + F_c \tag{8.2}$$

and

$$\frac{1}{l'} = \frac{1}{l} + \frac{1}{f_c} \tag{8.3}$$

where l is the distance of the point object and l' the distance of the line image.

For a toroidal surface both meridians have refractive power. The optical effect of a toroidal refracting surface is to create a toroidal wavefront very similar to that formed by the third-order aberration, astigmatism, described in Chapter 7 and there called oblique astigmatism to distinguish it from the effect due to toric and cylindrical lenses which we will now call **axial astigmatism**.

In the case of a lens with a toroidal surface having powers, for example $+6\,\mathrm{D}$ and $+4\,\mathrm{D}$, in its principal meridians, the optical effect results in the formation of two line images at their second principal foci. The same effect would result from using a

+4 D thin spherical lens in contact with a thin cylindrical lens of +2 D power. Furthermore, the same effect can be obtained by using a +6 D thin spherical lens in contact with a thin cylindrical lens of −2 D power. Yet again, the same effect could be obtained by using two cylindrical lenses of powers +6 D and +4 D with their cylinder axes at right angles. This last case is called **crossed cylinders**.

Figure 8.5 shows the two cylindrical lenses at right angles and the two line images formed. Because each lens has zero power in the power meridian of the other lens, the two line images are located independently of each other. Any changes in the power of the cylindrical lens forming image B_1' will not affect the location of image B_2'. The same is true when a toroidal surface is used to form two line images. The power values in the principal meridians independently determine the line image distances.

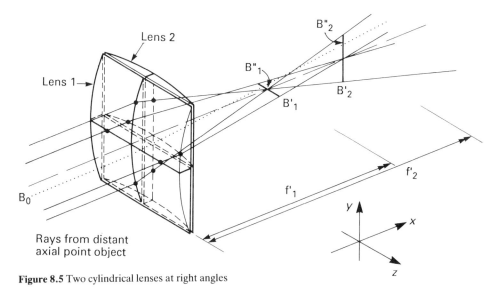

Figure 8.5 Two cylindrical lenses at right angles

Although the toroidal surface has a different *curvature* in oblique meridians, it does not have effective power in any meridian other than the principal meridians. This is because the line images are really caustics of the emerging wavefront. The off-set ray B_0, shown dotted, is in neither power meridian. It passes through both line images at the off-axis points B_1'' and B_2'' and so only these two line images exist.

A demonstration of this effect is provided by changing the orientation of one of the lenses in Figure 8.5 so that the cylinder axes are not at right angles. To calculate the optical effect it is helpful to remember that the contour map of a perfect spherical surface is a series of concentric circles (see Section 7.3) and the departure of such a surface from the plane surface through the pole is given by

$$S_s = \frac{y^2 + z^2}{2r} \tag{8.4}$$

The case of a cylindrical surface is given by

$$C_y = \frac{y^2}{2r} \tag{8.5}$$

when the cylinder axis is parallel to the z-axis (lens 1 in Figure 8.5) and by

$$C_z = \frac{z^2}{2r} \tag{8.6}$$

when the cylinder axis is parallel to the y-axis (lens 2 in Figure 8.5).

It is now easy to see that when two cylindrical surfaces are put together with their axes at right angles the total departure from the plane is given by

$$C_y + C_z = \frac{y^2}{2r} + \frac{z^2}{2r}$$

$$= \frac{y^2 + z^2}{2r} = S_s \tag{8.7}$$

which is a spherical surface if the two cylinders have the same radii of curvature. If the cylindrical curvatures are different, we have

$$C_y + C_z = \frac{y^2}{2r_1} + \frac{z^2}{2r_2} \tag{8.8}$$

which is an elliptical function used to describe third-order astigmatism in Chapter 7. Such an elliptical function can still be thought of as a spherical function plus a cylindrical function by splitting the steeper of the two curvatures into two to give:

$$C_y + C_z = \frac{y^2 + z^2}{2r_1} + \frac{z^2}{2} \left(\frac{1}{r_2} - \frac{1}{r_1} \right) \tag{8.9}$$

The general relationship is shown diagrammatically in Figure 8.6. Contour lines of equal sag which define cylindrical lenses are parallel lines spaced as

$$y = \sqrt{(2rC_y)} \qquad \text{and} \qquad z = \sqrt{(2rC_z)}$$

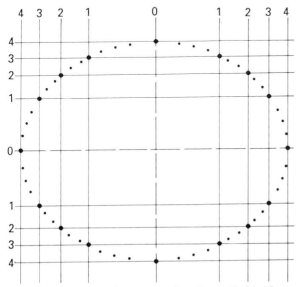

Figure 8.6 Diagrammatic representation of two cylindrical lenses crossed at right angles. The locus of equal sag (4 arbitrary units) is an ellipse. This diagram uses contours to indicate surface curvature in the same way as Figure 7.13

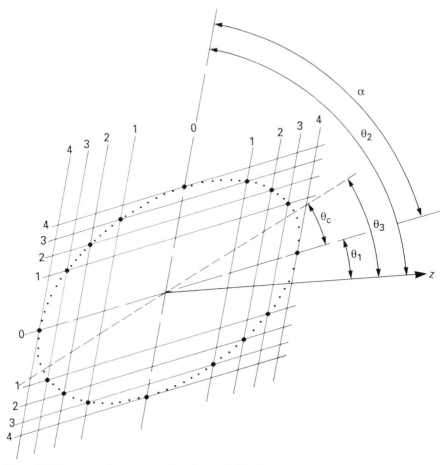

Figure 8.7 Diagrammatic representation of two cylindrical lenses crossed at an angle α. The locus of equal sag (4 arbitrary units) is an ellipse

from Equations 8.4 and 8.5. These are given arbitrary units in Figure 8.6, and the locations where these add to give a total of 4 units are marked and found to lie on an ellipse.

The problem becomes more difficult when one or both axes of the cylinders is not parallel to the Y or Z axis. Figure 8.7 shows two cylindrical lenses crossed obliquely at an angle α, and therefore at angles θ_1 and θ_2 from some arbitrary z-axis. They have different powers shown by the contours of one being spaced differently from the other. The points where the total departure from the plano is 2 units are shown by the dots and these lie on an ellipse at angle θ_3.

In coordinate geometry the rotation of axes by some angle θ means that for the new condition what was previously z must be replaced by $z \cos \theta + y \sin \theta$ and what was previously y by $z \sin \theta + y \cos \theta$. Because both the cylindrical lenses of Figure 8.7 have been rotated from being parallel with the z-axis, and their initial equations were as in Equation 8.6, they are now

$$C_1 = \frac{(z \sin \theta_1 - y \cos \theta_1)^2}{2r_1}$$

$$C_2 = \frac{(z \sin \theta_2 - y \cos \theta_2)^2}{2r_2}$$

(8.10)

The total departure from plano, the combined optical effect of the two lenses, is $C_1 + C_2$:

$$C_1 + C_2 = \frac{z^2 \sin^2 \theta_1}{2r_1} + \frac{2yz \sin \theta_1 \cos \theta_1}{2r_1} + \frac{y^2 \cos^2 \theta_1}{2r_1}$$
$$+ \frac{z^2 \sin^2 \theta_2}{2r_2} + \frac{2yz \sin \theta_2 \cos \theta_2}{2r_2} + \frac{y^2 \cos^2 \theta_2}{2r_2}$$

(8.11)

This is the general equation of an ellipse. It converts to a simpler equation if we choose the axes so that

$$\frac{2yz \sin \theta_1 \cos \theta_1}{2r_1} + \frac{2yz \sin \theta_2 \cos \theta_2}{2r_2} = 0$$

(8.12)

or

$$\frac{\sin 2\theta_1}{2r_1} + \frac{\sin 2\theta_2}{2r_2} = 0$$

(8.13)

We do not have complete choice of θ_1 and θ_2 because the angle between them is α. However, we can rotate the axes so that θ_1 is replaced by $\theta_1 - \theta_3$ and θ_2 by $\theta_2 - \theta_3$ and there will be a value of θ_3 for which Equation 8.13 is zero. Assuming that this is when $\theta_1 - \theta_3 = \alpha_1$ and $\theta_2 - \theta_3 = \alpha_2$, we can write Equation 8.11 as

$$C_1 + C_2 = \frac{z^2 \sin^2 \alpha_1}{2r_1} + \frac{z^2 \sin^2 \alpha_2}{2r_2} + \frac{y^2 \cos^2 \alpha_1}{2r_1} + \frac{y^2 \cos^2 \alpha_2}{2r_2}$$

(8.14)

This then is the optical effect at some unknown meridian angle with respect to some unknown axes! We now calculate the optical effect at the meridian angle (also unknown) which is at 90° to that of Equation 8.14. For this case *both* α_1 and α_2 change by 90°, therefore

$$(C_1 + C_2)_{90°} = \frac{z^2 \sin^2 (\alpha_1 + 90°)}{2r_1} + \frac{z^2 \sin^2 (\alpha_2 + 90°)}{2r_2}$$
$$+ \frac{y^2 \cos^2 (\alpha_1 + 90°)}{2r_1} + \frac{y^2 \cos^2 (\alpha_2 + 90°)}{2r_2}$$
$$= \frac{z^2 \cos^2 \alpha_1}{2r_1} + \frac{z^2 \cos^2 \alpha_2}{2r_2} + \frac{y^2 \sin^2 \alpha_1}{2r_1} + \frac{y^2 \sin^2 \alpha_2}{2r_2}$$

(8.15)

The surprising thing about Equations 8.14 and 8.15 is that if we add them together all the angles disappear because $\sin^2 \theta + \cos^2 \theta = 1$ for any value of θ. Therefore

$$C_1 + C_2 + (C_1 + C_2)_{90°} = \frac{z^2}{2r_1} + \frac{z^2}{2r_2} + \frac{y^2}{2r_1} + \frac{y^2}{2r_2}$$
$$= \frac{z^2 + y^2}{2r_1} + \frac{z^2 + y^2}{2r_2}$$

(8.16)

Because Equation 8.16 holds for any pair of meridians at right angles, the sum of their optical effects is a constant.

What the above reasoning shows is that any general toroidal surface has a curvature that varies with the chosen meridian but, if any two meridians at right angles are chosen, the sum of their curvatures is a constant for that surface. This arises, in brief, because the curvature is proportional to z^2; rotation of axes in trigonometry gives $(z \cos \theta)^2$ terms; $\cos(\theta + 90) = \sin \theta$; and $\sin^2 \theta + \cos^2 \theta = 1$, independent of angle. This result forces the maximum curvature to be at right angles to the minimum curvature. All toroidal surfaces have this feature, as have the off-axis parts of aspheric surfaces (Section 8.2). It is useful to refer to the sum of any orthogonal curvatures as the **total curvature** of the surface.

In ophthalmic optics it is normal to designate a toric lens, used for the correction of the eye's axial astigmatism, in terms of its spherical power, S, and its cylindrical power, C. If F_A and F_B are the maximum and minimum powers then, recognizing that optical power is proportional to curvature, we can say

$$C = F_A - F_B \tag{8.17}$$

$$S + (S + C) = F_A + F_B = F_{TOTAL} \tag{8.18}$$

The problem of the two cylindrical lenses crossed at an angle α between their cylinder axes in Figure 8.7 can make use of Equation 8.17. *Each* cylinder has a total power of F_1 and F_2, respectively, because their minimum powers are both zero. These total powers are independent of orientation and so the total power of the combination (thin lenses in contact) is $F_1 + F_2$ even though the two cylindrical lenses are not at right angles. The cylindrical power C is given by

$$C = \pm\sqrt{[(F_{TOTAL})^2 - 4F_1F_2 \sin^2 \alpha]} \tag{8.19}$$

and the angle θ_3 (from the F_1 cylinder axis) is given by

$$\tan \theta_3 = \frac{-F_1 + F_2 + C}{+F_1 + F_2 + C} \tan \alpha \tag{8.20}$$

Equation 8.20 is obtained by subtracting 8.15 from 8.14 and differentiating to find the maximum. Equation 8.19 is found by inserting the angle at which the maximum occurs into the difference of 8.14 minus 8.15. Much rearrangement is required.

Given the value of C from Equation 8.19, it is straightforward to obtain S from Equation 8.18. These two values determine the lens, although for ophthalmic purposes (Section 8.2) it is also necessary to specify the angle at which the cylinder axis must be orientated within the spectacle frame, so that it correctly balances the astigmatism of the eye.

8.2 Ophthalmic lenses with cylindrical power

Ophthalmic optics is a major user of toroidal surfaces and has developed a standard terminology. In the UK, for instance, the term *toric* should be used only for lenses and not for surfaces. *Toroidal* surfaces are defined as those with curvatures in their principal meridians which are arcs of circles. The surfaces of Figure 8.3 satisfy this but the distorted surfaces of Figure 8.1 do not. (See A. G. Bennett and R. B. Rabbetts, *Clinical Visual Optics*, Butterworths.)

While the human eye rarely exhibits astigmatism in the symmetrical manner of

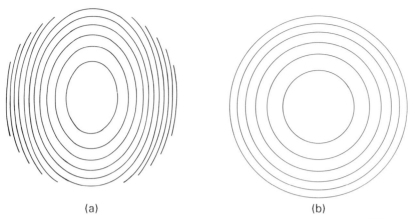

(a)   (b)

Figure 8.8 Contour plots of a toric lens: (a) with extra power in the horizontal meridian (cylinder axis vertical), compared with (b) a spherical lens

the toroidal and cylindrical surfaces of Section 8.1, the vision of many eyes can be improved by using a spectacle lens with a cylindrical component which corrects the cylindrical component of the generally irregularly aberrated eyes (see Chapter 15). Most spectacle lenses even if of zero spherical power normally have curved surfaces and so a spectacle lens with cylindrical power will have one toroidal surface. The contour plot of such a surface will be a series of ellipses as shown in Figure 8.8, although the form of practical ophthalmic torics departs somewhat from this.

Figure 8.9 shows the form of a pencil (or bundle) or rays which has come from a point object on the axis and passed through an astigmatic lens having positive power in both meridians. Ophthalmic lenses to correct myopic eyes would have

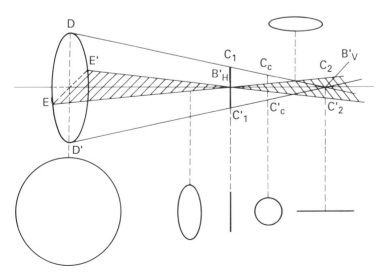

Figure 8.9 Astigmatic pencil

negative power and some ophthalmic lenses could have positive power in one principal meridian and negative power in the other. In Figure 8.9 the power axis of the positive cylindrical component is in the horizontal meridian EE', but if this were provided by a separate cylindrical lens the mechanical axis of the required cylinder would be in the vertical meridian parallel to DD'. Optometrists would say that this lens has a positive power cylinder axis in the vertical meridian (see H. Obstfeld, *Optics for Vision* and A. G. Bennett and R. B. Rabbetts, *Clinical Visual Optics*, for fuller treatments).

Light passing through the vertical section DD' in Figure 8.9 is affected only by the power of the spherical component and converges to B'_v; in sections parallel to DD' the light converges to points on a horizontal line $C_2C'_2$ passing through B'_v. In the horizontal section EE' the power is the sum of the powers of the spherical and cylindrical powers, and the light in this section converges to B'_H, sections parallel to EE' converging to points on a vertical line $C_1C'_1$ through B'_H. A pencil of this form, in which the light originating at a point passes through two perpendicular lines, is called an **astigmatic pencil**.

Cross-sections of such a pencil are shown in Figure 8.9. If the aperture of the lens is circular, the cross-section of the emergent pencil is at first an ellipse with its major axis parallel to the cylinder axis, narrowing down to a line parallel to the cylinder axis at B'_H. The cross-section then becomes circular at $C_cC'_c$, after which it becomes elliptical with the major axis perpendicular to the cylinder axis and degenerates into a line again at B'_v. The circular cross-section at $C_cC'_c$, where the pencil will have its smallest cross-sectional dimension, is known as the **circle of least confusion**.

The properties of the astigmatic pencil were investigated by the mathematician Sturm (1838), and the pencil of the form shown in Figure 8.9 is known as Sturm's conoid. The distance between the focal lines is known as the **interval of Sturm**.

The positions of the line foci are found by applying the conjugate foci expression $(L' = L + F)$ to each meridian in turn. To find the lengths of the line foci and the position and diameter of the circle of least confusion, it is convenient to represent the sections of the pencil in the two principal meridians in one plane, as in Figure 8.10, where the section in a horizontal meridian is shown full lines and in a vertical meridian by broken lines.

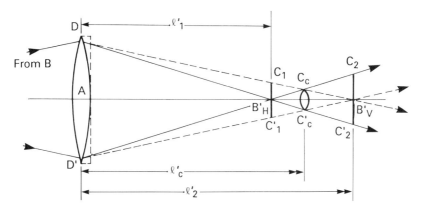

Figure 8.10 Astigmatic lens – positions and lengths of line foci

From Figure 8.10,

$$\frac{C_1C_1'}{DD'} = \frac{B_H'B_v'}{AB_v'} = \frac{l_2' - l_1'}{l_2'}$$

or

$$C_1C_1' = DD' \left(\frac{L_1' - L_2'}{L_1'}\right)$$

(8.21)

$$\frac{C_2C_2'}{DD'} = \frac{B_H'B_v'}{AB_H'} = \frac{l_2' - l_1'}{l_1'}$$

or

$$C_2C_2' = DD' \left(\frac{L_1' - L_2'}{L_2'}\right)$$

(8.22)

To find the circle of least confusion,

$$\frac{C_cC_c'}{DD'} = \frac{l_c' - l_1'}{l_1'} = \frac{l_2' - l_c'}{l_2'}$$

(8.23)

$$l_2'l_c' - l_1'l_2' = l_1'l_2' - l_1'l_c'$$

$$l_c' = \frac{2l_1'l_2'}{l_1' + l_2'}$$

or

$$L_c' = \tfrac{1}{2}(L_1' + L_2')$$

(8.24)

giving the position of the confusion circle.

Equation 8.23 may also be written

$$z = C_cC_c' = DD' \left(\frac{L_1' - L_2'}{L_1' + L_2'}\right)$$

(8.25)

where z is the diameter of the confusion circle.

The optical axis of a toric lens is a line passing through the centre of curvature of the spherical surface and perpendicular to the cylinder axis. In the case of a thin lens the point on the lens through which the optical axis passes is the optical centre.

As with a spherical lens, any ray passing through the optical centre will be undeviated and will be a chief ray from the object point from which the ray started. Since one ray from every point of the object passes undeviated through the toric lens, the image formed when the aperture is quite small will correspond in its general shape to the object, and it size may be found in the same way as with the image formed by a spherical lens. Each object point is, however, imaged as a line, parallel to one or other of the principal meridians, the length of the line being dependent upon the aperture of the lens and the distance between the two focal lines. The form of the image will therefore only agree with that of the object when the length of each line focus is quite small, compared with the size of the image.

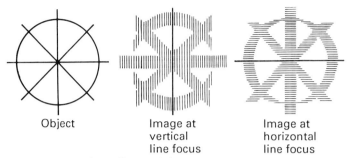

| Object | Image at vertical line focus | Image at horizontal line focus |

Figure 8.11 Image formed by a toric (astigmatic) lens

The nature of the image formed by a toric lens with its principal meridians horizontal and vertical is shown in Figure 8.11. It should be noticed that lines sharply defined in the image are perpendicular to the meridian focusing in the image plane. The form of the image produced by astigmatic beams is of considerable importance in the case of the astigmatic eye.

8.3 Aspheric surfaces

8.3.1 Approximate conicoids

The word *aspheric* strictly means 'not spherical' and therefore covers a very wide range of surfaces. In optics, however, its use is generally restricted to surfaces that are rotationally symmetrical about some axis. Non-rotational aspheric surfaces are described in Section 8.3.3. Aspheric surfaces that are rotationally symmetrical may be divided into various types and this section deals with those described mathematically within third-order theory as developed in Chapter 7. These surfaces approximate to conics which are described more fully in Section 8.3.2.

The study in Section 7.2 of the ways in which a wavefront could depart from the perfectly spherical was restricted to third-order theory and found five independent aberrations. One of these was oblique astigmatism and the wavefront surface was found to be toroidal (Section 7.4.1) in a similar way to the toroidal optical surfaces discussed in Section 8.1. The general equation of third-order theory (Equation 7.6), however, has only one term that is rotationally symmetrical about the axis. This is the first term used to describe spherical aberration. Because it has the expression $(y^2 + z^2)^2$ we can reduce this to the single variable ρ^4, which can rotate about the x-axis without any loss of accuracy.

Within the approximation called third-order theory, the equation of a sphere of radius r and passing through the origin is

$$x = \frac{\rho^2}{2r} + \frac{\rho^4}{8r^3} \tag{8.26}$$

To modify this in a rotationally symmetrical way and within third-order theory, the expression $A\rho^4$ is the only term available. We therefore have

$$x = \frac{\rho^2}{2r} + \frac{\rho^4}{8r^3} + A\rho^4 \tag{8.27}$$

where the first two terms on the right-hand side define a spherical surface (within third-order theory), and $A\rho^4$ is an aspherizing term.

The value of A now controls the amount by which we want to alter the spherical surface. This is easier to fit into the mathematics if we make $A = a/8r^3$. For any given surface, r is a constant so that we have merely to choose a different value for a than for A. If we now choose to make $a = -1$, we obtain

$$x = \frac{\rho^2}{2r} + \frac{\rho^4}{8r^3} - \frac{\rho^4}{8r^3} = \frac{\rho^2}{2r} \qquad (8.28)$$

which is the equation of a parabola or, because ρ is allowed to rotate about the x-axis, the equation of a paraboloidal surface. For all the sag calculations from Chapter 3 onwards, we have been approximating our spherical surfaces to paraboloids (in Equations 3.3a and 3.3b, for example). For other values of A we approximate to other types of conic surface – ellipsoids and hyperboloids, which are described more fully in Section 8.3.2. Unlike the spherical surface, all these aspheric surfaces have only one optical axis, that axis around which ρ has been rotated. Some indication of the extent of the aspheric effect can be seen from a simple numerical example. If r is regarded as a base spherical curve of, say, 84 mm radius (a +6 D surface in CR39 material) then, for $\rho = 10$ mm, the difference between a spherical surface and a parabola is given by

$$\frac{\rho^4}{8r^3} = \frac{10^4}{8 \times 84^3} = 0.002 \text{ mm} = 2 \text{ μm}$$

and at $\rho = 20$ mm this rises to 34 μm.

This can be seen in a different way using a contour map of the spherical base curve compared with a paraboloidal base curve as in Figure 8.12. The contours are circles because both surfaces are rotationally symmetrical, but when drawn at sag intervals of 1 mm it is seen that the paraboloidal surface generates contours of increasingly larger diameter than the sphere.

The most immediate effect of aspherizing a spherical surface in this way is on the amount of spherical aberration. In the case of a convex mirror the spherical aberration of Figure 8.13 renders the idea of a point image meaningless above

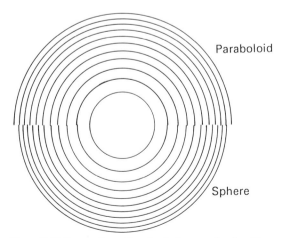

Paraboloid

Sphere

Figure 8.12 Comparison of a spherical curve of radius 84 mm with a paraboloid of the same central curvature. Equivalent to a 6 D base curve in CR39. Contours at 1 mm intervals

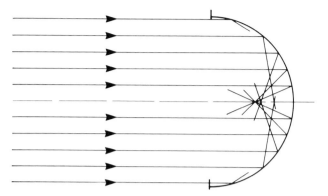

Figure 8.13 Spherical aberration of a spherical mirror

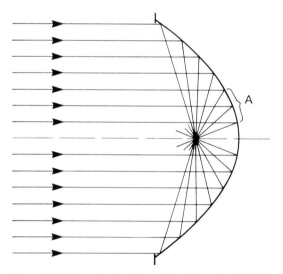

Figure 8.14 Zero spherical aberration of a paraboloid mirror

about $F/1$. All this aberration is completely corrected when the surface is made a paraboloid as shown in Figure 8.14. This is because the effective refractive index is -1 and equations similar to 7.18 but for a single spherical surface give a spherical aberration value equal to twice the difference between a sphere and a paraboloid. Equation 7.18 is subject, of course, to a conjugate factor, and a paraboloidal mirror is free of spherical aberration only for infinity conjugates. It is therefore used in telescopes and collimators. In the former its use is not a complete solution because it does suffer from coma and astigmatism and, in the absence of spherical aberration, the coma cannot be corrected by altering the stop position.

Off-axis paraboloidal mirrors are common in spectrophotometric and lens testing equipment. These use a portion such as A in Figure 8.14 of the mathematical surface which is offset from the axis. For an object on the axis at the focus, these mirrors provide an unobscured collimated beam which is not impeded by the object itself.

For a single thin lens one or other surface may be aspherized to correct third-order spherical aberration for a particular pair of conjugates. The aspherizing process can be seen as a thin shell of zero power which has a thickness t which varies according to the formula:

$$t = \frac{na\rho^4}{8r^3} \qquad (8.29)$$

where a is some constant and n is the refractive index of the shell. Such a shell will have its first surface sag given by

$$x_1 = \frac{\rho^2}{2r} + \frac{\rho^4}{8r^3}(1 + an) + \ldots \qquad (8.30)$$

and its second surface by

$$x_2 = \frac{\rho^2}{2r} + \frac{\rho^4}{8r^3} + \ldots \qquad (8.31)$$

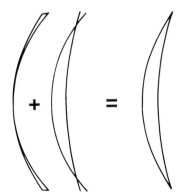

Figure 8.15 The aspheric 'shell' concept

Figure 8.15 shows the concept. Whatever the value of r, this thin shell has no basic power. The aspheric term gives it a thickness change that varies with ρ^4 and so introduces a third-order spherical aberration only. It has no *third-order* coma, astigmatism or distortion while the stop is at its surface. However, a thin lens may be chosen with a bending that gives zero coma for some specified conjugates. If it is then aspherized its spherical aberration will be corrected, giving an aplanatic system within the limits of third-order theory. Such a lens will not be corrected for chromatic aberration as described in Chapter 14 and is therefore of limited value compared with an achromatic doublet lens as described in Section 14.3.

With spectacle lenses the stop (at the centre of rotation of the eye) is displaced from the lens and the aspheric shell can be arranged to balance the oblique astigmatism of any meniscus lens by using its spherical aberration and Equation 7.53. A meniscus lens with two spherical surfaces can be given a bending that removes oblique astigmatism, but for lenses of high dioptric power the result becomes extremely meniscus and difficult to fit into a normal spectacle frame. In 1980, Jalie proposed a flatter series of lenses in which the residual oblique astigmatism was corrected by using a hyperboloidal surface for the front surface of

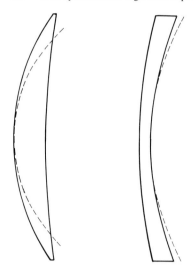

Figure 8.16 Aspheric spectacle lens with hyperboloid surfaces proposed by Jalie. The dotted curve indicates the equivalent spherical surface in each case

positive lenses and for the rear surface of negative lenses, with a resulting lens thickness as shown in Figure 8.16. Such a design improves the quality of vision and the cosmetic appearance of the lenses. Because the eyes can tolerate a level of astigmatism in the peripheral field some spectacle lens makers offer lenses in which the aspherizing is more than is needed optically. The residual astigmatism is now of the opposite sign but no larger than with a spherical lens of that bending. This gives a thinner lens of better appearance.

8.3.2 Accurate conicoids

The aspherizing described in the previous section was restricted to approximate conic aspherics using the single term:

$$\frac{a\rho^4}{8r^3}$$

and an analysis within third-order theory. If we now extend the polynimial expression beyond third-order theory we have an infinite number of terms for the sphere and an infinite number of terms for the aspherizing action:

$$
x = \frac{\rho^2}{2r} + \frac{\rho^4}{8r^3} + \frac{\rho^6}{16r^5} + \frac{5\rho^8}{128r^7} + \cdots
$$

for the sphere

$$
+ \frac{a\rho^4}{8r^3} + \frac{b\rho^6}{16r^5} + \frac{c\rho^8}{128r^4} + \cdots
$$

for the aspherizing action

(8.32)

Alternatively, we can use a general polynomial to describe the surface directly without reference at all to a sphere. This latter approach is needed for general aspherics as discussed in the next section. For surfaces only slightly different from a sphere, or which belong to the family of conic surfaces as described below, the 'sphere plus asphere' of Equation 8.27 is used.

The equation of a circle with the origin on the circle is

$$\rho^2 - 2rx + x^2 = 0 \tag{8.33}$$

and this has the polynomial form given in the first part of Equation 8.32. If Equation 8.33 is aspherized by including an extra variable p in front of the x^2 term we have

$$\rho^2 - 2rx + px^2 = 0 \tag{8.34}$$

This has the polynomial form:

$$x = \frac{\rho^2}{2r} + \frac{p\rho^4}{8r^3} + \frac{p^2\rho^6}{16r^5} + \frac{5p^3\rho^8}{128r^7} \cdots \tag{8.35}$$

or, if stated as a departure from a sphere:

$$x = \frac{\rho^2}{2r} + \frac{\rho^4}{8r^3} + \frac{\rho^6}{16r^5} + \frac{5\rho^8}{128r^7} + \cdots$$

$$+ \frac{(p-1)\rho^4}{8r^3} + \frac{(p^2-1)\rho^6}{16r^5} + \frac{5(p^3-1)\rho^8}{128r^7} + \cdots \tag{8.36}$$

Unlike the aspherizing expression in Equation 8.32, which has coefficients a, b, c, etc., which can take any value, Equation 8.36 has an aspherizing expression with a strictly controlled progression of coefficients, $p-1, p^2-1, p^3-1$, etc. If p is equal to 0, these terms cancel all the sphere terms except the first and we have a paraboloid. If p is positive (other than equal to 1, which gives a sphere) we have an ellipsoid, and if p is negative we have a hyperboloid. The cross-sections of these surfaces are shown in Figure 8.17.

The analytical equations of these surfaces are derived from Equation 8.34 by inserting the value of p. Using these formula it is possible to show that in reflection perfect imagery can be obtained for any pair of axial points using one of these surfaces. As shown in the last section, the paraboloid is needed when one of these points is at infinity. The ellipsoid is required for finite real conjugates, and the hyperboloid when one conjugate is virtual. The sphere belongs to this family because it provides perfect imagery when the axial object and image points are coincident. Figure 8.18 shows the sequence. In mathematical texts on two-dimensional coordinate geometry, the salient points of ellipses, parabolas and hyperbolae are referred to as foci and these are identical with the conjugates of these three-dimensional mirrors given in Figure 8.18.

With refraction the situation is more complicated. It is possible to define the required surface for perfect point imagery by calculating the location at which Snell's law will hold. Descartes (who defined Snell's law before Snell did!) calculated in this way and found that the surfaces were general aspherics and not conics. The term **Cartesian ovals** is applied to these. For the special case of reflection, these degenerate into the conic surfaces defined above. However, when the object point is at infinity these Cartesian ovals degenerate to conics even for refraction. When n' is greater than n an aberration-free point image is formed on axis by an ellipsoidal surface where one half of the surface refracts the light into the

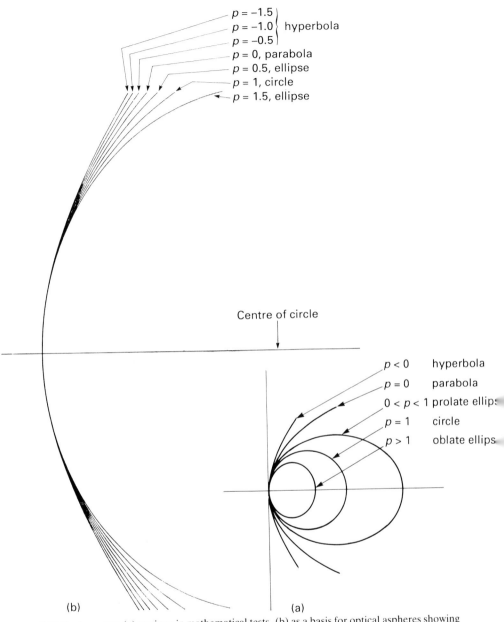

$p = -1.5$
$p = -1.0$ } hyperbola
$p = -0.5$
$p = 0$, parabola
$p = 0.5$, ellipse
$p = 1$, circle
$p = 1.5$, ellipse

Centre of circle

$p < 0$ hyperbola
$p = 0$ parabola
$0 < p < 1$ prolate ellips
$p = 1$ circle
$p > 1$ oblate ellips

(b)   (a)

Figure 8.17 Conic curves: (a) as given in mathematical tests, (b) as a basis for optical aspheres showing relatively small differences

focus of the other half. When n' is less than n, a hyperboloidal surface is needed, which refracts the light into the other focus. The form of each conic is dependent on the refractive indices but the general effect is shown in Figure 8.19.

Conics when properly calculated offer a method to obtain aberration-free images for these axial points. For off-axis points, image quality rapidly deteriorates owing

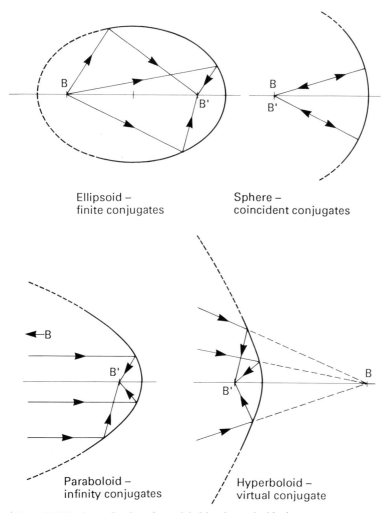

Ellipsoid –
finite conjugates

Sphere –
coincident conjugates

Paraboloid –
infinity conjugates

Hyperboloid –
virtual conjugate

Figure 8.18 Perfect reflection of an axial object by conicoid mirrors

to coma. Aspheric surfaces of these types have found application in condensing systems where elliptical reflectors are used to re-image sources as shown in Figure 8.20, and a pair of elliptical refracting surfaces, each at infinity conjugates, re-image a source at a finite distance as shown in Figure 8.21.

8.3.3 General

Section 8.3.2 shows that the restriction to the conicoid series of surfaces means that the coefficients of the polynomial terms are not independent of each other. If we now allow these to be completely independent as indicated by a, b, c, etc., in Equation 8.32, we no longer need to retain the basic sphere series at all. The general polynomial equation is:

$$x = \frac{\rho^2}{2r} + a\rho^4 + b\rho^6 + c\rho^8 + \ldots \tag{8.37}$$

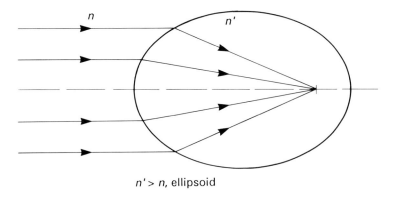

n' > n, ellipsoid

n > n', hyperboloid

Figure 8.19 Perfect refraction of a distant axial object by conicoid surfaces

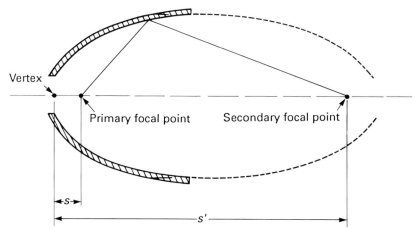

Vertex

Primary focal point   Secondary focal point

Figure 8.20 Ellipsoidal condensing reflector. For practical purposes the distances s and s' become more important than focal lengths (Adapted from Melles-Griot with permission)

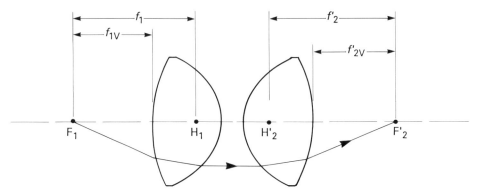

Figure 8.21 Aspheric condenser lenses. Facing surfaces are ellipsoidal. A source placed at F gives aberration-free collimated light after refraction by the first lens. The second lens does the same in reverse. For a different image size at F' aspheric lenses with different focal lengths are used while maintaining roughly collimated light between them

where r, a, b, c, etc., are all independent values, which can be made to fit any curve with a basic curvature r, although even this will not be apparent if a, b, c, etc., are large compared to $1/2r$. What is retained by Equation 8.37 is the rotational symmetry because ρ rotates about the x-axis.

A particularly straightforward example of a general rotationally symmetric aspheric is the Schmidt plate used to correct the spherical aberration of a spherical mirror. Because there is no restriction on the profile of the aspheric the correction is not restricted to third-order spherical aberration, all orders of the aberration being corrected. The principle is shown in Figure 8.22. Rather than design the system so that the corrected image is at the paraxial focus, a thinner corrector plate is required if an intermediate image location is chosen. This means that the rays at some value of ρ are undeviated by the plate while those above and below are deviated in opposite directions. This type of profile needs positive and negative values to the coefficients a, b, c, etc., in the polynomial. Figure 8.22 also indicates the plate shape needed (mainly a ρ^4 term for the third-order spherical aberration) if the system is corrected for the paraxial focus, but both profiles are considerably exaggerated. The value of this system is that a wide corrected field can be obtained because the plate is located at the centre of curvature of the mirror, which is also the stop of the system. However, the image has a curved field and the plate cannot correct all wavelengths of light to the same precision. Other methods of spherical mirror correction were invented after the Schmidt plate and these use spherical surfaces concentric to the centre of the mirror. The whole system is then concentric and has a very wide field of view. Quite often in optical development an aspheric is used in the early design and then found to be more expensive than using more (usually two) spherical surfaces.

Another approach to general aspherics uses a number of spherical surfaces which are then blended together. This has been used for moulded ophthalmic lenses in which the expense of aspheric surface manufacture is restricted to the mould. The manufacturing approach uses a number of spherical surfaces, usually four or five, which approximate to the required aspheric, usually a hyperbola, as shown in Figure 8.23. The intervening areas are removed by smoothing and polishing with a

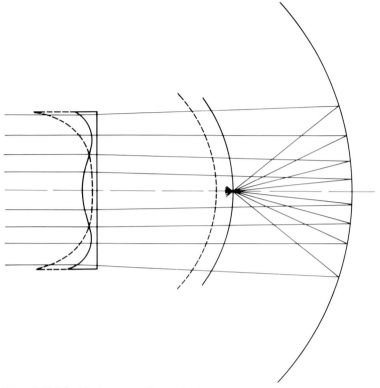

Figure 8.22 Schmidt plate correcting a spherical mirror. The plate shown corrects for an intermediate image position. For correction to the paraxial focus of the mirror (shown dotted) the thicker plate (shown dotted) would be needed. Both plates are shown with exaggerated thickness changes

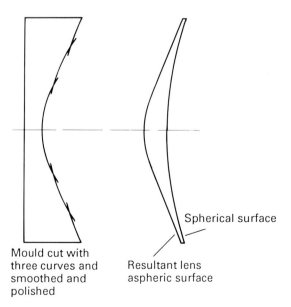

Spherical surface

Mould cut with
three curves and
smoothed and
polished

Resultant lens
aspheric surface

Figure 8.23 'Blended' lens manufacture

pliable tool which preferentially removes protruding areas. The optical quality of such a surface would be poor if the resultant (positive) lens were used at its full aperture. However, for ophthalmic lenses the quality in the peripheral field, although having some unevenness to its oblique astigmatism, generates a loss of visual acuity due to blur which is not greatly worse than that of a spherical lens of the same power. Such blended lenses are generally used at powers of about +10.00 to 14.00 D with a 4 or 5 D drop in power from the centre to the edge of the lens.

At even higher powers smooth lenses that make no attempt to correct the vision at wide field angles have been manufactured on the blended surface principle. Figure 8.24 shows the general form. In the region A the lens surface has become

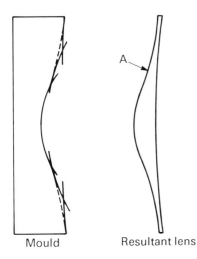

Mould   Resultant lens   **Figure 8.24** Blended aphakic lens

toroidal in the capstan form. Although there is still positive power in the sagittal meridian, the tangential meridian has negative power and considerable astigmatism is found. The blended lenses of Figures 8.23 and 8.24 were particularly useful for correcting **aphakia**, the condition of the eye after the crystalline lens has been removed because of cataract opacities. These days it is generally possible to insert an **intraocular lens** into the eye which compensates for the 16–22 D loss due to lens removal.

Although cataracts affect a small percentage of eyes, **presbyopia**, the ageing of the lens that reduces the ability to focus near objects, affects all eyes sooner or later. The use of segmented bifocal spectacles to compensate for this condition is described in Section 8.4. A blended solution involves the use of a progressive change in power between the upper part of the spectacle lens, for distance vision, and the lower part being more positive (or less negative if the overall lens correction is negative) for near vision such as reading. The optical problem can be succinctly stated using the contour diagrams discussed previously. In Figure 8.25 the upper part has a low positive power and the front curvature of the lens shows this in terms of widely spaced contours, while the lower portion needs more positive power indicated by the more closely spaced contours. In the **progressive lens** the need is to join up contours smoothly at each value without generating elliptical shapes, which were seen in Section 8.1 to denote toric surfaces with astigmatism.

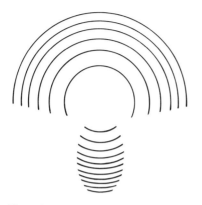 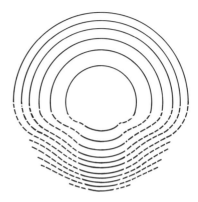

Figure 8.25 The progressive lens problem

The short answer is that it cannot be done. Whatever shape is chosen some measure of astigmatism and distortion must occur in the areas shown dotted in Figure 8.25. The greater the power change the worse these aberrations become. If, for the same power change, the 'blended' area is made larger, the aberrations become less, but then the useful areas of the lens are reduced. Various designs exist and the better designs are those that make the inevitable aberrations as smooth as possible and of a form to which the eye can adapt.

8.4 Segmented and Fresnel lenses

The commonest example of a segmented lens is the ubiquitous bifocal spectacle where a minor part or **segment** of the lens is given a different power by changing the curvature on the lens material. This same technique can be used with bifocal contact lenses, but less successfully. In general this method involves a sudden change in the slope of the lens surface, and sometimes an actual step. Because of the difference in power the deviation of light beams on each side of the segment edge is noticeably different and bifocal spectacle wearers have to become accustomed to a jump in the image. This can be minimized by making a **flat-top segment** so that the most used transition by the bifocal spectacle wearer from distance vision in the upper part of the lens to near reading vision in the segment, as shown in Figure 8.26, is close to the centre of the extra power of the reading segment. However, even in this case and when the segment boundary is an actual discontinuity in the lens surface, this image jump remains more noticeable than the segment edge. This is partly because the pupil of the eye is close to the lens and so the segment edge is well out of focus. When the segment has the same or nearly the same power as the rest of the lens the noticeability of the edge is very low.

The segmented lens for bifocal spectacles was invented in the eighteenth century probably by Benjamin Franklin and the image jump across the two lenses of his design (Figure 8.27) would be very low in the central region where the junction line passes through the optical centres of the two lenses.

The idea of segmented lenses having uniform power but much reduced bulk was suggested at about the same time, and developed later by Fresnel in 1819 for use in lighthouses. These lenses are now known as **Fresnel lenses**, but should not be confused with the Fresnel zone plate (Chapter 13) which utilizes diffractive

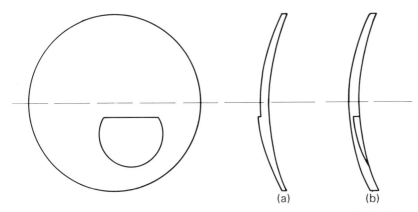

Figure 8.26 'Flat top' bifocal spectacle lenses for the right eye: (a) moulded, (b) fused

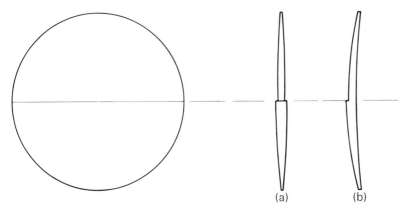

Figure 8.27 'Franklin' bifocal spectacle lenses: (a) as invented, (b) modern design

principles and has a very small discontinuity between the segments, of the order of the wavelength of light. The Fresnel lens is a refractive device in the same way as an ordinary lens. Figure 8.28 shows the basic concept by which the curved refracting surface of a plano-convex lens is allowed to collapse onto the plano surface while still retaining its localized curvature. The segments now become concentric zones and the narrower these are the thinner the lens can be. These lenses are usually made by cutting a master shape in metal using a diamond on a precision lathe and then moulding polymer replicas. The diamond must be rotated through a different angle for each zone and the lens can therefore simulate an aspheric curve without any more difficulty than a spheric curve. If the rings are close enough the curvature change across each zone can be ignored and flat grooves may be machined by a flat diamond. The technique spawns its own vocabulary and Figure 8.29 explains some of the terms used.

   At very high angles of refraction the relief surface intercepts the light and considerable scattering can occur. With very narrow groove widths diffractive effects scatter more light, as described in Chapter 13, because the adjacent grooves

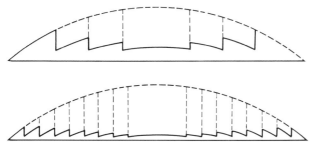

Figure 8.28 Fresnel lens concept

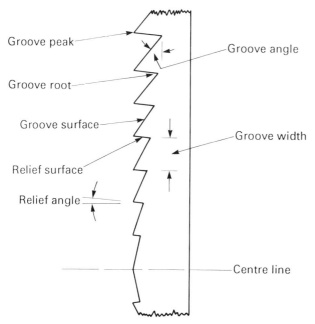

Figure 8.29 Terms and definitions for Fresnel lenses. Sometimes the word 'facet' is used in place of 'groove'

are incoherent with respect to each other. Nevertheless, the basic optics remains much the same as with ordinary lenses. In Figure 8.30, the refraction of the light at G is little different from what it would be if it had occurred at G', particularly when the thickness of the original lens is small compared with the object and image distances. In fact, first-order optics with these lenses remains the same because they are nearer to the 'thin' lens concept. The $L' = L + F$ equation applies, as do the equations for thin lens pairs.

With third-order aberrations some differences show up. In particular, there is an extra coma-like term, which interacts with field curvature so that the field curvature of a Fresnel lens is dependent on stop position and not fixed as indicated in Equation 7.53 for ordinary lenses.

Fresnel lenses are used in illumination systems in which their extreme apertures easily outweigh any light loss due to scattering. They are also used in some visual

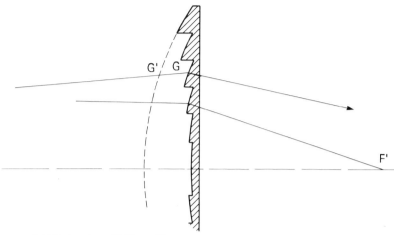

Figure 8.30 Refraction with Fresnel lenses

systems such as field lenses in vehicular rear-view systems and cameras, again in the viewing system. They have also been used as temporary stick-on lenses for vision correction where this is changing after an eye operation, particularly in the form of a prism.

Exercises

8.1   A thin lens has cylindrical surfaces of equal curvature, the axes of the cylinders being at right angles. Show that the effect of the lens is the same as that of a spherical lens with a curvature equal to either of the cylinders.

8.2   (a)   A point source is placed 0.5 m from a cylindrical lens of +5 D power with cylinder axis vertical and 40 mm diameter. Find the position, length and direction of the line focus.

   (b)   If a +3 D spherical lens is placed in contact with the cylindrical lens, what will be the positions and lengths of the line foci and the position and diameter of the circle of least confusion?

8.3   Two thin cylindrical lenses, one of power +5 D cylinder axis vertical and the other +4 D cylinder axis horizontal and 45 mm diameter, are placed in contact 0.5 m from a point source of light. Find the dimensions of the patch of light on a screen parallel to the plane of the lenses and (a) 0.2 m, (b) 0.4 m, (c) 0.6 m from the lenses.

8.4   A sphero-cylindrical lens has powers +5.0 D spherical/−2.0 D cylindrical, axis vertical. It is required to make a lens equivalent to this with one surface plane; what powers must be worked on the other surface? What type of surface is this? Give a diagram.

8.5   A lens made of glass of refractive index 1.523 has a toric surface the powers of which are +10 D in the vertical meridian and +6 D in the horizontal; the other surface is plane. Find the radii of curvature of the surfaces when a lens of the same power is made up in the crossed cylinder form.

8.6   A polished metal cylinder 2 in in diameter stands on a table with its axis vertical and at a distance of 10 ft from a window 6 ft broad. What would be the breadth of the image seen by reflection from the cylinder?

8.7   (a) A Fresnel lens made from PMMA ($n = 1.49$) has a facet width which has a
            constant value of 1 mm. If the lens is 40 mm in diameter with a focal length of
            100 mm calculate (using the thin prism equation) the facet angle for the 15th and
            20th facets assuming that the second surface is plano. What is the depth (peak to
            root distance, Figure 8.29) of the 16th and 20th facets?
       (b) If the facets now have a variable width given by the equation $y_n - y_{n-1} = \sqrt{20} \, (\sqrt{n}$
            $- \sqrt{(n - 1)})$ where n is the number of the facet counting the central facet as $y_1 -$
            y_0, the next as $y_2 - y_1$, etc., it will be found that again 20 facets just fit inside the
            40 mm diameter lens. Calculate again the facet angle and depth for the 16th and
            20th facets. Comment on the results.

Chapter 9

Light sources and the nature of light

So far we have considered that light can be represented by straight lines drawn on paper. Indeed this is why the name **geometrical optics** is applied to that branch of the subject. This description is adequate where we are concerned with the way light reacts with prisms, lenses and mirrors, but is no longer sufficient when we wish to understand how light reacts with itself or when it meets optical devices with very fine structure.

In these circumstances a more complete description is required, which includes the wave motion described in Chapters 1 and 2 and incorporates the way in which this wave motion is produced and propagated. This branch of the subject is called **physical optics**.

9.1 The dual nature of light

Our understanding of the propagation of light began some centuries earlier than our understanding of how it is produced, and the same order will be followed here. Very early experiments with burning-glasses showed light to be a form of energy. In the seventeenth century **Römer's** observations on the moons of Jupiter revealed that the times between their eclipses by the planet steadily increased and decreased as the earth orbited the sun. Calculations showed that these changes could be accounted for by the different distances the light had to travel if it had a velocity of about 190 000 miles per second.

This finite velocity means that the energy has to travel in some sort of self-contained packet across the intervening distance. The possible ways in which this might occur reduce to two. In the first case energy may be propagated as moving matter as in the case of energy received when a ship is hit by a projectile; or it may be propagated as a wave motion travelling through a continuous medium which does not move as a whole, as in the case of the movement of a ship caused by water movement arising at a distant disturbance.

Each of these methods has formed the basis of a theory of light and indeed the modern concept of radiant energy propagation is based on a duality that utilizes both ideas. A full description of these concepts is beyond the scope of this book and would be more than is needed to understand a large part of physical optics. The essential idea that light is propagated in discrete packets of energy whose actions *can be described* by a wave motion will be developed as required.

Isaac Newton considered the evidence of rectilinear propagation and the laws of reflection and refraction, and championed the **corpuscular theory**, which assumed that light consisted of exceedingly minute particles shot out by the source. Refraction was explained by assuming that the particles travelled faster in the denser medium. Although this sounds inherently unlikely, a greater problem was the partial reflection and partial refraction that was found to take place on glass–air surfaces and similar boundaries.

It would seem that a particle is either reflected or refracted and, if so, why should identical particles be treated differently? Newton was forced to assume that the boundaries between media were subject to 'fits of easy reflection and easy refraction', which were in turn determined by vibrations set up by the particles.

In this last hypothesis, Newton came surprisingly close to the modern concepts of the dual nature of light. However, the assumptions regarding the speed of the particles in the denser medium were the first to be challenged. Measurements made by Foucault and Michelson in the nineteenth century gave contrary results and the **wave theory** became totally accepted until the present century.

The original wave theory was propounded somewhat vaguely by both Grimaldi and Hooke, but to **Huygens** in 1690 is accorded the honour of the first written explanation of the principle described in Chapters 1 and 12. This was not altogether accepted, and its opponents maintained that waves on water and sound waves required a medium to carry them whereas light could pass through a vacuum. Huygens assumed the existence of a medium known as the *ether*, filling all space. All experiments have failed to detect the nature or even the existence of such a medium and modern theory regards it as a mathematical concept rather than an actuality.

A more pertinent objection to the wave theory was that waves could bend round objects while light gives sharp shadows and appears to travel in straight lines. It was left to the work of Young and Fresnel to show that light did bend round objects but that its vibrations were of such short wavelength that the effect could be seen only in special experiments. This will be discussed more fully later in this and in succeeding chapters. An added virtue of the wave theory was that if transverse vibration was assumed, instead of longitudinal as in sound, the phenomenon of polarization (Chapter 11) could be explained.

These successes ensured the acceptance of the wave theory of light throughout the whole of the last century, and **Maxwell** was able to integrate these ideas with those of electricity and magnetism. The observed actions of reflection and refraction, interference and diffraction, and polarization could all be explained.

In retrospect, the undermining of this theory began in the early nineteenth century when Fraunhofer discovered a dark line across the spectrum of the radiation from the sun. Later, when scientists applied Maxwell's equations to the radiation from hot bodies, no good agreement was found with the experimental facts.

After the discovery of the electron as the fundamental particle of electricity and the ability of light to eject these from metals (photoelectricity), further problems were seen to arise. It was found by experiment that although the numbers of electrons ejected were proportional to the intensity of the light, the energy of the individual electrons was only proportional to the frequency of vibration of the light. In other types of vibrating energy the energy flow is proportional to the amplitude, and it was difficult to see why the energy of the ejected electrons did not vary with this. The solution proposed by **Planck** depends on the idea of **harmonic oscillators**, which vibrate at a preferred frequency determined by their structure, their binding forces and boundary conditions. If their resonant frequency is v_0, they can emit and absorb energy at frequencies, v_0, $2v_0$, $3v_0$, etc., but at no frequencies less than v_0. These theories of Planck regarding the emission and absorption of radiant energy not as a continuous action but in discrete packets or **quanta** and Einstein's concept of **photons** carrying the concentrated energy of the light surrounded by a wave motion of finite length were found to be viable.

Further deductions from Einstein's theories led to the idea of stimulated emission, which culminated in the invention of the **laser** in 1960. Thus the history of man's ideas on the nature of light is a story of change and modification as new discoveries were made. We have no reason to suppose that the present concepts represent an ultimate description, and much of today's cosmological theories of creation, time and the universe demand greater and greater insight into the nature of light. At a rather lower level the next sections deal with the basic mathematics of simple harmonic motion followed by a longer review of Maxwell's electromagnetic theory and Planck's quantum theory leading to lasers, the most fundamental discovery/invention in twentieth-century optics.

9.2 Wave motion, simple harmonic motion

Because the original concept of a continuous wave motion goes so far in explaining the basic phenomena of interference and diffraction, it is necessary to study the fundamental characteristics of a simple wave motion. In any case such a study is a prerequisite to understanding the more complex theories to follow. If we observe the ripples on the surface of a pond, two things are at once evident; first, that any floating object executes an up-and-down motion without appreciably altering its

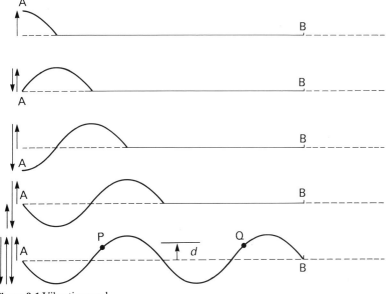

Figure 9.1 Vibrating cord

general position, and, secondly, that crests and troughs, places of maximum displacement of the surface above and below its normal position, are repeated at regular intervals. It is obvious that there is no general forward movement of the water as the ripples or waves continue to travel out from a disturbance.

Let AB in Figure 9.1 represent part of a flexible string. If a regular up-and-down motion is given to the end A, each particle will communicate its movement to the next particle in turn, and a wave will travel along the cord. The form taken up by the string after the start of the motion at A is shown for five particular times. Two particles such as P and Q, having the same displacement, d, and direction of motion, are said to be in the same **phase**, and the distance between two successive particles in the same phase is a **wavelength**. If the end A moves in such a manner that it retraces its path in regular intervals of time, its motion is said to be periodic, and the time between successive passages in the same direction through any point is termed the **period** or **periodic time** of the motion. From Figure 9.1 it will be evident that during the time that A makes one complete up-and-down vibration, the disturbance has travelled a distance equal to one wavelength; during two complete vibrations it will have travelled a distance of two wavelengths, and so on. Then if λ = wavelength, V = velocity at which disturbance travels, and T = period, we have

$$\lambda = VT \tag{9.1}$$

which is the fundamental equation of wave motion.

The number of vibrations taking place per second is termed the **frequency**, v, of the vibration, and

$$\lambda = \frac{V}{v} \tag{9.2}$$

Each vibration is commonly called a **cycle**, being that movement before the motion repeats itself. Frequency is therefore measured in cycles per second and this unit has now been given the name **hertz** (Hz), after the German physicist who showed that light and radiated heat are electromagnetic waves.

The maximum displacement from its normal position of a particle, i.e. when it is at the crest or trough of a wave, is termed the **amplitude**, a, of the wave.

The form of the wave will depend on the nature of the vibration of the individual particles, and each particle will be vibrating in the same way according to the periodic motion imparted by the source, such as the end of the cord in Figure 9.1. The motions of the individual particles in a wave motion may be very complex, but it can be shown that any complex vibration can be resolved into a number of simple to-and-fro motions of the type known as **simple harmonic motion** (SHM).

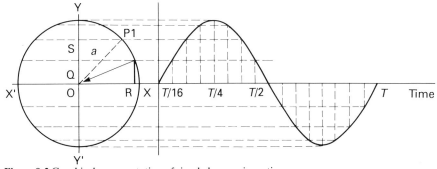

Figure 9.2 Graphical representation of simple harmonic motion

If the point P (Figure 9.2) moves around a circle with uniform velocity, the to-and-fro motion of the projection of P on *any* diameter is termed a simple harmonic motion; thus the point R moves between XOX' and the point Q between YOY'. The maximum displacement of Q and R is equal to a, the radius of the circular path of P. This is the **amplitude** of the wave form generated if Q is plotted against time as shown. The **phase** is the angle turned by OP from some arbitrary fixed point such as X. If y is the displacement of Q from O a time t after P was at X, then

$$y = a \sin \text{POX}$$

or

$$y = a \sin \omega t = a \sin \frac{2\pi t}{T}$$

where $\omega/2\pi$ is the frequency (ν), the number of cycles P completes in a second, or where T is the time for P to complete a cycle. Each revolution of P completes one cycle of simple harmonic motion. If t is measured from some other moment such as P at Y then we would have

$$y = a \cos \omega t = a \sin (\omega t + 90°)$$

or, more generally,

$$y = a \sin (\omega t + \delta) \tag{9.3}$$

where δ is the initial phase between the start location of P and some arbitrary location.

A simple harmonic motion is characteristic of the vibration in a straight line of a particle acted on by a restoring force linearly proportional to the displacement of the particle from the centre of attraction. An example is seen in the motion of a weight suspended from the end of a spring. Many of the phenomena of light, especially those of interference, diffraction and polarization, may be explained by considering the resultant effect as if a particle were being acted on simultaneously by two or more simple harmonic motions. The effects due to each motion at successive values of t are calculated and the resultant at each value of t is the algebraic sum of the separate effects at the corresponding times. This is the **principle of superposition** and has been shown to be true where the effect is small compared with the energies and separation distances of the particles affected. This proviso ensures the linearity of the restoring force described above. Different effects can occur with high-powered lasers where the energies are large, and in materials in which the binding forces are complex.

The resultant of a number of motions acting in the same straight line may be solved graphically by plotting each motion to the same scale and with the same coordinates, in the way shown in Figure 9.2. The curve of the resultant displacement is then the sum of the separate displacement curves. An example is shown in Figure 9.3, where the broken line is the curve of the resultant displacement due in two motions,

$$y_1 = a \sin \frac{2\pi t}{4}$$

and

$$y_2 = b \sin \frac{2\pi t}{8}$$

executed in the same vertical line.

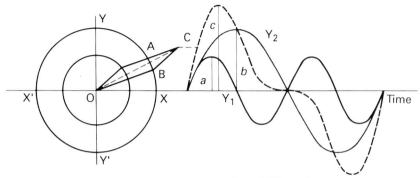

Figure 9.3 Composition of two simple harmonic motions of different frequencies

At any moment in time the displacement of the resultant can be found by drawing the lines OA and OB at the correct phase angles and constructing, by the method of the parallelogram of forces, the vector addition to obtain the resultant OC. As the vectors OA and OB rotate at different frequencies in this example, OC has a varying length which when projected on to YY′ produces the broken line wave motion.

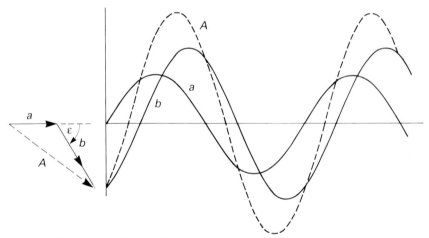

Figure 9.4 Composition of two simple harmonic motions of the same frequency

If the two motions are of the same frequency, the phase difference between them must remain constant. In Figure 9.4, the two motions have different amplitudes, a and b, but the resultant found by the parallelogram of forces has a constant amplitude, A, because the angle between a and b remains the same. Because a, b and A all rotate together it is easy to see that the frequency of the resultant is the same as that of the two composing motions, but it has its own phase angle.

From the triangle the resultant amplitude A is given by:

$$A^2 = a^2 + b^2 + 2ab \cos \epsilon \tag{9.4}$$

where ϵ is the phase difference between a and b. Note that A can be zero if a equals b and ϵ equals $180°$. It often occurs in optics that there are two or more motions to

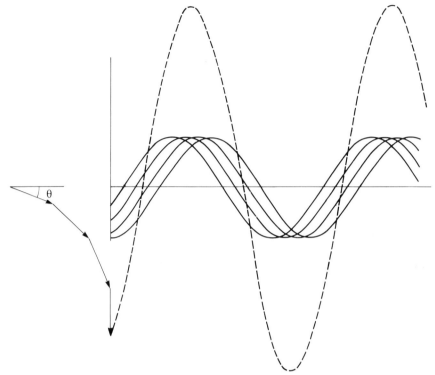

Figure 9.5 Composition of four simple harmonic motions of equal frequency and amplitude but progressively out of phase

be summed of equal or nearly equal amplitudes. Furthermore, the phase differences between them are often equal. Figure 9.5 shows four such motions, and the broken line representing the resultant emphasizes the large amplitude this can achieve. It may be shown, however, that the resultant is zero if the phase differences add up to 360°.

Although Figures 9.4 and 9.5 show only two and four motions of the same frequency, it is a general rule that any number of motions of the same frequency will yield a resultant of that frequency.

Figure 9.4 shows the case when the two motions are of equal frequency but differ in phase. We have

$$y_1 + y_2 = a \sin(\omega t + \gamma) + b \sin(\omega t + \delta) \tag{9.5}$$

where a and b are the amplitudes of the constituent motions and γ and δ are their phases. Expanding Equation 9.5 (using $\sin(A + B) = \sin A \cos B + \cos A \sin B$ twice) gives

$$a \sin \omega t \cos \gamma + a \cos \omega t \sin \gamma + b \sin \omega t \cos \delta + b \cos \omega t \sin \delta$$

$$= \sin \omega t \, (a \cos \gamma + b \cos \delta) + \cos \omega t \, (a \sin \gamma + b \sin \delta)$$

which may be set equal to

$$\sin \omega t \, (A \cos \epsilon) + \cos \omega t \, (A \sin \epsilon) = A \sin(\omega t + \epsilon) \tag{9.6}$$

which is a motion of the same frequency but different amplitude and phase.

We find A from

$$A^2 = (A \cos \epsilon)^2 + (A \sin \epsilon)^2$$
$$= (a \cos \gamma + b \cos \delta)^2 + (a \sin \gamma + b \sin \delta)^2$$
$$= a^2 + b^2 + 2ab \cos (\gamma - \delta) \tag{9.7}$$

and ϵ from

$$\tan \epsilon = \frac{a \sin \gamma + b \sin \delta}{a \cos \gamma + b \cos \delta} \tag{9.8}$$

When more than two motions are summed of equal amplitudes but consistently differing phases as often occurs in interference we have, for example,

$$a \sin (\omega t + \epsilon) + a \sin (\omega t + 2\epsilon) + a \sin (\omega t + 3\epsilon) + a \sin (\omega t + 4\epsilon)$$

$$= a(2 + 2 \cos \epsilon)^{1/2} \sin \left(\omega t + \frac{3\epsilon}{2} \right) + a(2 + 2 \cos \epsilon)^{1/2} \sin \left(\omega t + \frac{7\epsilon}{2} \right)$$

$$= a(2 + 2 \cos \epsilon)^{1/2} (2 + 2 \cos 2\epsilon)^{1/2} \sin \left(\omega t + \frac{5\epsilon}{2} \right) \tag{9.9}$$

This resultant is a motion of the same frequency as the component motions but whose amplitude and phase depend on ϵ. The amplitude is 4 when ϵ is zero and zero when ϵ is a multiple of $\pi/4$. The resultant for ϵ equal to $\pi/8$ is shown in Figure 9.5. If the value of ϵ is varied, the amplitude of the resultant varies between $4a$ and zero.

The resultant waveform of Figure 9.3 occurs when OA and OB produce the simple harmonic motion by projection on to the same vertical line YY'. If the projections on to XX' were used a similar waveform would result. If, however, we combine OA projected on to YY' and OB projected on to XX' a more complex result occurs. If the analysis is restricted to cases where the frequency of the two contributing motions is the same a result is obtained of great importance in the study of polarized light.

Suppose the two contributing waveforms are represented by

$$x = 11 \sin \left(\frac{2\pi t}{T} + \frac{\pi}{4} \right) \quad \text{and} \quad y = 7 \sin \left(\frac{2\pi t}{T} \right)$$

which are executed along XOX' and YOY' respectively of Figure 9.6. If two circles are drawn with radii proportional to 11 and 7 they can be divided into equal parts representing equal time intervals. As the two motions are of the same frequency the number of divisions is the same but the difference in phase is represented by the different starting points. Remembering that the resultant is derived from the components of the rotating vectors on *different* axes, these must be taken *before* the parallelogram of forces is drawn. The result of these at any instant is found to lie on the ellipse shown. In fact, the result of any two orthogonal SHMs of equal frequencies is an ellipse except for the special case where the phase difference is zero or π, when the result is a straight line. A further special case occurs when the phase difference is an odd multiple of $\pi/2$ and the amplitudes are also equal. These conditions result in a circular motion.

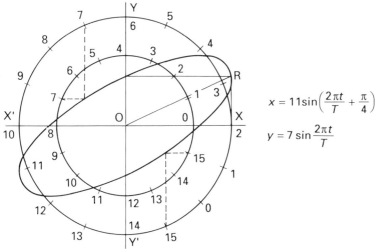

Figure 9.6 Two simple harmonic motions executed in directions at right angles

$$x = 11\sin\left(\frac{2\pi t}{T} + \frac{\pi}{4}\right)$$

$$y = 7\sin\frac{2\pi t}{T}$$

In more general terms we have

$$y = b \sin \omega t$$

$$x = a \sin (\omega t + \delta)$$

$$= a \sin \omega t \cos \delta + a \cos \omega t \sin \delta$$

$$\frac{x}{a} = \frac{y}{b} \cos \delta + \left(1 - \frac{y^2}{b^2}\right)^{1/2} \sin \delta$$

$$\left(\frac{x}{a} - \frac{y}{b} \cos \delta\right)^2 = \left(1 - \frac{y^2}{b^2}\right) \sin^2 \delta$$

$$\frac{x^2}{a^2} - \frac{2xy}{ab} \cos \delta + \frac{y^2}{b^2} \cos^2 \delta = \sin^2 \delta - \frac{y^2}{b^2} \sin^2 \delta$$

Therefore

$$\frac{x^2}{a^2} + \frac{y^2}{b^2} - \frac{2xy}{ab} \cos \delta = \sin^2 \delta \tag{9.10}$$

This is the equation of an ellipse the principal axes of which, like that in Figure 9.6, are not aligned with the x and y axes, but contained in the rectangle of sides $2a$ and $2b$ as shown in Figure 9.7. If δ is made equal to $\pi/2$ the principal axes do align and the familiar equation is obtained:

$$\frac{x^2}{a^2} + \frac{y^2}{b^2} = 1 \tag{9.11}$$

This becomes a circle when $a = b$, while Equation 9.10 does not, remaining an ellipse with its principal axis at 45° to the x and y axes. Thus for circular motion we need equal amplitudes, $a = b$ and $\delta = \pi/2$ or some multiple of this.

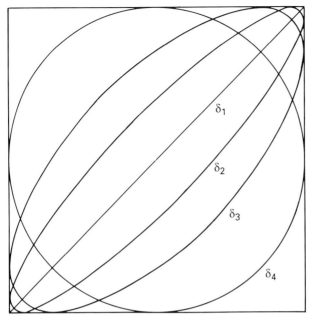

Figure 9.7 Composition at right angles of equal-amplitude, equal-frequency simple harmonic motions with phase difference. $\delta_1 = 0°$ (or 180°, 360°, etc.), $\delta_2 = 22.5°$ (or 157.5°, 202.5°, etc.), $\delta_3 = 45°$ (or 135°, 225°, etc.), $\delta_4 = 90°$ (or 270°, 450°, etc.)

On the other hand, when $\delta = 0$ or some multiple of π we have, from Equation 9.10,

$$\frac{x^2}{a^2} + \frac{y^2}{b^2} - \frac{2xy}{ab} = 0$$

that is,

$$\left(\frac{x}{a} - \frac{y}{b}\right)^2 = 0 \qquad \text{or} \qquad y = \frac{b}{a}x$$

which is the equation of a straight line for all values of a and b. The resultants for equal a and b but different phase differences are shown in Figure 9.7.

9.3 Wave trains, complex waveforms – superposition and analysis

It was seen in Section 9.2 that the form of a wave will depend on the motion imparted to the particles by the source. If this motion is simple harmonic, the simultaneous displacement of all the particles may be found using Equation 9.3. This applies in the case where particles are in actual motion and in the case where 'motion' is really that of a varying electromagnetic field.

Let us suppose that a source A (Figure 9.8) is executing a simple harmonic motion given by the expression

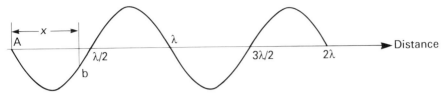

Figure 9.8 The form of a wave motion: a wave train

$$y = a \sin \frac{2\pi t}{T}$$

and that this is transmitted from one location to the next with a velocity V. Then a location b at a distance x from A commences its motion x/V seconds after that at A and its motion is therefore given by

$$y = a \sin \frac{2\pi}{T} \left(t - \frac{x}{V} \right) \tag{9.12}$$

As the wavelength λ is given by $\lambda = VT$, we have

$$y = a \sin 2\pi \left(\frac{t}{T} - \frac{x}{\lambda} \right) \tag{9.13}$$

This equation gives the complete form of any simple wave motion (provided $y = 0$ when x and t are zero), for by giving a constant value to x we have the equation of the motion at this point, and by giving any constant value to T the curve obtained gives the form of the train of waves at that instant. The curve so obtained (Figure 9.8) should not be confused with that in Figure 9.2, which shows the displacement at a *single* location for various values of T.

When two or more waves are superposed, the resultant displacement at any point may be found from the separate displacements by the methods of Section 9.2. As the resultant displacements will still be periodic, a new wave is formed. If the waves are travelling with equal velocities in the same direction and the vibrations are taking place in the same plane, the resultant waveform may be found graphically. Figure 9.9 shows the resultant waveform (full line) when two waves (broken lines) are superposed. The two waves of the same amplitude are travelling in the same direction with equal velocities and one has twice the wavelength of the other. Any

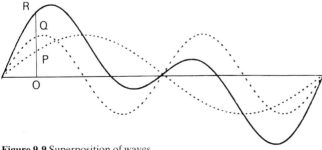

Figure 9.9 Superposition of waves

point on the resultant wave has a displacement OR equal to the algebraic sum of the separate displacements.

This process may be repeated for any number of simple waves and the most complex waveforms may be built up. By the same reasoning, the most complex waveforms can be expressed as a sum of a number of simple waves of simple harmonic form. The mathematics of this was developed by Fourier (1768–1830) and the full theory is beyond the scope of this book. A superficial description is given at the end of this section and the principle is readily understandable. It is applied to optical phenomena in Chapters 12 and 13.

The difference in phase at any instant between two locations in a simple waveform may be found from Equation 9.13. Thus, if x_1 and x_2 are the distances of the two locations from the source, the displacement of each is

$$y_1 = a \sin 2\pi \left(\frac{t}{T} - \frac{x_1}{\lambda} \right)$$

and

$$y_2 = a \sin 2\pi \left(\frac{t}{T} - \frac{x_2}{\lambda} \right)$$

Therefore the difference in phase at any instant

$$\epsilon = 2\pi \left(\frac{t}{T} - \frac{x_1}{\lambda} \right) - 2\pi \left(\frac{t}{T} - \frac{x_2}{\lambda} \right)$$

$$= \frac{2\pi}{\lambda} (x_2 - x_1)$$

$$= \frac{2\pi}{\lambda} \quad \text{(path difference)} \tag{9.14}$$

This will also express the difference in phase between two disturbances arriving at a single location from two sources at distances x_1 and x_2, provided the two sources are producing exactly similar motions. Expression 9.14 is of great importance in the phenomena of interference (Chapter 12).

Sections 9.2 and 9.3 have analysed the wave motion as a one-dimensional vibration. Chapter 1 loosely considered light as a wavefront moving out from a point source and its propagation in terms of wavelets generated by each point on the wave. The wave itself had travelled from the source in a given time and this time is the same for all parts of the wavefront. If we isolate different points on the wavefront they will all be in the same phase and the wavelets generated will be in phase. Thus, if exactly similar vibrations are travelling out uniformly in all directions in a homogeneous medium from the point source B (Figure 9.10), any sphere DAE having its centre at B will be a surface of equal phase generating wavelets of equal phase. The only way to obtain wavelets of different phase is to take them from points at different distances from B. Thus if the wave motion at J and K were redirected to F using, in the case of light, two semi-reflecting mirrors, for instance, the effect at F would be two motions of the same frequency but with a phase difference given by Equation 9.14.

A simplistic description of the synthesis and analysis of complex waveforms initiated by Fourier makes use of complementary pairs of graphs. In Figure 9.11,

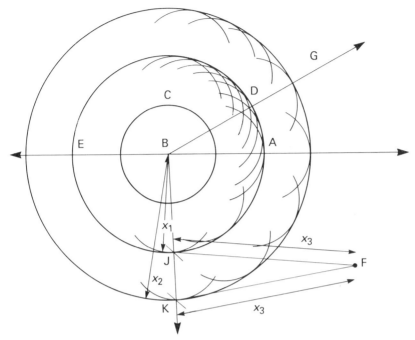

Figure 9.10 Light of differing phase superposed at F

the graphs on the left show the actual waveform plotted in either space or time. It does not matter to the analysis whether this represents height of a wave motion on water, pressure of a soundwave, or, in our case, amplitude of an electromagnetic field. The right graph of each pair shows the amounts of the different frequencies contained in the waveform. These graphs simplify the phase differences of the constituent frequencies so that in-phase frequencies are plotted to the right of zero frequency and out-of-phase frequencies are plotted to the left of zero frequency.

In Figure 9.11(a) the waveform is a sine wave, which is assumed to stretch from minus infinity to plus infinity. The complementary frequency plot of this shows just one single value where $v = V/\lambda$ in the case of a light wave. The amplitude, a, is the same in both cases. (Sometimes, for mathematical completeness, the negative out-of-phase component is also shown at $-v$.) Figure 9.11(b) shows the two-component waveform of Figure 9.3. Again the waveform is assumed to reach from minus infinity to plus infinity. Two component frequencies are now shown on the right-hand graph.

It is possible to create a perfect square wave by superposing sine waves of differing frequencies. However, this requires an infinite number of components at ×3, ×5, ×7, and higher frequencies although their amplitudes are diminishing as shown in Figure 9.11(c).

Similar complications set in when the sine wave is of restricted duration. In Figure 9.11(d) a single-frequency sine wave of only five cycles is shown. The frequency components not only now reach from minus infinity to plus infinity, but they are infinitely packed together so that only the envelope of their amplitudes can be drawn.

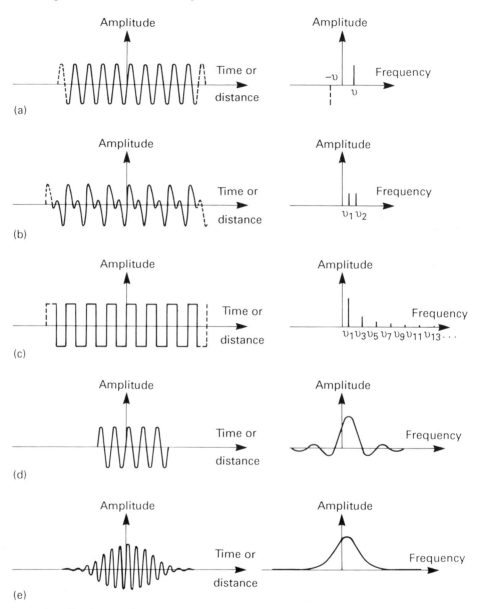

Figure 9.11 Waveforms and their frequency spectra (Fourier transform). See text

Of considerable interest to optical science is the wave train, which is largely a single frequency but with an amplitude that gradually increases and then reduces to form a packet or group of waves. This has frequency components, which are also packed closely together, but have a well restricted smooth envelope as shown in Figure 9.11(e). When the frequency spectrum of light emitted from specific materials is examined using the methods of Chapter 11 this shape of envelope is found and leads to the quantum theory described in Section 9.4.

By considering Figure 9.2 again, it can be seen that a particle in simple harmonic motion has, on passing through its mean position, its maximum velocity $2\pi a/T$, which is the velocity of the point P. Its velocity then diminishes as it recedes from its mean position, becoming zero as it reaches its maximum displacement. The energy of the particle, being all kinetic at the moment of maximum velocity, is given by $\frac{1}{2}mv^2$ where m is the mass and v the velocity. Thus energy is proportional to $(2\pi a/T)^2$ or, more importantly, to the amplitude squared, a^2. This again applies where the 'motion' is really that of a varying electromagnetic field. As such, it is found that I, the intensity of the light, represented by the wave motion, is proportional to the amplitude squared. This result is important in interference and diffraction.

9.4 Electromagnetic theory to quantum theory

For many years after the wave theory of light had become generally accepted it was supposed that the vibrating atoms of a luminous body set up waves in an elastic-solid medium that filled all space, and most of the more important phenomena of light have been satisfactorily explained on this theory.

In 1873, however, Clerk-Maxwell (1831–1879) put forward a new form of the wave theory in which light is considered to consist of alternating electromagnetic waves. This **electromagnetic theory**, which conceives an alternating condition of the medium in place of an oscillation of position of its particles as in the elastic-solid theory, satisfactorily explains most of the known facts about light and overcomes the difficulties met with in the older theory. Maxwell showed mathematically that the electromagnetic vibration would be transverse to its direction of travel and would travel in free space with a velocity of 3×10^8 m/s, which, as we have seen, agrees with the measured velocity of light.

About 1886 Hertz (1857–1894) succeeded in producing electromagnetic waves by purely electrical means, and this was a most important discovery in confirmation of Maxwell's theory. These waves, which are those now used for radar, television and radio, have many of the properties of light waves, but their wavelengths extend to many thousand times those of light. The discovery in 1895 of the Röntgen or X-rays by Röntgen (1845–1923) and later of the gamma rays emitted by radioactive substances, gave us another series of electromagnetic waves with the same characteristics as light waves, but of very much shorter wavelength.

Later researches have succeeded in filling the gaps that existed between the various groups of radiation and we now have a complete range of radiations, differing only in frequency, and therefore in wavelength, extending from the longest electrical waves of some thousands of metres wavelength, as used in radio, through the infrared or heat radiations, to the extremely small portion of the complete range, the visible spectrum, which we know as light, where the wavelengths vary from 7×10^{-4} mm in the red to 4×10^{-4} mm in the violet. Beyond the violet we have the still shorter wavelengths, the ultraviolet radiations, with wavelengths extending from that of the extreme violet down to about 14×10^{-6} mm, and beyond these again the X-rays, the wavelengths of which extend down to 6×10^{-9} mm. In the gamma rays the shortest wavelength found is about 6×10^{-10} mm. More recent work has shown the existence of radiations of very great penetrating power and having wavelengths as short as 8×10^{-12} mm. To these radiations the name cosmic rays has been given.

The various regions of this enormous range of radiations have been shown to possess many properties in common; they all travel at the same speed, and all give rise to certain phenomena, such as interference and diffraction. The different regions are, of course, produced by different sources and detected in different ways. The known range of electromagnetic waves is shown diagrammatically in Figure 9.12.

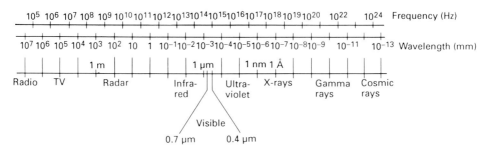

Figure 9.12 The electromagnetic spectrum

The wavelengths of light and other short-wavelength radiations are usually expressed in terms of one of the following units:

1. The **micrometre** (μm) which is equal to one-thousandth of a millimetre, 10^{-6} m (formerly called a micron).
2. The **nanometre** (nm), which is equal to 10^{-9} m or one-millionth of a millimetre (10 Å).
3. The **angstrom** (Å), which is equal to 10^{-10} m or a ten-millionth of a millimetre.

Thus the wavelength of the yellow light of a sodium lamp is 5890 Å = 589 nm = 0.589 μm = 0.000 589 mm. We shall adopt the nanometre in referring to wavelengths.

Turning now to matters of more common experience, a dark metal bar when heated becomes firstly a dull red colour and progressively brighter, the hotter it gets. Its colour also changes from red through yellow to white and possibly a bluish-white colour at the highest temperature. This indicates that the peak wavelength being emitted steadily moves from the long wavelength red towards the shorter wavelength blue as the metal has more heat energy available.

If, instead of a solid metal bar, a pure element such as sodium is heated in a flame, the light it gives out is an intense orange-yellow (in the case of sodium). The same colour appears when an electric current is passed through a sodium vapour as in a common type of street light. If the sodium vapour is replaced by mercury vapour, blue and green colours are generated to give a bluish-white appearance. Thus it is seen that particular colours are associated with particular elements.

The Huygens conception of light travelling out from a luminous point as a *continuous* wave surface can satisfactorily account for the observed facts of reflection, refraction, interference, diffraction and polarization. There are, however, phenomena concerned with the interchange of energy between radiation and matter, such as emission, absorption and the photoelectric effect, which cannot be explained in terms of the radiation being transmitted as a continuous wave surface.

At the beginning of the twentieth century, Planck successfully explained the experimental facts of the emission and absorption of energy by recognizing that the atoms involved in these processes were harmonic oscillators with resonant frequencies. As such, they can oscillate only at discrete frequency values, and these in turn define specific amounts of energy, which Planck called **quanta**.

Later Einstein extended the quantum theory to light, and considered that light travelled, not as a wave surface over which the energy was equally distributed, but in minute bundles of energy – **light quanta** or **photons** – the energy of which remains concentrated as they travel through space.

The wave theory is therefore valid in defining the general direction of radiated energy, but the interaction of this energy with matter requires this concentration to occur. Subsequently de Broglie postulated that wave effects must also be associated with every particle of matter and these control the statistics of the manner in which, from a large number of energized atoms, each one emits light in a random **spontaneous** manner. Years later this spontaneity was tamed with the invention of the laser. In the seventeenth century, Isaac Newton in his original experiments on the spectrum allowed light from the sun to pass through a small circular aperture into a darkened room and received the small circular patch of light on a white screen. When he put a prism in the beam the circular patch was broadened into a band of colour. Each of the colours which make up white light was deviated by a different amount by the prism. Unless the aperture is very small (and then the coloured band is very dim) the coloured patches overlap and the spectral colours are indistinct. To obtain a purer spectrum, the circular aperture is replaced by a slit aperture parallel to the apical edge of the prism. The illumination source can be imaged onto this slit and in turn this is collimated (in one plane) just before the prism as shown in Figure 9.13. The deviated output can then be viewed with a simple astronomical telescope. Such instruments, known as **spectroscopes**, have little value these days except as teaching apparatus. **Spectrophotometers**, which use a much more complicated array of optics and detectors, are very useful in chemical analysis. For optical purposes the basic system is converted into a **refractometer** as described in Section 11.2.

When the light emitted by energized gases is examined by the spectroscopic methods described above, a bewildering array of spectral lines is obtained. The simplest arrangement is found when the gas is hydrogen (Figure 9.14). This shows a series of lines which become closer together with decreasing wavelength in a very regular manner. These lines led to the establishment of the whole subject of atomic physics when Niels Bohr postulated a structure of matter in which lightweight

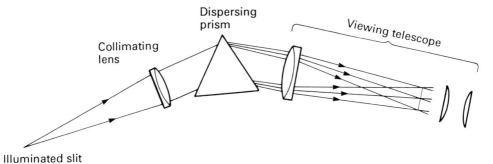

Figure 9.13 Optical arrangement of a spectroscope

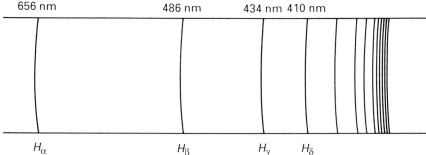

Figure 9.14 Emission spectrum of the hydrogen atom

electrons orbit a heavy nucleus. The electron orbits are restricted in number and each orbit requires a specific amount (or level) of energy in the electron that occupies it. Bohr postulated that light is emitted when an electron moves from a high energy level orbit to one having a lower energy level. The difference in energy is carried away by the emission of a single photon of light, the wavelength of which is defined by the equation

$$E = h\nu$$

where E is the energy difference, h is Planck's constant and ν is the frequency of the light.

This equation is of fundamental importance in physics and the constant h (6.63×10^{-13} joule seconds) is a fundamental constant of nature. Thus the energy radiated by a heated metal bar increases with increasing temperature as more atoms are emitting energy more frequently but the energy of each photon of light emitted depends solely on the change in energy of the electron actually emitting the light.

The particular photon and the electron energy levels are very closely related. This is evidenced by the fact that if the photon passes another electron in another atom in the lower of the energy levels there is a high probability that the photon will be absorbed by the electron which then jumps to the higher energy level.

Thus specific atoms when heated give out light which has a narrow spectral extent particularly when the atoms are in the form of a gas with little interaction between them. Each wave train has an envelope such as that given in Figure 9.11(e) and a spectral spread of frequency of the bell-like shape (also given in Figure 9.11(e)). If the atoms are overheated and the gas contained so that its pressure increases it is found that the frequency spread increases. This **pressure broadening** reduces the accuracy of measurements, particularly those associated with the measurement of refractive index (Section 11.2).

If the atoms of the gas are sufficiently heated but then placed in front of an incandescent source it is found that the light from the source of the same wavelength as that emitted by the gas is *absorbed* by the gas atoms. Although this energy is also re-emitted by these atoms, an analysis of the incandescent spectrum reveals dark lines across it because the absorbed and re-emitted wavelengths are generally sent in other directions. This was first noticed by Fraunhofer, who examined the spectrum of the sun and recognized that the dark lines on it were the **absorption spectra** of the hot gases near the surface of the sun absorbing the light from the hotter incandescent core. Fraunhofer gave letters to those lines he could identify with specific elements on earth and these are used in the definition of refractive indices of optical materials (Section 14.2).

9.5 Incandescent lamps, wavelength, colour and temperature

When a material gives out light because of its high temperature it is said to be **incandescent**. In the last century this was achieved by applying a gas flame on to a thin mesh of non-combustible material. The invention of electricity enabled Edison to heat thin metal wires to incandescent temperatures.

Tungsten-filament lamps are so called because the source of light is a fine filament of tungsten metal through which an electric current is passed. The resistive heating of the filament raises its temperature to a maximum of about 3500 K. At this temperature the tungsten would rapidly oxidize. To avoid this the filament is enclosed in a glass envelope which is evacuated or filled with an inert gas. Even with this precaution the tungsten is slowly evaporated from the filament on to the walls of the envelope, reducing its transmission and eventually ending the life of the lamp after a few hundred hours of use. Longer life can be obtained by reducing the power through the lamp (underrunning), but this reduces the colour temperature (see later) and also reduces the efficiency of the conversion of electrical power into light. At its design voltage a tungsten filament lamp will normally give between 15 and 25 lumens per watt. This is the total amount of light radiated by the lamp and represents an efficiency of conversion of electrical energy into visible energy of between 8 and 13%.

These efficiency figures improve with higher filament temperatures, but the increased evaporation of the filament material at higher temperatures delayed the

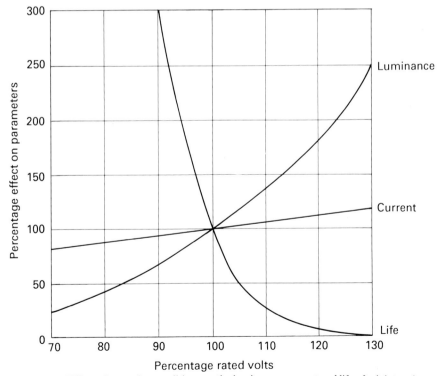

Figure 9.15 Effect of operating conditions on the luminance, current and life of miniature lamps

exploitation of this feature until the invention of the 'quartz-iodine' lamp (more correctly termed the tungsten-halogen lamp). In this lamp the inert gas contains a halogen, iodine or bromine, which has the property of redepositing evaporated tungsten on to the filament. This action requires a high bulb temperature, and as glass proved unsuitable quartz is used in a much smaller shape so that it becomes hotter, but the quartz does not play an active role.

Figure 9.15 shows the effect of changing the operating conditions of miniature lamps. It can be seen that the lifetime increases dramatically with underrunning, but with tungsten-halogen lamps the halogen cycle breaks down at lower temperatures. Overrunning and underrunning also change the colour temperatures of lamps as described below.

By methods described in Chapter 12 it is possible to measure experimentally the wavelength and hence, as the velocity is known, the frequency of any light. It is found that, when light is dispersed to form a spectrum, each portion of the spectrum consists of light of a particular wavelength and therefore frequency. Thus, in the case of a continuous spectrum of white light, there is a continuous range of wavelengths extending, in the portion visible to the eye, from about 390 nm at the violet end to 760 nm at the red end. We may therefore specify any *spectrum colour* or any position in a spectrum in terms of the wavelength or frequency of the light dispersed to that position (shown around the periphery of Plate 5).

The continuous spectrum of *white* light is usually obtained from an incandescent solid or liquid. The word incandescent means glowing with heat, and the hotter the source the more it glows. As well as becoming more intense, the source changes colour from red through white to blue-white. When the intensity for each wavelength is plotted for a perfect (or black-body) source the curve varies with temperature as in Figure 9.16.

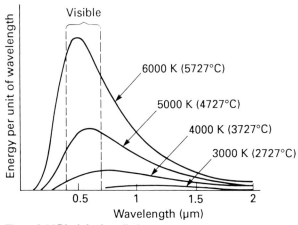

Figure 9.16 Black-body radiation

It is therefore possible to describe the red-white or blue-white colour seen in terms of the temperature and this is known as the colour temperature, usually given in kelvins. This need not be the same as the actual temperature; if a blue filter is placed in front of a tungsten lamp the red parts of its spectrum will be absorbed more than the blue and through the filter the colour temperature will appear to be higher even though the lamp itself remains the same. Due to atmospheric absorption the colour temperature of the sun is about 6000 K even though its actual temperature is higher than this. The colour temperature of most lamps ranges from 2000 to 4600 K. Standard lamps are manufactured to specific colour temperatures (Plate 5).

Each colour temperature can be stated in terms of its 'mired' value (microreciprocal degree) which is given by 10^6 divided by the colour temperature in kelvins. The reciprocal nature of this unit allows the filters described above to be assigned a mired shift value, which may be positive or negative. Lamps of different colour temperatures will be 'shifted' by the same number of mireds by a given filter. This analysis is used in photography, but care must be exercised as it assumes that the lamps are nearly black-body radiators. Some fluorescent lamps depart considerably from black-body behaviour, and photographic results show these differences.

Some typical colour temperatures are given in Table 9.1.

Table 9.1 Some typical colour temperatures

	Temperature (K)
Clear blue sky	12 000–25 000
Overcast sky	6500
Heavily overcast sky	6000
Candle	2000
Tungsten lamp filament	2600
Arc lamp crater	4000
Sun	5000–6000
Moon	3000–4000
Fluorescent lamp	Various – often departing strongly from black-body radiation

See also Table 10.1.

It is convenient to refer to light of any particular frequency or wavelength in terms of the colour sensation to which it gives rise when received by the normal eye, but it should be understood that colour is a purely visual sensation. Light, which we call white, consists of a continuous band of frequencies, as is seen from its spectrum, and this complex vibration gives rise to the sensation of white when received by the eye. Portions of this band of frequencies or a predominance of any portion of it give rise to the sensation of colour as distinct from white. Thus light of wavelengths from about 760 nm to about 620 nm gives the sensation of red, from about 580 nm to about 510 nm the sensation of green, and so on.

That the type of sensation is not specifically determined by the wavelength of the radiation can be shown in a number of ways. Thus the sensation of yellow as produced by a radiation of wavelength 589 nm may also be evoked by presenting to the eye a suitable mixture of pure red and pure green, that is, by light containing no true yellow radiation. Also the sensation of white will result from the mixture in suitable proportions of certain pairs of pure colours known as complementaries. This aspect of colour is dealt with in Section 10.13.

Any colour can be specified in terms of three variables, its **hue**, its saturation or **purity**, and its **luminosity**. The hue is the property that depends on the frequency of the light. The saturation depends on the amount of white present in addition to the light giving the hue; the less white light the more saturated the colour is said to be. Thus, for the same hue we have the intense saturated yellow of the spectrum, the paler yellows when this hue is mixed with white, and the various browns as the luminosity is reduced.

In the spectrum there are five very distinct hues; red, yellow, green, blue and violet. There is also purple, not found in the spectrum, but formed by a mixture of red and blue. The divisions of the spectrum given in Table 9.2 with their corresponding wavelengths were given by Abney (1844–1920).

Table 9.2 Abney divisions of the spectrum

	Wavelength (nm)
Violet	446 to end
Ultramarine	464–446
Blue	500–464
Blue-green	513–500
Green	578–513
Yellow	592–578
Orange	620–592
Red	End to 620

More recently, H. B. Tilton has suggested the adoption of equal hue bands each containing 10 just noticeable differences in wavelength for the colour-normal observer except for the violet (eight colour noticeable differences) and the red (15). The suggested names with their wavelength limits are given in Table 9.3.

Table 9.3 Tilton's equal-hue bands

	Wavelength (nm)
Violet	388–429
Indigo	429–458
Blue	458–481
Cyan	481–499
Turquoise	499–513
Green	513–528
Emerald	528–546
Chartreuse	546–561
Yellow	561–575
Amber	575–587
Ochre	587–599
Orange	599–610
Tangerine	610–622
Scarlet	622–636
Red	636–782

Thus there should be 143 just noticeable colours in the spectrum shown round the edge of Plate 5 but the quality of normal colour printing cannot achieve this.

9.6 Stimulated emission and resonant cavities

When an incandescent filament emits light we can imagine many atoms having their electrons raised to high energy levels by absorbing energy from the free electrons constituting the electric current through the filament. These electrons subsequently fall to lower levels and emit light. No specific colour is generated, as many different electrons and energy levels are involved. The actual moment when a particular electron emits light is not predictable as it does so on a *spontaneous* basis, although the time spent in the higher (or excited) level may be very short. However, for transparent sources such as gas discharge lamps a second process exists.

It is now known that if a photon of the same wavelength as that about to be emitted encounters an electron in the excited state a high probability of **stimulated emission** occurs. In this case the emitted photon is naturally of the same wavelength as the stimulating photon, but more importantly it is emitted in the same direction and in phase. However, this process is not very usual as there are many electrons in the lower state ready to absorb the stimulating photon. The excited electron may also emit spontaneously before the stimulating photon arrives. Thus, the process of stimulated emission, spontaneous emission and absorption can be seen as competing actions within the luminous volume. Very careful arrangements are needed if the stimulated emissions are to predominate over the normal luminescence of the material.

One approach is to ensure an over-abundance of electrons in the excited state. This can be achieved without upsetting the stimulation process by indirect **optical pumping**. With this method intense light is focused into the material so that its particular wavelength is absorbed by the electrons, which enter a higher energy level than that required. A proportion of these will then emit light spontaneously and descend to the required high energy level. From here they may spontaneously or by stimulation descend to the lower state from which they can be again raised by the incoming light beam. If this light is sufficiently intense it is possible to maintain more electrons in the high level than in the lower. This **inverted population** of electronic energy levels is essential in the design of **lasers**, but in itself it is not enough. Consider again the analogy of ripples on the surface of water used in Chapter 1. If the ripples are caused by a single stone dropped into the water, they are seen to have a definite duration. A circle of a finite number of ripples moves out from the point of impact. When these have passed no other ripples occur until another stone has been thrown in. If the second stone entered the water before the first set of ripples had died away it would be possible, by carefully judging the correct moment, to make the second set of ripples in phase with the first. Indeed a whole series of stones could be dropped at specific and regular intervals to generate a continuous ripple system moving out from the point of impact. If the stones were dropped on a random basis over a small area, the ripple system would be a complex superposition of several ripple trains even though they would have the same wavelength and speed of propagation.

The difference in these two cases is similar to that between **coherent** and **incoherent** light. The latter case is typified by the incandescent filament where the atoms are spontaneously emitting at random times. To make the atoms emit in a controlled sequence that can build up and maintain a single continuous wave motion, the inverted population must be stimulated into emission by a single controlling wave. This wave can be generated only in a **resonant cavity**. Resonance occurs in sound when a travelling wave is reflected at one end of an enclosed space and returns to the other end for a second reflection. If it now finds itself in phase with the remainder of the wave not yet reflected, the two superpose to give a greater amplitude. If they are out of phase, zero amplitude will result.

It is easy to show that the length of the cavity is the important dimension. For a fixed length l, any wave for which $n\lambda = 2l$ will resonate, n being an integer. If the material with an inverted population is contained in an optical resonant cavity a controlling light wave reflected at each end will stimulate emission from the material to build up its amplitude as it travels backwards and forwards. Because the wavelength of light is very small, the mirrors of the resonant cavity must be positioned and fixed to a very high precision.

9.7 Practical laser types

The first gas laser was built in 1961 by Javan, Bennett and Herriott at Bell Telephone Laboratories. It consisted of a tube, 100 cm long, filled with a mixture of helium and neon. At both ends of the tube, reflectors were fitted to make the resonant cavity. These were flat mirrors having a flexible coupling to the tube so that they could be adjusted parallel to each other to within a few seconds of arc. In the original model the discharge in the gas was generated by radio-frequency radiation from electrodes wrapped round the tube. Electrodes can also be used inside the tube between which a current can be passed through the gas. Figure 9.17 shows the general arrangement.

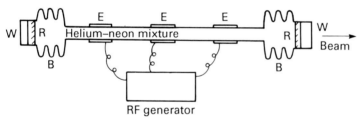

Figure 9.17 Layout of the first gas laser: W, window; R, reflector; B, bellows; E, electrode

The stimulated emission occurs in the neon, but the helium helps to absorb the electrical energy and generate the inverted population in the neon. When spontaneous emission occurs one of these photons is reflected normally by an end mirror and proceeds down the tube stimulating other excited electrons to emit. On reflection the same action occurs on the return journey and again after a further reflection.

When the stimulating wave passes over an excited electron the stimulated emission may not start until most of the wave has passed and need not occur at all. This means that after two reflections of the stimulating wave the stimulated action may still be in process and so the need to remain in phase is paramount. Energy is continually lost in the absorption of the stimulating wave by electrons in the lower energy levels and in the less than perfect reflection at the mirrors. A lasing action will occur and be maintained only if these losses are smaller than the gains due to stimulated emission. Once this has been achieved one of the mirrors may be made partially transparent so that the beam of coherent light emerges from the cavity. This also represents loss to the system.

In general terms the efficiency of the system is of the order of 0.3% and the energy radiated a few milliwatts – far less than a simple flash lamp. On the other hand the beam is extremely well collimated so that its luminous intensity is very high, making it a safety hazard as described in Section 9.8.

Since the first helium–neon laser, many other gases have been found to lase. This original combination has proved to be the most tractable, however, and low-powered types can now be purchased for a few tens of pounds while specially stable types are used in measuring systems. The light emitted is red, 632.8 nm.

Other electronic levels will also lase giving better efficiencies although the light emitted is infrared. For visual work a green colour is preferred and after some delay the argon ion laser was invented. Now a large number of gases are known to lase, giving a large choice of wavelengths.

In all cases the light emitted has an extremely narrow spread of wavelengths. The need to remain in phase after a reflection at each of the end mirrors means that the cavity length must contain an integral number of wavelengths. Whereas photons spontaneously emitted may have a small range of wavelengths, as the various atoms are influenced by collisions with other atoms which give rise to slight differences in the energy levels, photons in the light beam from a laser being related to the resonant cavity length are the most monochromatic known to man.

Although the HeNe laser described above was the world's first *continuous* source of stimulated radiation, it was preceded by some months by the pulsed ruby laser of **Maiman** at Hughes Research Laboratories. The major problem to be overcome was too great a loss of energy by spontaneous emission, non-optical losses and absorption in the end mirrors. Previous exploratory work by **Townes**, **Basov** and **Prokhorov**, for which they received the Nobel Prize, had shown the extent of these other means of energy dissipation compared with the required method.

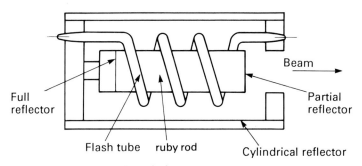

Figure 9.18 Layout of the first ruby laser

Maiman's solution (Figure 9.18) was to wrap a xenon flash tube around the working element, which was a cylinder of pink ruby containing 0.05% chromium. The end faces of the cylinder were polished flat and parallel and coated to provide the resonant cavity, although one face is only partially reflecting to allow the emergence of the laser beam. The flash lamp surrounding the ruby is operated by discharging a bank of capacitors across it. A large proportion of this energy is dissipated as heat, but a fraction is emitted as blue-green light, which is absorbed by the ruby. For a short time the ruby has an inverted population in its chromium atoms. With high-speed detectors it is possible to observe that spontaneous incoherent emission from the ruby starts radiating in all directions almost immediately, but after about 0.5 ms the coherent radiation emerges from the partially reflecting face of the ruby rod. This laser radiation may last for only a few milliseconds, but peak powers of the order of tens of kilowatts have been recorded.

Once again the precise wavelength emitted depends on the length of the rod, the resonant cavity. During the discharge this heats up and expands and it is possible to observe the change in wavelength due to this. The wavelength is about 694.3 nm, which is governed by the energy levels in the atoms of chromium known as the *active* element or *working* element. Again, other wavelengths can be found, although the red line first discovered is the most efficient.

An alternative lasing material is neodymium, which can be contained in a glass rod or crystal. This gives out light at 1.09 μm in the infrared part of the spectrum. Solid lasing materials such as these can now be made to lase continuously like gas lasers, but any defects in the material absorb energy and the solid can heat up and fracture. Later research turned to liquid lasers, in which any local hot spot can be healed by the movement of the liquid.

The population inversion so necessary to coherent light generation can be obtained in the junctions of semiconductor devices. Instead of inverting the population by putting energy into the working volume from other lamps, the direct passage of a current across the junction produces this effect. The excited electrons may emit light spontaneously and such devices are often called **light emitting diodes** or **LEDs**. If the ends of the junction region are plane and well polished a resonant cavity is formed as the refractive index of the material is so high that substantial reflectivities occur without any reflective coatings.

The most common material is gallium arsenide and this emits in the infrared part of the spectrum. LEDs are available that emit red and green light. The efficiency is about 50%, but large powers are difficult to obtain. For the infrared emitters a number of suitable detectors are available and such devices have been used on eye-movement experiments where the infrared beam is reflected off the sclera but not from the cornea. There is very little heat generated and at the low power levels that suffice, emitters and detectors can be mounted close to the eye on a spectacle frame without danger and without the subject being able to see the light as it is infrared.

9.8 Photometry of lasers – vision hazards

If plane mirrors define the resonant cavity of a laser, the emitted laser beam will be plane as far as diffraction effects allow (see Chapter 13). Plane mirrors can be difficult to adjust and often concave mirrors are used but with a further curved surface to give a parallel output beam having a divergence of about one-third of a minute of arc. This means that the total solid angle (see Section 10.2) containing the energy is about 10^{-8} steradians, while more common sources of light radiate into 4π steradians.

Because of the large loss of energy by other means which occur in laser systems the efficiency is usually low. Most small gas lasers produce between 1 and 30 mW of red light. In photometric terms this is equivalent to little more than 5 lumens. However, luminous intensity (see Chapter 10) is measured in lumens per steradian and as these few lumens are concentrated into such a narrow beam the intensity of the source seen looking along the beam is 500 million candela! Because this is in an almost parallel beam the eye focuses this source on to the retina. In all but the very weakest laser such an intensity will cause burn damage on the retina before the muscles of the eyelid can close to protect the eye. It is therefore very important that operating lasers are not directed towards observers or on to mirrors which could reflect the beam into unprotected eyes.

With pulsed lasers the hazard is even greater as the chance reflection of the pulse from a metal surface such as an optical bench or equipment case could direct sufficient energy into an eye to cause blindness. Workers in laser laboratories are normally supplied with protective goggles containing glass that absorbs the wavelengths at which the laser is operating. Other parts of the spectrum are transmitted so that the experimenter can see the apparatus. Protection is equally important for lasers generating ultraviolet or infrared light.

The use of lasers for interference and holography (Chapters 12 and 13) normally requires the laser beam to be divergent, and a mirror or lens is used for this purpose. This gives the beam a solid angle of about one-tenth of a steradian and the apparent intensity of the source is now only 500 candela – similar to tungsten filament lamps. Nevertheless, all lasers should be treated with the utmost respect and any examination of the beam is best done by allowing it to fall on a piece of white card or paper even if the result looks quite dim (see the calculation in Section 10.2).

9.9 Application of lasers

The potential of lasers was recognized from their beginning. Many uses were proposed, but as these did not come to fruition within a few years some cynicism was generated. It was said that lasers were 'a solution looking for a problem'. This has now been shown to be far from the truth. Two major uses of lasers occur in optical metrology and in holography, which are discussed in Chapters 12 and 13 respectively. Other uses involve one or more of the three most obvious attributes of this unique source:

1. Narrow wavelength range.
2. High collimation.
3. Large radiant energy density (high intensity).

Referring to the last attribute first, the use of pulsed lasers for photocoagulation in cases of retinal detachment is a case of a hazard being put to controlled use. This application often uses green light at 514 nm from the argon laser, but this and other ophthalmic applications sometimes use krypton lasers with wavelengths in the orange part of the spectrum.

The **excimer** laser is a recently developed type of gas laser which uses an *excited dimer* molecule for its inverted population. Dimer molecules are made up of atoms of a rare gas and halogen such as argon and fluorine. The molecules are short-lived but can be made very excited so that when stimulated emission occurs a lot of energy is released, giving pulses of short wavelength light in the ultraviolet region, in this case at 193 nm wavelength. The light produced is very quickly absorbed by most materials. This means that when this light is incident on a surface all the energy is absorbed by the first few molecular layers and these are then exploded away or **ablated** without any change to the underlying or adjacent material. The cutting and marking of polymers and photolithographic patterns using excimer lasers is now established, and current research is looking at the medical applications including the ablation of the living human cornea to give a different curvature and so correct defective vision. Other energy-based applications include the cutting of metals, glass, cloth and even the erasing of typewritten letters by the vaporization of material. The absorption of the laser energy can be so immediate that no burning is caused, although in the case of metal cutting some systems incorporate a jet of oxygen to cause combustion.

The continuous wave carbon dioxide laser is on the other side of the visible spectrum, having a wavelength of 10.6 µm. This is a very efficient system with efficiencies above 30%, and output powers of 10 kW of radiant energy are available. Their applications include welding, drilling, cutting and scribing, as well as surface treatment to harden the material.

Semiconductor lasers and semiconductor diodes emit mainly in the near infrared, where optical fibres can be made with very high transmission values. These lasers and diodes are now used in data communication systems.

The high degree of collimation available particularly from gas lasers allows aerial straight-edges to be created. A laser beam shone into a tunnelling or road grading machine can be used to ensure a straight line of travel. In setting up optical bench experiments a laser can be very useful, but stray reflections must be carefully avoided.

The narrow wavelength range permits lasers to be used as sources in various types of spectrophotometer (Chapter 11); of great value in the chemical and allied industries. Various laser systems are now in use for monitoring pollution.

The major attribute of lasers is, of course, their coherence as described in Section 9.6. The applications that use this are optical metrology and holography, which, as mentioned above, are dealt with in subsequent chapters; optical communications and material processing are outside the scope of this book.

9.10 Luminescent sources

Although a source of light gives out energy which it must obtain somehow, the method of heating used in incandescent sources is not the only way, nor the inverted populations in lasers. Substances called **phosphors** have the property of absorbing energy from electrons, electric fields, chemical reactions and even other light, particularly ultraviolet. They then re-emit this energy in the form of light. Phosphors are frequently sulphides and oxides or silicates and phosphates of such metals as zinc, calcium, magnesium, cadmium, tungsten and zirconium. They are used in such commonplace items as fluorescent lamps and television picture tubes and more complex devices such as X-ray machines and electron microscopes. They are all luminescent sources, but as they are often of large dimensions and flat the term luminescent screen is generally used.

Phosphors are used in fluorescent lamps to obtain a much broader range of colours than can be obtained with incandescent sources. They are also more efficient, with luminous efficiency values towards 90 lm/W and less heat radiation. The light is generated by a two-stage process as the electric current is discharged through a mercury vapour, which produces green and ultraviolet light. This ultraviolet light is then absorbed by the phosphor coating on the inside of the glass tube containing the gas and re-emitted as white light.

In television picture tubes the phosphor is selected not only for its colour and efficiency but also for its short response time. The light emission is stimulated by high-energy electrons striking the phosphor on the inside of the screen. As the picture changes it is important that the phosphor can start and stop emitting light very quickly as the electrons are switched on and off.

Lamps that cannot switch off at all use phosphors coated on the inside of glass containers, which are then filled with tritium gas. This is a radioactive isotope of hydrogen and emits low-energy beta particles (electrons), which are absorbed by the phosphor and re-emitted as light. These lamps are used in instrument panels and exit signs, as they need no external source of power.

Exercises

9.1   What is meant by the wavelength of light, and how is it related to its frequency of vibration and velocity of propagation?
9.2   State the fundamental equation of wave motion. Given the velocity of light in air as 3 $\times 10^8$ m/s, determine the wavelengths for red, yellow and blue light, the frequencies being

Red 395×10^{12} per second
Yellow 509×10^{12} per second
Blue 617×10^{12} per second

Give the wavelengths in angstrom units, nanometres, micrometres and inches.
9.3   A series of waves of wavelength 100 cm is travelling across a pond on which three corks A, B and C are floating at distances of 1.5, 2.25 and 3.8 m respectively from a fixed post. In which direction will each of the corks be moving when a trough of a wave is passing the fixed post? State also whether each cork is above or below its normal position.
9.4   What is meant by a simple harmonic motion? Define the terms period, amplitude and phase.
9.5   Plot the graph of the simple harmonic motion represented by

$$y = 3 \sin \frac{2\pi t}{8}$$

9.6   A particle is acted on simultaneously in the same straight line by two simple harmonic motions represented by

$$y = 3 \sin \frac{2\pi t}{6} \quad \text{and} \quad y = 5 \sin \left(\frac{2\pi t}{8} - \frac{\pi}{2} \right)$$

Plot a graph showing the resultant motion.
9.7   Find graphically the resultant motion of a particle acted upon by two perpendicular simple harmonic motions of equal period and amplitude, and differing in phase by:
(a) 0.
(b) $\pi/4$.
(c) $\pi/2$.
(d) $3\pi/4$.
(e) π.
(f) $3\pi/2$.
9.8   Repeat the graphical constructions of Exercise 9.7 for the case when one simple harmonic motion has twice the amplitude of the other.
9.9   A particle B executing a simple harmonic motion given by the expression $y = 8 \sin 6\pi t$ is sending out waves in a continuous medium travelling at 200 cm/s. Find the resultant displacement of a particle 150 cm from B, 1 s after the commencement of the vibration of B.

9.10 If a small float on the surface of a lake is seen to be oscillating up and down at 2.5 Hz, at what speed are the water waves travelling if their wavelength is seen to be 700 mm?

9.11 A transverse wave is described by the equation $y = 5.0 \sin(0.02\pi x + 4.0\pi t)$ where x and y are in millimetres and t is in seconds. Calculate:
(a) The frequency of the wave.
(b) The amplitude of the wave.
(c) The wavelength of the wave.

9.12 A particle executing simple harmonic motion given by the equation $y = 3 \sin(2t/6 + \alpha)$ is displaced by 2 units when $t = 0$. Find:
(a) The phase angle α when $t = 0$.
(b) The difference in phase between any two positions separated in time by 1 s.
(c) The time needed to reach a displacement of 2.5 from the start.

9.13 Four simple harmonic motions of the same amplitude and frequency are superimposed. If the phase difference between one motion and the next is the same, find the value of the phase difference for which the resultant effect is zero.

9.14 Describe fully a method of obtaining a pure spectrum. What difference will be seen between the spectra of:
(a) The sun.
(b) The electric arc.
(c) A tube containing incandescent hydrogen.

9.15 Sketch and explain the spectral radiation curves for a black body at various temperatures.
   With reference to the curves describe the change in appearance of a black body which would take place if its temperature were raised from cold to, say, 7000 K.

Chapter 10

Photometry and detectors

10.1 Amount of light – the visual response

In the study of geometrical optics developed in the previous chapters we were very concerned with the *direction* of light rays. In the study of **photometry** the *amount* of light is our first concern, although obviously the direction is important when the amount in different places is being calculated. In Section 1.1 the energy given out by the heated metal included heat as well as light. The heat energy does not stimulate the retina of the eye and is therefore not included in photometry, which is solely concerned with the visual sensation produced by the radiated energy.

The visual sensation produced will vary greatly in two respects with the wavelength of the radiation. It has already been indicated in Section 9.4 that different wavelengths give rise to the sensation of different colours, and this is considered more fully in Section 10.13. Also, the intensity of the sensation produced by equal amounts of emitted energy will vary throughout the visible spectrum. From the spectrum of white light (Plate 5) it will be evident that the brightest portion is situated in the yellowish-green and that the brightness falls off rapidly towards both the red and violet ends. Assuming a source that is emitting equal energy at all wavelengths and plotting the luminosity, that is the amount of visual sensation, against the wavelength, we obtain the relative luminosity or visibility curve shown in Figure 10.1. The form of this curve varies somewhat with different normal observers and also with different values of luminance, particularly when the luminance is low. The curve shown, which is the result of a large number of observations, has been adopted as the standard luminosity curve for ordinary values of luminance, that is over about $10 \, \mathrm{cd/m^2}$. It will be seen that the greatest visual effect occurs for a wavelength of 555 nm, in the yellowish-green portion of the spectrum. This curve is often called the V_λ **curve**. It is also called the **photopic curve** and is the relative sensitivity of the cones in the retina. Another curve, similar but shifted towards the blue, relates to vision at low luminance.

The eye is incapable of making an absolute measurement of the amount of light entering it; we can look at two sources and estimate that one appears 'brighter' than the other if there is sufficient difference between them, but cannot form a reliable judgment as to by how much they differ. The eye can, however, decide with a fair degree of accuracy whether two adjacent surfaces appear *equally* bright; and this is the basis of all practical visual photometric measurements.

According to the law of Weber (1834) the smallest perceptible difference of apparent brightness or luminosity is a constant fraction of the luminosity. This

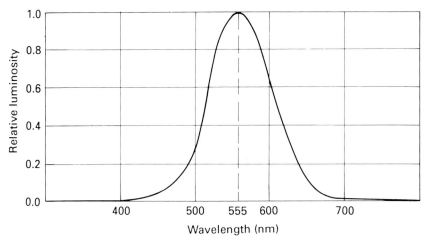

Figure 10.1 Relative luminosity or visibility curve

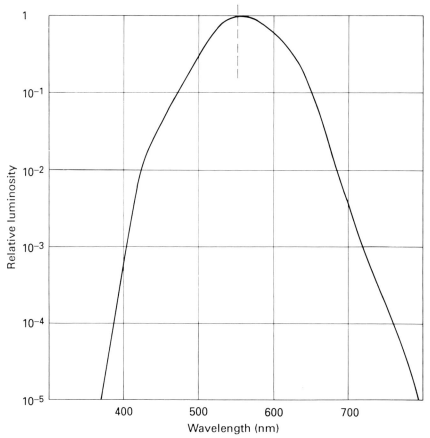

Figure 10.2 The overall visual response plotted with a logarithmic ordinate scale

fraction, known as **Fechner's fraction** (1858) is, over a large range of luminosities, about 1% and the eye can therefore distinguish between two adjacent surfaces that differ in luminance by this amount. This is a reasonable approximation to the curves of Figure 15.17.

In photometry we have to differentiate between the following quantities:

1. The amount of light emitted by a point source, the **luminous flux** and **luminous intensity**.
2. The amount of light received on a unit area of a surface in a given position; that is, the **illumination** or **illuminance** of the surface.
3. The amount of light *emitted* or *re-emitted* per unit area of a surface; that is the **luminance** of the surface. This was formerly called the brightness of the surface.

In discussing these measurements it should be remembered that in practical photometry it is rarely necessary to strive for accuracies better than ±1%, Fechner's fraction. The exposure meters on cameras, for instance, may only be accurate to ±10%. However, the range of the measurements can be over many orders of magnitude and it is often crucially important to eliminate *stray* light from the experimental area. A better description of the response of the eye is given in Figure 10.2, where a log scale is used in order to cover the large range of response.

The curves of Figures 10.1 and 10.2 can be approximated by the expression $\cos^4$ $(0.6\lambda - 333)$ where λ is the wavelength in nanometres, but for most of this chapter we will ignore the effects of wavelength and treat all light as if it were white or neutral.

As indicated above, photometry is restricted to radiant energy which stimulates the human retina to give a visual response. It is therefore a part of **radiometry**, which is the broader science concerned with radiant energy measurements throughout the ultraviolet, visible and infrared parts of the spectrum out to the longest wavelengths detectable. This energy as a radiant flux is measured in watts, which compares directly with luminous flux defined above, measured in lumens (defined in the next section). The concepts of photometry and radiometry are very similar. Most optics is concerned with visible light and this book will deal only with photometry to avoid confusion. (An article by D. A. Roberts in the Designers Handbook series in *Photonic Spectra*, April 1987, p. 59 gives a comparison between practical photometry and radiometry.) For practical comparisons it may be noted that *at the peak wavelength of 555 nm* shown in Figures 10.1 and 10.2, 1 W of radiant energy is equivalent to 685 lm. At ultraviolet and infrared wavelength, 1 W of radiant energy is equivalent to 0 lm. Strictly we should not refer to ultraviolet or infrared light. The correct descriptions are ultraviolet and infrared *radiation*.

10.2 Luminous flux and luminous intensity

There is a great difference in the amount of light emitted by different sources. Most practical sources are *incandescent* sources, which emit energy over a wide range of wavelengths, mainly as heat in the infrared part of the spectrum (see Chapter 9). This is wasted energy as far as the emission of light is concerned and the *luminous efficiency* (see Section 9.5) is very important to lamp manufacturers. In considering the amount of *light* given out by a source, the energy emitted must be evaluated or *weighted* according to its ability to stimulate visual sensation as given by the relative luminosity curve (Figure 10.1). The rate of flow of *light* from a source is the

luminous flux, Φ, of the source, and the unit of this is the **lumen** (lm), which is defined later.

In no practical source is the distribution of flux uniform in all directions; hence it is necessary to define the amount of light radiated with respect to a given direction. In no practical source is the amount of luminous flux uniform over all parts of the source; hence it is necessary to define the amount of light radiated with respect to a point on the source or to assume that the whole source is contained at a single point. This last assumption works very well in the majority of cases (see Sections 10.3 and 10.4), and the two requirements above can be met by considering the flux emitted by a point source into a cone of very minute solid angle constructed round a given direction. The concentration or density of this flux per unit solid angle is termed the **luminous intensity**, I, of the source in the given direction. The unit of luminous intensity is the **candela** (cd), formerly the candle, and luminous intensity is often referred to as **candle-power**. If Φ is the flux in lumens emitted within a cone of solid angle w steradians, theoretically in this case infinitely small, then $\Phi/w = I$ lm/sr $= I$ cd, in the given direction.

Note that the unit solid angle, or **steradian**, is the solid angle of a cone that, having its apex at the centre of a sphere, cuts off an area of the sphere's surface equal to the square of its radius (Figure 10.3). The value of any solid angle in steradians is equal to the area of the sphere's surface included in the angle divided by the square of the radius of the sphere. The surface area of a sphere of radius r being $4\pi r^2$, the solid angle for a complete sphere is 4π sr.

It can be shown that in a cone having a half plane angle θ the solid angle w is given by

$$w = 2\pi (1 - \cos \theta) = \pi \sin^2 \theta \text{ sr} \qquad \text{when } \theta \text{ is small} \tag{10.1}$$

While no source will be a point as required by the definition of luminous intensity, in many cases the size of the source is negligible at the distances at which it is used (see Section 10.3).

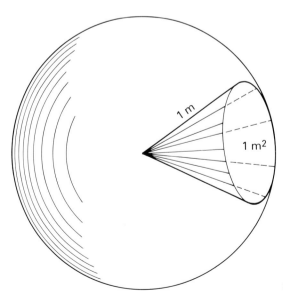

Figure 10.3 Unit solid angle: the steradian

Worked example   The output of an argon ion laser at 514 nm is 20 mW. Using the radiant equivalent value given in Section 10.1 and the relative luminosity curves, assess the luminous flux of this source. Repeat the calculation for a helium–neon laser of the same power. If both have a beam divergence of 0.025° calculate their luminous intensities.

The argon laser energy has a relative luminance (from Figure 10.1) of approximately 50% compared with the peak wavelength 555 nm for which the radiant equivalent value is 685 lm/W. The argon laser therefore has a luminous flux of

$$0.5 \times 685 \times 0.02 = 6.85 \, \text{lm} \qquad \text{(Argon)}$$

For the helium–neon laser the relative luminous efficiency (for $\lambda = 633 \, \mu m$) is about 20%. This laser therefore has a luminous flux of

$$0.2 \times 685 \times 0.02 = 2.74 \, \text{lm} \qquad \text{(HeNe)}$$

Note that neither of these values is particularly high.

If both lasers have a beam divergence of 0.025° the half-angle of the cone is 0.0125°, which by Equation 10.1 gives a solid angle, W, of 6×10^{-10} sr. The luminous intensity in candela or lumens per steradian is then $6.85/(6 \times 10^{-10}) \approx 10^{10}$ cd for the argon laser and 4×10^9 cd for the helium–neon laser.

These values are large and the safety precautions described in Section 9.8 are very necessary.

10.3 Illumination, illuminance

A surface receiving light is said to be illuminated, the **illuminance**, E, at any *point* of it being defined as the density of the luminous flux at that point, or the flux divided by the area of the surface, when the latter is uniformly illuminated. The metric unit of illuminance is the **lux** (lx) or **lumen per square metre** (lm/m²), the illumination of a surface, normal to the direction of the light, one metre from a source of one candle-power. The imperial unit is the **lumen per square foot** (lm/ft²), the illumination of a surface normal to the direction of the light one foot from a source of one candle-power. (These units were formerly called the metre-candle and foot-candle respectively.)

If Φ is the flux from a point source at the centre of a sphere of radius d, the illumination E on the surface of the sphere, from the definition, will be

$$E = \frac{\Phi}{\text{Area of sphere}} = \frac{\Phi}{4\pi d^2}$$

$\Phi/4\pi$ is the candle-power, I, and therefore

$$E = \frac{I}{d^2} \qquad (10.2)$$

This is the law of inverse squares (Section 1.3). It should be clearly understood that Equation 10.2 applies only to the illumination received on a surface *normal* to the direction of the incident light from a *point* source, where I is the candle-power of the source in the particular direction.

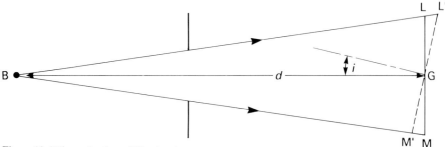

Figure 10.4 The cosine law of illumination

It can be shown by applying the inverse square law to each part of a source larger than a point that the error introduced by assuming it is a point source will be less than 1% if the distance, d, is greater than 10 times the largest dimension of the source.

Figure 10.4 represents the section of a pencil of light from a point source B of candle-power I falling on a surface LM perpendicular to the pencil. If Φ is the flux in the pencil of solid angle w, then the illuminance of the surface at distance d from the source is given by

$$E = \frac{\Phi}{\text{Area of LM}} = \frac{\Phi}{wd^2} = \frac{I}{d^2}$$

If the surface is rotated about G into a new position L'M' the same flux is distributed over the larger area, and the illumination of the surface is now

$$E' = \frac{\Phi}{\text{Area of L'M'}}$$

As the illumination refers to any one point on a surface, we may consider the area L'M' as very small compared with its distance from the source, and then

$$\text{Area of L'M'} = \frac{\text{Area of LM}}{\cos \text{LGL'}} = \frac{\text{Area of LM}}{\cos i}$$

where i is the angle of incidence.

Therefore

$$E' = \frac{\Phi \cos i}{\text{Area of LM}} = \frac{I}{d^2} \cos i \tag{10.3}$$

Thus, *the illuminance of a surface varies as the cosine of the angle of incidence*; this is known as the cosine law of illumination. The law of inverse squares and the cosine law are the two fundamental laws of photometry.

In practical applications of photometry we frequently require to find the illumination at different points on a plane surface, as for example the surface of a street, an illuminated test chart, etc. Figure 10.5 represents a source B of a candle-power I above a plane surface LM; the illuminance at any point G' on the surface will be

$$E = \frac{I_i \cos i}{d_i^2}$$

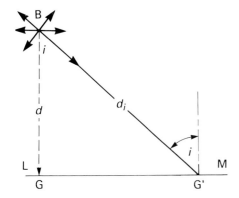

Figure 10.5 Illumination of a surface

where i is the angle of incidence and I_i the candle-power of the source in that direction. From which we have

$$E = \frac{I_i \cos^3 i}{d^2} \qquad (10.4)$$

$$= \frac{I_i d}{d_i^3} \qquad (10.5)$$

where d is the perpendicular distance of the source from the surface.

Worked example   A room 15 ft by 12 ft is illuminated by a 100 candle-power lamp suspended from the centre of the ceiling and 8 ft above the floor. Find the illumination on the floor (a) directly under the lamp and (b) in a corner of the room, assuming the candle-power to be 100 in both these directions and ignoring indirect light.

(a) $I = 100$ c.p. $d = 8$ ft

$$E = \frac{I}{d^2} = \frac{100}{64} = 1.56 \text{ lm/ft}$$

(b) Distance GG′ (Figure 10.5) from point on floor directly under lamp to corner of the room:

GG′ $= x = \sqrt{(7.5^2 + 6^2)} = \sqrt{(92.25)}$ ft

Distance d_i from lamp to corner:

$d_i = \sqrt{(d^2 + x^2)} = \sqrt{(64 + 92.25)} = 12.5$ ft

Hence, from Equation 10.5,

$$E = \frac{100 \times 8}{12.5^3} = 0.41 \text{ lm/ft}^2$$

Worked example   Light from a point source of 50 candle-power, which radiates uniformly in all directions, falls on a +1 D spherical lens of 50 mm aperture placed 500 mm from the source. What fraction of the total flux from the source is incident on the lens? What will be the illumination in lumens per square metre on a screen held 1 m beyond the lens, on the side opposite from the source, and perpendicular to the optical axis of the lens?

At 500 mm from the source the total flux is spread over a sphere of 500 mm radius, the area of which will be $4\pi \times 500^2$. The area of the lens aperture is $\pi \times 25^2$. Therefore the fraction of the total flux received by the lens will be

$$\frac{\pi \times 25^2}{4\pi \times 500^2} = \frac{1}{1600}$$

As the source is 500 mm from a +1 D lens a virtual image of the source is formed 1 m in front of the lens. Therefore the diameter of the patch of light on the screen will be twice the diameter of the lens and its area four times the area of the lens.

Total flux emitted from the source $= 4\pi I = 200\pi$ lm. Flux received by the lens is 1/1600 of total flux:

$$\frac{\pi}{8} \text{ lm}$$

Area of patch of light on screen $= \pi \times 50^2$ mm^2. Therefore, neglecting any loss of light at the lens,

$$\text{Illumination of screen} = \frac{\pi}{8 \times \pi \times 50^2}$$

$$= \frac{1}{20\,000} \text{ lm/mm}^2$$

$$= 50 \text{ lm/m}^2$$

$$= 50 \text{ lx}$$

Alternative solution:

$$\text{Illumination on lens} = \frac{I}{d^2}$$

$$= \frac{50}{500^2} \text{ lm/mm}^2$$

The area of the illuminated patch is four times that of the lens and therefore the illumination is one-quarter; hence

$$\text{Illumination of screen} = \frac{50}{500^2 \times 4}$$

$$= \frac{1}{20\,000} \text{ lm/mm}^2$$

$$= 50 \text{ lx}$$

10.4 Luminance

Illumination refers only to the amount of light *received* by a surface, regardless of the nature of that surface, but in many cases we are more concerned with the amount of light *emitted* in a given direction by a surface. The light emitted per unit area of a surface is the **luminance**, B, of the surface. It will depend, in the case of a non-self-luminous body, upon both the illumination and the fraction of the incident light diffusely reflected by the surface. This latter is known as the diffuse-reflection coefficient or **albedo** of the surface. Luminance may also refer to a diffusely transmitting substance, such as the diffusing globe of a lamp, or to an actual source, such as a flame or a fluorescent lamp.

In its perception of luminance the eye interprets the flux to be coming from an area projected at right angles to the direction of vision (the sun or a lamp surrounded by a spherical diffusing globe appears to the eye as a flat disc), and the luminance of a surface is defined as follows: *the luminance in a given direction of a surface emitting light is the luminous intensity measured in that direction divided by the area of this surface projected on a perpendicular to the direction considered.*

Luminance is expressed in candelas per unit area of surface. In metric units luminance is given in candelas per metre squared, which used to be called the **nit** (nt). Fortunately, this has dropped out of use and candelas per metre squared is the accepted SI term. In imperial units luminance is expressed in candelas per square foot.

Some typical approximate values of luminance are given in Table 10.1.

Table 10.1 Luminance of some common sources

	Luminance (cd/m^2)	
Clear blue sky	10^4	(or 3×10^2 mL)
Overcast sky	10^3	(or 3×10^2 mL)
Heavily overcast sky	4×10^3	
Candle	2×10^6	
Tungsten lamp filament	2×10^6 to 3×10^7	
Arc lamp crater	3×10^8	
Sun	1.6×10^9	
Moon	2.6×10^3	
Fluorescent lamp	10^4	

See also Table 9.1.

A figure of 1.6×10^6 cd/m^2 for the luminance of the sun may seem somewhat strange. As the sun is 96 million miles away the idea of a square metre of its surface may seem a little academic. If, however, a 1 m^2 aperture were placed far enough away from an observer so that the sun just filled it, the luminance of that square would be 1.6×10^9 cd or, remembering the definition of candelas in Section 10.2, its luminance would be 1.6×10^9 lm/sr. This luminance is not available over a steradian because an observer who moves no longer sees the sun through the aperture.

If, on the other hand, the aperture is moved further away from the observer, its luminance remains the same although the size of the image on the retina is reducing. This does not mean that the intensity of this retinal image is increasing, because the luminance is measured in lumens per steradian and the more distant the aperture the smaller fraction of a steradian is being subtended by the pupil of the observer's eye. The intensity (per unit area) of the retinal image therefore remains the same. *This intensity, with the full sun as quoted in the table, is, of course, at a dangerous level and retinal burns can easily occur.*

Table 10.2 Albedo (diffuse reflectance) of some common materials

Material	Percentage of incident light reflected (measured at 30° to the normal)
Magnesium oxide	96
Plaster of Paris	91
Matt white celluloid	80–85
White blotting paper	80–85
Ground opal glass	76
Foolscap paper	70
Black cloth	12
Lunar surface	7
Black velvet	4
3M's Nextel paint	2
Martin Marietta Black	0.5

Referring again to the table, it is seen that the luminance of the moon is also quoted. This is a non-self-luminous body and so its apparent luminance is dependent on its albedo. The albedo of the moon and some surfaces more close to hand is given in Table 10.2.

Unlike specular reflection, diffuse reflection (as its name implies) means that the light is reflected into a range of angles. Another method of expressing luminance assumes that the surface is a perfect diffuser, i.e. one that appears equally bright from whatever direction it is viewed. This will mean that the luminous intensity per unit area of apparent or projected area is constant for all angles. Since for a plane surface the projected area is proportional to the cosine of the angle between the direction of viewing and the normal to the surface, the luminous intensity of the surface must vary in the same proportion. This is stated in the *cosine law of emission* of Lambert (1727–1777) as follows: *for a perfectly diffusing surface the candle-power per unit area of the surface in any direction varies as the cosine of the angle between that direction and the normal to the surface, so that the surface appears equally bright whatever be the direction from which it is viewed.* No surface completely satisfies this requirement, but a few, such as a coating of magnesium oxide, sand-blasted opal glass and scraped plaster of Paris, have diffusion closely approaching the ideal. Many surfaces closely resemble the perfect diffuser when the angle of incidence is small and the direction of viewing is nearly normal. Such surfaces are said to be 'matt'.

Treating the surface as a perfect diffuser, luminance is expressed in terms of the total luminous flux in lumens emitted by a unit area (actual not projected) of surface. A surface emitting or reflecting $1 \, lm/cm^2$ has a luminance of one **lambert**, but a luminance of one-thousandth of this value, the **millilambert** (mL), is a more generally useful unit. A unit that is now outmoded is the **foot-lambert**, the luminance of a surface emitting a flux of $1 \, lm/ft^2$. One foot-lambert equals 1.076 mL. The corresponding metric unit is the **apostilb** (asb), the square metre being the unit of area. Millilambert values for the sky are given in Table 10.1.

It can be shown that for a perfect diffusing surface the luminance in lumens per unit area is π times the brightness in candelas per unit area, i.e. a surface having a luminance of $1 \, cd/ft^2$ has a luminance of π foot-lamberts.

As the illumination, E, of a surface is the flux in lumens received by a unit area of the surface and of this flux a fraction, r, the reflection factor, is reflected, the luminance of an illuminated perfectly diffusing surface will be given by rE lumens per unit area or rE/π candelas per unit area.

A further measurement of luminance, recognizing that photometry is based on visual parameters, makes allowance for the variations in diameter of the pupil of the eye. The concept of **retinal illuminance** from an extended source assumes that the retina is a smooth surface. The value is given by the equation:

$$E = \frac{B t \alpha \cos \theta}{k} \tag{10.6}$$

where B is the luminance in the direction of viewing (cd/m^2), t the transmittance of the eye, α the area of the pupil (m^2), θ the angle of incidence of the principal ray through the eye, and k a constant equal to the area of the retinal image (m^2) divided by the solid angle of the visual field (sr).

Usually, $\cos \theta$ is unity and t and k are constants of such values that it turns out that the retinal illumination can be found by multiplying the luminance of the scene in candelas per metre squared by the area of the pupil in square millimetres:

$$E(\text{Td}) = B(cd/m^2) \times \alpha(mm^2) \tag{10.7}$$

This gives the retinal illumination E in units called **trolands** (Td).

10.5 Standard sources of light

Because the measurements of photometric quantities are largely comparisons between sources it is necessary to define a standard source in the same way that a metre or yard is a standard length. The original standard of luminous intensity was the **international candle**. This, despite careful specification, was far from constant in its intensity, and its colour differed considerably from that of modern sources. The candle as originally specified is now only of historical interest, although the term *candle-power* still survives. The candle was replaced by various flame lamps, the chief of which was the Vernon Harcourt lamp burning pentane vapour with a hollow cylindrical flame and having an intensity of 10 candle-power. Later the standard was maintained with a series of specially constructed electric carbon filament lamps kept at the standardizing laboratories of the various countries.

In 1948 an entirely new primary standard was adopted. This consists of a very small cylinder containing molten platinum. The surface of the platinum at the temperature of the 'freezing point' of platinum has a constant luminance. One sixtieth of the luminous intensity from $1\,cm^2$ of this surface is the candela, which now replaces the candle as the unit of luminous intensity.

As working standards, specially constructed electric filament lamps are now generally used. In these the filament is mounted in a single plane perpendicular to the photometer bench. The glass bulb is considerably larger than would be used in the ordinary way in order to minimize the blackening caused by the deposition of metal particles on the glass, and the filament is 'aged' by being run for at least $100\,h$ before being standardized. In use the lamp is run at a constant current and voltage slightly below that at which it was 'aged'.

10.6 Comparison photometers

Photometry is concerned with the visual effects of emitted radiations so the primary instrument of measurement is the human eye. As this can only judge differences in luminous intensity qualitatively it must be assisted if quantitative results are to be obtained. The method adopted consists in arranging that the two sources produce equal luminance on two similar adjacent surfaces, each surface being illuminated by only one of the sources. A simple arrangement is shown in Figure 10.6.

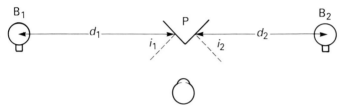

Figure 10.6 The photometer principle

The diffusing surfaces at P lie at equal angles to the light paths, but can both be seen simultaneously. If E_1 and E_2 are the illuminations on the two surfaces produced by sources B_1 and B_2, whose candle powers are I_1 and I_2 respectively, then

$$E_1 = \frac{I_1 \cos i_1}{d_1^2}$$

and

$$E_2 = \frac{I_2 \cos i_2}{d_2^2}$$

where i_1, i_2 are the angles of incidence of the light on the surfaces. If d_1 or d_2 is changed the eye can observe the surface and judge when they are equally bright; that is, when no difference exists and the system is *balanced*. Under these circumstances,

$$\frac{I_1}{d_1^2} = \frac{I_2}{d_2^2} \qquad \text{as } i_1 = i_2 \tag{10.8}$$

Thus the relative candle-powers of the two sources are given by the ratio of d_1^2 to d_2^2.

For the judgement of equality to be as accurate as possible, the two surfaces must be seen simultaneously and have a sharp dividing edge. Various methods have been devised to obtain these conditions. An early arrangement merely consisted of a rod in front of a screen. When the two sources are properly located, the two shadows on the screen are just touching and are equally dense. The ratio of intensities is then given by the ratio of the square of the source-to-screen distances as in Equation 10.8.

A more complex system replaces the V-screen of Figure 10.6 with a screen of opaque white paper normal to the line between the lamps, having at its centre a spot made translucent by treating the paper with oil or wax. Unequal illumination will show a dark central spot on a bright surround or vice versa. Although the precise derivation of the brightness is a little complex, the balance position when the spot disappears still occurs when the illuminations are equal and Equation 10.8 applies.

A more complex and sensitive approach is that of the **Lummer–Brodhun photometer** shown in Figure 10.7. The light from each source passes through an aperture in the metal case and falls on either side of a white diffusing screen S; this is formed of plaster of Paris or of two sheets of ground white opal glass with a plate of metal between them. Two right-angled totally reflecting prisms P_1 and P_2 reflect the diffused light from the screen surfaces to the comparison prisms P_3 and P_4. Prism P_3 has its hypotenuse face ground to a curve except for a small circular area in the centre, which is plane and polished, and this is pressed into optical contact with the plane hypotenuse of prism P_4.

Thus light from the left-hand side of the screen reaches the eye via the central part of P_3P_4, while the surrounding area receives light from the right-hand side. The edge of the central area may be sharp but will disappear when a balance is obtained as before. A more complex working of prism P_3 can produce a more complex pattern to be erased at balance. One such pattern is given in Figure 10.7, and systems incorporating this are known as **contrast photometers**.

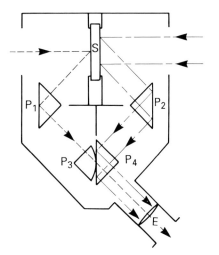
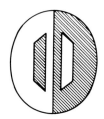

Figure 10.7 The Lummer–Brodhun photometer and the contrast pattern commonly used between P_3 and P_4

These systems give adequate accuracies provided the two sources are of identical colour. When a colour difference is apparent an alternative method can be used in which the two illuminated surfaces are viewed sequentially rather than side by side. Any difference in illumination now shows up as a flicker and the basis of this method is that flicker due to colour difference disappears at a lower speed of alternation than flicker due to luminosity difference so that the judgement of equal luminance is not affected by the presence of a colour difference.

Such a system, due to **Guild**, is shown in Figure 10.8. D is a disc which can be rotated at various speeds. D, T and S are coated with diffusing material such as magnesium oxide. The eye is placed at E and sees the inside of the tube T uniformly illuminated to 2.5 foot-lamberts while the hole A subtends an angle of 2°.

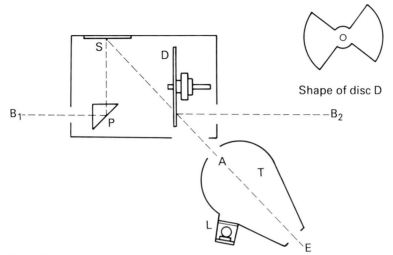

Shape of disc D

Figure 10.8 The Guild flicker photometer

As the disc is rotated, the eye sees the aperture field alternately illuminated by the light from S and D, and when the disc is rotating slowly the field will appear to flicker even when the luminance at S and D is equal. On increasing the speed of rotation the flicker becomes less pronounced and at a high enough speed will disappear. The test source B_2 and the photometer are fixed and measurements are made by moving the comparison lamp, B_1, while reducing the speed of rotation of the disc, until flicker almost disappears. The luminance of S and D is then equal and the relative intensities of B_1 and B_2 can be found as with other photometers.

To obviate any difference there might be between the reflection factors of the two surfaces of the photometer and any effects of stray light, a substitution method is generally employed. In this a third lamp, the comparison lamp, which need not be of known candle-power, is used with the standard and the test lamp, i.e. the lamp to be tested. Balance is first obtained between the comparison lamp and the test lamp. The distance between the comparison lamp and the photometer is then kept fixed and the standard substituted for the test lamp. By moving the standard lamp, balance is again obtained. Thus the same side of the photometer is illuminated in turn by both standard and test lamps. Then if I_1 and I_2 are the candle-powers of test lamp and standard respectively and d_1 and d_2 their respective distances from the photometer when balance is obtained, Equation 10.8 may be used.

10.7 Photoelectric photometers

The limitations of comparison photometers are never more evident than when a simple portable instrument is required for measuring the luminous intensity of actual lamps in offices, on roadways, etc. These need to be measured because they have reflectors which may become dirty or misaligned. Photoelectronic devices offer a solution provided the detector:

1. Responds to light in the same way as the eye.

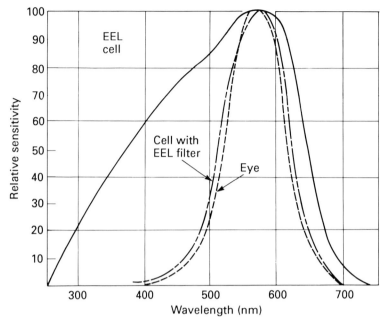

Figure 10.9 Spectral response of EEL cells (Courtesy of Diffusion Systems Ltd)

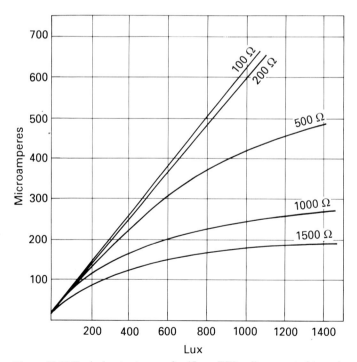

Figure 10.10 Typical output curves for 45 mm EEL cells connected to various external resistances (Courtesy of Diffusion Systems Ltd)

2. Has an output proportional to the luminous intensity.
3. Does not change from day to day.

Very few devices satisfy the first requirement and the nearest is the **selenium barrier-layer photocell**. The response of this device to light of different wavelengths is shown in Figure 10.9, compared with the visual response. If a correction filter that absorbs ultraviolet light is placed in front of the cell, the response can be made very close to that of the eye.

The cell produces a current through a resistance connected across it and the variation with light falling on it is shown in Figure 10.10. It can be seen that this is linear for low values of the resistance and a microammeter in the circuit could be calibrated directly in foot-candles or lux.

In manufacture, pure molten selenium is poured on to a steel base plate. A very thin transparent metal layer is spluttered over the selenium to form a conducting surface. The light reaches the selenium through the metal layer and the current is generated between this layer and the base plate. Rather like standard sources of light, these cells are 'aged' after manufacture by exposing them to a very strong light under short-circuit conditions. This helps to ensure constant values to the curves of Figure 10.10 throughout their life.

Figures 10.9 and 10.10 refer to products of Evans Electroselenium Ltd (now Diffusion Systems Ltd) and are reproduced with their permission.

(See also Section 10.10.)

10.8 Photometry of screens

In the section on illumination the concept of the perfectly diffusing surface was discussed and defined as a surface which appears equally bright from all directions. If such a surface is used as a screen for the projection of a colour transparency or cine-film, the projected image would appear equally bright in all directions. Most surfaces are not perfect and return the light in a partially specular reflection. As the audience tends to sit between the screen and the projector this can give a brighter image. Provided no objectionable high spots occur this can be a good thing.

The ratio by which a given screen appears brighter than a perfectly diffusing surface with the same illumination is often referred to as the **gain** of the screen in a particular direction. It is important to realize that there is no gain in the sense of extra light being generated. The additional brightness in a given direction is obtained at the expense of less light in another direction. The reflectivity of the screen can affect the apparent gain particularly in the case of a dirty surface.

Curves of the gain of typical screens are given in Figure 10.11. The curve for retroreflective screens can reach gain figures above 1000% and these have application in road signs and vehicle number plates where the driver sits within a few degrees of the projection axis of the headlamps.

10.9 The photometry of sources, integrating spheres

Luminous intensity of candle-power refers to the quantity of light emitted by the source in one given direction. The standard sources described have their intensities specified in a particular direction. In ordinary commercial work, however, the

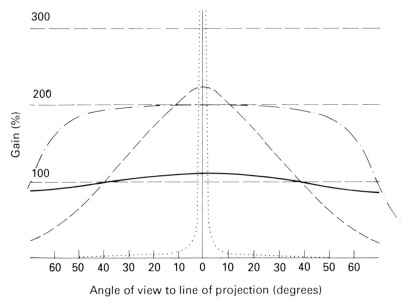

Figure 10.11 Gain of various screens: – – glass beaded; · · · · · retroreflective; — matt white; – · – aluminium-coated lenticular

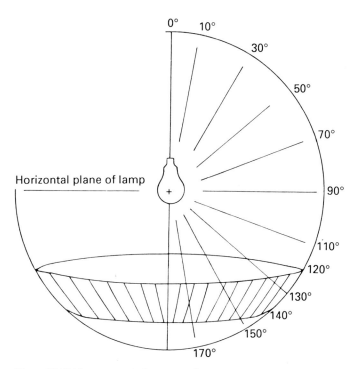

Horizontal plane of lamp

Figure 10.12 Measurement of mean candle-power

given candle-power is usually an average of the intensities in all directions, when it is known as the **mean spherical candle-power**.

The determination of this value for a lamp or lighting system may be done by one of two methods. The intensity may vary in all directions although a proportion of systems may be symmetrical about a vertical axis. In this case the total luminous flux may be measured by supporting the lamp in its operating position and measuring the intensity at different vertical angles as in Figure 10.12.

If the luminous intensity is measured at equal angles as shown, the results cannot simply be averaged as the light emitted near the horizontal plane of the lamp is also being emitted over a considerable solid angle all round the fitment. Similarly the light measured directly above or below the lamp is being emitted nowhere else.

This geometrical feature leads to the adoption of zone factors, which are proportional to the area swept out by the region around the angle chosen. These zone factors are given by the difference in the cosines of the *bounding* angles multiplied by 2π. For zones of 20° the factors are given in Table 10.3.

Table 10.3 Zone factors

Angle of measurement (degrees)	Zone (degrees)	Zone factor
10	0–20	0.378
30	20–40	1.091
50	40–60	1.671
70	60–80	2.051
90	80–100	2.182
110	100–120	2.051
130	120–140	1.671
150	140–160	1.091
170	160–180	0.378

The total of each luminous intensity reading multiplied by its zone factor gives the total flux, Φ, from the lamp. When this is divided by 4π the mean spherical candle-power is obtained in lumens per steradian or candela. If the zone factors in Table 10.3 are summed, the result is 4π.

As a more approximate method, a series of angles may be taken which in themselves give zones of equal solid angle. These are known as **Russell angles** and those giving six equal zones would be 33.6°, 60°, 80.4°, 99.6°, 120° and 140.4°. The average of luminous intensity values obtained at these angles gives the mean spherical candle-power directly.

When the source has large variations in its intensity in different *horizontal* directions, the same approach may be used but the lamp rotated completely about its vertical axis for each reading so that an average reading may be obtained. This is sometimes referred to as the **mean horizontal candle-power**. The values obtained in this manner may be used in the same way as for the symmetrical source described above.

An alternative method for total flux measurement involves the use of an **integrating sphere**. This device is a large sphere with a diffusely reflecting inner surface. The theory of its use depends on the application of the cosine law of illumination to the geometry of a sphere from which it can be shown that the surface illumination on the sphere wall *due to reflected light* is proportional to the total flux of a source inside. No direct light from the source must reach the point at which the illumination is measured.

This means that baffles must be inserted in the sphere and a hole cut in the wall to allow the illumination to be measured. These detract from the accuracy of measurement, as does the presence of fittings inside the sphere. If the test lamp and standard source can be introduced successively into the same place within the sphere for measurements, these inaccuracies can be minimized.

10.10 Measurement of illumination and luminance

The photometers described in Sections 10.6 and 10.7 are used to measure the candle-power of sources, although they do this by measuring the illumination due to a particular source at a given distance. If more than one source is present (other than the standard), the illumination due to both will be measured. The comparison photometers are rather too cumbersome and the photoelectric system is commonly used for this purpose, being placed on desktops and roadway surfaces to measure the illumination due to many sources. Indeed it is not capable of differentiating between the sources.

If the photoelectric device is used as a camera exposure meter, it measures the total light reaching the camera either from the scene or from some restricted area near the centre. No account can be taken of whether this contains a few bright sources or an even distribution, and exposure errors can be caused by this. The device essentially measures the illumination on the camera due to the scene and not the luminance of the surfaces comprising the scene.

A more sophisticated instrument is the **Spectra-Pritchard photometer** (Figure 10.13). This uses the eye solely to define the surface to be measured. The eye views the scene via a mirror having a hole in it which therefore appears as a black dot to the observer. The light from the scene is imaged on to the mirror so that the black dot is sharp and the light passing through the hole is coming only from the part of

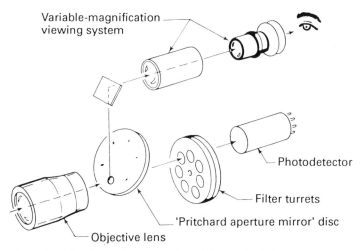

Figure 10.13 Aperture mirror disc system of the Spectra-Pritchard Photometer, model 1980, manufactured by the Photo Research Division of the Koll Morgan Corporation

the scene covered by the black dot. This transmitted light reaches a specially calibrated photodetector via a series of filters. The mirror has holes of different sizes which can be rotated into the observed scene while the filters have a considerable range of optical densities so that a very versatile broad-range instrument is made.

10.11 The action of filters

Unless the light is travelling in a complete vacuum, there is a continuous reduction in its intensity due to the media through which it passes. This reduction may be caused by absorption such as occurs with dark glasses, scattering such as occurs with ground glass, or reflection such as occurs with glass surfaces. In principle there is no difference between reflection and scattering, but all three processes reduce the amount of light transmitted from one place to another. Even the atmosphere has finite values of absorption and scattering. These vary considerably with meteorological conditions, but the transmitted light is unlikely to be better than 95% over a kilometre.

For optical and ophthalmic glass the absorption value is considerably more, but a further loss occurs at the surfaces of any sample due to reflection. Thus the absorption of most optical glasses is below 2% per centimetre but curves of the transmission of 1 cm samples often show values of about 90% because of the 4% reflection loss at each surface. This subject is dealt with in Sections 11.4 and 11.5.

10.12 Contrast

The contrast pattern in the Lummer–Brodhun photometer was used in a *null* method; that is, the instrument was in correct adjustment when a pattern could not be seen. Provided the pattern subtends a reasonable size and is reasonably bright, the eye can detect a contrast difference somewhat better than the 1% value of Fechner's fraction. It is found that the ability of the eye to just detect a small step in luminance, the **threshold contrast** of the eye, is very dependent on background luminance once this is below 100 Td, and also varies with the type of pattern, a single edge being less easily seen than a series of edges or bars (see Chapter 15).

The word *contrast* has been used in different ways and care must be taken when comparing authors. If B_t is the luminance of the object (or target) and B_b the luminance of the background, contrast is defined as

$$C_d = \frac{B_b - B_t}{B_b} \tag{10.9}$$

for objects darker than the background, and

$$C_l = \frac{B_t - B_b}{B_b} \tag{10.10}$$

for objects brighter than the background. When repetitive object patterns are used (see Section 14.7) the term **contrast modulation** is more generally used, which is defined as

$$C_m = \frac{|B_t - B_b|}{B_t + B_b} \tag{10.11}$$

The modulus of the difference means that only positive values of C_m occur, as is the case with all the other definitions. Strictly, contrast modulation should be applied only to sinusoidally varying patterns, but it is generally used for any repetitive pattern. It does not matter whether the bright parts or dark parts are used for B_t.

When an object is viewed from greater and greater distances, not only does its apparent size diminish and its luminance reduce because of absorption by the atmosphere, but its contrast is reduced due to scattering. Both these latter effects are very variable with weather conditions. (The book *Vision through the Atmosphere* by W. E. K. Middleton is a well known text on this subject.) The **visibility** or **meteorological range**, V, is a measurement of the greatest distance at which a large black object can be seen against the sky background.

The absorption coefficient and scattering coefficient of the atmosphere are often combined together in the **extinction coefficient**, σ. We can obtain an *approximation* of this for visible light from $\sigma = 3.9/V$, where σ is in kilometres^{-1}, if V is in kilometres.

The contrast C_r at range r is given by

$$C_r = C_0 e^{-\sigma r}$$

where C_0 is the contrast at zero range. Both C_0 and C_r are defined as in Equations 10.9 and 10.10, and the range must be in kilometres if σ is in kilometres^{-1}.

10.13 Colour mixing and measurement – colorimetry

When a mixture of light of two or more colours reaches the eye, new colour sensations are produced, and the eye is incapable of recognizing the simple colours contained in the mixture. Light of different colours may be mixed in a number of simple ways. Two or more projectors with coloured filters may be arranged to project overlapping patches of coloured light on a white screen. A disc may be coloured with sectors of the colours to be mixed, and on quickly rotating the disc the colours blend and produce the sensation due to the mixture. An improvement on this method is the Maxwell colour disc. Three discs coloured say red, green and blue-violet, and having a radial slot cut in each, as in Figure 10.14, may be fitted together to form a single disc, with three sectors, the angles of which can be varied by sliding one disc over the others.

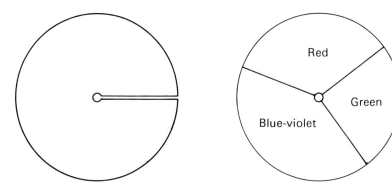

Figure 10.14 Maxwell's discs

The results of adding colours may be stated in a general way as follows:

1. Red + green + blue-violet = white
2. Red + green = yellow
3. Green + blue-violet = green-blue (peacock blue)
4. Red + blue-violet = purple

Therefore

5. Yellow + blue-violet = white
6. Green + purple = white

By altering the proportions of the colours one to another, all the hues are produced; thus, starting with red and adding increasing quantities of green, the hue changes from red to orange-red, orange and yellow and then through the greenish-yellows to green. Adding increasing amounts of blue-violet to green the hue changes from green through the various hues of green-blue and blue to blue-violet. In the same way, adding blue-violet to red gives magenta, the various purples and violet.

It is generally more useful, however, to study the effects of mixing purer colours than can be used in the above methods. Three suitably chosen spectrum colours, usually red, green and blue, when mixed in certain proportions, produce the *sensation* of white, and any colour sensation can be matched in *hue* by varying proportions of the three primaries or **matching stimuli**. Although any colour can be matched in hue, a complete match is not always possible because the colour produced by the mixture may be less saturated than the colour to be matched, the *test colour*. To obtain a complete match the test colour will have to be desaturated by the required amount by adding to it either a known amount of white or, as can be shown to produce the same effect, a certain amount of one of the matching stimuli.

Any colour can therefore be expressed in terms of the three matching stimuli, R, G, and B, and this is the basis of **colorimetry**, a subject of considerable importance in industry. A number of instruments, known as **colorimeters**, have been devised for colour measurement, and these differ mainly in the methods by which the matching stimuli are produced, mixed and varied in intensity. Coloured light, desaturated if necessary, from the specimen under test fills one-half of the observation field of the instrument and the mixture of the matching stimuli the other half.

In some instruments, such as the Wright colorimeter, three narrow selected bands of the actual spectrum are mixed together in varying proportions, but in others the matching stimuli are obtained by the use of three colour filters giving highly saturated colours. This enables a simpler instrument, more adapted for industrial use, to be produced. An example of an instrument using colour filters is the Donaldson colorimeter (Figure 10.15). The light source, L, is a 250 W projection lamp; this is run at 90% of its rated voltage, which gives a longer life and more uniform light output. The light is received on the plate A in which are three rectangular apertures, R, G and B, the dimensions of which can be independently varied by shutters sliding in grooves. Each shutter is operated by a separate rack and pinion, and the extent of the opening is read on a scale on the shutter. Three colour filters are mounted behind the apertures.

An image of the lamp filament in each of the three colours is formed by means of the lens C on the aperture D in the integrating sphere S. This sphere has a diameter

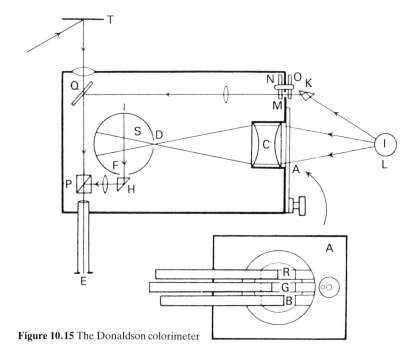

Figure 10.15 The Donaldson colorimeter

of 6 inches and its interior surface is silver plated and coated with magnesium oxide. By repeated reflections in the sphere the light from the three filters is completely mixed and this mixed light from the sphere wall at I passes through a second aperture F in the sphere. The light from F after reflection by the right angle prism H is reflected by the photometer prism P to the eye at E. The eye sees one-half of the field evenly illuminated by light which is a mixture of the light passing through the three filters and the proportion of the three component colours is adjusted by means of the shutters in A. Light from the test surface T fills the other half of the field.

When it is necessary to desaturate the test colour, light from the source is deflected by the prism K through a small aperture M fitted with a diffusing screen. A disc O carries three filters similar to the three main filters which can be brought in turn in line with the aperture, and N is a circular neutral wedge which can be rotated over the aperture to regulate the amount of desaturating stimulus. The desaturating light is then reflected into the field of the test colour by the plane parallel glass plate Q.

When a given colour has been matched by the mixture of the three matching stimuli its colour can be expressed in the form of an equation, as was first suggested by Clerk-Maxwell in 1856.

Thus $C = r(R) + g(G) + b(B)$, where C is the colour and the numerical coefficients r, g and b represent the amounts of the red, green and blue stimuli required for the match. The units in which the matching stimuli are measured are arbitrarily chosen so that an equal number of each stimulus will match a standard white when added together. These units are call **trichromatic units** and are units of *colour* as distinct from units of *light*. In the **unit trichromatic equation** the

trichromatic coefficients r, g and b will always add up to unity; thus, for standard white,

$$1W \equiv \tfrac{1}{3}(R) + \tfrac{1}{3}(G) + \tfrac{1}{3}(B)$$

For some colour, say a yellowish-green, the equation might be

$$C \equiv 0.3(R) + 0.6(G) + 0.1(B)$$

The coefficient 1.0 for the test colour in trichromatic equations is usually omitted. In the case of a colour, such as a pure spectrum green, which must be desaturated before a match can be obtained, the equation might read

$$C \equiv 0.480(R) + 0.543(G) - 0.023(B)$$

or

$$C + 0.023(B) \equiv 0.480(R) + 0.543(G)$$

that is, when the given amount of blue is added to the colour a match can be obtained by the mixture of appropriate amounts of red and green.

The fact that most colour sensations could be produced by the mixture of three suitably chosen primaries led to the formulation of the **trichromatic theory** of colour vision. This theory, first put forward by Young in 1802 and later elaborated by Maxwell and Helmholtz, is usually known as the Young–Helmholtz theory, and supposes the existence in the retina of three types of receptors, which when stimulated give rise to the sensations of red, green and blue respectively.

Plate 5 shows a colour reproduction of the colour triangle. The edge colours of this shape represent the spectral hues while the central 0.33, 0.33, 0.33 point is pure white. A straight line from this point to the edge represents the locus of saturations from 0 to 100% at that hue. It is seen that the values for which $r = 1$ or $g = 1$ or $b = 1$ are three different colours with saturations above 100%. They are therefore not real colours and cannot be duplicated by any dyes. Essentially they are theoretical points, which allow the whole of the observable colour diagram to lie within the positive axes of the graph.

A plot of colour temperatures may also be added to this diagram, and this is shown on Plate 5 (see Section 9.5).

Any two colours that when added together give the *sensation* of white are known as **complementary colours**; thus, as was seen above, yellow and blue-violet are complementaries, as are also green and purple. In many cases pairs of pure spectral hues, that is monochromatic light, having appropriate relative luminosities may be complementaries. Some of these complementary pairs of wavelengths with their required relative luminosities as determined by Sinden (1923) are shown in Table 10.4.

Table 10.4 Wavelengths and relative luminosities of complementary colours

Complementaries							
λ_1 (nm)	650	609	586	578.5	574	573	570.5
λ_2 (nm)	496	493.5	487.5	480.5	472	466.5	443
Relative luminosities							
L_1	42.1	53.4	73.8	85.5	92.2	94.2	97.5
L_2	57.9	47.6	26.2	14.5	8.0	5.8	2.5

The colour of most objects is due to selective absorption. When white light falls on a coloured opaque substance, some of it passes into the substance, and certain frequencies are diffusely reflected close to the surface. A red object, for example, illuminated with white light, is reflecting some of the white light from the surface, and for a short distance inside the surface is diffusely reflecting the red and possibly also the orange and yellow, while the green, blue and violet are absorbed. The same object illuminated with green or blue light will appear almost black, as except for a small amount from the surface it is reflecting no light of this colour. Coloured transparent substances, coloured glasses, dyes, etc., owe their colour to a similar cause, a red glass transmitting only red light and absorbing light of other colours.

The colour of an object will obviously depend on the colour of the illuminating light, and in speaking of the colour of any object we are referring to its colour when illuminated with white light.

The colours obtained by reflection and transmission are usually far from pure, and examined with a spectroscope will be seen to contain a band of frequencies on either side of the predominant hue. There is also usually a considerable amount of white light reflected from the surface, and if this is diffusely reflected and mixes with the coloured light from below the surface, the colour will be less saturated than when the surface is polished, and the white light is reflected in a definite direction. This effect is seen in the difference in colour in polished and unpolished wood and marble.

The effects of mixing coloured substances, such as pigments, or of combining colour filters – dyes, coloured glasses, etc. – are quite different from those obtained in adding coloured lights. It is well known, for example, that the mixture of blue and yellow watercolours gives green, whereas we have seen that the addition of blue and yellow light gives the sensation of white. The explanation is simple: in the case of the pigments, the blue particles, when illuminated with white light, are absorbing the longer wavelengths and reflecting the violet, blue and green; the particles of yellow pigment are absorbing the violet and blue and reflecting the green, yellow and red. A mixture of the two pigments will therefore absorb or subtract everything from the white light but the green which is reflected. If we place a red filter in one lantern and a green filter in another and project two overlapping patches on a white screen, the additive mixture produces the sensation of yellow, but if the two filters are placed together in one lantern, no light passes through the combination, as the light transmitted by one filter is absorbed by the other.

A great range of colours can be obtained by the mixture of pigments or dyes of three carefully chosen colours; these are known as the **subtractive primaries**, as distinct from the **additive primaries** dealt with earlier. The subtractive primaries are magenta, yellow and green-blue (called cyan). These colours are in themselves mixtures, being composed of broad bands of the spectrum. They are such that, of white light, the magenta reflects all but green, the yellow all but blue and violet, and the green-blue all but red and yellow. The general results of additive and subtractive mixtures are shown diagrammatically in Figure 10.16.

The subtractive primaries are the colours used in three-colour printing. Three photographs are taken, one through a red, one through a green and one through a blue filter. Prints are made from the three negatives in colours complementary to those of the filters, and the three prints are superposed. As the red filter transmits only red light, the dark portions of a print from the negative taken through the red filter represent absence of red, and a print from this negative is therefore made in white minus red, i.e. cyan. Similarly the print from the negative taken through the

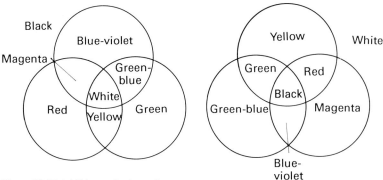

Figure 10.16 Additive and subtractive primaries

green filter must be made in magenta, and the print from the negative taken through the blue filter must be made in yellow. If the colours of the filters and inks are carefully adjusted the superposed prints can give excellent reproduction of the colours of the original object. Both the additive and subtractive systems of colour mixing have been used in colour photography but all the present day processes are subtractive.

When light passes through a space filled with minute particles, some of it is scattered, chiefly by diffraction. The percentage of light so scattered is found to be inversely proportional to the fourth power of its wavelength, so that the scattered light will consist mainly of the shorter wavelengths, the blue and the violet. This is known as the **Tyndall effect** after Tyndall (1820–1893), who investigated the effect experimentally.

The blue of the sky is due to this selective scattering. A certain amount of the shorter wavelengths of the sun's light is scattered by the particles of dust and water and the molecules of gas in the earth's atmosphere, while the longer wavelengths are transmitted. The increasing redness of the sun, as it gets nearer the horizon, is due to the same cause. As the light is then travelling through a great thickness of atmosphere more of the shorter wavelengths are scattered, and the remaining transmitted light is therefore redder. The light scattered in this way is also polarized (Section 11.8).

Exercises

10.1   Given that the illumination of the earth in full moonlight is $0.02\,lm/ft^2$ and that the distance of the moon is 235 000 miles, find its candle-power.

10.2   How does the intensity of illumination at a point on a screen vary (a) with the distance of the point from the source, (b) with the angle of incidence of the light?

A small source of 100 candle-power is suspended 5 ft above a horizontal table. Draw a curve showing how the intensity of illumination on a table produced by direct light varies along a straight line passing directly under the source.

10.3   Distinguish source intensity and intensity of illumination. How can the latter be measured and in what units is it expressed?

A surface receives light normally from a source at a distance of 2.82 m. If the source is moved closer so that the distance is only 2 m, through what angle must the surface be turned to reduce the illumination to its original value?

10.4   A point source of light of 20 cd is situated 50 cm from a plane mirror and on a normal to its centre. If the mirror reflects 90% of the incident light, find the illumination on a screen 3 m from and parallel to the mirror.

10.5   A small source of 30 cd is placed 10 cm from a +5 D lens of 10 cm diameter. Find the illumination on a screen 1 m from the lens, neglecting losses by reflection or absorption at the lens. What change in the illumination will be produced by reducing the aperture of the lens to 5 cm diameter?

10.6   Light from a lens 40 mm in diameter is converging to a point 75 cm from the lens. Find the diameter of the pencil at 15, 30 and 80 cm from the lens. Compare the illumination on a screen in these three positions.

10.7   State Lambert's law on the variation of the emission of light with direction.

A perfectly diffusing incandescent surface A of area 1 mm^2 and brightness 20 cd/mm^2 is set up parallel to and 50 cm from a screen, the line AB being normal to the screen. Find: (a) the luminous intensity of the surface A in candles along a direction AC inclined 30° to the normal AB, (b) the illumination in lumens per square metre on the screen (i) at B, (ii) at C (the point where AC cuts the screen).

10.8   Two lamps, one of 10 cd, are placed at opposite ends of a 2 m bench and a screen placed between them is found to be equally illuminated on its two sides when 80 cm from the 10 cd lamp. Find the power of the other lamp. What form should the screen take in order that the illumination of the two sides may be easily compared?

10.9   In comparing two sources of light by means of a photometer, it is necessary to arrange that:

(a)   The photometer head is on the line joining sources (axis).
(b)   All extraneous light is excluded.
(c)   Both sides of screen are equally clean.
(d)   Both sides of screen are equally inclined to axis of photometer.

Explain *clearly* the need for each of these precautions.

10.10   State and briefly discuss Lambert's law on the emission of light. A small source of light of area 2 mm^2 is set up 20 mm from a + 60 D condensing lens of diameter 40 mm. The luminous flux falling on the lens from the source is 40 lm. Find the luminance of the source in candelas per square millimetre and, assuming no losses, the illumination of the image of the source in lumens per square metre.

10.11   Two lamps of 30 and 20 cd respectively are placed 2 m apart. Find the positions on a line joining them where a screen would be equally illuminated by these two lamps. How would you determine these positions experimentally?

10.12   Explain the meaning of the term optical density as applied to an absorbing medium. Illustrate by finding the optical density of a filter which transmits 20% of the light flux incident upon it.

What will be the transparency (i.e. percentage of light transmitted) and optical density of a filter of the same material and twice the thickness?

10.13   A photometer was placed between two lamps 2 m apart and was found to balance when 80 cm from one of the lamps. Where must it be placed to again balance if a tinted glass absorbing 55% of the light is placed in front of the brighter lamp?

10.14   State Lambert's cosine law of emission. Explain what is meant by a perfectly diffusing surface.

The luminance of a flat perfectly diffusing surface of area 2 mm^2 is 25 cd/mm^2. What is its luminous intensity along (a) the normal direction, (b) a direction inclined at 60° to the normal?

If the light leaving the surface normally falls on a circular screen 1 cm in diameter and 50 cm from the surface, what amount of luminous flux, expressed in lumens, will fall on the screen?

10.15   A *uniform diffuser* is a surface having the same luminance (L) in all directions. Show that, for such a source of light, the luminous intensity (I) in any direction varies as the cosine of the angle between that direction and the normal to the surface.

Such a source, small in area, is suspended parallel to and 2 m above a horizontal table. The illumination on the table at a point A vertically beneath the source is 5 lm/mm^2. Find the illumination on the table at a point B 1 m from A.

10.16 (a) Light from a point source of 50 cd enters an eye of effective pupil diameter 3 mm, the source being 50 cm from the eye. What amount of luminous flux, in lumens, enters the eye?

(b) Calculate the illumination on the ground midway between two lampposts 100 yards apart and 16 ft high, each lamp being of 450 cd in the direction considered. Express your result in lumens per square foot.

10.17 Two small lamps, A and B, give equal illuminations on the two sides of a photometer head when their distances from it are in the ratio 2:5. A sheet of glass is then placed in front of lamp B and it is found that equality of illumination is again obtained when the distances A and B are in the ratio 6:5. Find the percentage of light transmitted by the glass.

From this figure calculate the optical density of the glass.

10.18 Two pieces of glass, one blue and the other yellow, are superposed. What colour is seen? If the two pieces are placed side by side in a lantern to illuminate a screen, will the effect be the same? Give reasons in each case.

10.19 How would you specify a given colour to reproduce it from a knowledge of the specification?

10.20 Explain why it is possible with two pure spectrum colours suitably chosen (say a red and a green) to match daylight containing the whole range of spectrum colours. What does *match* mean here? Could a picture be painted successfully by such a light?

10.21 The letters on a poster printed in red ink disappear altogether when looked at through a red glass. So does the lettering on a poster printed in yellow ink when seen by gaslight. Explain these results.

10.22 Explain each of the following:

(a) Yellow and blue paints mixed together produce green paint, but yellow and blue lights projected on to a screen produce white light.

(b) Smoke from the end of a burning cigarette appears blue whereas smoke from the mouth of the smoker appears grey.

(c) Materials with bright saturated colours can be obtained in the red-orange-yellow range, but in the green-blue range the colours may be saturated but not bright.

Chapter 11

Optical materials: interaction of light with matter

11.1 Dispersion

Having very briefly considered matter as a source of light and equally briefly reviewed the detection of light by organic and inorganic materials, we now turn to the interaction of light with optical materials. Obvious optical materials are glass and plastics from which lenses may be made, metallic materials such as silver or aluminium from which mirrors may be made, scattering materials for screens, and coloured materials for filters. In a general sense all materials are 'optical' as every material has some effect on visible light. Even the air in the atmosphere has a scattering, absorbing and refractive action. In the simplest type of optical material, optical action is constant throughout its volume. Such materials have already been defined in Section 1.2 as **homogeneous**. Another restriction on 'simple' optical materials is that they are **isotropic**, which means that they have the same effect no matter what direction the light is travelling through them. Inhomogeneous and anisotropic materials will be discussed later in this chapter.

All materials have an optical action which changes as the wavelength of the light is changed. The interaction of light with materials is always dependent on the spacings (and binding energies) of the atomic and molecular components of the material. Sometimes when these spacings are very small with respect to the wavelength the optical effect has only a very small wavelength dependence. At visible wavelength most optical materials have values which change by between 1% and 10% over the extent of the spectrum from red to blue.

Chapter 1 defined the refractive index of a material as the ratio of the velocity of light in vacuum divided by the velocity of the light in the material. Although the velocity of light in the vacuum appears constant and the same for all wavelengths, in materials the velocity is different for different wavelengths. This means that a multi-wavelength flash of white light travelling down an optical fibre, for instance, will steadily separate into its various components with, usually, the blue light coming last. This separation is called **dispersion** and it means that the refractive index values which have been used through the earlier chapters of this book must, for the same material, be adjusted after reference to the actual wavelength of the light used. There is no accurate theoretical method for calculating these values and so they are measured on instruments called **refractometers**. In the teaching laboratory it is instructive to use the theory of refraction through a prism (Section 2.8) on a **spectrometer** and this method will be described first.

11.2 Spectrometers and refractometers

To measure the angle through which light of various colours is deviated by a prism, the instrument used (for teaching purposes) is the **spectrometer** (Figure 11.1). The arrangement of the instrument is shown in plan in Figure 11.2. Its essential parts are a collimator, an astronomical telescope, a table for carrying a prism, and a divided circle. The collimator, with an adjustable vertical slit, is mounted with its axis directed towards the axis of the divided circle and is usually fixed. The telescope, which has a Ramsden eyepiece with crosslines at its focus, rotates in a horizontal plane about the axis of the circle and its position is read on the circle by means of a vernier. The prism table, which is provided with levelling screws, also

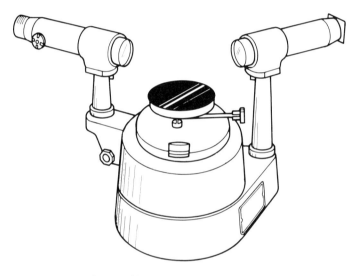

Figure 11.1 A modern teaching spectrometer (Courtesy of PTI Co. Ltd)

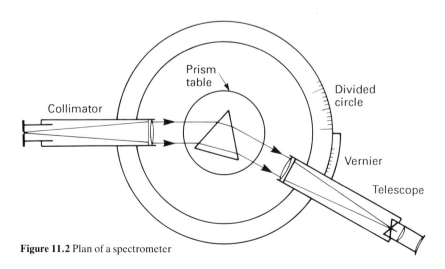

Figure 11.2 Plan of a spectrometer

rotates about the axis of the circle, and in some instruments its position can likewise be read on the circle by means of a vernier. In adjusting the spectrometer the telescope eyepiece is first focused on the crosslines and the telescope then focused on a distant object. The slit of the collimator is illuminated and, the telescope having been brought into line with the collimator, the latter is adjusted by moving the slit in or out until a sharp image of the slit is formed in the plane of the crosslines of the eyepiece.

In order to use Equation 11.1 given below it is necessary to measure the angle of the prism. This may be measured on the spectrometer by either of the following methods:

1. The prism is placed on the table of the spectrometer with its refracting faces vertical and the angle to be measured facing the collimator (Figure 11.3). A certain amount of light is reflected from the two faces of the prism, and the angle between the two reflected beams is measured by reading the position of the telescope when an image of the slit is received on the crosslines, first by reflection at the face AC and secondly at the face AE. It can easily be proved that the angle between these two reflected beams is equal to twice the angle of the prism.

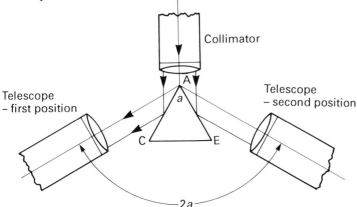

Figure 11.3 Measurement of the angle of a prism

2. The telescope is fixed at an angle of roughly 90° with the collimator (Figure 11.4). The prism is rotated until an image of the slit formed by reflection at the face AE is received on the crosslines, and the position of the prism table is read on the circle. The prism table is then rotated until the image of the slit reflected from the face AC falls on the crosslines, and the reading of the table again taken. The angle through which the table has turned will then be equal to 180° $-a$, where a is the angle of the prism.

To determine the refractive index of the prism using Equation 11.1, it will be necessary to measure the angle of minimum deviation; this may be done as follows. The collimator slit is illuminated by a source giving the light for which the refractive index is wanted, and the telescope is brought into line with the collimator, so that an image of the slit falls on the crosslines. The reading of the telescope in this position is taken. The prism, which will usually have a refracting angle of about 60°, is then placed on the table with one refracting face making an angle of about 45°

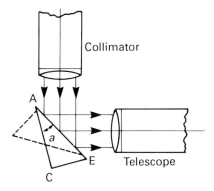

Figure 11.4 Measurement of the angle of a prism

with the beam from the collimator (Figure 11.5). The telescope is turned until an image of the slit is seen. To find the position of minimum deviation, the prism is slowly rotated in the direction that causes the image to move towards the first position of the telescope, i.e. towards the undeviated direction. Following the image with the telescope a position is found where the image commences to move in the opposite direction, and this is therefore the direction of minimum deviation.

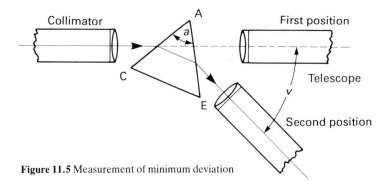

Figure 11.5 Measurement of minimum deviation

The image of the slit in this position is brought on to the crosslines, and the reading of the telescope again taken. The difference in the readings gives the angle of minimum deviation v, and the refractive index is found from Equation 11.1.

$$n = \frac{\sin\left(\dfrac{a + v}{2}\right)}{\sin\left(\dfrac{a}{2}\right)} \tag{11.1}$$

(from 2.18)

Although the refractive index of a material may be found by grinding and polishing it into a prism shape and using it on a spectrometer to give an angle of minimum deviation, the method is time consuming and expensive as a high degree of flatness is required on the incident and emergent faces for good accuracy. The methods described above are insufficiently accurate when the precise value of index is required at different wavelengths so that achromatic lenses (Chapter 14) may be designed and manufactured. For this purpose the Hilger–Chance V-block

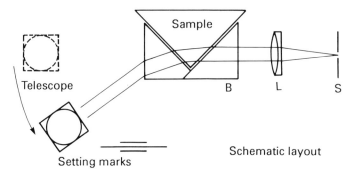

Figure 11.6 Hilger–Chance V-block refractometer

refractometer was developed. A side view of this instrument is shown in Figure 11.6.

The V-shaped block, B, is made as indicated of two prisms of medium-index glass. The angle between the faces of the V is exactly 90° and the sample of material is ground and polished to about 90°. The quality of the polish and the accuracy of the angle are not of great importance as a film of liquid is placed between the sample and the block to ensure optical continuity. Monochromatic light from the source, S, is collimated by the lens, L, and passes through the loaded V-block, which forms a three-component prism system. The deviation of the emergent beam is measured by the horizontal telescope, T, which can be rotated as shown. This deviation depends on the refractive index of the sample and may be in either direction. In use the instrument is usually calibrated with samples of known materials.

The setting system received special attention during the design and utilizes two short parallel lines in the eyepiece of the telescope. These can be set equidistant about the image of the slit seen in the telescope as in Figure 11.6. The eye is very sensitive to any non-symmetry and maximum accuracy can be achieved (Section 6.6.4). This instrument is commonly used for quality checking by optical and ophthalmic glass manufacturers.

11.3 Transmitting optical materials

In practical terms transmitting materials such as glass and water can be easily distinguished from absorbing materials such as wood, metals and bitumen. In fact all transmitting materials also absorb and this is particularly evident when large thicknesses are involved. There is no light in the depths of the oceans and light transmission through a metre thickness of ordinary window glass would be much less than 50%. For light transmitting optical fibres for communications very special manufacturing techniques yield a 50% drop over distances of 10 km or more. Nevertheless they still absorb some light.

On the other hand all absorbing materials will transmit light if they are thin enough. This is particularly evident with metals, which can be vacuum deposited (see Chapter 12) on to the surface of optically transmitting materials. This section is concerned with materials that may be used in optical systems and for thicknesses up to 100 mm or so having absorption values of only a few per cent. In Section 11.7 it will be seen that other losses occur due to some light being reflected at the surfaces but the predominant effect is that of light transmission. The main parameter of such materials is their refractive index and their dispersion, the extent to which the refractive index changes with wavelength.

For practically all transparent materials, the dispersion is such that the refractive index is greater for light of shorter wavelength. This means that all optical components such as lenses and prisms deviate blue light more than red light. To correct this **chromatic aberration** it is usual to have a pair of lenses or prisms made of *different* materials but in an **achromatic** combination. The details of this approach are given in Chapter 14 but for this to be done accurately it is clearly necessary to know the refractive indices and dispersions of the available materials to an equal degree of accuracy. There are many types of optical glass available from glass manufacturers. Optical plastics, now becoming more widely used, suffer because there are relatively few types.

For convenience the value of the refractive index of a material for a wavelength near the centre of the visible spectrum is known as its **mean refractive index**. This used to be the orange spectral line put out by the sodium atom, the Fraunhofer D line (see Section 9.4) but this is in fact a line pair and although easy to obtain in the laboratory it has been superseded by the nearby helium d line, which allows more accurate measurements. The helium d line has a wavelength of 587.56 nm, which is some distance from the peak of the visible response curve. Glass manufacturers therefore also quote the refractive index for the mercury e line, which has a wavelength of 546.07 nm, closer to the peak visual response at 555 nm.

The definition of dispersion has also changed over the years. In its simplest form it is merely the difference in refractive index between the blue hydrogen F line at 486.13 nm and the red hydrogen C line at 656.28 nm. Some manufacturers also quote the difference between the blue cadmium F′ line at 479.99 nm and the red cadmium C′ line at 643.85 nm. Further details are given in Chapter 14.

The mean dispersion of a glass is usually greater for glasses of high mean refractive index, but the increase in mean dispersion is not proportional to the increase in mean refractive index. This is illustrated by the figures for the deviation of the C, D and F lines produced by two prisms, one of crown glass and the other of flint glass, the refracting angles being such that the minimum deviation for the D line is equal in each case (Table 11.1).

It will be seen that, while the mean deviation, v_D, produced by the two prisms is equal, the dispersion between the F and C lines, $v_F - v_C$, for the flint prism is

Table 11.1

	n_C	n_D	n_F
Hard crown glass	1.5150	1.5175	1.5235
Dense flint glass	1.6176	1.6225	1.6349

	v_C	v_D	v_F	$v_F - v_C$
Crown prism: $a = 60°$	38° 31′	38° 42′	39° 14′	0° 43′
Flint prism: $a = 52° 6′$	38° 27′	38° 42′	39° 42′	1° 15′

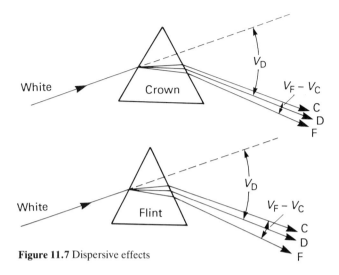

Figure 11.7 Dispersive effects

almost double that for the crown. The effect is shown diagrammatically in Figure 11.7, the dispersion being exaggerated.

From Equation 2.20, the deviation of a prism of small refracting angle for the D line is given by

$$v_D = (n_D - 1)a$$

Similarly,

$$v_F = (n_F - 1)a$$

$$v_C = (n_C - 1)a$$

The dispersion between the F and C lines is therefore

$$v_F - v_C = (n_F - n_C)a = \frac{n_F - n_C}{n_D - 1} \cdot v_D$$

This ratio of the dispersion of the C and F lines to the mean deviation

$$\frac{v_F - v_C}{v_D} = \omega = \frac{n_F - n_C}{n_D - 1} \tag{11.2}$$

is called the **dispersive power** or **relative dispersion** of the medium.

In the calculation of achromatic lenses and prisms the reciprocal of dispersive power is often used. This is called the **constringence**, but it is also referred to as the **Abbe number** or **V value** after the formula

$$V = \frac{n_D - 1}{n_F - n_C} = V_D \tag{11.3}$$

The variety of spectral lines referred to earlier give two other definitions:

$$V_d = \frac{n_d - 1}{n_F - n_C} \tag{11.4}$$

and

$$V_e = \frac{n_e - 1}{n_{F'} - n_{C'}} \tag{11.5}$$

Partial dispersion values can also be defined as the index difference between the mean value and that at another wavelength. These are important because the mean refractive index is never centrally placed between the index values for red and blue.

In optical design work (described in Chapter 14), computer programs store the actual refractive index values for all the glass types for eight or more wavelengths or else calculate them as required from stored coefficients used in a polynomial equation of wavelength, **Cauchy's formula**. However, concepts of mean refractive index and V value are important in the understanding of optical systems and in the appreciation of optical systems and of different optical materials.

At the time of the introduction of the achromatic lens in 1733, the only optical glasses available were the ordinary crown and flint types. During the first half of the nineteenth century systematic efforts to extend the range of available types of glass were made by Fraunhofer and Guinand in Germany and by Faraday, Harcourt and Stokes in England, but the great advance in this direction was made by Schott and Abbe, who in 1885 issued the first catalogue of the Jena glasses. This contained particulars of a number of glasses, the dispersive powers and partial dispersions of which differed considerably from those of earlier types of crown and flint.

Many more types of glass are now available and optical designers have a wide choice. Optical design, particularly that of achromatic doublets, is discussed in Chapter 14. Figure 14.2 (p. 473) shows a diagram of glass types plotted by refractive index and reciprocal dispersion (V value). The double line somewhat arbitrarily divides the types into crowns and flints. Each area shown is commonly referred to by the descriptions indicated, but for specific glass types six-figure

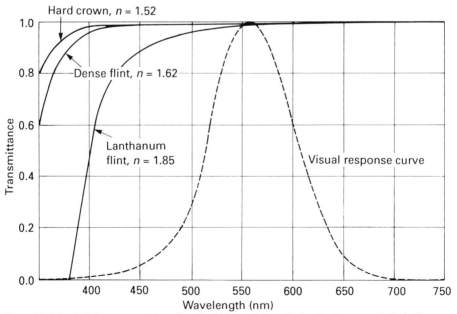

Figure 11.8 Optical glass transmission. Sample thickness 25 mm (reflection losses not included)

reference numbers are also used. These are derived from the first three significant figures of $n_d - 1$ and the first three significant figures of the V value. Thus glass type HC 524592 has a refractive index of 1.52400 and a V value of 59.21.

This glass, like all glasses, requires a glass former and a series of glass-modifying materials, which control the characteristics of the glass. Hard crown, 524592, may be taken as a starting point. For most of the optical diagram the glass former is silica and in the case of hard crown the modifiers comprise oxides of calcium, sodium, potassium, barium and magnesium. In order to move to the lower left part of the diagram, part of the silica is replaced by phosphorus or fluorine. To move to the right, that is from crown to flint, requires the introduction of lead oxide at the expense of some of the silica and alkali oxides. This was a very early change as it gave a glass composition that would melt more easily, the index change being a by-product.

Lead is a heavy element and increasing amounts of lead oxide gave increasingly heavier glasses with higher indices. The terms dense flint, extra dense flint, etc., mean what they say. With about 50% or more of the constituent materials being lead oxide the EDF and DEDF types have densities of $4-5\,g/cm^3$. Recently flint glasses have been introduced in which the lead oxide is replaced by titanium oxide, with significant reductions in density to below $3\,g/cm^3$; little more than crown glass.

If some of the lead oxide is replaced by barium oxide, high index is achieved without higher dispersion. Barium oxide can also be introduced into crown (lead-free) glass types to give high index. For even further increase in index, elements from the rare earth series have been added to the barium crowns and flints. As the principal rare earth element is lanthanum, they are commonly called lanthana crowns and lanthana flints. Such a two-dimensional plot of glass types does not cover all the features. For instance, the addition of boron to flint glass gives a higher index but with a relative partial dispersion increased in the red.

As an alternative to glass, organic polymers (plastics) have advantages and disadvantages. Although they can be formed into lenses by moulding, which is cheaper than grinding and polishing, the accuracy of curve is not as good. They have about half the weight of the equivalent glass lens, but their resistance to scratching is usually lower, although abrasion-resistant coatings are now available. The homogeneity attainable with plastics is considerably poorer than that with optical glass, and the choice of index and V value is restricted to two locations for readily obtainable plastics (Table 11.2).

Apart from spectacle lenses, optical plastics are used in cheap cameras and binoculars and also in some display systems where weight is important. The limitation to two main types is a very serious restriction on lens design and the extra

Table 11.2 Refractiveindex and V value of some plastics

	Refractive index	V value		
Polycarbonate	1.58	29.5		} Flint glass
Polystyrene	1.59	30.5		} equivalents
Polymethyl methacrylate	1.49	57.5	(Perspex acrylic)	} Crown glass
Allyldiglycol carbonate (CR39)	1.50	56	(CR39)	} equivalents

elements required to give equal performance often offset the weight and cost advantage.

Most glasses have very high transmission (between 450 and 1000 nm), although some of the high index types tend to absorb in the blue. When tabulating the optical transmission of a sample, it is important to state the thickness and also whether or not the losses due to reflection at the input and output (polished) surfaces are included. In Figure 11.8 the samples are 25 mm thick and the reflection losses are not included. Optical plastics are generally worse than this not only because they absorb more but also because they tend to scatter the incident light. The term **transmittance** is defined as the ratio of light out to light in:

$$T = I'/I \tag{11.6}$$

where I is the intensity of the incident light and I' the intensity of the transmitted light.

When reflection losses are ignored, the term **internal transmittance** is sometimes used.

11.4 Absorbing optical materials

The absorption of light by the bound atoms and molecules of a solid material and by the atoms and molecules of liquids and gases occurs in many ways. The quantum theory of light, which links together the concepts of wave motion and discrete packets of energy, photons, has a general theory to account for all these ways. This theory is beyond the scope of this book but some broad distinctions can be made. One type of encounter between a photon and an atom (or molecule) is called an **elastic collision**. The photon is very briefly trapped by the atom and flies off in a new direction, rather like a sling-shot, without loss or gain of energy. This is called **scattering** and is considered more fully in Section 11.8. Because a large number of atoms and photons are involved with even a weak light beam these processes are essentially random and obey the statistical laws of chance.

Absorption occurs when the photon happens to collide more directly with the atom and loses its energy to the atom, usually to its electrons. The photon therefore ceases to exist and so the light beam is destroyed in a progressive way once it has entered the material. The rate at which this occurs throughout the thickness of a material is apparently non-linear, but in fact follows a simple rule. If, for instance, half the light is absorbed in the first millimetre of the material, then half of the *remaining* light will be absorbed by the second millimetre, and so on. Thus the transmittance of the material against thickness will be as shown in Table 11.3.

The equation that describes this type of effect is:

$$I' = Ie^{-\alpha t} \tag{11.7}$$

(see Section 11.5), where I and I' are the incident and transmitted intensities, t is the thickness and α is the **absorption coefficient**, which will be given in units of

Table 11.3

Thickness (mm)	1	2	3	4	5	6
Transmittance (I'/I)	0.5	0.25	0.125	0.0625	0.03125	0.015625

reciprocal length such as mm^{-1}. This equation does not account for any reflection losses at the surfaces, and in a real sample these would need to be taken into account.

If the absorption coefficient of Equation 11.7 is obtained experimentally by measuring samples and allowing for reflection losses, the value obtained for α includes the effects of scattering. This is generally not the case if α is obtained theoretically. In coloured materials such as dyes the value of α is strongly dependent on wavelength. This is called **selective absorption** and is generally due to resonance effects between the energy content of photons of a given wavelength and the energy difference between stable atomic or molecular motions. In all materials the refractive index, scattering, selective absorption, reflection and transmission are all linked together in the quantum theory. The effect of light on conducting materials such as metals is generally absorbing because they have free electrons, which can absorb over the range of energy values associated with visible light. Most transmitting materials are insulators or **dielectrics**, in which the electrons and atoms are bound so that they absorb only energies for which the photon wavelengths are outside the visible spectrum. Anisotropic effects (including scattering and reflection) lead to polarization of the light, and these are described later in this chapter.

The selective effects of absorption complicate the calculations when absorbing materials are used together with selective emitters and detectors. The photometric calculations must be done at sufficiently close intervals over the range of wavelengths used. Selective absorption usually occurs because there is some resonance between the light of a given frequency and the oscillation modes allowed in the atoms of the material by the quantum theory. This is the same as saying that the energy of the photon matches the energy differences of the electron orbits. This match does not have to be exact and the usual shape of selective absorption is shown in Figure 11.9. This gives the transmission for a given thickness, t, of

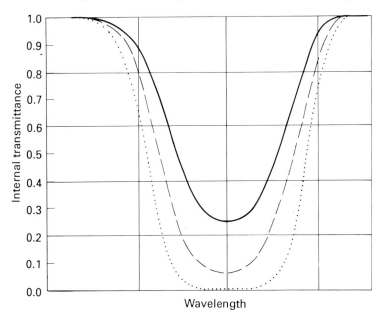

Figure 11.9 Absorption filters: —— single thickness; – – – double thickness; · · · · quadruple thickness

material and if the thickness is increased the shape shows a *multiplicative* change. Thus the dotted curve for twice the thickness shows that the transmission value for each wavelength has changed to be the square of transmission at thickness t.

Materials that strongly absorb a part of the spectrum are used as dyes to colour fabric and paint, etc. Some of these materials can be introduced into otherwise clear plastics to give **absorption filters**.

In optics, absorption filters are more often used to select certain wavelengths of light by allowing transmission between two absorption areas. Figure 11.10 shows a typical result using adjacent absorption bands from the same or two different materials. When the thickness is increased, say by ×2, the selection bandwidth reduces but the peak transmission also reduces. This is a limiting problem with absorbing filters and better results can be obtained with interference effects. **Interference filters** are described in Chapter 12.

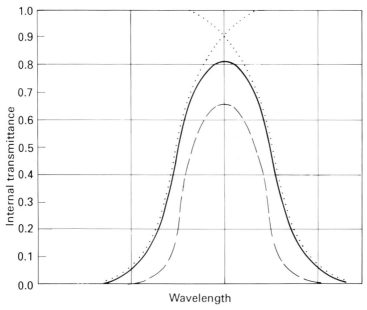

Figure 11.10 Bandpass filter: · · · · component absorption filters; —— single thickness effect; – – – double thickness effect

11.5 Practical filters, photochromic materials

Although transmittance values may vary considerably with wavelength (Section 11.4), filters can be obtained with very little transmittance change with wavelength. Such filters are neutral to the colour of the light passing through them and are known as **neutral density filters**. If two such filters have transmittance values of 0.4 and 0.6, the transmittance of both used together is found by **multiplying** these values, i.e. 24%. This fact has led to the adoption of the concept of **optical density**, which is the $\log_{10}$ of the reciprocal of the transmitted fraction. Thus a transmission of 50% means that half the light is transmitted so the density is $\log_{10}(1/0.5)$, i.e. 0.3. This is very useful when the transmittance values are very low as the density values increase as in Table 11.4.

Table 11.4

Optical density	0.1	0.25	0.4	0.5	0.75	1.0	2.0	3.0	4.0
Transmittance	0.80	0.56	0.40	0.32	0.18	0.10	0.01	0.001	0.0001

The usual advantage of a logarithmic scale is found in that when two filters of density 0.5 and 0.25 are used together they give a density of 0.75 by simple addition. In transmittance terms this is $0.32 \times 0.56 = 0.18$ or 18%. This calculation is correct only when the spectral absorptions are identical for the two filters. Otherwise it must be done at all wavelengths.

The multiplying of the transmittance values for filters in combination is an outcome of **Lambert's law**, which states that equal paths in the same homogeneous absorbing medium absorb equal fractions of the light entering them. Thus light entering a small section of the medium of thickness dt will have its intensity reduced from I to $I - dI$ so that we can say that

$$\frac{dI}{I} = - \alpha \, dt \tag{11.8}$$

where α is a constant called the **absorption coefficient**. Integrating this, the intensity at any value of t is given by

$$\log I_t = - \alpha t + C \tag{11.9}$$

where C is the constant of integration. When t is zero, the intensity is the incident intensity, I, so that $C = \log I$. Thus, the emergent intensity, I', from a filter of absorption coefficient α and thickness t is given by

$$I' = I e^{-\alpha t} \tag{11.10}$$

Its transmittance is therefore given by

$$T = \frac{I'}{I} = e^{-\alpha t} \tag{11.11}$$

from which it may easily be shown that an increase in thickness by a factor m gives a new transmittance value T' given by

$$T' = T^m \tag{11.12}$$

Equation 11.10 is the same as Equation 11.7.

For filters that change their transmittance with wavelength a graph may be drawn such as Figure 11.10, but optical science is often very interested in high transmittance values (for maximum efficiency) or low transmittance values (to ensure that light is blocked). It would be more useful to plot optical density values, as these expand the detail of the graph at the low transmittance values. However, this does not help at the high transmittance regions. The 'solution' is to plot the reciprocal of the density values on a log scale. This is known as a log–log scale, but the actual y-axis position is given by

$$\log \left(\frac{1}{\log \left(\frac{1}{T} \right)} \right)$$

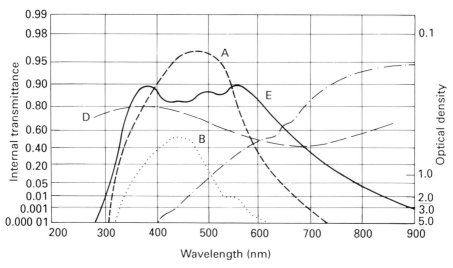

Figure 11.11 Selectively absorbing filters: (A,B) bandpass, (C,D) colour conversion, (E) heat absorbing

The actual values of T are indicated on the y-axis, and Figure 11.11 is an example of this type of graph. This is useful when different thicknesses of the same material are being considered, as the whole curve can be moved vertically. It is not so easy to obtain the effect of combinations of filters having different curves. This would be served best by graphs of optical density with wavelength, but these are rarely available. However, computer programs able to multiply values in one table with the corresponding values in another table are now commonplace (spreadsheets).

Curves C and D of Figure 11.11 represent colour conversion filters. The use of temperature to describe whiteness of a light source as if it were an incandescent source was described in Section 9.5. It was also indicated that by filtering out some of the red light the effective colour temperature could be increased, while filtering out some of the blue light causes the apparent colour temperature to be decreased. In colorimetry and visual psychophysics colour temperature can be expressed in **micro reciprocal degrees**, or **mireds**. This is simply the value of the colour temperature in kelvins, divided into 10^6.

This means that the sun, with a colour temperature of 5000 K, has a mired value of about 200. A filter such as C of Figure 11.11 has a mired shift value of $+120$ and, if the sun is viewed through this filter, it has a new mired value of 320, which is equivalent to a colour temperature of 3125 K. Thus, mired filters with positive values reduce the colour temperature while those that transmit more in the blue have a negative mired value. Filter D in Figure 11.11 has a mired value of -60 and *raises* the colour temperature. The usefulness of the mired method arises because while filter D raises 1000 K to 1063 K it raises 5000 K to 7143 K.

Filter E of Figure 11.11 is a **heat absorbing** filter. These are particularly useful in slide projectors where the infrared energy content of the projector lamp would damage the photographic slide. A heat absorbing filter traps this energy and raises its own temperature, which is then cooled by the fan. These are normally glass filters and in extreme situations need to be made from tempered (toughened) glass so that the range of temperature change does not shatter the glass. Heat *reflecting* filters (hot mirrors) are considered in Chapter 12.

All the above filters are passive and do not change their characteristics no matter how much or how little light they are absorbing. A range of active filters is called photochromic filters because they change their characteristics as the incident energy changes. Although some of this change is in the colour of the filters, most of it is of a neutral effect. Materials that exhibit photochromic effects have been used in data storage and holographic recording, but their most common use is as variable sunglasses.

The material used has the property of darkening to a sunglass tint in strong sunlight and fading to almost clear in dusk and dark conditions, rather like a reversible photograph. The photographic process works because crystals of silver halide are soluble in developers while silver is not. The action of exposing the photosensitive material is that of breaking the molecular bond between the silver and the chlorine, bromine or iodine. The latter diffuse away, leaving the silver to record the image. Silver halides are transparent substances while silver is not.

The same principle is used in photochromic glass except that the silver and halide, when separated by the activating light, move apart by only a very short distance and recombine shortly afterwards. The time of recombination depends on temperature – at low temperatures there is little thermal motion to bring the ions into contact. At too great a temperature (above 400°C) the ions diffuse into the glass to give permanent darkening. At normal temperatures a state of dynamic equilibrium is achieved when the rate of separation is equal to the rate of recombination.

The activating light is more effective when in the ultraviolet and blue end of the spectrum. Usually, a mixture of halides is used, bromide and iodide being sensitive to yellow light as well as blue and ultraviolet.

Although this ultraviolet and blue light needs to be absorbed to drive the reaction, the absorption spectra of photochromic glass (that is, its appearance as a filter or tint) is a different spectral curve, usually with a peak at 500 nm, giving a grey to brown appearance. This is affected by the size of the silver halide crystals used. If these are too small (<5 nm), light of 500 nm wavelength is not affected. If they are above about 500 nm the crystallites scatter the light and the glass becomes opalescent. Choice of crystal types, size and concentration, glass type and sensitizing dopants such as copper are all factors influencing the final product.

The most important parameters are the working colour, density and speed of response both to darkening and to fading. Figure 11.12(a) shows the transmission curves for the Reactolite Rapide* when clear and when exposed to bright sunlight as found at latitudes of about 60°. This particular material has a very fast reaction time and Figure 11.12(b) shows the changes in transmittance at 550 nm when the activating light is suddenly increased from zero to full sunlight and then back to zero.

As indicated on these graphs, these figures apply to a specific thickness and temperature. If the thickness is increased the change in transmittance in the fully exposed condition does not follow the density rule given at the beginning of this section because the darkening near the front of the lens reduces the activating light intensity – and hence the darkening – in the rear portions of the lens. Table 11.5 and the graphs allow 8% loss in transmittance due to reflections at the surfaces.

---

* Reactolite Rapide is a trade mark of Pilkington PLC.

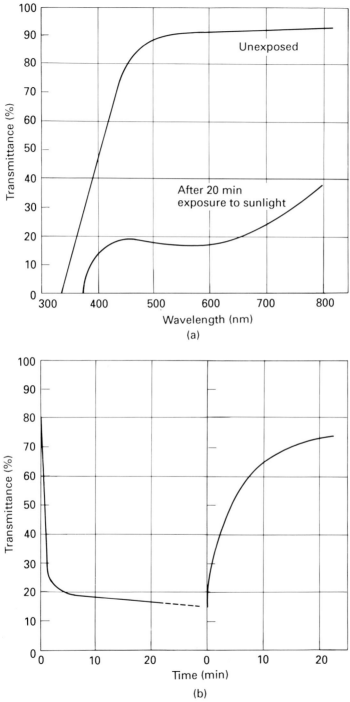

Figure 11.12 The salient parameters of a photochromic glass – Reactolite Rapide. These curves relate to glass 2 mm thick at 25°C: (a) spectral transmittance, (b) darkening and fading

Table 11.5 Transmittance values for various thicknesses of Reactolite Rapide at 25°C

Thickness (mm)	Transmittance (%)	
	Fully faded	Fully darkened
1	90	28
2	88	16
3	86	10.5
4	84	8
5	83.5	6

The effect of temperature was indicated previously. In Table 11.5 the fully exposed transmittance of 2 mm material would change to 11% at 15°C and 20% at 30°C.

11.6 Inhomogeneous optical materials

Inhomogeneity in optical materials is generally unwelcome. The need to have accurately predictable light rays deviated only at material boundaries in turn dictates that each optical material should have an accurately consistent refractive index everywhere within its volume. Badly melted glasses contain *straie*, which are local changes in refractive index, and these are of no use for optical work. Changes in index due to stressing the material can be useful in models of mechanical structures as a design tool, but the effect of stress in toughened (tempered) glass must be unobservable in normal use. Liquids and gases can show local variations, particularly when they contain temperature gradients. A mirage in the desert occurs not because the atmosphere separates out into layers of different temperature and refractive index but because there is a *continuous* change in index (reducing) towards the ground. Gas lenses, of very long focal length, can also be made by spinning a hollow tube. Minor pressure variations from the still, axial air to the air being in laminar flow near the tube's internal surface create refractive index changes from centre to edge and therefore make a lens. Solid materials exhibiting the same effect are described below.

When laying down thin films of material using vacuum deposition it is possible to produce very thin layers of differing refractive index. As described in Chapter 12, it is also possible to deposit two materials simultaneously. If the ratio of their deposition rates is steadily changed a filter may be built up which has a *continuous* change in index throughout its thickness. A light beam incident on this material at an angle will be deviated according to the law of refraction, but then continue to be deviated in a *continuous curved* fashion as it traverses the index gradient.

The precision needed to make useful larger optical elements out of **GRIN (graded index** or **gradient index)** material is very high. This technology was first used in optical fibres where the lateral dimension is small and then in lenses for coupling optical fibres together. Latterly, linear arrays of fibre-like lenses have been used in desktop photocopiers, where the system needs to be very compact.

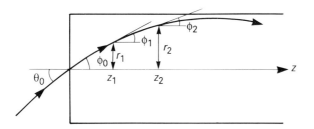

Figure 11.13 Meridional ray travelling through a gradient index lens

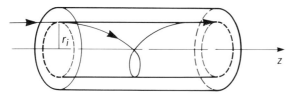

Figure 11.14 Skew ray travelling through a gradient index lens

The Gradient Index Laboratory at Rochester University, USA, has taken the lead in this technology and manufactured binocular objectives and eyepieces out of GRIN glasses. Figure 11.13 shows the passage of a light ray through a thick GRIN lens where the gradient is such that the highest refractive index is along the central axis and diminishes by r^2 towards the periphery. Figure 11.14 shows the complex helical path of a skew ray.

So far this technology has not been used in spectacle lenses, probably because currently obtainable index differences require a fairly thick lens to give appreciable optical power. Spectacle lens design would undoubtedly benefit if bigger index gradients were available, as this gives an extra degree of freedom to the lens designer.

The natural lens of the eye, however, does have an appreciable gradient index. Figure 11.15 shows a laser beam traversing, in this case, a goldfish lens and the distinctive curved path can be seen.

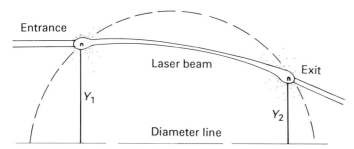

Figure 11.15 Laser beam traversing the nearly spherical lens of the goldfish showing the curved path due to gradient refractive index (Adapted from Axelrod *et al.*, *Vision Research*, **28**, 57–65, 1988)

11.7 Anisotropic materials – polarization

In Section 11.4 the concept of resonant interaction between the oscillating energy we call light and the energy absorption characteristics of the bound atoms and molecules in a solid was briefly described. The complex geometry of the molecular structure of most solids means that the characteristics of the atom for a vibration in one orientation is not likely to be the same as in another orientation. Light is a transverse vibration and this means that, within the same beam of light, vibrations may exist in all orientations. Different vibrations may therefore react differently when entering a solid.

Most optical materials such as glass are **amorphous solids**. This means that the molecules are arranged in a random fashion. Whatever crystalline structure exists is not repeated over many molecules. This means that a light beam traversing the glass undergoes an effect which averages out for all orientations of the vibration. Large transparent crystalline solids are a different matter. They react differently to vibrations of different orientation and are called **anisotropic**. This results in polarization effects.

The action of polarizing sunglasses is that of reducing reflected sunlight more than other light in the general scene. If two such sunglass lenses are put together, it is possible, by rotating one about their common perpendicular axis, to vary the total amount of light passing through them (Figure 11.16). This phenomenon requires that the light between the two lenses has a preferred orientation which the second lens can accept or reject. In the full theory of light, the wave motion used to describe it has a transverse varying electric field and transverse varying magnetic field which are orthogonal. Usually light is unpolarized, that is, it contains a great many wave motions with their vibrations at all angles (although each electric vector and its associated magnetic vector are always at right angles). The polarizing sunglass lens selects a particular orientation, and allows this to be transmitted while absorbing the rest (see Figure 11.16).

One form of polarized light is **linearly polarized light**, which occurs when the light has its electric field varying in a single fixed orientation. The term 'plane of polarization' was originally used to describe the orientation of the magnetic vector but (of course) it has sometimes been used referring to the electric vector. To avoid this confusion the term **plane of vibration** is now used and this refers to the electric

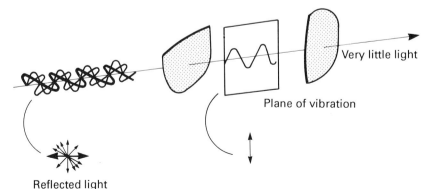

Figure 11.16 The action of polarizing sunglasses. The reflected light is predominantly horizontal vibration (see text)

vector, being that plane containing the electric vector and direction of propagation of the light. The term 'plane of polarization' should not be used. In the paragraphs that follow we will think of light in terms of its electric field vector only.

The action of the sunglass lens described above is not that of selecting a particular plane of orientation from among *all* others. The action of a light beam polarized in one plane of vibration is not restricted to that plane as it is possible to calculate its components in other planes. Thus, a disturbance *a* orientated at an angle θ to the *x*-axis has a component $a \cos \theta$ in the *x*-direction and $a \sin \theta$ in the *y*-direction. For disturbances in all orientations between $0°$ and 2π, the intensity will be $2\pi a^2$. For the components of these disturbances in the *x*-direction the intensity will be

$$\int_0^{2\pi} a^2 \cos^2 \theta \, d\theta \; = \; \pi a^2 \qquad (11.13)$$

while, in the *y*-direction,

$$\int_0^{2\pi} a^2 \sin^2 \theta \, d\theta \; = \; \pi a^2 \qquad (11.14)$$

Thus the intensities are divided equally between the orthogonal orientations. The perfect linear polarizer would transmit 50% of the incident **unpolarized light** in which the planes of vibration are randomly orientated. Quite often the in-coming light is not completely random in this respect and exhibits **partial polarization**. The more monochromatic light becomes the more likely it is to be polarized in some way. Strictly, monochromatic light always exhibits some form of polarization as it has insufficient wave trains to obtain randomness during normal observation times. When unpolarized light is divided into two orthogonal linearly polarized beams it is found that the two beams are incoherent and no interference effects can be obtained between them. This remains the case even when their planes of vibration are subsequently made parallel.

An optical system that generates polarized from unpolarized light is called a **polarizer**. When used to examine polarized or partially polarized light it is called an **analyser**. When two polarizers are used orthogonally as in Figure 11.16, they are said to be **crossed**.

The earliest form of polarizers were **dichroic crystals** like tourmaline. This family of substances has crystals with long narrow lattice structures, which orientate parallel to each other. This means that the material has an action on any light beam going through it that depends on the direction of the beam and on the orientation of the plane of vibration with the axis of the crystals. The material is **anisotropic**. It is useful to define a direction within the crystal where the effect *does not* occur – this is known as the **optic axis**. If a plate is cut from a tourmaline crystal with the optic axis parallel to the surfaces of the plate, an unpolarized light beam passing through the plate will have one component of its electric vectors in the optic axis and the other perpendicular to it. In tourmaline the electric vector perpendicular to the optic axis is very strongly absorbed while the other is less so. The emergent beam is thus strongly polarized provided the plate is a few millimetres thick. These effects are very colour dependent and so tourmaline changes colour depending on the angle from which it is viewed; hence the term *dichroic*, which merely means two-coloured. If the plate was cut so that the light passing through it was going *along* the optic axis, no effect would occur because the absorption would be the same for the two components.

Other crystals exhibit similar properties. The earliest form of Polaroid used large numbers of dichroic herapathite crystals aligned on a plastic sheet. Once again the unpolarized light has its electric vector component strongly absorbed when it is perpendicular to the optic axis. This selective absorption occurs because some of the electrons in the crystal are partially free to move in a direction perpendicular to the optic axis but not at right angles. The electric vectors of the incident light beam try to stimulate electric currents in the material but only those perpendicular to the optic axis succeed and the resultant electron energy is absorbed into the lattice.

An alternative to long crystals is long parallel wires in the form of a grid. When the incident unpolarized beam reaches the grid the components of the electric vectors that are parallel to the wires induce currents in the wires by driving the conduction electrons along the wires. These conduction electrons lose energy either by collision with the lattice, in which case the light is absorbed, or by reradiating in the backwards direction, in which case the light is reflected. The components of the electric vectors which are perpendicular to the grid are not able to generate electric currents across the narrow wires (if they are sufficiently narrow) and are therefore transmitted (Figure 11.17).

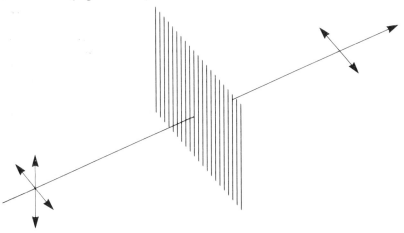

Figure 11.17 The action of a grid polarizer

Note that, when electric vectors are considered, the component *against* the gaps in the grating is transmitted. The efficiency of this device as a polarizer is critically dependent on the ratio of the wavelength of the incident light to the wire spacing. Figure 11.18 shows that d, the wire spacing, needs to be at least four times smaller than the wavelength of the light.

The most common linear polarizer is not the dichroic Polaroid described above but a later development, which is a molecular analogue to the wire grid. This E-sheet Polaroid is formed from a sheet of polyvinyl alcohol which is heated and stretched in *one* direction. This process aligns its long hydrocarbon molecules. When the sheet is dipped into an iodine solution the iodine attaches to the long molecules, forming long aligned conducting chains, which act like wires. These are separated by distances of molecular dimensions so the grid polarizes in the visible although it is less good at the blue end of the spectrum.

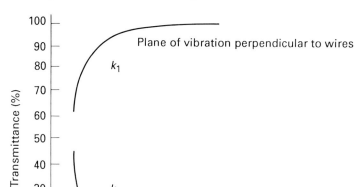

Figure 11.18 The polarizing effect of a wire grid

Birefringence is another property of anisotropic crystals in which the electrons can move with different amounts of freedom in different directions. An example of this is calcite or Iceland spar, which is a crystalline form of calcium carbonate ($CaCO_3$). This cleaves readily into the form of rhombohedra or rhombs in the form shown in Figure 11.19. All the sides are parallelograms of angles 78° and 102°. All the corners except two contain both angles. The two 'blunt' corners contain angles

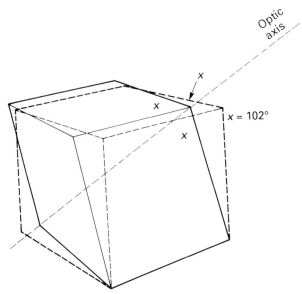

Figure 11.19 Rhombic calcite crystal showing blunt corner, optic axis and reference cube (dashed). The top and lower faces of the reference cube are co-planar with the rhomboid faces. $x = 102°$

of 102° only. The two blunt corners in the equally sided crystal shown are joined by an axis of *symmetry* which is an optic axis. In this direction the crystal appears symmetrical and therefore apparently isotropic. All other directions are asymmetric and the crystal is anisotropic along these.

The anisotropy takes the form of *two* refractive indices. When a *near* object, such as a small hole in a card, is viewed through a rhomb of calcite, two images are seen and on rotating the rhomb about the direction of view one image is seen to remain stationary while the other moves round it. The line joining the two images will always be parallel to the optic axis direction. If the blunt corners are cut off by plane faces so that it is possible to view the object along this direction it will be found that there is then only a single image and therefore no double refraction of the light travelling along the optic axis.

Because calcite has only one direction of zero effect it belongs to a class called **uniaxial** crystals. The two types of ray that give rise to double imaging are called **ordinary** rays (o-rays) and **extraordinary** rays (e-rays). Figure 11.20 shows how a beam of unpolarized light falling normally onto one surface of the rhomb is doubly refracted into two beams. The ordinary ray behaves as if the material were isotropic, as in the case of glass. The extraordinary ray behaves quite differently, changing its direction on entering the crystal, even though the incident light is normal to the surface. It is found that these rays are linearly polarized in planes orthogonal to each other.

By measuring the angle of refraction of the ordinary and extraordinary rays for different angles of incidence we find that $\sin i/\sin i'$ is a constant for the ordinary ray but varies with angle of incidence for the extraordinary ray. Hence we conclude that the velocity of light and the refractive index of the crystal are constant in all directions for the ordinary ray but differ in different directions for the extraordinary ray. Only along the optic axis of the crystal will the two velocities be equal.

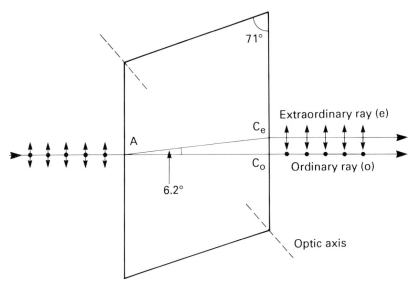

Figure 11.20 Double refraction by calcite of an unpolarized beam incident normally in the section containing the optic axis

The principle of wavelets developed by Huygens (see Section 1.2) and used later by Fresnel (Section 11.8) may also be applied to the propagation of wavefronts in anisotropic media. In uniaxial crystals such as calcite, quartz, tourmaline and ice, the wavefront associated with the ordinary ray is constructed from wavelets having the same velocity in all directions; that is, they are spherical wavelets. For extraordinary rays the wavelets are ellipsoids of revolution which coincide with the sphere of the ordinary wavelets along the optic axis, where their velocities are equal. Figure 11.21 shows that the velocity of the extraordinary wavelets may be smaller or larger than that of the ordinary ray, giving rise to the terms positive

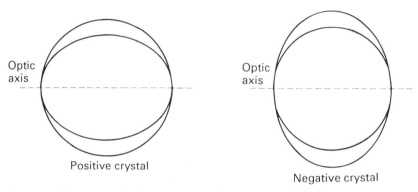

Figure 11.21 Wavelet sections for uniaxial crystals

uniaxial crystal and negative uniaxial crystal as shown. The general sphere–ellipsoid wavelet shape for a positive uniaxial crystal is shown in Figure 11.22, which also shows the wavelet shape for a **biaxial** crystal; that is, a crystal having two axes. These wavelets are both ellipsoids with four distortions on them. The inner ellipsoid is pinched outwards while the outer ellipsoid is dimpled in to coincide in the same plane as the optic axes although not coincident with them as the axes define identical velocities and *directions*. Strongly biaxial crystals are not often used and will not be dealt with here.

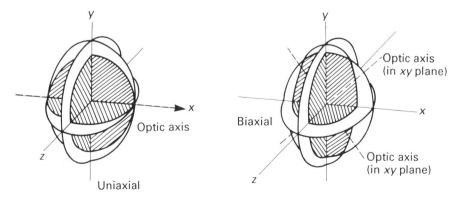

Figure 11.22 Crystal wavelets

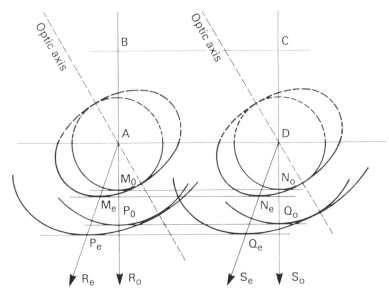

Figure 11.23 Huygens' construction for ordinary (o) and extraordinary (e) wavefronts and ray directions

In Figure 11.23, parallel light is shown incident normally on one face of a crystal of calcite, the optic axis of which is in the plane of the paper. When the plane wavefront meets the surface AD, wavelets of the form described above travel into the crystal from each point on the surface; those originating at A and D are shown in the figure. The refracted wavefronts of the ordinary light M_oN_o and P_oQ_o are tangential to the spheres, while those of the extraordinary light M_eN_e and P_eO_e are tangential to the ellipsoids. It is seen that, in this case, the two sets of wavefronts are parallel to one another, but the rays of the *extraordinary beam are not perpendicular to the wavefronts*, and may not always lie in the plane of incidence; the extraordinary beam is therefore deviated from the normal. The rays of the ordinary beam are perpendicular to the wavefronts as in an isotropic medium. Other examples with the incident light reaching the surface at various angles should be solved by similar graphical constructions. It will be obvious that when the light is incident in the direction of the optic axis the ordinary and extraordinary beams coincide, since the spheres and ellipsoids are in contact along this direction.

An important case occurs when the optic axis is perpendicular to the plane of incidence and parallel to the face of the crystal. In the graphical construction of this (Figure 11.24), as the optic axis is perpendicular to the plane of the paper, the sections of both ordinary and extraordinary wavelets will be circles. Therefore, in this plane, the velocity of the extraordinary light is the same in all directions, and the value sin i/sin i' is a constant. For this reason the refractive index n_e of a crystal for the extraordinary beam is defined as the ratio of the velocity of light in air to the velocity of the extraordinary beam in a plane perpendicular to the optic axis. The values of the ordinary and extraordinary refractive indices for the D line of a few of the more important uniaxial crystals are given in Table 11.6. The **linear birefringence** of uniaxial crystals is defined as the difference between the extraordinary and ordinary refractive indices, $n_e - n_o$. Few naturally occurring crystals have a linear birefringence greater than that of calcite (negative).

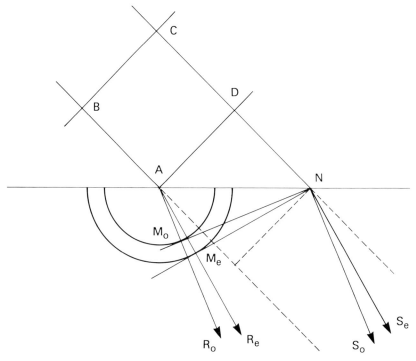

Figure 11.24 Huygens' construction for double refraction. Optic axis perpendicular to the plane of incidence

Table 11.6 Indices for birefringent crystals

	n_o	n_e
Positive crystals		
Quartz	1.5442	1.5533
Ice	1.309	1.313
Mica (slightly biaxial)	1.561	1.594
Negative crystals		
Calcite	1.6585	1.4864
Tourmaline	1.669	1.638
Sodium nitrate	1.5874	1.5361

As shown in Figure 11.20, the two emergent beams of light are found to be linearly polarized in orthogonal directions. The ordinary ray has a vibration orientation that is *always* perpendicular to the optics axis. On the other hand, the extraordinary ray plane of vibration contains the transmitted ray and the optic axis. This plane is sometimes called a **principal section** (Figures 11.20 and 11.23 are principal sections). The polarization is complete; that is, each of the two beams is completely linearly polarized. Even with the high linear birefringence of calcite, the angle between the beams is not very large so that some extra technique is needed to separate them.

In the **Nicol prism**, the crystal is cut diagonally and recemented together with low-index optical cement. The angles are chosen so that the ordinary ray is totally internally reflected while the extraordinary ray is transmitted. The **Glan–Foucault prism** uses the same idea, but with the two components differently orientated. The **Wollaston double-image prism** also uses two components arranged so that both beams are transmitted, but the angle between them is maximized. This prism is shown in Figure 11.25. The first component prism is cut from birefringent material so that its optic axis is parallel to the entance surface, AC, and lies in the prism section, while the second prism has its optic axis parallel to the exit surface but perpendicular to the prism section.

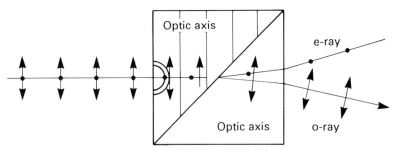

Figure 11.25 Wollaston double-image prism polarizer in calcite. The rays are designated with regard to the second prism

On entering the first prism the light divides into an e-ray and o-ray, travelling along the same path but at different velocities, experiencing different refractive indices. On reaching the angled interface, which must be well cemented or in very good optical contact to avoid total internal reflection, the designations of the two rays exchange. Because the optic axes of the two prisms are perpendicular to each other, the o-ray of the first prism becomes the e-ray of the second and vice versa. Therefore, one ray sees an interface between n_o and n_e while the other ray sees n_e to n_o. Thus, one ray is refracted towards the normal to the interface while the other is refracted away from it. Each ray undergoes a further refraction at the final surface.

In this way, considerable angular separation of the two beams can be obtained, while the dispersion of one prism is almost neutralized by the other. Prisms of this kind have been used for doubling the image in measuring instruments, where the fact that the light is polarized (as in Figure 11.25) is of no importance; for example, the Javal–Schoitz ophthalmometer or keratometer.

The first component of the Wollaston prism is a birefringent crystal cut with its incident face parallel to the optic axis. In this direction the crystal displays the maximum difference between the o-ray and the e-ray refractive index and, therefore, light velocity and phase. If a parallel-side plate of the crystal is cut in this way, the two rays emerge without change in direction but with part of the energy *retarded* behind the other part. No effect of this can be seen when the two beams are incoherent, as the summation of the two random vibrations remains random.

If the incident beam is linearly polarized, the two emergent beams will be coherent but the electric vector of one will be delayed with respect to the other. These are the conditions described in Section 9.2. The resulting emergent light depends on the relative amplitudes and relative phase of the o-ray and e-ray. The relative amplitudes depend on the orientation of the plane of vibration of the incident beam with the optic axis of the retarder, and the relative phase $\Delta\varphi$ depends on the index difference and thickness, t, of the retarder:

$$\Delta\varphi = \frac{2\pi}{\lambda_0}(n_o - n_e)t \qquad (11.15)$$

where λ_0 is the wavelength in vacuum and n_o and n_e are the refractive indices with respect to vacuum. In most cases the resultant will be an ellipse, as was found with Equation 9.10 (Section 9.2), and indeed **elliptically polarized** light may be regarded as the general form of polarized light, with **linearly polarized** and **circularly polarized** being special cases.

This latter case occurs when the amplitudes are equal and the relative phase difference is 90°, i.e. $\pi/2$. A retarder of this thickness is called a **quarter-wave plate**:

$$\frac{\lambda_0}{4} = (n_o - n_e)t \qquad (11.16)$$

Clearly it is not possible to make a simple plate retarder that is a quarter-wave plate for all wavelengths. Figure 11.26 shows how the form of the polarization changes as the light passes through a retarder for linearly polarized incident light with a direction of vibration at 45° to the optic axis (giving equal components) and at 30°. It is seen that a plate of double thickness produces linearly polarized light of a different orientation. Such a **half-wave plate** introduces a phase difference of 180° between the components and so for any orientation with respect to the optic axis the emergent beam has the reverse orientation, as shown.

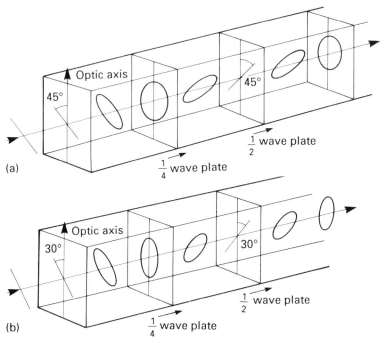

Figure 11.26 Passage of polarized light through a retarder: (a) incident plane of vibration at 45° to the optic axis, (b) incident plane of vibration at 30° to the optic axis

Circularly polarized light may be produced from unpolarized light by mounting a linear polarizer in front of a quarter-wave plate with its plane of transmitted vibration at 45° to the optic axis of the plate. Such a combination is called a **circular polarizer** and it is evident that it only works for light incident from one side.

11.8 Polarization by scattering and reflection

This section deals with polarization effects that do not depend on asymmetric crystal structures. Here the asymmetry needed is provided by the geometry of the situation. Light is scattered by atoms, molecules and particles that can vibrate in resonance or near-resonance with the frequency of the light. The interaction is generally between the light energy and the electron cloud in the atom or molecule. The characteristics of the electron clouds at the frequency of the incident light are very variable in the case of solids and these determine whether the light is reflected, refracted or absorbed.

In the case of gases the most likely interaction is elastic collision, which leads to **scattering**. The nearer the light frequency to resonance, the greater the scattering. Most gases resonate in the ultraviolet and therefore the atmosphere scatters more at the blue end of the spectrum, giving rise to the blue sky and, because the blue is scattered *out* of the beam, to red sunsets.

If we consider a beam of unpolarized light reaching a scattering medium at A (Figure 11.27), the near-resonant vibrations in the atoms of the media are transverse to the incident direction. Consider an observer at B whose awareness of the vibrations at A is by means of the scattered light. It is apparent from the geometry that there are no vibrations in the direction of the incident beam – the scattered light is therefore linearly polarized (and similarly at C).

If an analyser is used to view a patch of blue sky at about 90° to the sun, the polarization of the scattered light can be seen by rotating it. Multiple scattering,

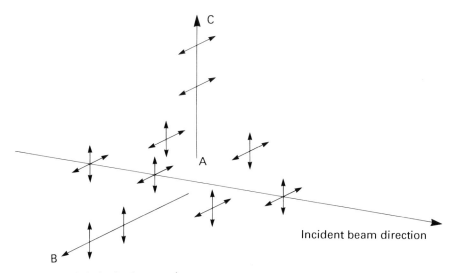

Figure 11.27 Polarization by scattering

that is further scattering of the once scattered light, tends to depolarize the beam but 70–80% polarization can be found in favourable circumstances.

The geometry of Figure 11.27 is also applicable when light is reflected. In Figure 11.28 the angle of incidence of the unpolarized beam approaching a dense medium has been chosen so that the angle between the light refracted and the light reflected is 90°. The reflection and refraction takes place in the first few molecular layers of the dense material and viewed from the position C the light has no vibration in the BD direction so that only light of the polarization shown is reflected; akin to the scattering situation. This has its electric vector perpendicular to the plane of incidence, that is, the plane containing the incident ray and the normal to the surface. The other plane of vibration would be parallel to this plane. The two orientations have been given many symbols. Probably the most common are p and s, derived from the German 'parallel' and 'senkrecht'. (Others are TM, TE; ∥, ⊥; and l, r.) The light reaching C is the s type, having its electric vector vibrating in a plane perpendicular to the plane of incidence. The angle i_b is known as the **Brewster angle** or polarizing angle.

From Figure 11.28,

$$n \sin i = n' \sin i' = n' \cos i \tag{11.17}$$

if $i + i' = 90°$. Therefore

$$\tan i_b = \frac{n'}{n} \tag{11.18}$$

This is the angle at which the polarization of the reflected beam is the most complete. At other angles some polarization occurs in accordance with equations that can be derived from a consideration of the electric and magnetic vectors at the point of refraction/reflection. There must be no abrupt changes or *discontinuities* in

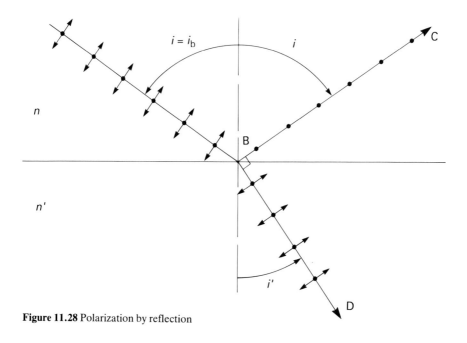

Figure 11.28 Polarization by reflection

these fields. The reasoning of the foregoing is a simplification of the E field conditions. Different conditions apply to the magnetic field, which yield the same result. **Fresnel's equations** were derived by him to give the relative *amplitudes* of the reflected and refracted beams in the two principal planes of vibration. When these are squared we obtain the relative intensities which for reflection are:

$$\left(\frac{I_R}{I_I}\right)_p = \frac{\tan^2 (i - i')}{\tan^2 (i + i')}$$

$$= \text{Reflection coefficient, } R_p \tag{11.19}$$

$$\left(\frac{I_R}{I_I}\right)_s = \frac{\sin^2 (i - i')}{\sin^2 (i + i')}$$

$$= \text{Reflection coefficient, } R_s \tag{11.20}$$

It is apparent that Equation 11.19 gives $R_p = 0$ at the Brewster angle given by Equation 11.18. For normal incidence, $i = i' = 0$ and the equations become indeterminate. However, if we expand the sine terms, and remembering that $\sin = \tan$ at small angles, we have

$$R_p = R_s = \left(\frac{\sin i \cos i' - \cos i \sin i'}{\sin i \cos i' + \cos i \sin i'}\right)^2 \tag{11.21}$$

Using Snell's law and the fact that cosines go to unity for small angles,

$$R_p = R_s = \left(\frac{n' - n}{n' + n}\right)^2 \tag{11.22}$$

When i exceeds the critical angle, these simplified equations cannot be directly applied and a more comprehensive treatment including the relative phases of the vibrations is needed. (See Longhurst, *Geometrical and Physical Optics*, Longman (1973) or Hecht and Zajac, *Optics*, Addison-Wesley (1987).)

The graphs of Figure 11.29 show the reflection coefficients for three refractive indices and for external or internal reflection. The intensities of the transmitted beams are found by the coefficients

$$T_p = 1 - R_p \tag{11.23}$$

and

$$T_s = 1 - R_s \tag{11.24}$$

assuming no absorption and limiting the measurements to air (i.e. after two surfaces).

Where there is absorption in the material the reflected light as well as the transmitted light is affected. Most importantly the reflectivity for the p beam does not fall to zero, although the general shape of the curves remains the same. With metallic coatings for mirrors, unwanted polarization effects are normally not too serious, but with multilayer dielectric coatings as described in Section 12.8 serious problems can occur at non-normal angles.

On the other hand, if polarization effects are required, this method may be used. With a single surface of water the reflected light is more than 80% polarized over a range of incidence angles from about 40° to 70°, which explains the value of polarizing spectacles in marine or rain conditions (see Figure 11.16). The axes of

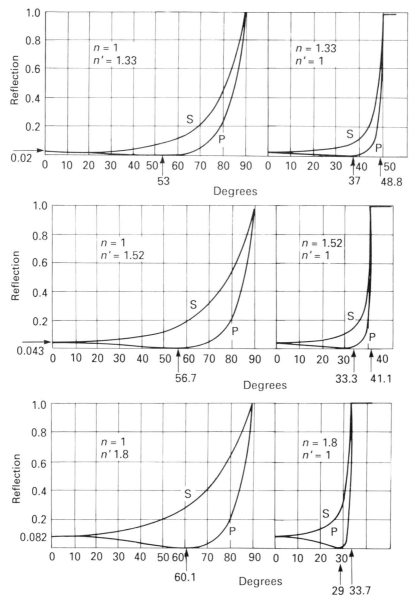

Figure 11.29 Reflection coefficients at boundaries of non-absorbing media

any polarizer can be found by looking through it at a scene reflected from a liquid or glass surface near the Brewster angle and rotating the polarizer for minimum transmission.

With a parallel plate of glass two surfaces are in action. The internal and external Brewster angles are linked by Snell's law and so maximum polarization occurs at both surfaces for the same beam angle. The reflection of the s-polarization at this angle rises from about 7.5% for the single surface to 13% for the plate ($n = 1.52$).

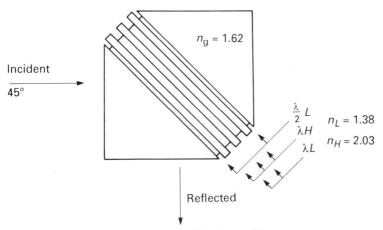

Figure 11.30 Design of a seven-layer polarizing beamsplitter

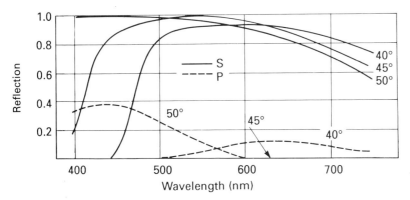

Figure 11.31 Performance *versus* (external) angle of incidence. Design wavelength 540 nm (Design by Dr P.B. Clapham, National Physical Laboratory, used with permission and acknowledged with thanks)

For two plates it is 20%, and for four 28%. When a large number of plates are stacked together with small air gaps between, a maximum reflection of about 35% may be obtained. The transmitted beam also becomes more polarized.

This concept may be realized using optical thin films whose refractive index with respect to a glass cube may be chosen to give a Brewster angle of 45° for a chosen wavelength. At other angles the effect is less complete, as it is at other wavelengths. Figures 11.30 and 11.31 show the design and performance of such a polarizing beamsplitter.

11.9 Definitions and typical values

See R. J. King, *Polarized Light: Definitions and Nomenclature*, NPL Report MOM25, National Physical Laboratory, Teddington, England (July 1977), on which this section is based, with permission.

11.9.1 Degree of polarization

A beam of partially polarized light may be considered as comprising a completely polarized component of intensity I_a and a completely unpolarized component of intensity I_b. The **degree of polarization**, P, is defined by

$$P = \frac{I_a}{I_a + I_b} \tag{11.25}$$

assuming no coherence between the beams.

11.9.2 Principal transmittances and ratios

When the direction of vibration of an incident linearly polarized beam is orientated so that the transmittance of a polarizer is a maximum, the ratio of transmitted to incident intensities is defined as the **major principal transmittance**, k_1. This direction of vibration also defines the **transmission axis** of the polarizer.

When the orientation is such as to give a minimum transmittance, the ratio of the intensities is defined as the **minor principal transmittance**, k_2. The **principal transmittance** ratio is defined as k_1/k_2 while the **extinction ratio** is the inverse of this.

11.9.3 Polarizance

In the particular case in which the incident light is unpolarized, the degree of polarization produced by a polarizer is sometimes called **polarizance**. In terms of principal transmittances, the polarizance is equal to

$$\frac{k_1 - k_2}{k_1 + k_2} \tag{11.26}$$

If the incident light is partially polarized, the degree of polarization produced by the polarizer will not be equal to its polarizance.

11.9.4 Extinction ratio

The inverse of the principal transmittance ratio is defined as the **extinction ratio**, k_2/k_1. If two identical polarizers are used together, the transmittance when their transmission axes are parallel is

$$\tfrac{1}{2}k_1 k_1 + \tfrac{1}{2}k_2 k_2 \simeq \frac{k_1^2}{2}$$

when k_2 is small. When the axes are crossed the transmission is

$$\tfrac{1}{2}k_1 k_2 + \tfrac{1}{2}k_1 k_2 = k_1 k_2$$

This crossed transmittance is often referred to as the extinction and the ratio

$$\frac{k_1 k_2}{k_1^2/2} = \frac{2k_2}{k_1} \tag{11.27}$$

is sometimes (erroneously) called the extinction ratio.

When polarizers are used with visible light the values of k_1, k_2, etc., must be weighted to match the visual response. The performance of polarizers in the infrared and ultraviolet will be different. When a polarizer is used with unpolarized light the total visual transmittance is

$$k_v = \frac{k_1 + k_2}{2}$$

which is usually very nearly equal to $k_1/2$. For sheet polaroid the type numbers used refer to the value of k_v. Types HN-38, HN-32 and HN-22 have k_v values of approximately 0.38, 0.32 and 0.22 respectively. The transmittances of crossed pairs $(2k_2/k_1)$ of the above types are about 5×10^{-4}, 5×10^{-5} and 5×10^{-6} respectively.

The best Glan–Thompson prisms have an extinction ratio (k_2/k_1) less than 1×10^{-6}. The extinction ratio of a single reflection may not be as small as the zero value of the curves of Figure 11.29 might suggest. In practice the incident beam would show some angular spread and the surface will not be perfect as regards cleanliness and structure. For a thin-film polarizer designed along the lines of Figure 11.30 an extinction of about 10^{-3} in transmission is found at the design angle and wavelength.

The degree of polarization of a single reflection varies with incident angle and refractive index as shown in Figure 11.32. Naturally, these curves apply to specular reflections only. Light diffusely reflected from textured surfaces has a much lower degree of polarization, but this light contains information which is normally required. The value of polarizing sunglasses in reducing glare from specular reflections can be judged from Figure 11.32.

When light is transmitted by an efficient linear polarizer the extinction ratio is at least 10^{-4} and so the degree of polarization may be taken as unity. Subsequent components of this linearly polarized beam will be coherent and so we must work with amplitudes. If a second linear polarizer is inserted in this beam with its transmission axis orientated at angle θ to the first, the light transmitted by the system will be a maximum when θ is zero and a minimum when θ is 90°. At intermediate angles the second polarizer selects the component amplitude $a_1 \cos \theta$, where a_1 is proportional to $\sqrt{k_1}$ for the first polarizer. The transmitted intensity is then proportional to $k_1 k_1' \cos^2 \theta$, where k_1' applies to the second polarizer. If the incident light is unpolarized a factor ½ must be applied as k_1 is the principal transmittance for polarized incident light. Thus, the system transmittance is given by

$$\frac{k_1 k_1'}{2} \cos^2 \theta$$

If now a retarder is introduced between the linear polarizers, the situation is complicated to the extent that the optic axis of the retarder must be defined with respect to the transmission axes of the polarizers, and the phase difference, ϕ, introduced by it must be known. If the angle between the first polarizer and retarder is α and between the retarder and second polarizer is β we may proceed as follows.

The light reaching the retarder has amplitude components $a \cos \alpha$ and $a \sin \alpha$ with respect to its optic axis. These components will have further components $ab \cos \alpha \cos \beta$ and $ab \sin \alpha \sin \beta$ with respect to the transmission axis of the second

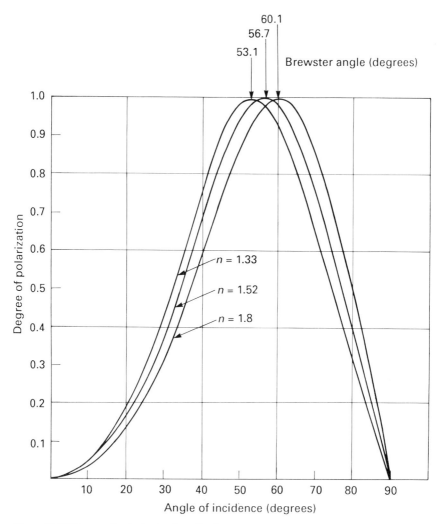

Figure 11.32 Degree of polarization by reflection

polarizer, but they will differ in phase by ϕ by the action of the retarder. The transmitted disturbance is therefore

$$A' = ab \cos \alpha \cos \beta \, [\sin \omega t] + ab \sin \alpha \sin \beta \, [\sin (\omega t + \phi)] \qquad (11.28)$$

where these are not orthogonal components but both in the transmission axis of the second polarizer.

By Equation 9.7,

$$I' = a^2 b^2 \cos^2 \alpha \cos^2 \beta + a^2 b^2 \sin^2 \alpha \sin^2 \beta + 2a^2 b^2 \sin \alpha \cos \alpha \sin \beta \cos \beta \cos \phi$$

$$= a^2 b^2 \cos^2 (\alpha - \beta) - a^2 b^2 \sin 2\alpha \sin 2\beta \sin^2 \frac{\phi}{2} \qquad (11.29)$$

because we have calculated with respect to the transmission axes of both polarizers.

The transmittance of the system is given by

$$k_1 k_1' \left[\cos(\alpha - \beta) - \sin 2\alpha \sin 2\beta \sin^2 \frac{\phi}{2} \right] \tag{11.30}$$

Obviously the problem becomes much more complex when k_2 is not negligible, and when more complicated systems are to be analysed. A number of different methods have been invented to cope with this. Each one requires the application of a very specific set of rules under stringent conditions in an empirical procedure designed as an analogue computing scheme. The **Poincaré sphere** uses a sphere as a map with linearly polarized light (of varying orientation) around the equator and circularly polarized light at the poles. Everywhere else is elliptically polarized and calculations are made by changing position. **Jones' vectors** and **Stokes' vectors** are similar approaches with complementary advantages and disadvantages. In matrix form the former are 2 × 2 while the latter use 4 × 4 matrices (Mueller matrix). The form of these have been calculated and tabulated in W. A. Shurcliff, *Polarized Light*, (1962) Harvard/Oxford.

11.10 Applications of polarized light

The most obvious use, in the form of polarizing sun spectacles, has been covered in Section 11.7 and their value noted in Section 11.9. The polarizing effect by reflection can also be used in the study of surface structure. This is particularly useful on materials that absorb light but which can be optically polished. The reflection of linearly or circularly polarized light is accompanied by a change in phase of the components giving an elliptically polarized reflected beam. Instruments called **ellipsometers** are used and can measure surface effects a small fraction of a wavelength thick.

Many organic tissues have polarizing properties and a polarizing microscope is used to study these. This is not very different from an ordinary microscope but has provision for polarizers and analysers together with a rotatable mount for the stage. The lenses used must be particularly free from strain.

If a normally isotropic glass plate is compressed it is found to act like a negative uniaxial crystal. The optic axis is in the direction of the stress. This means that for light incident perpendicular to this direction the glass acts like a retardation plate. This stress induced birefringence is called **photoelasticity**. For most purposes it is found that the birefringence is linearly proportional to the stress, the constant of proportionality being the **stress-optical coefficient**.

The effect of applied stresses may be seen by placing the material between linear polarizers, or between circular polarizers. Models of proposed structures may be built and loaded so that the distribution of the strain can be seen. (Stress is the applied load; strain is the response of the material.) With optical glass a similar method may be used to check the absence of strain for good homogeneity and isotropism. On the other hand glass, which is very strong in compression, may be toughened by generating zones of tension and compression so that the compressive region is at all the surfaces while the central bulk is in tension. This is done either by quickly cooling the outer surfaces with a cold air blast or by chemical means.

The human eye can just detect the polarization of light. This was first noticed in 1844 by Haidinger and the phenomenon is known as **Haidinger's brush**. When a sheet of linear polarizer is held between the observer and a uniform white background, the brush appears if the polarizer is quickly rotated through 90°. This faint pattern is yellow and double-ended like two cones stuck point to point. It subtends an angle of about 2°. Its orientation depends on the plane of vibration of the polarizer and as it fades away in a few seconds it is generally thought to be a fatigue effect. If the polarizer is rotated again it reappears and can be made to rotate if the polarizer is slowly rotated. A similar but fainter effect can be found with circularly polarized light. The phenomenon is confined to blue light – if this is excluded the brush fails to appear – so that the yellow occurs where the blue response has been fatigued.

Not all light reaching the retina is absorbed by it. The techniques of retinoscopy and ophthalmoscopy would be impossible if this were the case. Although less blue light than yellow and red (and infrared) light is reflected, the blue reflection is mainly specular while for the longer wavelengths the reflection is mainly diffuse. This can be shown by using linearly polarized light which retains its polarization with specular reflection. The location of the reflecting surfaces in the retina of the living eye is still open to argument. It seems clear that the longer wavelengths are reflected from more than one place while the blue has a single plane of reflection. Wavelengths shorter than 400 nm are hardly reflected at all.

Some materials, such as quartz, are not only birefringent, they also exhibit **optical activity**. When it is cut as a parallel plate with its optic axis perpendicular to the surfaces it has the ability to rotate the plane of vibration of incident linearly polarized light. Thus, if a plate of quartz cut in this way is inserted between a pair of polarizers crossed to give extinction, it will allow light to be transmitted. If the second polarizer is rotated, a new position of extinction can be found. It follows, therefore, that the light leaving the quartz plate is linearly polarized but the plane of the vibrations has been rotated.

With a given substance, the angle through which the plane of vibration is rotated is proportional to the thickness of the plate and, with a given thickness of plate, this angle varies inversely with the wavelength, being approximately proportional to λ^{-2}. Hence, with white light, the extinction described above will be incomplete and different colours will be transmitted in turn as the analyser is rotated.

Some materials are birefringent without being optically active (e.g. calcite), while others are optically active without being birefringent (e.g. liquids, particularly sugar solutions). The rotation may be right-handed (positive) or left-handed (negative), defined as the required movement of the analyser to regain extinction as seen by an observer looking from it towards the polarizer. Fused quartz (fused silica) has no birefringence or optical activity.

Exercises

11.1   Describe fully a method of obtaining a pure spectrum. What difference will be seen between the spectra of:
(a) The sun.
(b) The electric arc.
(c) A tube containing incandescent hydrogen.

11.2   Explain clearly how you would use a spectrometer to determine the refractive index of the glass of a prism. Give the necessary theory and describe the experimental details.

11.3   What is meant by the dispersive power of glass? Describe carefully a practical method of measuring this value.

11.4   In measuring a prism with refracting angle of 60° 4′ on the spectrometer the following values were obtained for the minimum deviation for different colours:

Red   (C Line)   38° 23′
Yellow   (D Line)   38° 37′
Blue   (F Line)   39° 8′

Find the dispersive power of the glass. What kind of glass is this?

11.5   Manufacturers' lists of optical glasses give particulars of refractive index, dispersive power and partial dispersions. Briefly explain these quantities.

11.6   A 60° prism is made of flint glass having refractive indices $n_C = 1.615$, $n_D = 1.620$, and $n_F = 1.632$. Find the angle between the emergent light corresponding to the C and F lines when a beam of white light is incident on the prism in the direction of minimum deviation for the D line.

11.7   Discuss the occurrence of the colour in the case of:
(a) The blue sky.
(b) A piece of blue glass.
(c) A piece of blue paper.

11.8   Table 11.7 gives the refractive indices of two glasses for the stated Fraunhofer lines. Explain the meaning of the term 'irrational dispersion', using the quoted figures to illustrate your answer. Find the dispersive powers and the relative partial dispersions of the two glasses.

Table 11.7

	C	D	F	G
Crown	1.506	1.509	1.514	1.523
Flint	1.749	1.757	1.776	1.812

11.9   The face AC of a prism ABC is silvered. A ray from an object is incident on the face AB and after two reflections, first as the silvered face AC and then at the face AB, it emerges from the base BC. Determine what relation must exist between the prism angles at A and B for there to be no chromatism.

11.10 Describe the advantages and disadvantages of using plastics instead of glass in an optical system.

11.11 Sketch and explain the spectral radiation curves for a black body at various temperatures.
    With reference to the curves describe the change in appearance of a black body which would take place if its temperature were raised from cold to, say, 7000 K.
    Explain what is meant by 'colour temperature' and 'mired value'. What does the term 'mired' signify when applied to a filter?

11.12 Describe two experiments which show that light is propagated as a transverse wave motion. How is the law of rectilinear propagation explained on the wave theory?

11.13 Explain what is meant by plane polarized light and describe two methods by which it can be produced.

11.14 Show that, when $i + i' = 90°$, $\tan i = n$. What is the difference between the incident and reflected light for this particular angle of incidence?

11.15 Using Fresnel's law of reflection plot the curves showing the variation of the intensity of the reflected light with variation of the angle of incidence when light is reflected from the surface of a medium of refractive index 1.6, for vibrations taking place (a) parallel (b) perpendicular to the plane of incidence. From the curves determine the polarizing angle.

11.16   The angle of incidence of a parallel pencil of white light on the plane polished surface of a block of glass of refractive index 1.760 is 60° 24′. What will be the angle between the reflected beam and the beam refracted into the glass?

Explain carefully any differences there will be in the nature of the incident, the reflected and the refracted light.

11.17   Explain what is meant by the term 'polarizing angle' when used in connection with light incident in air on the surface of a plane glass plate. Show that if the refractive index of the plate is n the tangent of the polarizing angle is equal to n.

When polarizing lenses are used in sunglasses they are orientated to accept light polarized in a particular direction. What is the direction normally chosen and why is it chosen?

A man uses a piece of Polaroid to minimize the glare reflected from the still surface of a lake. At what angle of view relative to the surface will he obtain the best extinction of the glare?

11.18   Explain what is meant by a quarter-wave plate. If the refractive indices of mica are 1.561 (ordinary) and 1.594 (extraordinary), what must be the thickness of a film of mica to produce a quarter-wave plate for sodium light? ($\lambda = 589$ nm.)

11.19   (a)   Explain how circularly polarized light may be produced by a quarter-wave plate.
(b)   Explain how a half-wave plate may be used to give any desired rotation of the plane of vibration of linearly polarized light.

11.20   A beam of plane polarized light of wavelength 540 nm falls normally on a thin plate of quartz cut so that the optic axis lies in the surface. If the ordinary and extraordinary refractive indices are respectively 1.544 and 1.553, find for what thickness of the plate the difference in phase between the ordinary and extraordinary beams is π rad on emergence from the plate.

11.21   Plane waves of light are incident obliquely on the surface of a calcite crystal. Give a clear diagram showing the passage of light within the crystal when the optic axis lies perpendicular to the plane of incidence and parallel to the surface of the crystal.

In what important respect do the ordinary and extraordinary beams emerging from the crystal differ from one another?

A ray of unpolarized light falls on a calcite crystal, the optic axis of which is parallel to the surface. The angle of incidence is 34° and the plane of incidence coincides with the principal section of the crystal. Find the angles of refraction of the o and e rays for sodium light. ($\lambda = 600$ nm), $n_o = 1.658$, $n_e = 1.486$.)

[*Hint*: Regard the normal as the x-axis. The line $y = mx + c$ touches the ellipse

$$\frac{x^2}{a^2} + \frac{y^2}{b^2} = 1$$

when $m^2 a^2 = c^2$.]

Chapter 12

Interference and optical films

The phenomenon of interference arises directly out of the wave theory of light. According to the principle of superposition of simple harmonic motions developed in Chapter 9, interference should occur whenever two or more waves meet. That this is not so in practice is due mainly to the departure of ordinary light from a *continuous* wave motion. To obtain steady interference effects with ordinary light sources we have to satisfy rather stringent conditions. On the other hand the light from a laser source approximates much more closely to a continuous wave motion and workers with these light sources will testify that interference phenomena occur so easily that they can become a nuisance.

If, however, the correct effects can be isolated, they form a very precise method of measurement, being able to assess accurately changes in dimension which are less than one thousandth of the wavelength of the light used. For green light this is 0.5 nm or 5×10^{-10} m (2×10^{-8} inches). This branch of optics is known as **optical metrology** or **nanotechnology** and an outline of this technology will be given in this chapter.

On the other hand, if optical components can be made with thicknesses less than the wavelength of light, the interference phenomena obtained can be used to modify reflection coefficients and obtain colour filters, very high transmission elements and polarizing effects. This branch of optics is known as **optical thin films** or **optical coatings** and this chapter will conclude with a review of the theory and applications of this subject.

12.1 Coherence and incoherence

Two sets of waves of equal amplitude and frequency spreading out from point sources will neutralize the effects of one another in certain places and will give increased disturbance in others. This can be shown visually by generating two sets of ripples on the surface of mercury. If two small points are attached to a tuning fork and arranged so that they just dip into the mercury, two exactly similar sets of ripples will be created when the tuning fork is set into vibration. Each ripple system will consist of a series of concentric circles. The appearance will be as in Figure 12.1, where B_1 and B_2 are the sources and the crests of the waves are represented by solid lines and the troughs by broken lines.

Along the lines marked with crosses the crest of one wave coincides with the trough of another, and since the amplitudes are equal the displacements will be

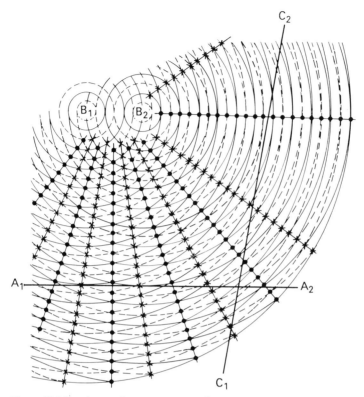

Figure 12.1 Interference between two sets of waves

neutralized in accordance with the principle of superposition. Along the lines marked with circles the crests coincide and the troughs coincide so that the resultant disturbance has twice the amplitude. As the energy of the wave motion is proportional to the square of the amplitude there will be four times the energy of a single source along these lines and thus there is no loss of total energy from the two sources due to interference, but only a redistribution.

If a barrier were erected at A_1–A_2, it would receive no waves where the crossed lines meet it and double amplitude waves at the place where the circled lines meet it. In moving along the barrier from A_1 to A_2, one receives doubled and zero effects in a cyclic fashion. Similarly a barrier along C_1–C_2 would show the same effect but with the different regions of energy more spread out. Their location is crucially dependent on the distance between B_1 and B_2 *measured in terms of the wavelength of the ripples*.

As has been indicated earlier, two ordinary sources of light placed in positions such as B_1 and B_2 do not produce visible interference effects. If they did we should expect to find on a screen at A_1–A_2 bright patches where the crests coincided and darkness, because of zero amplitude, in the regions where a crest always coincides with a trough. These patches would stay much the same on the screen above and below the plane of the page and would be seen as alternately bright and dark **fringes**. If either of the two sources varied in phase with respect to the other source these lines would move. With any two ordinary sources of light, the relative phase between them is constantly changing in a very fast random manner. The fringes formed are therefore blurred out to a uniform illumination. When the source has a finite size there is an equally random relationship in the phase between one part and another.

In Section 9.6 the concept of coherent sources was developed and the various forms of laser described. However, interference phenomena do not require such complete coherence and indeed fringe patterns due to interference were being studied long before the laser was invented. This was achieved with largely incoherent sources because the presence of stationary fringes on screen A_1–A_2 of Figure 12.1 depends only on obtaining a constant phase difference between sources. In other words it does not matter if the phase of B_1 changes as long as that of B_2 changes in an identical manner. In terms of superposition of waves at A_1 the wave motion from B_1 and B_2 can vary from moment to moment provided that a crest from B_1 always coincides with a trough from B_2 and vice versa. Such a relationship between B_1 and B_2 is termed **mutual coherence**.

In all aspects of coherence it should be remembered that a continuous single-frequency simple harmonic motion cannot be perfectly achieved, nor can complete identity between two sources. It is proper therefore to speak of **degrees of coherence** which are less than 100%. In general terms the visibility of the fringe patterns described will vary with the degree of coherence obtained. However, some fading of the pattern can usually be tolerated.

In mathematical terms it may be stated that for two coherent sources the intensity on the screen is found by adding the respective amplitudes and squaring the result; whereas for incoherent sources the final intensity is found by adding the respective intensities. The latter method can never give darkness as the intensities are always positive being the square of the amplitudes discussed in Sections 9.2 and 9.3.

12.2 Interference with incoherent and coherent sources

The methods of interference with incoherent sources are aimed at obtaining a sufficient degree of mutual coherence between two sources. This is done by making the two sources both dependent on the same source. For instance, the reflection of light in a mirror provides an additional source apparently behind the mirror surface. It is possible to obtain interference between an incoherent source and its reflection provided there is a region where the two sets of waves overlap and provided the distance between the source and its reflection is not greater than a few wavelengths. These conditions are easier to achieve if two reflected sources are used and Newton's rings (Section 12.6) is a particular case which constitutes the earliest study of interference phenomena.

The next example was devised by Young in the early nineteenth century. He illuminated a small aperture so that it became effectively a single small source. A short distance from this he placed a screen containing two small apertures which were illuminated only by the first aperture. These two apertures now constitute mutually coherent sources and, provided that the illuminating light is reasonably monochromatic, interference fringes may be observed on a screen in front of the two apertures as A_1–A_2 in Figure 12.1. The fringes are very faint as nearly all the original light has been obscured. Better results are obtained if narrow parallel slits are used in place of the three apertures.

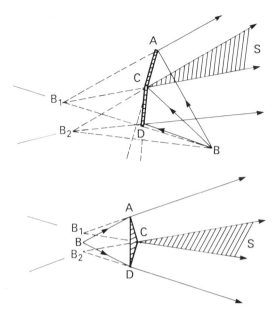

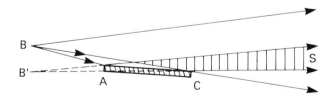

Figure 12.2 Fresnel's mirrors and bi-prism

Much better use of the available light was made by **Fresnel** in his **double-mirror** and **bi-prism** methods. These are most easily explained in Figure 12.2. The single source B is in each case converted to two sources, B_1 and B_2, by reflection or refraction at AC and CD. So that the apparent sources are close together the angle ACD of the bi-prism and between the mirrors must be in the region of 179°. The interference takes place in the regions S where the beams overlap.

Considerably brighter fringes may be obtained if B is an illuminated narrow slit parallel to the join of the mirrors or refracting edge of the prism. This gives clear parallel fringes in the region S with a regular spacing which may be measured experimentally using a travelling microscope. Fringes may be found anywhere in the shaded region S and are said to be non-localized.

Lloyd's single mirror method is different from the above as it uses only one reflected source. In Figure 12.3 the illuminated slit B is placed parallel and close to the surface of a plane mirror (or black glass) and the reflected source B_1 interferes with the original light waves in the shaded region S.

Figure 12.3 Lloyd's mirror

The mutual coherence of the interfering sources described above arises from the dependence of each source on a single original source. Although the precise phase and structure of the wavefront from the real source may be changing many times a second, the dependent sources are changing also. When there is a phase difference between the two sources it means that the precise parts of the waves that are interfering left the original source at different times. But the original source is changing many times a second and so, if the time difference becomes too large, the two waves are no longer mutually coherent as the source changes between the times of their emission. The allowed time difference is called the **coherence time** of the source and when multiplied by the velocity of light gives a **coherence length** which is a more useful parameter.

This is related to how monochromatic the source is. If the source contains wavelengths λ_1 to λ_2 then, as k of Equation 12.1 increases, x can equal $k\lambda_1$ and $(k + \frac{1}{2}) \lambda_2$ and the fringe pattern fades away, with λ_2 filling in the dark bands of the λ_1 pattern. With an incandescent source followed by a filter the coherence length may be only a few micrometres. With a spectral line source such as a mercury lamp a coherence length of 200–300 mm may be achieved. With continuous wave lasers source coherence over many metres may be readily obtained and special stabilizing methods can extend this to many kilometres.

The above discussion takes no account of the finite size of sources, which effectively reduces the coherence length. With lasers the length of the resonant cavity has an effect on their coherence, and fringes with highest visibility are generally obtained when the path difference is a multiple of the cavity length.

In Figure 12.4, two similar lasers are arranged as interference sources B_1 and B_2. Their normally collimated output is caused to diverge by lenses L_1 and L_2, so that an interference pattern is formed at S on the screen. With ordinary laboratory lasers no fringes will be seen. This is because the mutual coherence between the

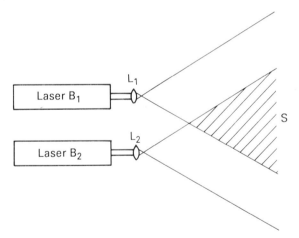

Figure 12.4 Interference with lasers – no fringes are seen

two lasers is limited by the random phase changes, which limit their individual coherence length to a few kilometres at best. Although this is a considerable distance, the velocity of light is such that the *coherence time* is only a few microseconds. Thus the interference pattern is only stationary over this period of time and moves in a random manner. It is therefore not possible to see fringes in this way, although special methods have shown they exist.

However, most laser interferometer systems use single lasers and beam splitting methods to achieve measurements with long path differences or provide a multiplicity of interfering sources to give very fine fringes for high accuracy. These are discussed later in this chapter.

12.3 The form of interference fringes

Referring again to Figure 12.1, the presence of a bright or dark fringe at a given point on A_1–A_2 depends on whether the waves arrive in or out of phase. If in Figure 12.5 B_1 and B_2 represent two mutually coherent line sources emitting waves of equal amplitude, frequency and phase, the waves will arrive in phase at point S on the screen as the distances B_1S and B_2S are equal if CS perpendicularly bisects B_1B_2. S is therefore a bright fringe.

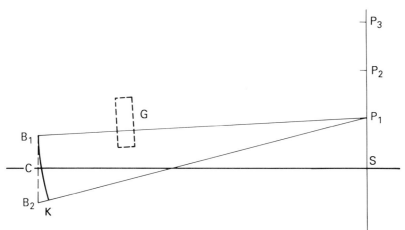

Figure 12.5 Width of interference fringes

For a point P_1 to one side of S the paths B_1P_1 and B_2P_1 will be unequal and the waves will arrive out of phase. If the phase difference is equivalent to one-half wavelength a dark fringe will result as crests and troughs are arriving together. A bright fringe will result at P_2 if the phase difference by then is equivalent to one whole wavelength. At P_3 the difference may have increased to one and a half wavelengths and so a dark fringe results. Therefore on either side of the bright fringe at S there will be a series of bright and dark fringes as in Plate 1. The bright fringes occur when the path difference is $k\lambda$ and the dark fringes when it is $(k + \frac{1}{2})\lambda$, where k is any integer.

The path difference for P_1 is found by constructing a circle centre P_1 through B_1. Then, as B_1P_1 and KP_1 are equal, the inequality is due to B_2K. Normally B_1B_2 is small compared with the distance to the screen if there is to be any appreciable separation of the fringes. Thus, B_1K approximates to a straight line perpendicular to P_1C and B_2K. Representing B_1B_2 by b, SP_1 by x and CS, which is approximately equal to CP_1, by d, we have, by similar triangles, B_1B_2K and $B_1P_1B_2$, *for the centre of a dark fringe*:

$$x = \frac{d}{b}(k + \frac{1}{2})\lambda \qquad\qquad (12.1)$$

for the centre of a bright fringe:

$$x = \frac{d}{b}(k)\lambda \qquad\qquad (12.2)$$

and the distances between centres of consecutive bright or dark bands will be

$$y = \frac{d}{b} \lambda \tag{12.3}$$

It is thus seen that the spacing depends on the wavelength of the light used. If a source of more than one wavelength is used a number of interference bands will be produced which will mix in and out of step because of their different fringe spacings. The central band at S will remain white as all wavelengths have zero phase difference.

If, in Figure 12.5, a block of glass is inserted into one beam only, a displacement of the fringe pattern will result. It is assumed that the frequency of any light vibration remains constant whatever medium it is travelling in. As the velocity changes, by the definition of refractive index (see Section 1.10), the wavelength inside a medium of index n will be $\lambda_n = \lambda/n$ where λ is the wavelength *in vacuo* or in air. The calculation of phase difference due to a path difference in glass must take account of the different wavelength. Thus Equation 9.14 becomes

Phase difference, $\epsilon = \dfrac{2\pi}{\lambda_n} \times$ Path difference

$$= \frac{2\pi n}{\lambda} \times \text{Path difference} \tag{12.4}$$

Referring again to Figure 12.5, the effect of a block glass of index n and thickness t is to remove a patch difference due to air and insert one due to glass. The resulting change in the phase difference is

$$\frac{2\pi n t}{\lambda} - \frac{2\pi t}{\lambda} = (n - 1) \frac{2\pi t}{\lambda}$$

If the displacement of the central fringe due to the insertion of the glass is m fringes this represents a phase change of $2\pi m$ and so

$$(n - 1)t = m\lambda \tag{12.5}$$

The product nt is referred to as the **optical thickness** of the glass block. Optical thickness, nt, relates to the difference in phase caused by an optical material (Section 1.11). **Apparent thickness**, t/n, relates to imaging through an optical material (Section 2.5).

In the case of Lloyd's mirror (Section 12.2) it is found by experiment that the fringes produced are not as predicted by Equations 12.1 and 12.2. The fringe pattern is displaced by one-half fringe and this is due to the reflected source B_1 being π out of phase with the original source. This occurs due to a phase change of π occurring on reflection at the mirror and is considered further in Section 12.6.

From Equations 12.1 and 12.3 the wavelength of the light can be determined by measuring the distances x or y, b and d in any of the methods described in Section 12.2. Conversely, if the wavelength is known, variations in b and d will show in the spacing of the fringes.

The interference fringes will not be sharply defined, because the resultant amplitude at different points on the screen gradually decreases as the waves get more and more out of step, until a minimum is reached, and then gradually increases to a maximum as the waves again come into step. The variation in

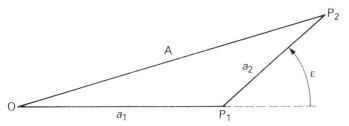

Figure 12.6 Resultant amplitude

intensity at different points on the screen may be found by applying the graphical method of Section 9.2. If a_1 and a_2 (Figure 12.6) are the amplitudes of the wave motion from b_1 and b_2, respectively, arriving at a general point P with a phase difference of ϵ, the resultant amplitude A at the screen is the vector addition of a_1 and a_2.

Analytically,

$$A \sin \left(\frac{2\pi t}{T} + \beta \right) = a_1 \sin \frac{2\pi t}{T} + a_2 \sin \left(\frac{2\pi t}{T} + \epsilon \right) \tag{12.6}$$

where β is the phase of the resultant wave motion.

By vectors (Figure 12.6) for the case where $a_1 = a_2 = a$,

$$A^2 = 2a^2 (1 + \cos \epsilon) = 4a^2 \cos^2 \frac{\epsilon}{2} \tag{12.7}$$

or

$$A^2 = 4a^2 \cos^2 \left(\frac{2\pi \times \text{Path difference}}{\lambda} \right) \tag{12.8}$$

It should be clear from Equation 12.8 and Figure 12.6 that A will be a maximum when the phase difference, ϵ, is 0, 2π, 4π, etc., and a minimum when ϵ is π, 3π, 5π, etc. The intensity of the fringes varies as A^2 and so is proportional to the cosine squared of the path difference. These fringes are usually produced when two sources are used and are known as *cosine squared* fringes and have an intensity profile as in Figure 12.7.

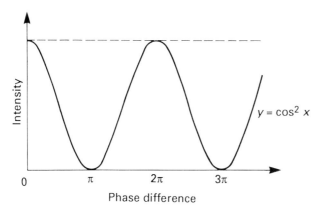

Figure 12.7 Intensity profile of two-beam fringes

The central fringes in Plate 9(b) are cosine squared fringes, although in this case they are a little overexposed so that the peripheral fringes can be seen. If the number of interfering beams is increased to four, the fringes become sharper and extra fringes appear between them as in Plate 9(c). The intensity profile of these is the square of the amplitude value calculated in Section 9.2 for four equal interacting motions (Equation 9.9).

From that expression the overall resultant may be evaluated for a range of ϵ (assuming a equal to unity). The profile obtained will show the subsidiary fringes of about $\frac{1}{16}$ the intensity of the main fringes. Outside the central three peaks of Plate 9, diffraction effects become evident.

12.4 Multiple beam interference

The cosine-squared fringes of Plate 1 are the result of two interfering beams. When measuring their position or spacing it is difficult to locate the centre of a fringe accurately, whether it is bright or dark. Using photodetectors in arrays or at some specific distance apart it is possible to monitor the phase and determine the fringe position to within 0.001 times their spacing as well as to display electronically sharpened images on a video screen. An optical method for obtaining sharper fringes uses multiple interfering beams. These have been employed since 1897 and the advent of the laser widened their application, although the electronic methods referred to above now allow improved accuracies with two-beam interference and multiple-beam interference thereby moving the latter to the more esoteric applications. The two general examples of multiple-beam interference are the Fabry–Perot etalon and the diffraction grating (Chapter 13).

In multiple-beam methods the multiple effective sources produce waves which are out of phase by multiples of the phase difference between the first two beams. This means that Equation 12.7 becomes

$$A \sin\left(\frac{2\pi t}{T} + \beta\right) = a_1 \sin\frac{2\pi t}{T} + a_2 \sin\left(\frac{2\pi t}{T} + \epsilon\right) + a_3 \sin\left(\frac{2\pi t}{T} + 2\epsilon\right)$$

$$+ a_4 \sin\left(\frac{2\pi t}{T} + 3\epsilon\right) + a_5 \sin\left(\frac{2\pi t}{T} + 4\epsilon\right) \quad \text{etc.}$$

The amplitudes a_1, a_2, a_3, etc., are generally reducing where multiple reflections are concerned. The graphical calculation of the resultant appears as in Figure 12.8, where the resulting amplitudes A_1 and A_2 can be compared for a relatively small change in the incremental phase difference from ϵ_1 to ϵ_2.

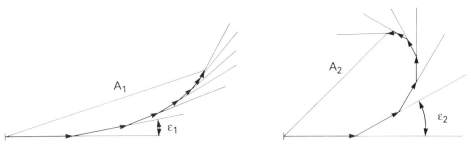

Figure 12.8 Vector addition for seven beams

An analytical result can be obtained when the values of a_1, a_2, a_3, etc., have a simple relationship such as a general reduction for multiple reflections. This uses a factor E, which is dependent on the reflection coefficients being greater for higher reflectivities, which gives more beams with slowly decreasing amplitudes. Figure 12.9 shows the effect of increasing the value of E for the transmitted beams of a Fabry–Perot etalon as described below. The reflected beams show the inverse intensity profile and examples of these are given in Plates 10(a) and 10(b), the latter showing some imperfection of the mirror surfaces.

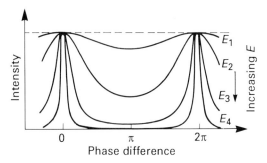

Figure 12.9 Intensity profile of multiple-beam fringes

Large path differences between interfering beams make heavier demands on the coherence of the light source, particularly if the path difference is traversed many times as in multiple-beam interference. The availability of lasers as sources has stimulated considerable developments in this type, although the Fabry–Perot was first conceived in the last century. This interferometer consists of nothing more than two glass or quartz plates with very plane surfaces held parallel a small distance apart by a hollow cylinder of invar or silica as in Figure 12.10. When the separation is fixed, as is normally the case, the instrument is called a **Fabry–Perot etalon**. The inner surfaces of the plates are coated to give very high reflectivities and the plates themselves are made slightly prismatic to avoid any interference effects inside them.

The prime purpose of the device is as a form of spectrometer. When the light from a monochromatic source is collimated and passes through the mirrors the multiple reflections result in very fine circular fringes, shaped as the sharpest in Figure 12.9, so that the fine circles are bright against a dark background. Any other wavelengths present in the source are easily seen even if their wavelengths differ by considerably less than 0.1 nm from each other.

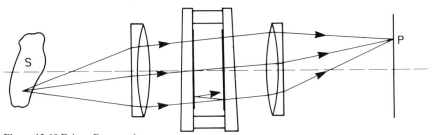

Figure 12.10 Fabry–Perot etalon

12.5 Interference by partial reflection

In Figure 12.1 the interference effects of two sources were shown as generating bright and dark regions on screens at A_1–A_2 and C_1–C_2. The methods of interference considered so far have formed apparent sources side by side to the screen as indicated by A_1–A_2 of Figure 12.1. However, interference effects can also occur when the sources are effectively one behind the other, as for screen C_1–C_2. Apparent sources displaced along the line of viewing are readily generated by partial reflection. (The distinction 'division of wavefront' for the methods of Section 12.2 and 'division of amplitude' for those of this section and following is commonly used.)

The best known phenomenon due to the interference of light is the brilliant colouring produced when white light is reflected from a thin film of a transparent substance. Examples of this are seen in the soap bubble, a patch of oil on a wet road, and a very thin film of air between two pieces of glass. Unlike the effects of interference previously described, which require the use of a small source, the interference colours of a thin film are best seen when the film is illuminated by an extended source such as the sky.

This action may be set up in the laboratory without the two surfaces being in close proximity. The Michelson interferometer is shown in Figure 12.11. This uses the partially reflecting element A to divide the light from source S into two mutually coherent beams which are reflected by M_1 and M_2.

The partially reflecting element is often called a beamsplitter from this first action, but it is also able to recombine the beams reflected by M_1 and M_2. Although some light is lost back to the source, the appearance to the eye at E is that the two reflecting surfaces appear to be very close together or even superimposed at M_1 and M_2'.

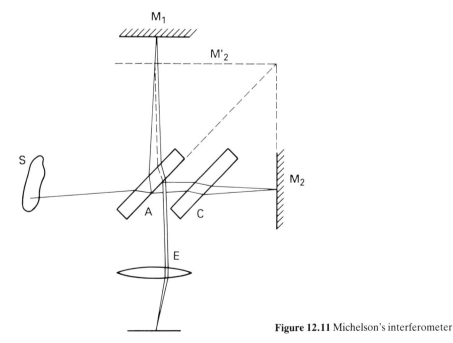

Figure 12.11 Michelson's interferometer

The light to M_1 passes twice through the glass support for the partial reflecting surface. Chromatic errors due to this may be cancelled out by inserting an identical block of glass to compensate at C. When the positions of M_1 and M_2 are such that the light paths are identical, the two effective sources superimpose. With a small angle between the mirrors, interference fringes can be seen even with white light. The different wavelengths produce fringe patterns with spacing proportional to λ (Equation 12.3), but with a central fringe coincident for all colours where the phase difference is zero for all colours. On either side a few highly coloured fringes occur which rapidly merge to an even illumination as the fringe patterns of the different wavelengths overlap. This action with white light can be very useful in positively identifying the point of effective coincidence of the mirrors, and was used by Michelson to relate the wavelength of various monochromatic light sources to the unit of length, the standard metre.

The Michelson interferometer shown if Figure 12.11 was modified by **Twyman** and **Green** to form a very useful testing instrument for optical shop use. When using collimated light from a monochromatic source the system can be considered to be a Fizeau interferometer when the reflected image of M_2 is close to M_1 but with the two beams separately available. The system as arranged in Figure 12.12 can be used for testing prisms or optical windows. If mirror M_2 is replaced by a mirror under test, that too may be interferometrically compared with M_1. For lenses the arrangement shown in Figure 12.12(b) may be used, where L is a lens producing a convergent beam. The test surface, M, is placed so that the beam is returned along its own path and therefore is recollimated by L to interfere with the beam from M_1. A range of different radii of curvature may be measured by this system, and concave surfaces by replacing L with a divergent lens.

The results are seen as fringe patterns in which departure from circles or straight lines indicates a less-than-perfect component. The interpretation of the patterns can be difficult and interferometric testing is generally reserved for only the highest quality of optical component.

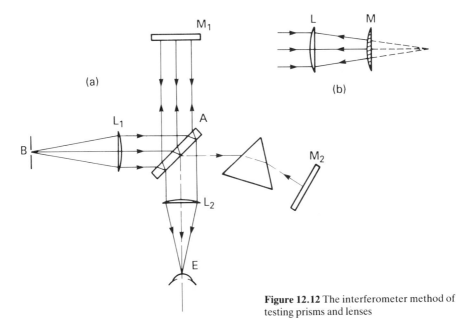

Figure 12.12 The interferometer method of testing prisms and lenses

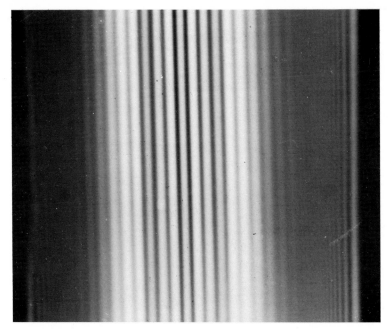

Plate 1 Fringes formed when slit, illuminated with monochromatic light, is viewed via a Fresnel bi-prism (see Section 12.3)

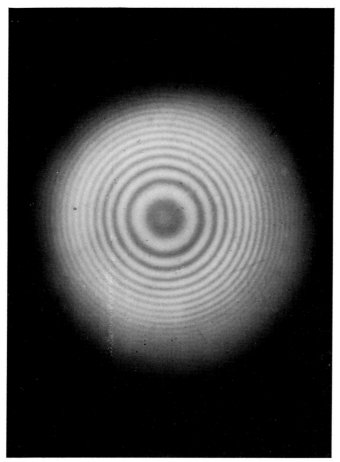

Plate 2 Newton's rings with monochromatic light. The light centre to the ring system is due to the two glasses not being quite in contact (see Section 12.6)

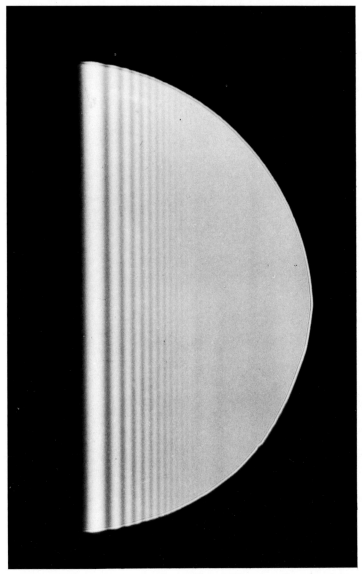

Plate 3 Shadow of a straight edge showing Fresnel diffraction effects (see Section 13.4)

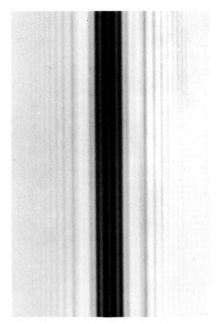

Plate 4 Shadow of a narrow wire showing Fresnel diffraction effects (see Section 13.4)

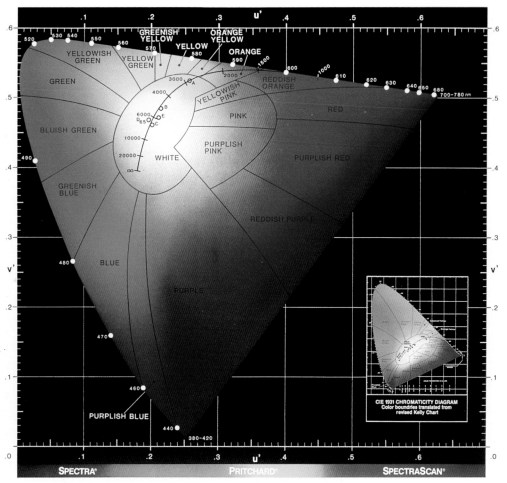

Plate 5 The CIE (1976 UCS) chromaticity diagram. The letters A to E indicate the colour and colour temperature values (in kelvins) of the standard illuminants along the black body curve. Spectral wavelengths (in micrometres) are shown along the periphery of the diagram. Reprinted by courtesy of Photo Research – a division of the Kollmorgan Company (see Section 10.13)

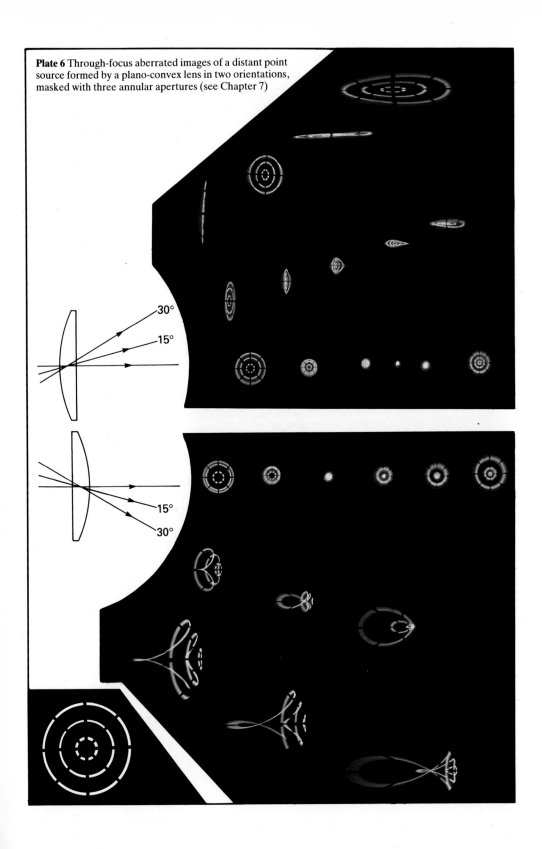

Plate 6 Through-focus aberrated images of a distant point source formed by a plano-convex lens in two orientations, masked with three annular apertures (see Chapter 7)

30°
15°

15°
30°

Plate 7 Photograph showing diffracted light forming a bright spot at the centre of a circular shadow (see Section 13.3.2)

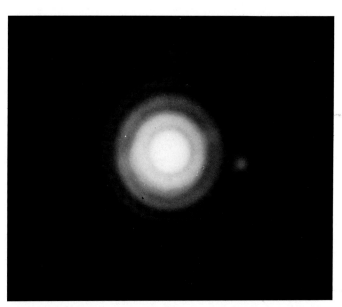

Plate 8 Airy disc. Pattern of light formed by a perfect optical system of circular aperture as the image of a point object (see Section 13.5)

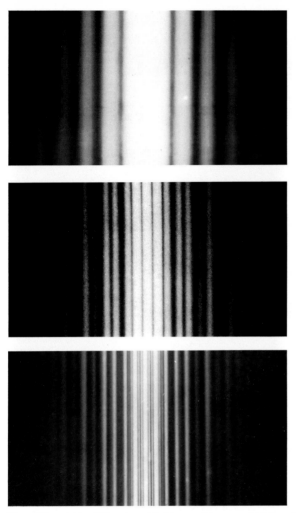

Plate 9 Fraunhofer diffraction with (a) one, (b) two and (c) four slits (see Sections 12.3, 13.5 and 13.7)

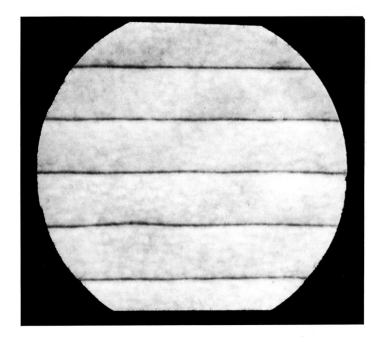

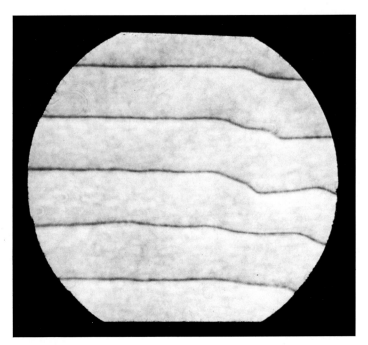

Plate 10 (a) Multiple-beam interference fringes (see Section 12.4); (b) multiple-beam interference fringes showing region of inhomogeneity (see Section 12.4). Reproduced by courtesy of Pilkington P-E Ltd

Plate 11 Double-swept contrast/frequency grating. In this plate the contrast modulation of the grating changes by a factor of 10 (one log unit) for every 60 mm along the y-axis although variations in printing can affect this. The angular frequency values along the x-axis are correct for arm's length viewing at 600 mm (see Section 15.9). Reproduced by courtesy of J. G. Robson and F. W. Campbell

H S Z D S N

C K R Z V R

N D C O S K

Plate 12 A small scale Pelli-Robson Letter Sensitivity Chart. It should be noted that the contrast levels of the letters in this reproduction of the chart are not exactly those on the full-size chart (see Section 15.9). Copyright D. G. Pelli; reproduced by courtesy of Metropia Ltd.

12.6 Thin films and Newton's rings

In Figure 12.13, light from an extended monochromatic source is incident upon a thin transparent film. A proportion of this light is reflected at each surface and the eye superimposes those coming from the same point on the film.

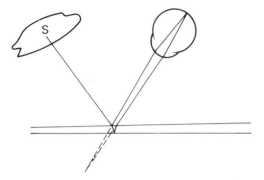

Figure 12.13 Interference with a thin film

The actual ray bundles to the eye may be separated slightly before the action of the film and eye superimposes them on the retina, but when the film is very thin this separation is very small and the bundles are largely mutually coherent. As they interfere the eye sees coloured fringes apparently localized in the plane of the film. The phase difference causing the fringes comes from the path difference between the light reflected from the top surface of the film and that which enters the film and is reflected at the lower surface. The colours arise because different wavelengths generate different phase differences from the same path difference. Thus one area of the film may generate a bright fringe for red-orange light but a dark fringe for blue-violet.

If a soap film is caught across a wire it may be supported vertically in a glass container to protect it from draughts. The film gradually becomes wedge-shaped, due to gravity, and if illuminated with white light the reflection from it is seen to be crossed by broad horizontal bands of colour. After a time the upper region will become very thin and a perfectly black band will be formed at the upper edge. The same effect can be seen when two carefully cleaned glass plates are pressed into contact; where the air film between them is very thin compared with the wavelength of light a black patch will appear and the plates are said to be in **optical contact** at that place.

This dark region shows that there is a phase difference between the two reflected waves for all colours. This cannot be due to the path difference, which is very small and would produce various phase differences depending on wavelength. From electromagnetic theory it can be shown that a phase change of π occurs when a wave is reflected at a surface between a rarer and a denser medium. If the reflection is from a boundary between a denser and a rarer medium there is no change of phase. The two beams being reflected from the glass plates in optical contact come into each of these categories.

If crown glass is used for one plate and flint glass for the other the intervening space may be filled with a liquid of refractive index between the two glass types, such as oil of sassafras. Under similar conditions of optical contact a bright patch is now seen as the light is being reflected from the same sort of boundary at each surface.

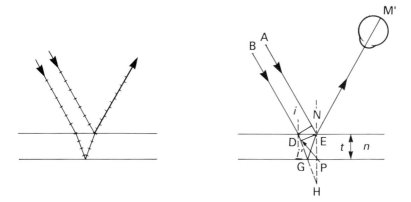

Figure 12.14 Reflection from a thin film. Conditions for a light or dark band

The path difference due to one beam traversing the film before reflection while the other is reflected from the first surface is required to be known so that the phase difference for a particular wavelength may be calculated. In Figure 12.14, the difference in path length for the two beams which are subsequently superimposed by the eye is found by constructing DN perpendicular to the incident beams. There is then no path difference before that line and none after the point E where they combine. The difference then appears to be between DGE and NE. However, the path NE does not lie in a medium of the same refractive index as DGE. The difference can most easily be accounted for by constructing PE perpendicular to DG. Then, as shown in Chapter 1, the time taken to travel along NE is the same as for DP.

Thus the waves are still in phase at P and E. The distance PGE can then be shown, for a parallel film of thickness t and index n, to be given by $2t \cos i'$, using the construction to H. The phase difference due to this is given by Equation 9.14 to be

$$\frac{4\pi nt \cos i'}{\lambda}$$

Remembering the phase difference due to reflection it is seen that for films bound by similar media a *dark fringe* will occur when

$$\frac{4\pi nt \cos i'}{\lambda} = 2\pi k$$

or

$$2nt \cos i' = k\lambda \tag{12.9}$$

and a *bright fringe* when

$$\frac{4\pi nt \cos i'}{\lambda} = 2\pi(k + \tfrac{1}{2})$$

or

$$2nt \cos i' = (k + \tfrac{1}{2})\lambda \tag{12.10}$$

The dark fringes so calculated will not be completely dark unless the reflections from the two surfaces produce interfering beams of equal amplitude. The fraction of incident light reflected at normal incidence at the boundary between media of index n_1 and n_2 is given by Fresnel's expression (Equation 11.22):

$$r = \left(\frac{n_1 - n_2}{n_1 + n_2}\right)^2$$

For a thin soap film in air or a thin film of air bounded by glass of identical type, the intensity of the lower beam cannot equal that of the upper in Figure 12.14 as this suffers two refractions as well as a reflection. However, there are other beams due to further reflections which provide interfering beams at larger phase differences. These give a multiple beam effect and provide for zero intensity in the dark bands. If the reflectivity of the two surfaces is increased a larger number of interfering beams occur, giving the sharp dark fringes shown in Figure 12.9.

Newton examined the effects produced on reflection from a thin film of air, by placing a convex lens of shallow spherical curve in contact with a glass plate (Figure 12.15). The enclosed air film will then be extremely thin at the point of contact and

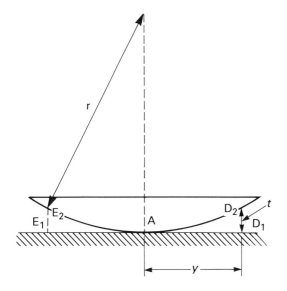

Figure 12.15 Newton's rings

will gradually increase in thickness with increased distance from this point, lines of equal thickness being circles with the point of contact as centre. Illuminated with monochromatic light, the film, as seen by reflection, will therefore have a black spot at the point of contact, surrounded by concentric alternate light and dark rings. These are known as **Newton's rings** and a photograph of these is shown in Plate 2. If the film is illuminated with white light the rings will of course be coloured with colours which blur together seven or eight rings from the centre.

Up to that ring, the colours appear in the following order from the centre outwards:

1st order   Black, blue, white, yellow, red
2nd order   Violet, blue, green, yellow, red
3rd order   Purple, blue, green, yellow, red

4th order   Green, red
5th order   Greenish-blue, red
6th order   Greenish-blue, pale red
7th order   Greenish-blue, reddish-white

This is known as **Newton's scale of colours** From Equation 12.9, the thickness t of the air film for any dark ring is

$$t = \frac{k\lambda}{2 \cos i'}$$

where k is the number of the ring ($k = 0$ for the central dark patch). If r is the radius of curvature of the lens surface and y the radius of the kth dark ring, then as t is small:

$$r = \frac{y^2}{2t}$$

(from Equation 3.3a), hence

$$r = \frac{y^2}{k\lambda} \tag{12.11}$$

for normally incident light when $\cos i' = 1$.

If the diameters of different rings are measured using a travelling microscope, it is possible to assess the curvature of the lens. This may be in error if the lens and plate are not in contact so a difference method is preferred using the expression

$$r = \frac{y_1^2 - y_2^2}{(k_1 - k_2)\lambda} \tag{12.12}$$

where y_1 and y_2 are the half-diameters of the k_1th and k_2th ring respectively.

If, instead of a lens, we use two flat glass plates with a piece of thin paper between them at one edge, we then have a thin wedge of air which produces **Fizeau fringes**. Lines of equal thickness of the air film are generally parallel to the edges of the glass plates which are in contact. Thus when the arrangement is illuminated with monochromatic light a series of dark fringes will be seen. If both plates have plano surfaces these fringes will be straight and equally spaced. With normal illumination the spacing is given by x where

$$x = \frac{\lambda}{2\alpha} \tag{12.13}$$

where α, the angle between the plates, is assumed to be very small. This expression may be calculated in a similar manner as Equation 12.11.

In both cases the interference pattern is said to comprise **fringes of equal thickness**. Any variation from flatness of the plates or sphericity of the lens will change the thickness of the air gap and this will show in the fringe pattern. With extended source illumination the fringes are localized in the plane of the film. For thin films, changes in the angle of viewing across the lens or plates only introduces a fraction of a fringe error in the contouring action as $1 - \cos\theta$ is less than 2% up to 10° from normal. The method constitutes a test of lens surfaces.

In optical component fabrication it is important to know the precise curvature of any surface and also its figure; that is, any variation in curvature over the surface. A traditional method, introduced by Fraunhofer, uses **test plates** and is based on Newton's rings. Two glass plates are ground and polished to the exact curve required by the test surface, one being convex and the other concave. The accuracy of the curves is tested by means of an accurate spherometer. By placing the plates in contact and observing the interference pattern from the enclosed air film, any departure from the spherical can be seen. If both surfaces are spherical and of equal curvature, a uniform illumination is seen over the whole surface.

In testing a lens surface during manufacture the test plate of the opposite curvature is placed in contact with the lens surface, both surfaces being perfectly clean. If the lens is spherical but not of the correct curvature, circular fringes will be seen, while any departure from spherical will show up in the shape of the rings. The rings do not immediately show whether the air film is thicker at the centre or at the edge, but in the latter case the lens can be rocked, causing the fringe centre to move while in the former a slight pressure in the centre of the lens will cause the fringes to move outwards.

In spite of precautions it is all too easy to scratch the lens surface when a contact method such as this is used. Non-contact interferometers are frequently used with laser illumination so that the tested and reference surfaces can be well separated.

12.7 Single-layer antireflection coatings

The interference effects of thin films as typified by Newton's rings and Fizeau interference have a usefulness beyond that of optical shop measurement. The particular shape, spacing and order of the fringes convey information on the film thickness, so it is possible by controlling that thickness to obtain a particular level of light reflectance which is uniform over a lens or other optical component if the film thickness is uniform. This effect was first observed by Fraunhofer as reducing the reflection at glass surfaces when a thin film or tarnish developed on exposure to the atmosphere of the glass used in lenses. Artificial methods of obtaining the tarnish were based on chemical etching and developed by Dennis Taylor in the early years of the century. Modern methods are based on the evaporation of particular substances in a high vacuum, so that no air molecules impede transfer to the optical surface.

For a single film to give zero reflection it is necessary to create the conditions for a dark fringe all over the surface. However, a further condition exists as the principle of superposition requires that the interfering beams must be of equal amplitude if they are to cancel out completely.

The fraction of the *amplitude* of the incident light reflected at normal incidence from a surface between two media of refractive indices, n and n', is given by Fresnel's expression (Equation 11.22):

$$\frac{n - n'}{n + n'} \tag{12.14}$$

When a film is deposited onto glass, at the first surface of the film $n = 1$ (air) and if n_f is the refractive index of the film, the fraction of the *amplitude* of the incident light reflected will be

$$\frac{1 - n_f}{1 + n_f} \tag{12.15}$$

At the second surface of the film in contact with the glass of refractive index n_g the fraction of the *amplitude* of the light reflected will be

$$\frac{n_f - n_g}{n_f + n_g} \tag{12.16}$$

Hence if the *intensities* of the light reflected from each surface of the film are to be equal

$$\left(\frac{1 - n_f}{1 + n_f}\right)^2 = \left(\frac{n_f - n_g}{n_f + n_g}\right)^2 \tag{12.17}$$

from which we have $n_f = \sqrt{n_g}$ or the refractive index of the film should be equal to the square root of the refractive index of the glass. As reflection takes place each time at a surface between a rarer and a denser medium, there will be no difference in phase between the two beams due to reflection alone. Therefore, in order that there shall be extinction of a particular wavelength at normal incidence, the thickness of the film must be such that

$$n_f t = \left(k + \frac{1}{2}\right) \frac{\lambda}{2} \qquad \text{where } k \text{ is an integer}$$

That is, $n_f t$, the optical thickness, must be an odd number of quarter wavelengths. Usually only one is applied and this is known as a **quarter-wave coating**.

Not all substances evaporate easily or deposit on a clean glass surface to form a hard durable film. Table 12.1 gives a short list of commonly used materials with their refractive indices.

Table 12.1 Refractive indices of some materials commonly used as optical films

		Refractive index
Magnesium fluoride	MgF_2	1.38
Silicon dioxide	SiO_2	1.45
Aluminium oxide	Al_2O_3	1.65
Silicon monoxide	SiO	2.0
Zinc sulphide	ZnS	2.3
Titanium dioxide	TiO_2	2.35

For optical crown glass extinction requires a film of index $\sqrt{1.52} = 1.233$, but the lowest useful film index is that of magnesium fluoride, which has a refractive index of 1.38. Using Expression 12.15, we find the amplitude reflected at the air/film interface is 0.159 (ignoring signs). Using Expression 12.16, we find that the amplitude reflected at the film/glass interface is 0.048. These amplitudes are in antiphase for a quarter-wave coating and so the resultant amplitude is given by $0.159 - 0.048 = 0.111$. The square of this value gives the fraction of the intensity reflected, which is 0.012 or 1.2%. Although this is not zero it is a considerable improvement over the 4% of the untreated glass. This latter value is found from Equation 12.14 using $n = 1$ and $n' = 1.52$ and squaring.

Although the index of magnesium fluoride is too high for spectacle crown glass, for higher index glass types the performance of the single layer coating improves. With a glass index of 1.9 it is theoretically possible to obtain zero reflectance as $\sqrt{1.9}$ = 1.38.

These figures have been calculated for normal incidence and are effective for the wavelength at which the refractive indices are measured and for which the coating has a quarter-wave optical thickness. For other wavelengths the result is less good so it is general practice to choose a design wavelength near to the centre of the visual range, such as 510 nm. This means that most light is reflected in the red and blue extremes of the spectrum, which gives rise to the purplish tinge on coated lenses and the old term **bloomed lenses**. (These films for anti-reflection are now frequently called **AR coatings**.) At angles other than normal incidence the optical thickness is affected by the cos i term and the Fresnel expressions change. Yet again any departure from the design wavelength gives a poorer result overall. Figure 12.16 shows the reflectance values obtained at different angles of incidence for different wavelengths.

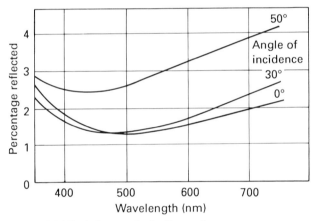

Figure 12.16 Single-layer antireflection coating

The value of antireflection coatings is two-fold. In the first instance the reduction in reflection gives an equivalent increase in transmission as the absorption of the film is negligible. This may not seem large for a single surface but for a lens system having six separate components there are 12 surfaces at which light is lost. If each surface transmits only 96%, the total transmission of the lens is $(0.96)^{12} = 61\%$. For coated surfaces transmitting 98.8%, the total transmission is 87%.

The second advantage of coated surfaces is that the 40% portion of the light reflected in the example above is reflected about inside the lens and can end up in the image plane as stray light or flare spots. With coated lenses these flare spots are much reduced and are typically purple in colour.

The formation of a very uniform layer of material with a precise thickness in the range 0.1–1.0 μm on an optical component is not a simple process. The now traditional method comprises electrically heating the material in a small container of tantalum, so that hot molecules of the material are ejected, and holding the object to be coated in the path of these ejected molecules, which lose their heat as they strike the cold object and adhere to its surface. If this were done in an ordinary room, most substances would react with the atmosphere and even those molecules ejected would be impeded by the molecules of air in their path. This means that the process must be done in a vacuum and values of about 10^{-5} Torr are commonly used.

As some materials liquefy in the boat before evaporation the arrangement inside the vacuum chamber usually suspends the articles to be coated, commonly called **substrates**, over the point of evaporation so that they lie on the surface of a sphere. The density of the vapour stream follows the inverse square law and so it is necessary to have all the substrates to be coated at the same distance. If a large flat substrate is being coated or special accuracy is needed a rotating plate of metal may be positioned in the stream with a shape such that the stream is interrupted longer in the centre than at the edge, and the propensity to deposit a greater thickness in the centre of the substrate is eliminated.

The material is evaporated at a rate depending on the temperature of the boat, but this is too crude a relationship to give good control. If it is required to lay down a film with a quarter-wave thickness, the increasing thickness must be monitored during evaporation and the vapour stream interrupted as soon as the value is correct. This is done by continuously measuring the value of reflectivity from a particular substrate near the centre of the stream and having an iris diaphragm which can be closed when the reflectivity is at a minimum.

12.8 Multilayer coatings

The disadvantage of single-layer antireflection coatings lies in the non-availability of materials of suitable refractive index for glass types of index lower than 1.9. A greater degree of freedom is found in design if two layers are used. Although a second coating material index may be even more incorrect, it now becomes possible to arrange for the larger amplitudes reflected to cancel to a more nearly zero value, thereby obtaining a lower reflected intensity.

If, for example, a layer of refractive index 1.7 such as lead fluoride is interposed between the glass (1.52) and the magnesium fluoride (1.38) the three amplitudes of reflection are:

Air/MgF$_2$ 0.17
MgF$_2$/PbF$_2$ 0.105
PbF$_2$/glass   0.06

If we arrange for the last two to be in antiphase with the first, the result is 0.005, giving a negligible reflectance value at the design wavelength. It should be remembered that the reflectance between air/MgF$_2$ and MgF$_2$/PbF$_2$ will undergo a change of phase of π, but that between PbF$_2$/glass will not, so that two quarter-wave layers as in Figure 12.17 will give the required phases.

The result for various wavelengths (Figure 12.17) shows a complete extinction of reflection at the design wavelength, but the reflectivity rises more sharply than for a single-layer coating. It is possible with two layers to design not for zero reflection at the design wavelength but to obtain extinction at two other wavelengths where the vector diagram forms a triangle but still gives a zero result.

If three-layer systems are considered, the number of degrees of freedom becomes still greater. This allows other constraints to be invoked. The durability of the coating may be considered important, and the material may be chosen with this

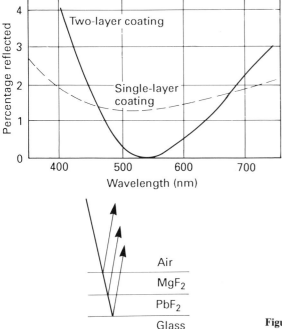

Figure 12.17 Two-layer antireflection coating

in mind. There is also the possibility that the coating can be 'tuned' for different substrate refractive indices by changing only the thicknesses of the layers and not the materials used. This means that the optical coating department of an optical company may meet most requirements without having to handle numerous coating materials. The actual materials and thicknesses used in these top-quality coatings tend to be proprietary information to the companies employing them.

The narrower spectral response of the two-layer antireflection coating and the steeper sides of the three-layer types are related to the fine fringes found with multiple-beam interferometry. In general, the larger the number of interfering beams, the sharper can be the effects on reflected or transmitted intensities for varying thickness or wavelength. This means that narrow spectral filters can be constructed with 10, 20 or even 40 layers of alternating materials so that the reflected light beams are alternately out of phase at a particular wavelength. This means that the transmission of the system is quite high (in the region of 60–70%) for that wavelength. However, for wavelengths on either side of this the reflecting beams no longer cancel and high (99%) reflectance values can be obtained. The width of the transmitted light at the half-intensity points may be 5 nm or less with 20 layers. These systems are known as **bandpass filters**. Other transmission bands may exist at other wavelengths, but these can be cut out using a blocking filter (either multilayer or absorbing). Figure 12.18 shows a bandpass filter designed for use in colour television cameras for selecting the blue channel. The red transmission is cut out using an absorbing filter.

Other multilayer designs give high transmission for wavelengths longer than a certain value and high reflectance for those shorter or vice versa. Such systems are known as **edge filters**. Both types are widely used in spectrophotometric

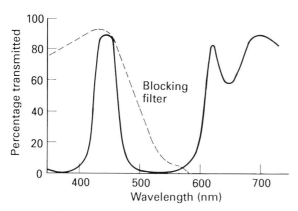

Figure 12.18 Bandpass filter (blue) for a colour television camera tube

instruments to assist in providing monochromatic beams, and in pollution-monitoring equipment where fixed wavelength beams are required.

The edge filter has found another application in the design of illumination systems in controlling the heat output of incandescent lamps. The heat is largely given out in the near infrared part of the spectrum and would severely damage film material in high-powered projectors. Between the lamp and the film is placed an edge filter known as a **hot mirror**, which reflects the infrared wavelengths while allowing the visible light to pass through. Often lamps in these systems have reflectors behind them to concentrate their output on to the film. These reflectors often have edge filters deposited on them, which reflect the visible light and allow the infrared to pass through. They are naturally known as **cold mirrors**, but the deeply curved shape of the reflector often makes a uniform coating difficult to achieve.

All multilayer designs are very sensitive to angle of inclination. Temperature changes cause the layers to change in thickness and some critical designs must be held at constant temperature. Filters can be designed to work at non-normal angles of incidence, but a problem arises in that the reflectivities of the interfaces vary with the polarization of the light. This can be useful when polarizing beamsplitters are required (Section 11.8).

One of the easiest materials to evaporate is aluminium, and when deposited on to a polished glass surface it forms a very good mirror both for light incident first on the coating and for light that passes first through the glass. Reflecting components for use in the former fashion are called **front-reflection mirrors**. Their advantage lies in having no chromatic effects as the glass is not used optically and also that wavelengths in the ultraviolet can be used which would otherwise be absorbed in the glass. Reflectance values of 90% are obtainable with freshly deposited aluminium.

Disadvantages stem from the tarnishing of the aluminium and that it is easily scratched. To avoid this, an *overcoat* of silicon monoxide is evaporated on top of the aluminium, although this causes a loss in reflectivity, particularly for blue light.

In spectrophotometers the light beams are generally focused by mirrors as they require no adjustment for variation of wavelength. If a number of mirrors are used in sequence, however, a reflectance of only 90% on each can rapidly reduce the total amount of light finally reaching the detectors in the same way that transmission values less than 100% for air/glass surfaces reduce the overall transmission of a lens. Increased reflectivity up to 99% can be obtained with pairs of half-wave layers on top of the aluminium. Here we are trying to obtain increased or *enhanced* reflection, and the refractive indices of the layers are chosen to be as far apart as practicable. If magnesium fluoride (1.38) is deposited next to the aluminium and then zinc sulphide (2.35), the total reflectivity increases from 92 to 97%. A further pair of layers gives 99%. This increase, as is typical with multilayer systems, is achieved only over limited wavelength ranges.

12.9 Optical coatings and interference – pertinent questions

In the case of a single-layer antireflection coating the explanation of its action may well be as follows. The light wave approaches the first surface and because of the difference in refractive index across the interface a small fraction is reflected. On reaching the second interface a further fraction is reflected which interferes with the first reflected wave giving zero (or reduced) reflectivity.

Many students have asked the question: how is it that the light on reaching the first surface knows of the presence of the second and is not reflected? The answer to this may be given in various degrees of erudition and mathematical rigour. However, the student is reminded of the points made in Section 9.4. The wave train builds up at the start in a gradual manner so that many wave crests have passed before full amplitude is reached. Thus between each half-wave (the phase difference of the 'reflected' portions) there is very little difference even during the early waves. Thus it is not a case that the reflection at the first surface has to be annulled in some way but rather that it is not allowed to form as the wave train reaches and crosses the interface system.

This whole question becomes less than hypothetical when the modern theory of light is adopted. This states that the wave motion associated with light is no more than a useful description of the fact that any given photon need not react with matter in a standard manner. It has a range of reactions available to it and the wave motion calculations are in fact determining the probability of particular events. Over 50 years ago at the University of Cambridge the Young's slits experiment was carried out with such a weak light source that most of the time only one photon was on its way from the source to the photographic plate which was the screen. Yet after an exposure of some months the interference pattern was found to be on the plate. Recent experiments with a laser source and sensitive photon detectors have shown that the expected pattern slowly built up from the apparently random arrival of each photon.

Thus the wave motion is associated with each photon and not with the beam as a whole. In other words, we can say that photons are components of light whereas waves are a description of it.

Exercises

12.1 Explain the conditions necessary for interference to occur between two beams of light, and describe two methods of producing these conditions.

12.2 The effect at a given point of two beams of mutually coherent light depends on the difference in phase between the two disturbances when they reach the point. Show that the difference in phase can be expressed in terms of the path difference by the relation

$$\text{Phase difference} = \frac{2\pi}{\lambda} \times \text{Path difference}$$

12.3   Two point sources emitting disturbances of 6 mm wavelength in a horizontal plane are 2 cm apart. Show in a diagram the lines along which there is no difference in phase and along which there are differences in phase of 1, 2 and 3 wavelengths.

12.4   Describe a laboratory method of obtaining interference fringes in a manner based on Young's experiment.
Explain what measurement should be made and the calculation necessary to enable the wavelength of the light to be determined from the experiment.

12.5   A narrow slit illuminated with sodium light ($\lambda = 589$ nm) is placed 15 cm from a bi-prism of angle 179° ($n = 1.5$). Find the separation of the dark interference bands on a screen 1 m from the prism.

12.6   A Fresnel bi-prism ($n = 1.5$) with angles of 0.5° at its two edges is used to produce interference. It is placed 10 cm in front of a narrow illuminated slit and the interference fringes on a screen 1 m from the prism are found to have a separation of 0.8 mm. What is the wavelength of light used? Explain what would happen if a very thin sheet of glass were placed in the path of the light from one half of the prism.

12.7   Two narrow slits 0.5 mm apart are illuminated with light of 600 nm wavelength, forming interference fringes on a screen 1 m from the slits.
(a) How far apart (centre to centre) are the dark bands in the pattern?
(b) A thin film 0.1 mm in thickness and of index 1.6 is placed over one slit. How far, and in which direction, are the fringes displaced on the screen?

12.8   Light from a narrow slit passes through two parallel slits 0.2 mm apart. The interference bands on a screen 100 cm away are found to be 3.29 mm apart. What is the wavelength of the light used? Describe a method other than the above for determining the wavelength of light.

12.9   Bi-prism fringes are produced with sodium light ($\lambda = 589$ nm). A soap film ($n = 1.33$) is placed in the path of one of the interfering beams and the central bright band moves to the position previously occupied by the third. What is the thickness of the film?

12.10   Explain carefully the reason for the colours seen in thin films.
A convex lens is placed on a plane glass surface and illuminated with sodium light incident normally; if the diameter of the tenth black ring is 1.5 cm, find the curvature of the surface (wavelength = 589 nm).

12.11   Newton's rings are formed between a plane surface of glass and a lens. The diameter of the third black ring is 1 cm when sodium light ($\lambda = 589$ nm) is used at such an angle that the light passes through the air film at an angle of 30° to the normal. Find the radius of curvature of the lens.

12.12   Newton's rings are formed in the case of an air film between an upper surface of radius r and a plane surface, normal incidence. Calculate the positions of the first six bright and dark rings for light of the three wavelengths, namely 680, 589 and 450 nm. Plot the results by smooth (sine) curves showing the variation of light intensity from the centre outwards and hence deduce approximately the first series of Newton's colour scale.

12.13   Explain the production of Newton's rings by the light *transmitted* by a thin transparent film. Why are these rings of complementary colour to those formed by reflection from the same thickness of film?

12.14   Explain the difficulty in obtaining interference fringes by reflection at a film of appreciable thickness when the source emits white light. How is the difficulty affected by using
(a) Light from a sodium lamp?
(b) Light from a HeNe laser?

12.15   Two plane plates of glass are in contact at one edge and are separated at a point 20 cm from that edge by a wire of 0.05 mm diameter. What will be the width between the dark interference fringes formed when light of wavelength 589 nm falls normally on the air film enclosed between the plates?

12.16 Give a brief account, with an explanatory diagram, of the optical arrangement of a Michelson interferometer. Good fringes were observed with such an instrument with monochromatic light; when the movable mirror was shifted 0.015 mm, a shift of 50 fringes was observed. What is the wavelength of the light used?

12.17 A Michelson interferometer is adjusted to give the brightest possible fringes with a source of sodium light, which is emitting light of two wavelengths (589.0 and 589.6 nm). On moving one of the mirrors a position is found where the fringes disappear, the maximum of one set falling on the minimum of the other. How far has the mirror been moved from its original position?

12.18 Interference fringes are formed by an arragnement such as a Fresnel bi-prism. When a thin flake of glass is introduced into one of the two interfering beams, the fringes are laterally displaced. Explain, with a diagram, why this is so and in which direction the displacement occurs.

   If the refractive index of the glass flake is 1.5, the wavelength of the light used is 600 nm and the central bright band moves to the position previously occupied by the fifth, calculate the thickness of the glass.

12.19 Antireflection coatings are applied to the surfaces of lenses and optical parts to reduce the amount of reflected light and increase the transmitted light. Explain in detail how this effect is brought about, making clear the role of the thickness and refractive index of the film. Give a diagram.

12.20 Explain what is meant by a front-reflection mirror. How can a reflectivity of 98% be obtained?

12.21 Give an account of the method by which a very thin layer of material can be put on to a piece of glass.

12.22 Show that the optical path difference between light reflected from the top surface and that undergoing a single reflection at the bottom surface of a parallel-sided film is $2\mu t \cos r$, where μ and t are the refractive index and thickness of the film respectively and r is the angle of refraction.

   When fringes are formed by interference in a thin film, explain why the condition is that $2\mu t \cos r = n\lambda$ for a dark fringe when fringes are viewed in reflection, but for fringes viewed in transmission this condition is that for a bright fringe.

   In a Newton's rings experiment the lower surface of the lens is not spherical but conical; contact is made with the glass plate over a very small flat area made by grinding flat the point or apex of the very shallow cone. What is the shape of the fringes formed and how are they spaced? When counting fringes radially outwards and viewing normally it is found that there are 10 fringes in a distance of 1 cm. What is the angle between the conical surface and the flat if the wavelength of the light used is 600 nm?

12.23 A simple interference filter is designed, using an air film bounded by parallel reflecting surfaces, to have a peak transmission at 546 nm. Calculate the thickness of film required for normal incident light and determine the shift in the wavelength of maximum transmission when light is incident at an angle of (a) 10° and (b) 30° on the filter.

12.24 Magnesium fluoride of refractive index 1.38 is coated on to heavy flint glass of refractive index 1.7 to produce a non-reflecting surface for a wavelength of 500 nm. What thickness of coating is required? Why will magnesium fluoride not produce such an effective non-reflecting surface if it is coated on to glass of refractive index 1.5?

12.25 A film of index 1.4 is coated on to a glass of index 1.6 to give minimum reflection for a wavelength of 500 nm at normal incidence. Calculate the thickness of the film and the effective reflection coefficients for wavelengths of 500, 400 and 600 nm. Deduce the probable colour appearance of normally incident light reflected from the surface, assuming no change in index with wavelength.

Chapter 13

Diffraction: wavefronts and images

13.1 Wavefront optics

Chapters 9 and 12 developed the concept that although the energy of light is propagated as photons the action of these can be most easily understood in terms of wave motion. From an ordinary light source this wave motion is extremely complex as the motion comprises many waves in a random arrangement. The laser is a special source where the motion approaches that of a simple single wave. With wave motions such as water waves and sound it is clear that the waves bend round obstacles such as breakwaters and buildings, and the discovery that the same is true of light held the wave theory of light in a position of pre-eminence for over a century.

This bending or **diffraction** of light has an important bearing on how precisely we can use light to measure size as the edges of shadows and images through lenses are found to have non-sharp forms with fringes. These effects will be described for incoherent sources and later for coherent sources, where the fringing results in less precision. When these more pronounced effects can be utilized fully, the promise of **holography** can be realized as described in Section 13.10.

Previously we have made use of Huygens' principle in which every point on a wavefront is assumed to be the centre of a system of secondary waves or wavelets and the new wavefront is the common tangent of the secondary waves. Fresnel, by considering the mutual interference that takes place between these wavelets, succeeded in giving an explanation of diffraction effects. A more general theory was developed by Kirchhoff and applied to optical images by Fraunhofer and Lord Rayleigh. Abbe applied it to microscope images where the different parts of the object being illuminated by the same source can become partially coherent although a full understanding of imagery by coherent and partially coherent sources has only been developed within the last 30 years, notably by Born and Wolf. The simplified theory presented in the following sections assumes that the illuminating source is monochromatic *and uses very non-rigorous mathematics!*

13.2 Fresnel zones – spherical waves

Fresnel's treatment of diffraction was based on the idea of secondary wavelets emanating from all parts of the wavefront. Figure 13.1 shows a wavefront arising from a point source at S. Fresnel maintained that the effect at P could be found

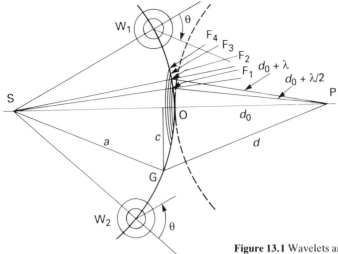

Figure 13.1 Wavelets and Fresnel zones

either by allowing this wavefront to proceed until it reaches P or by dividing the wavefront into small areas so that each area can be assumed to generate a wavelet, as W_1 and W_2. Each of these wavelets then produces an effect at P which can be summed according to the principle of superposition to give the total effect of the wavefront at P. Obviously these two methods should yield identical results, but the second will allow us to calculate the effect at P when parts of the wavefront are obscured by edges or apertures.

In his analysis Fresnel assumed that the *amplitude* of the effect at P of any wavelet is:

1. Proportional to the amplitude of that part of the wavefront.
2. Proportional to the area of the element of the wavefront generating the wavelet.
3. Inversely proportional to the distance of the wavelet centre to P. (This is the inverse square law when intensities are considered.)
4. Subject to an obliquity factor related to the angle θ, which is zero when θ is 90° and rises to a maximum when θ is 0°. There is no wavelet in the backwards direction.

This simple theory encounters some difficulties of interpretation and justification, particularly in the last assumption. However, a precise equation for this obliquity factor was not required in Fresnel's analysis as the major effect in the addition of the effects at P is the differences in relative phase of the disturbance due to each wavelet. This is correctly dealt with by this theory and remains an important contribution of Fresnel to the science of optics.

Referring to Figure 13.1 we consider the shortest distance from the wavefront to P to be d_o (= OP). Other parts of the wavefront will cause disturbances at P which are out of phase with that produced by the nearest part (at O) due to the longer routes giving a path difference. This may be most easily analysed by constructing a sphere centre P with a radius $d_o + \lambda/2$. This sphere cuts the wavefront in a circle centred on SP passing through F_1 on the wavefront.

Within this circle the phase of the disturbances arriving at P varies between 0 and π. Another circle can be constructed using a sphere centre P of radius $d_o + \lambda$. This circle is also centred on SP and passes through F_2. The disturbances arriving at P from the annulus between these circles have phases varying between π and 2π.

Further annuli or zones can be constructed, each having an effect at P varying by 180° of phase. These hypothetical zones are known as **half-period zones** or **Fresnel zones**. It should be realized that the size of these zones will clearly depend on the wavelength of the light and the distance of the point P.

The reader may easily prove that, when the wavefront is spherical or plane and the wavelength is small compared with the distance d_0, the areas of the zones are all very nearly equal to $\pi d_0 \lambda$, and the radii of their boundaries are proportional to the square roots of the natural numbers. With the short wavelength of light, these half-period zones will be very small. If d_0 is 500 mm and the light of wavelength 6 × 10^{-4} mm the area of each zone is 0.942 mm^2 and the radius of the first zone is 0.548 mm assuming a plane wavefront.

As the areas of these zones are equal they may be assumed to be sending out the same number of wavelets from equal elemental areas, but as the distance of P and also the obliquity are increasing as we go out from the central zone, the amplitude at P due to the zones is gradually diminishing from the central zone outwards. From the way in which the zones have been constructed it follows that the effect at P from any one zone is exactly opposite in phase from that of an adjacent zone. It is necessary therefore to find the resultant amplitude arising from a number of superimposed effects of very gradually decreasing amplitude and between every consecutive two of which there is a phase difference of π.

Let the amplitude at P due to successive zones be denoted by a_1, a_2, a_3, etc. As the average phase of the vibration due to adjacent zones differs by π, alternate amplitudes, a_2, a_4, etc., can be given negative values to show that the displacement is in an opposite direction to that of a_1, a_3, etc. Then the total amplitude at P will be

$$A = a_1 - a_2 + a_3 - a_4 + a_5 \ldots$$

As the amplitudes due to consecutive zones, although gradually decreasing, are nearly equal, we may say that

$$a_2 = \frac{a_1 + a_3}{2} \qquad a_4 = \frac{a_3 + a_5}{2} \qquad \text{etc.}$$

Then we may write A in the form

$$A = \frac{a_1}{2} + \left(\frac{a_1}{2} - a_2 + \frac{a_3}{2}\right) + \left(\frac{a_3}{2} - a_4 + \frac{a_5}{2}\right) + \ldots$$

where each term in brackets is equal to zero. The expression, if carried sufficiently far that the amplitude due to the outer zones is negligible, becomes

$$A = \frac{a_1}{2}$$

Hence, when a sufficiently large number of zones is considered, the total amplitude at a point, such as P, due to the wavelets from all points on a wave front is equal to half the amplitude due to the wavelets from the central half-period zone alone.

We have determined this result by dealing with the amplitude and phase of the effect due to each half-period zone taken as a whole. Actually, there is a continuous change in phase from 0 to π across the zone. This can be most easily analysed by

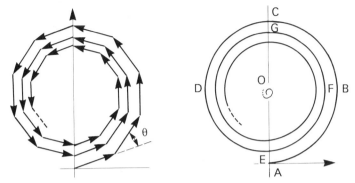

Figure 13.2 Vector addition of amplitudes – spherical wavefront

using the graphical method of Section 9.2. For this purpose each zone may be subdivided into smaller rings having equal areas but differing slightly in phase. The amplitude of these zones will therefore decrease only gradually due to distance and obliquity factors.

The graphical summation is shown in Figure 13.2, first for finite zones having equal phase differences and then in the limit as these zones become infinitely small. In this case the first half-period zone is represented by the arc ABC, which is very nearly a semicircle, and the resultant amplitude is given by the length AC.

In the same way the second half-period zone is given by CDE and the resultant amplitude due to both zones is the very small value AE. If the construction is continued for a large number of zones the curve is seen to be a spiral and approaches O. Thus the resultant for all the zones from the wavefront is seen to equal AO which is closely equal to AC/2 as shown before.

Referring to any general point G on the wavefront the path difference of the effect at P is given approximately as

$$\text{Path difference} = \frac{c^2}{2a} + \frac{c^2}{2d} = c^2 \left(\frac{a + d}{2ad} \right)$$

using the sag formula in Figure 13.1. This gives a phase difference of

$$\delta = \pi \left(\frac{a + d}{ad\lambda} \right) c^2$$

which is sufficiently accurate as c is usually very small compared with a and d.

The use of half-period Fresnel zones is helpful in so far as it allows the major diffraction effects to be explained. Later in this chapter the analysis will use full-period zones to describe the effects of *diffractive lenses* or *kinoforms* (Section 13.8).

13.3 Diffraction effects

It has been shown that in the case of light the half-period zones will be very small, and a large number will be contained in quite a small area about the central point O. The effective portion of the wavefront may therefore be considered as confined to this small area, as the effect of the outer zones will be negligibly small compared with that of a large number at the centre.

An object ordinarily considered small nevertheless has dimensions that are large compared with the wavelength of light and will cover a considerable number of zones; the resultant amplitude of the vibrations at a point P (Figure 13.2) behind it from the remaining unscreened zones of the wavefront will consequently be inappreciable, the effect at P being *as if* the light travelled from the source *along the straight line OP*. Thus, except that there will be diffraction effects around the edge of the shadow, of a nature indicated in the first paragraph and to be investigated below, the statement that light travels in straight lines is seen to be approximately true; and it is quite legitimate in practice to accept the rectilinear propagation of light as a law upon which to base the actions of optical instruments and to employ therefore the idea of 'rays' of light in geometrical optics.

The theoretical results obtained by considering the wavefront as in Section 13.2 are confirmed by the results obtained by experiment.

13.3.1 Small circular aperture

Light from a distant point source is passed through a perfectly circular aperture of 1 or 2 mm diameter, and the light is received on a screen, or better by means of an eyepiece. As the area of the half-period zones will depend on the distance d of the screen or eyepiece, the number of zones contained in the aperture can be varied by varying the distance d. At a certain considerable distance the aperture contains only one zone, and the illumination at the centre of the light patch is a maximum, since $A = a_1$. As the eyepiece is moved towards the aperture until two zones are included, the centre becomes dark, as now $A = a_1 + a_2 = 0$ (approximately). Moving still closer until three zones are included, the centre again becomes bright, since $A = a_1 + a_2 + a_3 = a_1$ (approximately). In this way a series of alternate light and dark centres is found as the eyepiece is moved towards the aperture.

13.3.2 Small circular obstacle

One of the chief objections to the theory of Fresnel at the time of its publication was that put forward by Poisson, who showed that, according to the theory, there should be little loss of illumination at the centre of the shadow of a *small* circular object. As the areas of the zones are approximately equal, a small circular obstacle intercepting a few central zones should have little effect on the total disturbance reaching the point P from the whole wavefront, and there should be a bright spot at the centre of the shadow. Arago (1786–1853) showed by experiment that this was actually the case.

The experiment may be carried out as follows. A circular opaque object, such as a smooth edged coin or a polished steel ball about 10 mm in diameter, is suspended in the path of the light from a pinhole aperture 2–3 m away. An eyepiece is mounted in the centre of the shadow at about an equal distance from the object, and a small bright spot of light will be seen in the field (Plate 7). (About 20 half-period zones are covered in this case.) Removing the object will make little difference in the brightness of the spot. It is important that the object is perfectly circular and has smooth edges, as otherwise irregular portions of a number of zones will be exposed and the appearance confused.

If the eye, placed in the centre of the shadow, views the object without the eyepiece, the edge of the object is seen as a brilliant luminous ring, showing that the light entering the shadow is travelling as if originating at the edge of the obstacle. An interesting case of the same kind can often be seen in mountainous districts. If one is just within the shadow of a fairly near mountain or hill before the sun has risen over the edge or just after it has set, trees on the skyline appear to be lit with an intense brilliance, while birds and even flies, far too small to be seen at such a distance in the ordinary way, appear as brilliant points of light.

13.3.3 The zone plate

An interesting confirmation of Fresnel's theory is provided by the device known as a **zone plate**. It follows from our consideration of the effect of the various half-period zones (Section 13.2) that, if the disturbance from alternate zones can be prevented from reaching the point P, the disturbance from the remaining zones, since it all arrives in the same phase, will add up and produce a greatly increased illumination at this point. To construct a zone plate a series of concentric circles with radii proportional to the square roots of the natural numbers are drawn on white paper, and the alternate rings blackened. The diagram is then photographed, considerably reduced, on to a glass plate so that the rings are alternately transparent and opaque. The rings on such a plate will correspond to half-period zones for a certain distance of the point P, depending of course on their size; and if the zone plate is set up at the correct distance in front of a screen, there will be a concentration of light on the screen corresponding to each point on the object. The zone plate thus behaves in much the same way as a positive lens. Zone plates of any 'focal length' can, of course, be made by reproducing the diagram to the required size. Each zone plate has more than one focal length as concentrations of light can occur at other distances where the path difference between neighbouring transparent rings is 2λ, 3λ, 4λ, etc., giving progressively shorter f'.

A considerable increase in the concentration of the light could be effected if, instead of stopping out alternate zones, the phases of the wavelets from these zones were changed by π. The disturbances from *all* zones would then arrive at P with the same phase.

This can be done using photographic plates of dichromated gelatin. This material can be dissolved away when unexposed, but the exposed areas remain to provide a transparent area giving a path difference from the clear areas. With a path difference of π, a **phase zone plate** is produced. These devices may be considered as diffractive optical elements, and are described in more detail in Section 13.8.

13.4 Fresnel diffraction

Many of the diffraction phenomena are best seen when the source is in the form of a very narrow slit. In such a case the wavefronts may be considered as cylindrical, and the resultant amplitude along a line through a point, such as P, and parallel to the slit can best be found by dividing the wavefront into strips rather than zones. The construction of these half-period elements is similar to that of the half-period zones. If d is the distance of P from the nearest point of the wavefront, the distances of P from the outer edges of successive half-period elements will be $d + \lambda/2$, $d + 2\lambda/2$, $d + 3\lambda/2$ and so on.

The length of these strips being equal, their areas, unlike those of the half-period zones, decrease rapidly at first but more slowly as the distance from the centre

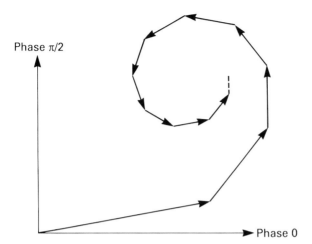

Phase π/2

Phase 0

Figure 13.3 Vector addition of amplitudes – cylindrical wavefront

increases, the more outer strips being practically equal in area. The amplitudes due to these outer strips, being approximately equal but of alternate phase, annul one another, and the effect of the whole wavefront is due to a few central elements. The effects of these central strips are not however equal, as was the case with zones, which had equal areas.

Because, for a cylindrical wavefront, the areas of the half-period strips are very unequal near to the axis we find that when the graphical method of Section 9.2 is applied the spiral effect of Section 13.2 still occurs but with a shift of centre. If narrow strips of equal phase difference are plotted as in Figure 13.3 the amplitudes of the first few strips are large enough to give an offset before the spiral effect occurs. This can be reduced to a smooth curve when infinitely small strips are considered, and the effects of those strips on the other side of the central strip show as a spiral to the opposite quadrant. The total curve, known as **Cornu's spiral**, is given in Figure 13.4.

In this curve, the amplitudes emitted by each of the elemental strips are laid end to end but as the lengths of the strips are equal, distance along the spiral is related to distance across the wavefront. The resultant amplitude at the point P may be found from a line joining the ends of the available part of the spiral, the length of the line being proportional to the amplitude. Any part of this spiral has a slope given by the phase of the particular strip of the wavefront.

Thus, from Section 13.2, the phase is

$$\delta = \frac{\pi}{\lambda} \left(\frac{a + d}{ad\lambda} \right) c^2 = \frac{\pi}{2} v^2$$

where v has been defined as

$$c \left(\frac{2(a + d)}{ad\lambda} \right)^{1/2}$$

It can be shown that this is the distance from the origin along the curve to the point having a phase difference δ. To obtain the equation of the curve it is seen that for a short element dv of it the slope dx/dv = cos δ and the slope dy/dv = sin δ.

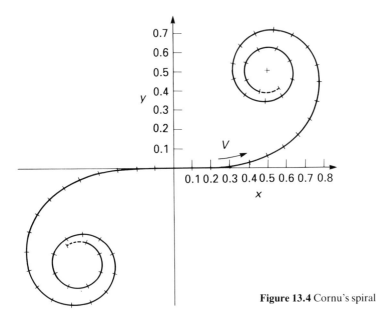

Figure 13.4 Cornu's spiral

These give

$$dx = \cos \delta\, dv = \cos \frac{\pi v^2}{2}\, dv$$

$$dy = \sin \delta\, dv = \sin \frac{\pi v^2}{2}\, dv$$

The coordinates of a particular point are then given by

$$x = \int_0^v \cos \frac{\pi v^2}{2}\, dv$$

$$y = \int_0^v \sin \frac{\pi v^2}{2}\, dv$$

These are known as **Fresnel's integrals**, which cannot be solved as they stand, but yield infinite series which can be evaluated for specific values to give a table of the coordinates of the curve. In reality the equations of Fresnel were formulated first from a more rigorous analysis of his theory. The curve was developed later by Cornu in 1874 as an elegant description of the mathematics. The Cornu spiral (Figure 13.4) describes the action of a cylindrical wavefront at a given point P. Each part of the spiral relates to a part of the wavefront. The part at the origin of the graph is given by that part of the wavefront nearest to P which is the part intersected by the line SP at W (Figure 13.5). If all the wavefront has an effect at P the total amplitude of this effect is given by the line connecting the centres of the spirals. If some of the wavefront is obscured the total effect at P is given by a straight line joining the ends of that part of the curve related to the unobscured portion. Thus when a straight edge, X–X (Figure 13.5(a)), parallel to the source of

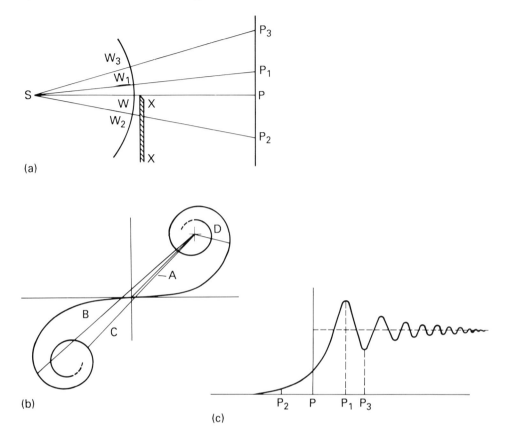

Figure 13.5 Diffraction at a straight edge: (a) geometry, (b) theory, (c) resultant

the cylindrical wavefront obscures part of the wavefront, the effect at P is given by the line A on the Cornu spiral of Figure 13.5(b). When squared, this gives the intensity for the position P as shown in Figure 13.5(c).

For other points such as P_1 the nearest part of the wavefront becomes W_1 so that the unobscured part of the spiral is larger, giving the line B (Figure 13.5(b)) as the resultant amplitude. This is longer than A and the greater intensity is plotted in Figure 13.5(c).

For P_3 an even greater length of the curve is used, but as this includes nearly a full turn of the lower spiral the resultant C is shorter than B. As more points are considered the amount of the curve used travels round the lower spiral so that the straight-line resultants oscillate in length. This generates the wave-like structure to the graph of Figure 13.5(c), which shows as fringes when photographed with sufficient magnification (Plate 3).

If points inside the geometrical shadow are considered (i.e. below P in Figure 13.5(a)), the amount of the spiral used reduces. For example, the point P_2 gives the resultant D, which gives the intensity shown. Beyond P_2 the resultant straight line joins the centre of the upper spiral to points on the spiral giving a steadily reducing intensity, with no fringes, although the phase angle is varying rapidly.

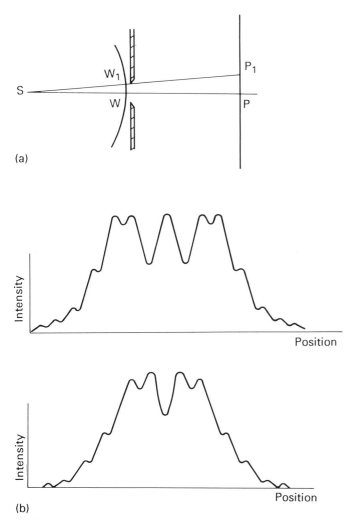

Figure 13.6 Fresnel diffraction effects at a slit

Because the upper part of the wavefront in Figure 13.5(a) is completely unobscured, the resultant straight lines all start at the centre of the upper spiral. If now a slit is considered as in Figure 13.6(a), this fixed point no longer occurs. The width of the slit determines the length of the curve to be used to obtain the resultant amplitude but the location of this on the spiral is dependent on the position of the point on the screen. For example, the point P on the screen uses a length of the spiral symmetrical about the origin while the point P_1 uses a length having one end at the origin. The resultant patterns show fringes but these are critically dependent on the apparent width of the slit, that is, the length of the spiral used. Two examples are shown in Figure 13.6(b) for narrow and broader slits. If instead of a narrow slit a narrow wire is placed in front of the cylindrical wavefront, again parallel to its source, a similar effect is obtained and this is shown in Plate 4.

13.5 Fraunhofer diffraction

The effect of apertures on wavefronts has been discussed so far with the resultant pattern being received as a shadow on a screen. When the source of the wavefront is imaged by a lens system on to the screen we find that the image has a structure also, due to the lens imaging the diffracted light. In this case there is no need to compare the spherical or cylindrical wavefront with a spherical or cylindrical reference surface centred at each point on the screen. Any lens that images the source on to the screen converts the divergent wavefront into a converging one centred on the image point. For locations on the screen near to this point the comparison is between spherical or cylindrical surfaces which are tilted with respect to each other. The comparison becomes simpler if we assume that the diffracting aperture operates in parallel light between two lenses as in Figure 13.7. Any case in which the screen is conjugate with the source is, however, known as **Fraunhofer diffraction**, and the images formed in optical instruments such as telescopes and the optical system of the eye are Fraunhofer diffraction patterns. The incident light does not need to be parallel but the instrument must be correctly focused.

In Figure 13.7, collimated light from a narrow slit S passes through the aperture, CD, and is imaged by the lens L_2 on to a screen. To develop the one-dimensional case, the aperture will be considered as the cross-section of a narrow slit parallel to the source slit S. According to geometrical optics a sharp image of the source should be formed at F′ in the focal plane of L_2, but close inspection in an experimental arrangement will show that a diffraction pattern is formed on the screen consisting of a central bright band bordered by fringes of rapidly reducing intensity. As with Fresnel diffraction these fringes are coloured if a white-light source is used.

Considering first the point F′ we note that this is the focus formed by L_2 of the plane wavefront emerging from L_1. If we consider the plane wavefront at CED, the wavelets arising from all points on it will produce effects over most of the forward direction. Those that travel parallel to the lens axis and are focused at F′ will all have the same path length as the lens creates the converging wavefront by making all the ray paths equal. These effects, therefore, arrive at F′ with the same phase and this point, which is the centre of the pattern, is thus always a maximum and bright. In this important respect, the pattern will differ from that with Fresnel diffraction, where the centre may be sometimes bright and sometimes dark.

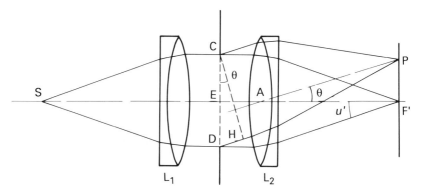

Figure 13.7 Fraunhofer diffraction: single rectangular aperture. CD = b, AF = F' and PF = h'

For other points in the focal plane the lens will effectively scan the wavelets, taking into account their relative phases, over a plane tilted with respect to the axis. For the point P in Figure 13.7, the ray paths CP and HP are equal, as are the paths from all points on the line CH. With a single-slit source at S, the only wavelets available are in phase along the line CD. If they are compounded along the line CH there will be an increasing phase difference moving from C to H. If the aperture CD is divided into narrow strip elements of equal width parallel to the length of the slit, then the amplitudes due to each strip are equal and there will be a constant difference in phase between successive elements. This means that the graphical summation of these effects gives Figure 13.8(a), which in the limit reduces to the arc of a circle, Figure 13.8(b). As different points in the focal plane are considered, the rate at which the phase difference increases will change and will give different curvatures to the arc. As the same total area of wavefront is being considered, the total amplitude and therefore the length of the arc remains constant (neglecting the obliquity factor for small angles) but the resultant amplitude, A, varies considerably, reducing to zero at times as shown in Figure 13.8(c).

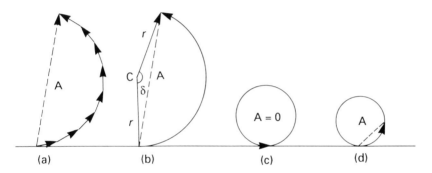

Figure 13.8 Vector addition for Fraunhofer diffraction

The resultant amplitude may be calculated more easily than under Fresnel conditions. In the simplest case, Figure 13.8(b), the arc of the contributing vectors has an overall phase difference δ, which is also the angle subtended at the centre, C, of the arc. If the length of the arc is l, the radius, r, is given by l/δ. The triangle gives the equation $A/2r = \sin \delta/2$, so that

$$A = 2r \sin \delta/2 = \frac{2l \sin \delta/2}{\delta}$$

or, if the amplitude for zero phase change is taken as unity (when $A = l = 1$),

$$A = \frac{\sin \delta/2}{\delta/2} \tag{13.1}$$

The overall phase difference, δ, is generated by the path difference DH (Figure 13.7) so that

$$\delta = \frac{2\pi DH}{\lambda} = \frac{2\pi b \sin \theta}{\lambda} = \frac{2\pi b h'}{\lambda f'} \tag{13.2}$$

where the angles $\angle$DCH and $\angle$PAF' are equal for small values. Thus the equation for the amplitude becomes

$$A = \frac{\sin\left(\dfrac{\pi b \sin \theta}{\lambda}\right)}{\left(\dfrac{\pi b \sin \theta}{\lambda}\right)} \qquad (13.3)$$

while the intensity is the square of this. The function $(\sin x)/x$ is known as the **sinc function**.

The full circles in the graphical addition diagrams occur when $\delta = 2\pi$, 4π, 6π, etc., and give zero amplitude. This is also shown in Equation 13.3, by putting the numerator equal to zero, which occurs when $(\pi b \sin \theta)/\lambda = \delta/2 = k\pi$. Thus the locations of P for zero intensity are given by $\sin \theta = k\lambda/b$, or

$$h' = \frac{k\lambda f'}{b} = \frac{k\lambda}{2 \sin u'} \qquad (13.4)$$

The first dark band occurs at $\lambda f'/b$ so that the width of the central bright fringe is seen to be $2\lambda f'/b$, which varies inversely with b.

The locations of the maxima are not so straightforward. The opposite phase value, $\delta/2 = (k + \frac{1}{2})\pi$, gives

$$\sin \theta = \frac{(k + \frac{1}{2})\lambda}{b} \qquad (13.5)$$

but this gives only the *approximate* positions as the denominator of Equation 13.3 has an effect. By differentiating Equation 13.1 and setting to zero we have $\tan \delta/2 = \delta/2$ for the maxima. This gives $\delta/2 = 0$, 1.43π, 2.46π, 3.47π, . . . which are different from $(k + \frac{1}{2})\pi$ although approaching it for the further locations. The amplitudes at these maxima may be calculated by putting these values of $\delta/2$ into Equation 13.1, which was obtained by assuming the central amplitude to be unity. (In Equation 13.1, $\sin \delta/2 = 0$, and $\delta/2 = 0$ for the central maximum. As the sine term approaches zero more slowly than the angle, the value of unity may be assumed.)

The values of maximum amplitude and intensity for the first three bright fringes are given in Table 13.1, along with the phase differences, δ, across the slit, which generate the maxima and minima.

Table 13.1

Phase difference (δ)	Amplitude	Intensity	Fringe no.
0	1	1	Zero
2π	0	0	Minimum
2.86π	0.217	0.047	1
4π	0	0	Minimum
4.92π	0.125	0.017	2
6π	0	0	Minimum
6.94π	0.091	0.008	3
8π	0	0	Minimum

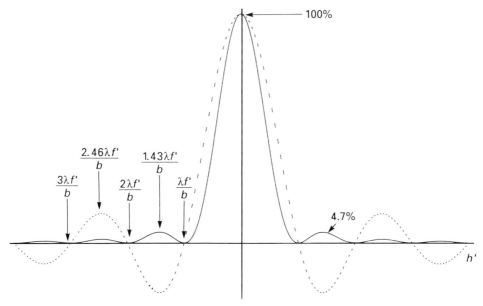

Figure 13.9 Amplitude (– – –) and intensity (——) curves for Fraunhofer diffraction from a slit. This is a line spread function

It is thus seen that the central maximum is of much higher intensity than the fringes and contains most of the light. The curves for amplitude and intensity against phase difference are given in Figure 13.9, and their appearance is shown in Plate 9(a).

The above analysis has been restricted to one dimension, CD, of the rectangular aperture between the lenses. If the other dimension, say CB, is considered, it can be shown that this produces a fringe pattern orthogonal to the first, where the locations of the maxima and minima are determined by the size of CB. The amplitude is given by

$$A(y) = \frac{\sin\left(\frac{\pi a \sin \varphi}{\lambda}\right)}{\left(\frac{\pi a \sin \varphi}{\lambda}\right)} \qquad (13.6)$$

and the total pattern by

$$A(xy) = \frac{\sin\left(\frac{\pi b \sin \theta}{\lambda}\right)}{\left(\frac{\pi b \sin \theta}{\lambda}\right)} \times \frac{\sin\left(\frac{\pi a \sin \varphi}{\lambda}\right)}{\left(\frac{\pi a \sin \varphi}{\lambda}\right)} \qquad (13.7)$$

Thus, when one pattern gives zero amplitude, the total must also be zero so that the pattern has zero intensities along two sets of lines parallel to the sides of the rectangular aperture. Within the rectangles formed by these lines the intensity rises to a maximum.

The most common aperture in optical instruments has a circular rather than a rectangular shape. When this is divided into strips to calculate the effect at off-axis points, their lengths are no longer constant so that the graphical addition is no longer a circle. This makes the summation more difficult because a double integral is involved. This can be reduced to a single integral by working in polar coordinates because the circular aperture has circular symmetry and so the pattern has circular symmetry taking the form of a bright central disc surrounded by a series of fading circular rings (Plate 8). As with the slit fringes, the diameters of these rings are inversely proportional to the diameter of the aperture and are unrelated to the size of the source, although this must be small for best fringe visibility.

With apertures usually found in lenses, the rings will be so small as to be visible only when the pattern is viewed with magnification. Also, unless the lens is of high quality, its aberrations will mask the diffraction effects. When a lens has residual aberrations that are negligible compared with the diffraction effects, it is said to be **diffraction limited**. With poor lenses, the pattern may always be artificially enlarged by placing a small aperture over the lens.

The mathematical calculation referred to above involves the use of Bessel functions. These are similar to the more familiar trigonometrical functions, sine and cosine, oscillating either side of zero, but Bessel functions show a steady decline in their amplitude. The one most similar to a sine function is called the Bessel function of the first kind of order one and is indicated by $J_1(x)$. In Chapter 7 the polynomial series for sine was given as

$$\sin x = x - \frac{x^3}{6} + \frac{x^5}{120} - \frac{x^7}{5040} + \ldots$$

The polynomial for J_1 is

$$J_1(x) = \frac{x}{2} - \frac{x^3}{16} + \frac{x^5}{384} - \frac{x^7}{18432} + \ldots$$

Although a few terms of the sine series sufficed for aberrational analysis, many terms are needed to accurately calculate the sinc function of Equation 13.3 and for the similar Bessel function equation for circular apertures. This is

$$A = \frac{2J_1(\delta/2)}{\delta/2} = \frac{2J_1(\pi b \sin \theta)}{\frac{(\pi b \sin \theta)}{\lambda}} \tag{13.8}$$

where b is now the *diameter* of the circular aperture and θ the angle to a circle on the screen, centre F' and passing through P (as in Figure 13.7).

The function $J_1(\delta/2)$ gives zero values for particular values of $\delta/2$ showing that the graphical summation of the amplitudes from the strips still gives a closed circuit at particular phase differences. However, the location of these zeros and the maxima between them is no longer open to the simple analysis of Equations 13.4 and 13.5. The calculated values for the first three bright rings are given in Table 13.2. The fourth column shows the total intensity in each ring, and it may be deduced that about 87% of the available light appears in the central disc. The radius of this disc can be considered to be within the radius, h', of the first dark ring. From Equation 13.2 and Table 13.2, we have that

$$\delta = 2.44\pi = \frac{2\pi b \sin \theta}{\lambda} = \frac{2\pi b h'}{\lambda f'} \tag{13.9}$$

Table 13.2

Phase difference (δ)	Amplitude	Intensity	Total fringe intensity	Fringe no.
0	1	1	1	Zero
2.44π	0	0		Minimum
3.27π	0.132	0.0175	0.084	1
4.47π	0	0		Minimum
5.36π	0.065	0.0042	0.033	2
6.48π	0	0		Minimum
7.40π	0.040	0.0016	0.018	3
8.48π	0	0		Minimum

so that

$$h' = \frac{1.22\lambda f'}{b} = \frac{0.61\lambda'}{\sin u'} \tag{13.10}$$

where u' is the aperture angle (Section 5.7) on the image side and λ' the wavelength in the image space. This is usually, but not always, the same as that in the object space; it is not so in the eye or with the immersion microscope objective.

This disc of light may be considered as the image of a point as formed by the lens. It is known as the **Airy disc**, after the astronomer Airy who first investigated (in 1834) the distribution of light at the focus of a lens. The variation in intensity is shown in Figure 13.10 and a photograph of the pattern is given in Plate 8.

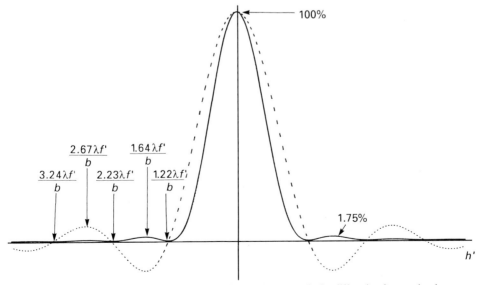

Figure 13.10 Amplitude (– – –) and intensity (——) curves for Fraunhofer diffraction from a circular aperture – the Airy disc. This is a point spread function

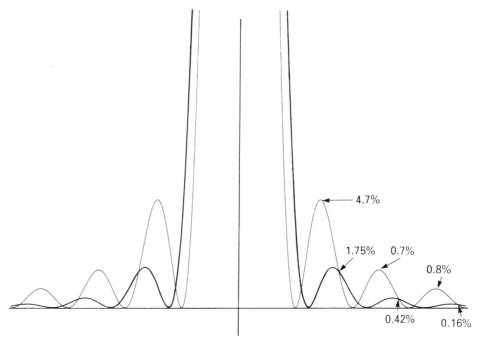

Figure 13.11 Comparison of diffraction patterns from rectangular (——) and circular (——) apertures

It is thus seen that, whereas in geometrical optics we assume that a perfect lens will bring light from a luminous point to a *point* image, actually the image will be a disc of finite diameter depending largely on the wavelength of the light and the aperture of the lens. While the diameter of the Airy disc will be extremely small, it will affect the quality of images formed by lenses as described in Section 13.6.

The comparison of rectangular aperture and circular aperture diffraction patterns is shown in Figure 13.11 where the width of the rectangle and the diameter of the circle are equal. It can be seen that while the circular aperture has a wider central area the intensities of the secondary maxima are less.

13.6 Point images and limiting resolution

Suppose B and Q (Figure 13.12) are two distant luminous points, such as two stars, subtending an angle w at a lens; then their images B′ and Q′ subtend the same angle at the second nodal point of the lens, and the separation of the images depends on this angle and on the focal length of the lens. Each image will consist of a diffraction pattern and when the separation of the images is large compared with the diameter of the Airy discs the distribution of light is as shown in Figure 13.12.

Under these conditions there will be no difficulty in seeing the two images as separate, that is, in 'resolving' the objects. As the angle w is reduced the images come closer together, while the diameter of the discs remains the same, and when the angle is reduced below a certain value the two discs will overlap to such an extent that the eye sees them as a single patch and is no longer able to interpret the image as that of two object points. The smallest value of the angle w for which the

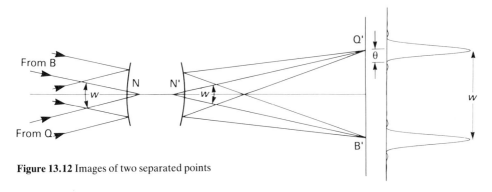

Figure 13.12 Images of two separated points

images of two points can be detected as double is the **limit of resolution** or **resolving power** and is of very great importance in instruments, such as the eye, the telescope and the microscope, where objects with fine detail or point objects separated by small angles, such as double stars, are concerned.

In order that an expression may be obtained for the limit of resolution, it is necessary to adopt some standard for the separation of the Airy discs, such that the two object points can be resolved. It is usually assumed that with a perfect optical system two points can be resolved if the centre of one diffraction pattern falls on the first dark ring of the other, that is, the angle w equals the angle θ, and the distribution curve is as shown in Figure 13.13. Hence it follows from Equation 13.9

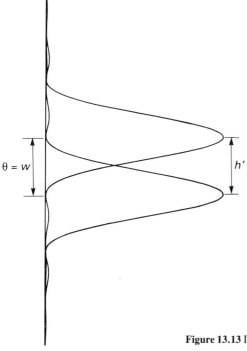

Figure 13.13 Intensity curves of two point images just resolved

that the smallest angle subtended by two point objects that can just be resolved, the limit of resolution, is

$$w = 1.22 \frac{\lambda}{b} \quad \text{or} \quad h' = \frac{0.61\lambda'}{\sin u'} \tag{13.11}$$

It is seen therefore that when an instrument, such as a telescope or microscope, is required to have a high resolving power its aperture must be as large as possible.

The effect of aperture on resolution may be easily demonstrated by the following simple experiment. A piece of fine wire gauze is held in front of a brightly illuminated background and slowly brought up towards the eye until the position is found at which it is just possible to distinguish the separate wires. If now a card with a very small pin-hole is placed before the eye, the separate wires will no longer be visible, and it will be necessary to bring the gauze much closer, so that the spaces between the wires subtend a much greater angle at the eye, before the separate wires are again seen.

In the foregoing consideration of the nature of the diffraction pattern at a focus and hence the resolving power of an instrument, it has been assumed that the optical lengths of all rays coming to the focus are equal; that is, that the emergent wavefront is perfectly spherical. Such perfection cannot be attained in practice, but Lord Rayleigh (1842–1919) showed in 1878 that the conditions would be satisfied if the difference between the longest and shortest paths to the focus did not exceed one-quarter of a wavelength.

Worked example   A telescope with an objective of 40 mm diameter is focused on an object 1000 m away; what is the smallest detail that could be visible with the instrument? If the eye can comfortably resolve two points subtending an angle of 100″, what magnifying power is required for the telescope in order that the resolved detail shall be comfortably seen? (Take the wavelength of light as 550 nm.)

From Equation 13.11, the angle subtended by smallest detail resolved

$$w = \frac{1.22\lambda}{b} = \frac{1.22 \times 0.000\,55}{40} = 0.000\,016\,8 \text{ rad}$$

As the angle is small, $w = h/l$, where h is the size of the object and l its distance. Then

$$\frac{h}{1000} = 0.000\,016\,8$$

and so

$$h = 0.0168\,\text{m} = 16.8\,\text{mm}$$

The image of this object formed by the telescope is required to subtend an angle of $100\,\text{s} = 100/206\,000$ rad.

Therefore the required magnifying power is

$$\frac{100}{206\,000 \times 0.000\,016\,8} = 29$$

In considering the resolving power of a microscope it will be convenient to consider the cones of light entering the objective from object points. From Equation 13.9, the smallest distance between two image points that can be resolved, $h' = 0.61\lambda'/\sin u'$. Assuming that the objective satisfies the sine condition (see Equation 7.45), i.e. $nh \sin u = n'h' \sin u'$, the minimum resolvable interval in the object space is

$$h = \frac{n'h' \sin u'}{n \sin u} = \frac{0.61 \, n'\lambda'}{n \sin u} = \frac{0.61\lambda}{n \sin u}$$

where λ is the wavelength in air.

As proposed by Abbe, $n \sin u$ is known as the **numerical aperture** of the objective and is represented by NA.

In the foregoing discussion the resultant of the curves in Figure 13.13 has been obtained by adding the intensities of the two diffraction patterns and it has therefore been assumed that the two luminous points are incoherent. This is by far the most usual case. If, however, the two sources are coherent or have a high degree of mutual coherence, we must add the amplitudes of their respective patterns before squaring to obtain the intensity. When this is done it is found that, at the just resolved position for incoherent sources, the pattern for coherent sources is not resolved. A comparison between the two patterns is given in Figure 13.14 where the curve for two coherent sources also assumes that they are in phase.

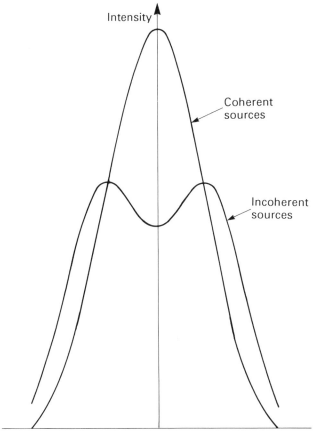

Intensity

Coherent
sources

Incoherent
sources

Figure 13.14 Resultant intensity curves for close objects

For two coherent sources in antiphase the resultant intensity pattern is easily resolved.

In the case where a telescope is being used to view two stars there is no doubt that the analysis of the diffraction patterns will be on an incoherent basis. The case of a microscope is more difficult, however, as the objects on the stage are illuminated by a single light source, which although incoherent can give rise to partially coherent conditions for objects close together, when their wavefronts will tend to be in phase.

13.7 Diffraction gratings

If in place of the slit CD (Figure 13.7) we have two equal narrow slits parallel to one another, each of these slits by itself would give rise to a diffraction pattern of the form described, the various maxima and minima due to one slit occupying the same position in the focal plane as those due to the other slit, if the incident light is parallel. There will, however, be interference between the light from the two slits. Since portions of the same wavefront pass through each slit, we may consider the slits as collections of coherent sources in the same way as the virtual sources of the double mirror or bi-prism experiment (Chapter 12).

If the width b of the slits is equal, the amplitude at the centre of the pattern will be twice, and the intensity therefore four times, that for a single slit. At any positions in the focal plane such that the difference in path C_2G_2 (Figure 13.15) of light from corresponding points on the two slits is an odd number of half wavelengths, the intensity drops to zero, since the disturbance from any point in one slit annuls that from a corresponding point in the other. Thus a series of dark

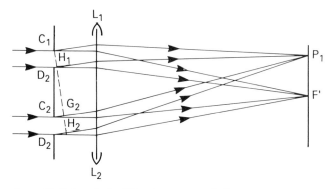

Figure 13.15 Fraunhofer diffraction: two equal slits

interference bands is superimposed on the diffraction pattern. The separation of these bands will depend on the separation, $b + c$, of corresponding edges of the two slits, c being the width of the opaque space, and a number may lie in the central maximum of the diffraction pattern. For example, when $b + c = 2b$, the new minima are half the distance apart of those for a single slit of width b.

Between these minima there will be maxima at positions where the difference in path C_2G_2 is an even number of half wavelengths. The resultant amplitude at these positions is the sum of the amplitudes due to each slit along the corresponding directions, and the intensity is therefore four times that at the same position with a single slit. Thus the maxima will be absent when they fall in positions corresponding to the minima of the diffraction pattern of the single slit, for in these positions the intensity due to each slit is zero. The intensity curve for two slits is shown in Figure 13.16, and the appearance in Plate 9.

The profile of the interference is that given for two-beam interference in Section 12.3:

$$A^2 = 4a^2 \cos^2 \epsilon/2$$

(from Equation 12.7), where a is the amplitude of the interfering beams and ϵ is the phase difference between them. The profile of the diffraction pattern is

$$A^2 = \frac{\sin^2 \delta/2}{\delta/2}$$

(from Equation 13.1), where δ is the phase difference across each slit. The resultant profile is found from the product of these. However, both ϵ and δ are dependent on the angle θ, subtended by P_1 (Figure 13.15):

$$\epsilon = \frac{2\pi(b + c) \sin \theta}{\lambda}$$

$$\delta = \frac{2\pi b \sin \theta}{\lambda}$$

so that the total intensity profile is given by

$$I = 4a^2 \left[\frac{\sin \left(\frac{\pi b \sin \theta}{\lambda} \right)}{\left(\frac{\pi b \sin \theta}{\lambda} \right)} \right]^2 \left[\cos \left(\frac{\pi(b + c) \sin \theta}{\lambda} \right) \right]^2 \tag{13.12}$$

which has a peak value four times that of a single slit.

As the first diffraction zero and the third interference zero coincide in the two-slit profile of Figure 13.16 it can be shown from Equation 13.12 that $c = 1.5b$. If $c = 2b$ the zero of the diffraction pattern falls on the third interference maximum, which is then virtually absent from the pattern although two very low intensity fringes mark the place.

When four equally spaced slits are considered, the interference becomes multiple-beam but without the reducing amplitudes assumed in Section 12.4. Thus, although the fringes become sharper, their profile is more akin to the diffraction patterns of single wide slits, which gave circular graphical summation curves in Section 13.5. More simply, principal maxima occur in positions such that the difference in path between the light from corresponding edges of adjacent slits is an even number of half wavelengths, as the amplitudes along these directions from each slit add up. The intensity of these maxima will therefore be 16 times the intensity at the same positions due to a single slit; that is, N^2 times, where N is the number of slits. Thus, as before, the maxima are absent where the positions in which they should fall coincide with the positions of minima in the diffraction pattern from a single slit.

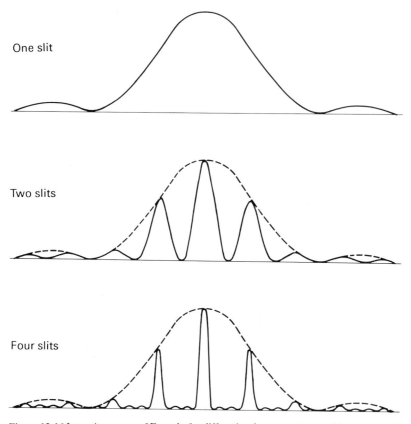

Figure 13.16 Intensity curves of Fraunhofer diffraction from one, two and four narrow slits. The heights of the curves for two and four slits are on a much smaller scale than those of the single slit

Dark bands are formed in positions such that the path difference between the light from corresponding edges of adjacent slits is an odd number of half wavelengths, in which case the vibrations from any point in one slit are annulled by those from a corresponding point in an adjacent slit. There are, in addition, dark bands in directions such that the path difference from corresponding points on adjacent slits differs by an odd number of quarter wavelengths, for in this case the path difference between alternate slits is an odd number of half wavelengths, and the vibrations from the first slit cancel those from the third, and the vibrations from the second cancel those from the fourth. There are thus three minima between every two principal maxima. Subsidiary maxima occur between every two minima but their intensity is very much less than that of the principal maxima. The intensity curve for four slits is of the form shown in Figure 13.16, and a photograph of the pattern is shown in Plate 9.

In the same way it may be shown that with any number of equally placed slits there will always be $N - 1$ dark bands and therefore $N - 2$ subsidiary maxima between every two principal maxima. When the number of slits is very large the intensity of the subsidiary maxima becomes so small compared with that of the principal maxima that the subsidiary maxima will not be seen. The diffraction pattern then consists of the principal maxima in the form of very narrow bright lines, which may be considered as images of the slit source, these being separated by comparatively broad dark spaces. These principal maxima will still, of course, be absent where their positions would correspond to the minima of the pattern for a single slit.

As with multiple-beam interference, the sharper fringes associated with the four-slit pattern of the previous section means that the measurement of their position can be made more precisely. In general terms, the more slits the sharper the fringes, which leads to one of the most important applications of diffraction, the **diffraction grating**. This consists, in its simplest form, of a large number of extremely narrow equal parallel slits separated by equal opaque spaces that are usually of the same width as the slits. The first diffraction gratings, made by Fraunhofer, about 1820, were formed of very fine wires closely and equally spaced. Later Rowland (1848–1901) produced gratings by ruling fine lines close together on the surface of glass with a fine diamond point. In these the diamond scratch may be considered as the opaque space between the transparent slits.

Most gratings in common use have rulings of between 10 000 and 20 000 lines per inch (periods $b + c$ of 2.5–1.25 µm). These are replicated from the original ruling by casting a thin film of polymer, which can then be peeled off the original and mounted on a glass plate. Slight mistakes in the ruling give rise to ghost images and a more common form of grating now in use is made by forming narrow interference fringes in a photoresistive material, which when developed gives a profile resembling the interference pattern. These are called **holographic gratings** but interference or photoresist gratings would be a better name. They give very clean images without ghosts but cannot be blazed (see later) as efficiently as ruled gratings.

From what has been said earlier, it will be clear that when parallel monochromatic light from a distant slit or from the slit of a collimator passes through a diffraction grating and is focused by a lens on a screen or by a telescope, a central bright image of the slit is formed together with fainter bands on either side of it separated by dark spaces. As the light is passing through such a great number of slits, and as these are so extremely narrow and close together, the various maxima will be sharply defined and widely separated.

The positions of the maxima, except the central one, are dependent on the wavelength, and therefore if the light is white or any other compound light the maxima for different wavelengths occupy different positions, and a spectrum is formed. The different spectra corresponding to the various maxima from the centre outwards are termed the first, second, third, etc., order spectra respectively. Some order spectra will be absent when their positions would correspond to the minima of the diffraction pattern for the single slit, as shown in Figure 13.16.

Figure 13.17 represents the section of a grating on which parallel light is incident normally. The diffracted light is received by lens L and imaged on to its focal plane. The intensity at point P is governed by two effects: the diffraction of the light from each individual slit which would, if the light from the slits were incoherent, give rise to a single diffraction pattern in the focal plane; and the multiple interference effects between light from all the slits if this is mutually coherent. When illuminated by a parallel beam from a collimated slit parallel to the slits of the grating this mutual coherence is normally very high, but the overall intensity of the interference pattern is given by the diffraction from each slit as in the two- and four-slit cases shown earlier.

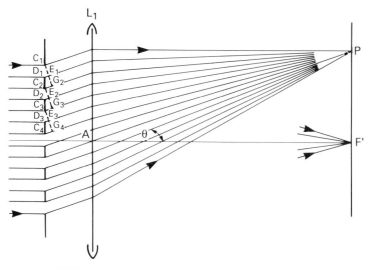

Figure 13.17 Diffraction grating

If the width of the slits is given by b, the diffracted amplitude from each slit at P is

$$a = \frac{\sin \delta/2}{\delta/2} \tag{13.13}$$

where δ is the phase difference and the amplitude at F′ is assumed to be unity (Section 13.5). δ is given by the path difference between the sides of each slit, D_1E_1, D_2E_2, etc. (Figure 13.17) which are identical, and so

$$\delta = \frac{2\pi b \sin \theta}{\lambda} \tag{13.14}$$

The interference effect of all the slits can be calculated in a similar manner to that used in Section 13.5 but with some important differences. The phase difference, ϵ, between each slit is given by the path differences C_2G_2, C_3G_3, etc., which are equal and so

$$\epsilon = \frac{2\pi(b + c) \sin \theta}{\lambda} \tag{13.15}$$

where c is the opaque space between the slits. The distance $b + c$, that is, the width of one slit and one opaque space, is known as the **grating interval** or **period**.

Although the phase of the light from each slit varies across the slit, the phase difference between respective parts of slits is a constant and so we can show the phase of each slit as a single value in the summation. The length of each slit in a rectangular grating is also constant and so the amplitudes are equal. Thus the graphical summation is as shown in Figure 13.18, where ϵ and the length, a, of each vector are constants. This is very similar to Figure 13.18, but the presence of the opaque strips means that the polygon never becomes a circle. The calculation of the resultant amplitude follows from constructing bisectors of the angles between the vectors. As the polygon is regular, these all intersect at C. The angle subtended by

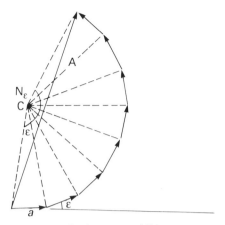

Figure 13.18 Grating vector addition

each line at C is ϵ, and the length of each bisector may be designated by r. The resultant A subtends the angle $N\epsilon$ at C where N is the number of slits. Then, from the figure,

$$A = 2r \sin N \epsilon/2$$

$$a = 2r \sin \epsilon/2$$

Hence

$$A = \frac{a \sin N \epsilon/2}{\sin \epsilon/2} \tag{13.16}$$

When Equation 13.13 is inserted for a, the equation becomes

$$A = \frac{\sin \delta/2}{\delta/2} \times \frac{\sin N \epsilon/2}{\sin \epsilon/2} \tag{13.17}$$

where the amplitude at F' is taken as unity.

The expression for intensity is, of course, the square of this and writing in the full terms for δ and ϵ we have

$$I = \left[\frac{\sin \left(\dfrac{\pi b \sin \theta}{\lambda} \right)}{\left(\dfrac{\pi b \sin \theta}{\lambda} \right)} \right]^2 \left[\frac{\sin \left(\dfrac{N\pi (b + c) \sin \theta}{\lambda} \right)}{\sin \left(\dfrac{\pi (b + c) \sin \theta}{\lambda} \right)} \right]^2 \tag{13.18}$$

Putting $N = 1$, 2 and 4 gives the curves of Figure 13.16, where the overall intensity is controlled by the first term and the sharpening of the fringes by the second term. This latter expression is a maximum when $\epsilon/2 = k\pi$ at which points its value equals N (k any integer). (Although both $\sin Nx$ and $\sin x$ become zero for $x = k\pi$, the sines become very nearly equal to the angles when x approaches $k\pi$ and it may be shown that the quotient equals N.)

This means that, for the maxima,

$$\frac{\pi(b + c) \sin \theta}{\lambda} = k\pi$$

or

$$(b + c) \sin \theta = k\lambda \tag{13.19}$$

In words, this means that the path differences C_2G_2, C_3G_3 in Figure 13.17 are equal to a whole number of wavelengths. The first, second, third, etc., order maxima occur in positions such that C_2G_2 is equal to one, two, three, etc., whole wavelengths respectively.

In contrast with the spectrum formed by a prism, the spectrum formed by a diffraction grating has the different colours directed into maxima whose angle θ is proportional to the wavelength. Thus the spectrum formed by one grating will be identical to that formed by another except in scale, which will depend on the grating interval. The diffraction spectrum, for this reason, is often termed a **normal spectrum**. For values of k greater than 2 the maxima from different orders may overlap for different wavelengths as the spread of the spectrum is proportional to the order. When attempting to separate two wavelengths close together it is preferable to use as high an order as possible as the width of the maxima does not change markedly with the order. However, the intensity of light in the higher orders is extremely small due to the overall diffraction pattern from a narrow slit. When the incident light is not at normal incidence it can be shown that Equation 13.19 becomes

$$(b + c)(\sin \theta + \sin i) = k\lambda \tag{13.20}$$

where i is the angle of incidence.

It is possible to rule transmission gratings so that each slit is in fact a prism. This has the effect of tilting the overall diffraction pattern away from the zero order and aligning its maximum intensity with one of the higher orders to one side. Such gratings are known as **blazed gratings**. It is also possible to produce reflection gratings where the slits become long narrow mirrors, which can be tilted to give the blazed effect. These can be made by vacuum coating the replica gratings described earlier. A reflection grating can also be made on a polished concave surface so that no lenses are required to give sharp spectra, the light being focused by the concave mirror. Gratings on which the rulings are made into circles form diffractive lenses and these are described in Section 13.8.

13.8 Diffractive optics, the diffractive contact lens

The diffraction grating, described above, is normally used for *dispersing* light and is used extensively in spectrophotometry. However, this optical element also *deviates* light and, like a prism, can be used for bending a light beam as well as splitting it into its component wavelengths. A comparison between a diffraction grating and and ophthalmic prism is given diagrammatically in Figure 13.19. If the grating is blazed into the first order, then for light at or near to a given wavelength most of the intensity is found in this first order and the other orders are no more intense than the stray beams reflected at the surfaces of a normal, uncoated ophthalmic

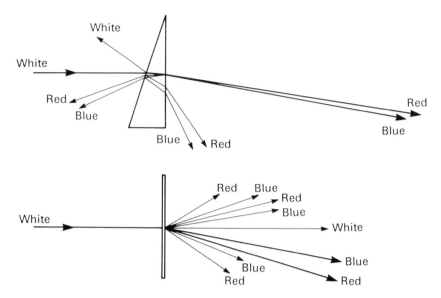

Figure 13.19 Comparison between a refracting ophthalmic prism and a blazed diffraction grating. Between each red and blue ray is a full spectrum of colours due to dispersion, which has been exaggerated for the prism

prism. The bigger difference is that the dispersion is greater and the opposite way to that of a refracting prism.

In Section 11.3, the dispersive effect of optical materials was studied and the concept of constringence given as

$$V = \frac{n_D - 1}{n_F - n_C} = \frac{n_D - 1}{(n_F - 1) - (n_C - 1)} \quad (11.3) \qquad\qquad (13.21)$$

With a diffraction grating the deviation is directly proportional to wavelength. When it is considered as a prism the apical angle, a, and refractive index, n, of the refracting prism are replaced by the grating interval, $b + c$, and the wavelength of the light, λ. The deviation, v, for the first order is then given by

$$v = \frac{\lambda}{b + c} \text{ rad} \qquad\qquad (13.22)$$

or

$$v = \left(\frac{1}{b + c}\right) \lambda$$

which compares with

$$v = a(n - 1) \qquad\qquad (2.20)$$

The dispersive effect of the refracting prism is proportional to $n - 1$, changing with wavelength as in the constringency equation. With glass prisms this is $1/V$, or about 2% of the prism deviation for ophthalmic glass, where $V = 58$.

When the diffraction grating is considered as a prism the dispersive effect (into the first order), using the same wavelengths as Section 11.3, depends on

$$V = \frac{587}{486 - 656} = -3.45 \tag{13.23}$$

and so $1/V = -29\%$.

This is the opposite direction and much larger than the dispersive effect with most glass prisms, as shown in Figure 13.19. At any appreciable value of prism dioptres such devices would be useless for vision work because the eye is very sensitive to this lateral spread of colour.

In Figure 4.17 it was shown that a series of refractive prisms could be equated to a lens. In the same way a series of diffractive prisms can be likened to a diffractive lens. In order to provide greater deviation at the edge of the lens the grating lines must be ruled closer than at the centre.

For a spherical lens action this ruling turns out to be a series of circles which are closer together nearer the periphery. The diffractive element which uses this effect is the *zone plate* described in Section 13.3. In the original device invented by Lord Rayleigh in 1871 alternate half-period zones were made opaque to the light as shown in Figure 13.20, which shows the appearance and the cross-section. This is equivalent to the amplitude grating described in Section 13.7. Such a device is a multiple-power lens having many focal lengths corresponding to the different diffraction orders (positive and negative) of the amplitude grating.

The *phase zone plate*, often called a *phase plate*, was a development from the amplitude zone plate in the same way as the phase diffraction grating – alternate

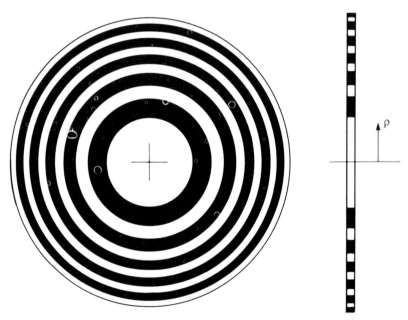

Figure 13.20 Amplitude zone plate. Just over 5½ full zones are shown. The inner half of each zone is transparent and the outer half is opaque as shown in the cross-section

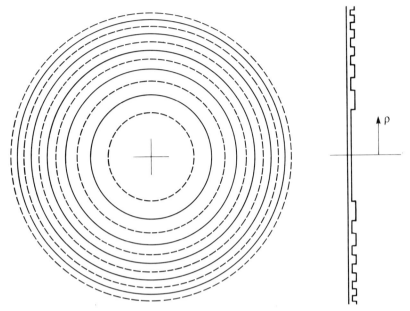

Figure 13.21 Symmetrical phase plate. Just over 5½ full zones are shown. The inner half of each zone delays the phase of the light by a different extent than the outer half of each zone as shown diagrammatically in the cross-section. The whole area is transparent

half-period zones are given a π change in phase. Such a device is more efficient with the light but still produces multiple images at positive and negative focal lengths. Figure 13.21 shows the appearance and profile in cross-section.

Taking the first order image as the reference image we can define this as the basic focal length for these devices. This image is formed from a wavefront that has one wavelength difference between each *full* zone. Using the centre of the whole pattern and the *outer* edge of each full zone, these wavelength differences will add constructively at a point F′ when there is exactly one wavelength difference between them. This is given approximately by the sag formula, $s = \rho^2/2r$, applied to each zone.

$$s_1 - s_0 = s_2 - s_1 = \text{other zones} = \lambda$$

$$= \frac{\rho_1^2 - \rho_0^2}{2f} = \frac{\rho_2^2 - \rho_1^2}{2f} = \text{other zones} \qquad (13.24)$$

so that

$$\rho_1^2 - \rho_0^2 = \rho_2^2 - \rho_1^2 = \text{other zones} = 2f\lambda \qquad (13.25)$$

When $\rho_0 = 0$ (and $s_0 = 0$) for the centre of the zone system,

$$\rho_1^2 = 2f\lambda \qquad \text{or} \qquad \rho_1 = \sqrt{(2f\lambda)} \qquad (13.26)$$

Then

$$\rho_2^2 - \rho_0^2 = (\rho_2^2 - \rho_1^2) + (\rho_1^2 - \rho_0^2) = 2(2f\lambda)$$

and so

$$\rho_2 = \sqrt{[2(2f\lambda)]} = \sqrt{2}[\sqrt{(2f\lambda)}]$$
$$\rho_3 = \sqrt{[3(2f\lambda)]} = \sqrt{3}[\sqrt{(2f\lambda)}]$$

and the nth zone radius is given by

$$\rho_n = \sqrt{[n(2f\lambda)]} = \sqrt{n}[\sqrt{(2f\lambda)}] \tag{13.27}$$
$$= \sqrt{n} \text{ times the radius of the first zone}$$

Thus the rings shown in Figures 13.20 and 13.21 have radii proportional to the *square root* of 1, 2, 3, 4, etc. Taking the first zone as a reference, the focal length, f, of a device for which $\rho_1 = 0.5$ mm will be given by

$$f = \frac{\rho_1^2}{2\lambda} = \frac{0.25}{0.0011} = 227 \text{ mm} \tag{13.28}$$

for a wavelength of 555 nm.

In this way an optical element of approximately 4 D power is generated by a full-zone system of 1, 1.4, 1.73, 2, 2.24, 2.5, 2.65, 2.8 mm diameters. However, for other wavelengths, such as blue light of 400 nm, this reference focal length will be 312 mm and for red light of 650 nm this reference focal length will be 192 mm. Furthermore, the other orders of diffraction will generate shorter focal lengths and also negative focal lengths of the same magnitudes because the optical system has a symmetrical profile.

A further development is the **blazed zone plate**, which again is a parallel concept to the blazed diffraction grating. As an optical element this device is sometimes

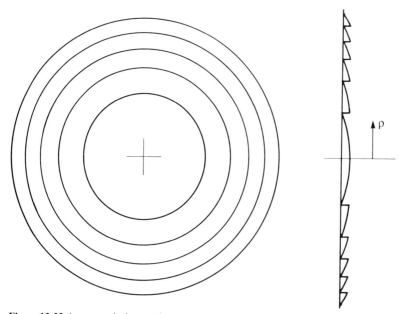

Figure 13.22 Asymmetrical zone plate. Just over five full zones are shown

called a **kinoform**. Here the zones must be considered as full-period zones and the phase shift changed in a continuous asymmetrical manner as shown in Figure 13.22. With such a device the light of that wavelength, λ_0, for which the step at the end of each zone provides a phase shift of 2π will be entirely directed into the first order image. However, other wavelengths will have some energy into other orders and, for the same device, light with a sufficiently shorter wavelength so that it encounters a phase shift of 4π will be directed entirely into the second order.

The change in focal length with wavelength is the same with this device as for the amplitude zone plate and the phase zone plate. This means that the focal length by Equation 13.28 for the shorter wavelength will be longer than that found for λ_0. This shorter wavelength λ_1 will, in general, be equal to $\lambda_0/2$ unless there are some dispersive effects in the optical materials used to generate the phase difference. Ignoring these we find that the shorter wavelength has twice the focal length of λ_0, when calculated for the λ_1 first-order image. However, all the light for this shorter wavelength is directed into the second-order image, which has half the focal length of its first-order image. Thus, the actual image for the λ_1 wavelength is at the same location as the first-order λ_0 image. Most people find this confusing.

The large change in optical power with wavelength as well as the division of light into more than one order which are intrinsic features of diffractive elements effectively limit their application to optical systems operating in monochromatic light. A non-obvious application using 'white' light across the visible spectrum is diffractive bifocal contact lenses. Refractive bifocal contact lenses normally use distinct areas or segments of the lens to provide different focal powers for distance and near vision. Often these segments interact with the pupil of the eye to give a variable quality of vision. Using diffraction it is possible to divide the incident light into more than one image at every point in the aperture. With the amplitude zone plate and the symmetrical phase zone plate multiple images are generated with both positive and negative power. With the asymmetrical kinoform or blazed zone plate it is possible to direct most of the light within the visible spectrum into a zero power and one positive power image.

This requires a diffractive kinoform, which has a phase shift so that λ_0, the wavelength at which the phase shift is 2π, lies outside the visible spectrum in the ultraviolet region. This means that for visible light near the peak of the visual response curve almost all energy will be split between the first-order and zero-order (no power) images. With visible red light the split will be biased towards the zero order and with visible blue light biased towards the first order. Nevertheless, the first-order image will provide a reading add, the intensity of which is independent of pupil size. This first-order image suffers from chromatic aberration so that for an add of +2.0 D for green light, the add for red light will be approximately +2.3 D and for blue light +1.7 D. However, the eye has an intrinsic chromatic aberration of nearly 1.0 D *in the opposite sense*, so that for the reading image this aberration is partially corrected in the same manner as an achromatic doublet described in Chapter 14.

Figure 13.23 shows the generation of multiple wavefronts associated with amplitude zone plates, phase plates and diffractive bifocal contact lenses using the asymmetric principle. Most contact lens wearers require some refractive power in their contact lenses to correct their distance vision, and this is undisturbed by the zero-order image of the diffractive power. The add power, being the difference between the overall power for distance and the overall power for near, is provided entirely by the diffractive effect.

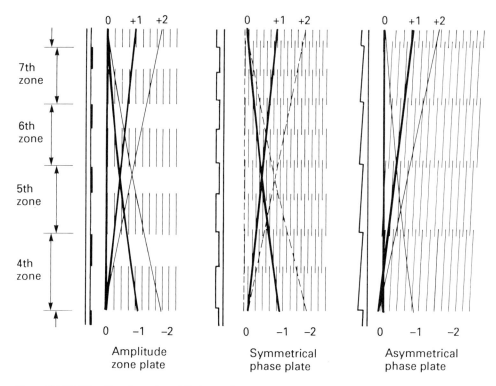

Figure 13.23 Diffractive elements and their wavefronts

13.9 Grating images and the optical transfer function

The diffraction gratings described in the last section were considered as elements in a beam of light but we may also investigate the problem of imaging such very fine objects. This was first considered by Abbe (1904) when deriving his theory of the microscope. In this, he took a transmission grating illuminated by a parallel beam as an object for lens L (Figure 13.24), to image on to a screen. The advantage of using a grating as an object lies in being able to apply the results of interference and diffraction previously considered. As this parallel illuminating beam is coherent the grating will give first and higher order beams as shown in the figure.

If the imaging lens L is of limited aperture not all these diffracted beams will pass through it. Provided that the first orders are accepted by this lens, it will then form three point images in its focal plane. These images act as coherent sources to provide interference effects in the final image plane, which constitute the image of the grating. If the first-order diffracted beams fail to pass through the lens, no image of the grating is seen as the zero-order 'source' at F′ has nothing with which to interfere. As the object grating is made finer, the angle of the first-order beam increases so that at some frequency of lines per inch the image of the grating is lost. With coherent illumination therefore it is found that the image of the grating quite quickly fades away as gratings of finer and finer structure are used as objects.

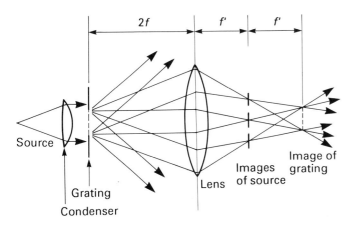

Figure 13.24 Grating images

In the study of this fading it is normal to define the fineness of the grating in terms of its **spatial frequency**, that is the number of slits in a given length. This is commonly measured in cycles per millimetre or line-pairs per millimetre. Only if the grating has a sine-wave variation in transmission can it be said to have one frequency, according to Fourier's theory, and even then the lack of *negative* transmission makes this less than exact. However, it is often a sufficient approximation to consider line/space gratings as having just one spatial frequency.

When such a grating is imaged by a lens it is found that diffraction and aberrations combine to reduce the clarity of the lines in the image. The ratio between the contrast of the image and the contrast of the object is generally called the **optical transfer function (OTF)** or **modulation transfer function (MTF)** of the lens (Chapter 14). This parameter is usually assumed to be unity at coarse grating frequencies and a graph of the variation in modulation against frequency for coherent illumination will look like curve A of Figure 13.25.

From the Abbe concept that the first order must be imaged by the lens we have that

$$(b + c) \sin \delta = \lambda$$

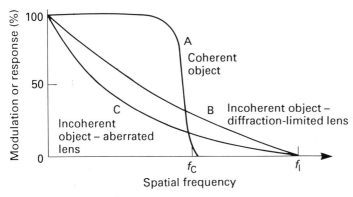

Figure 13.25 Image quality curves

(from Equation 13.19), where $b + c$ is the period of the grating and θ may be set equal to u (Section 5.7), the acceptance angle of the lens. So, with coherent light, we find that the maximum grating frequency ($f_C = 1/(b + c)$) that can be resolved is given by

$$\frac{1}{(b + c)} = f_C = \frac{\sin u}{\lambda} = \frac{NA}{\lambda} \tag{13.29}$$

The above has assumed that the light from the individual slits is mutually coherent. This is much the case with a microscope, but if the grating is self-luminous or illuminated by having an extended source imaged on to it the calculation becomes somewhat different and rather more complex. An idea can be gained from the minimum separation distance h' for just resolved points given by Equation 13.11 and (for rectangular apertures) by Equation 13.14. Although not exactly correct, the grating frequency may be regarded as equal to $1/h'$ and the maximum resolved frequency, f_I, is then $2 \sin u'/\lambda$ in the image plane or $2 \sin u/\lambda$ in the object plane. However, with incoherent light the change of modulation with frequency is much more gradual and the curves B and C of Figure 13.24 show typical results for a perfect diffraction-limited circular lens and a lens with aberrations respectively.

Various systems are now available for measuring the optical transfer function of lenses by presenting various gratings to the lens under test and measuring electronically the modulation of the image (see Section 14.8). When, as often happens in very sophisticated systems, the objects are not completely coherent or incoherent the calculation and measurement of the resolving action of lenses is very difficult.

13.10 Holography

The basic concepts of holography were developed by Dr Dennis Gabor in 1948, long before the laser was invented. Using mercury isotope lamps as his coherent sources he endeavoured to improve the resolution of the electron microscope and called his approach '*microscopy of reconstructed wavefronts*'. This description is the essence of the holographic process. Normal photography recreates an object in the light and dark tones of an emulsion, but Gabor appreciated that when an observer views a real object he receives only the wavefront of the scattered light leaving that object and reaching the eyes. All the optical information must be in the phases, intensities and wavelengths of that wavefront, which are extremely complex even for simple objects. To record this wavefront for later reconstruction is normally impossible.

In holography the simplification of coherent monochromatic illumination is essential, and even then it is not possible to record directly the phase values of the wavefront as these are not seen by optical detectors such as photographic emulsions and image tubes. In order to record phase effects it is necessary to add a coherent reference beam into the system, which will interfere with the scattered wavefront so that differences in phase will show up as differences in intensity.

Leith and Upatnieks in 1961 introduced this reference beam at a fairly large angle to the scattered beam. This simple move creates very fine interference fringes in the plane of the photographic plate. When this is exposed and processed and returned to its original position the reference beam alone 'reconstructs' the

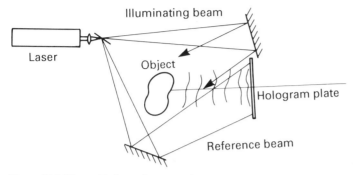

Figure 13.26 Fresnel holography – manufacture

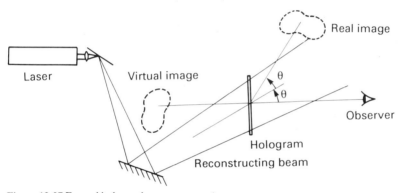

Figure 13.27 Fresnel holography – reconstruction

scattered beam from the object by diffraction between the beam and the processed interference fringes. A number of 'orders' of image are created as with a diffraction grating. Some are real images while some are virtual. The first-order virtual can be seen by an observer located as in Figure 13.27 as an image having considerable clarity of detail and shading and in three dimensions. The angle of the reference beam means that the straight-through light and the 'opposite' image are out of the field of view.

The three-dimensional aspect of the reconstruction is inherent in re-creating the wavefront rather than a flat representation of the object. The eyes use only two small areas of the wavefront at any one time. As the head is moved other areas are used and the object is seen 'in the round'. The wavefront is only reconstructed over the area of the interference fringes on the photographic plate so that the reconstructed object is visible only through the plate, which acts as a window to the scene. The limited power of laser sources has restricted the size of hologram objects to a few feet, although with special techniques holograms of rooms have been made. It is necessary to prevent incoherent illumination reaching the object or the plate.

Most holograms are made in darkrooms and it is generally necessary to mount the components on a vibration-free surface. If the exposed hologram is moved or tilted with respect to its original position in the reference beam the reconstruction distorts and detail is lost. If a reference beam of different wavelength is used the object may still be seen but at a magnification given by the ratio of the wavelengths. These holograms are called **Fresnel holograms**. As might be expected it is possible to produce **Fraunhofer holograms**. These are made when the photographic plate is distant from the object compared to its size either in actuality or because a lens is used to collimate the scattered light from the object. Figure 13.28 shows a general arrangement using a lens. In the special case where the recording plane and the object are at the principal foci of the lens the hologram so formed is called a **Fourier-transform hologram**. This is because the intensity in the hologram at a given distance from the axis is found to be related to the amount of a given spatial frequency in the object. This can be used to process transparencies so that particular spatial frequencies are enhanced and others rejected – a form of optical processing (Section 13.11).

Another form of hologram can be made when the lens of Figure 13.28 is positioned to form an image of the object close to the plane of the hologram. Such **focused holograms** are more tolerant on coherence, especially in reconstruction, and are used in display applications. If a thick emulsion is used on the photographic plate it is possible by virtue of interference effects within the emulsion to obtain a hologram that responds differently to different colours in the reconstruction beam, thus allowing colour holograms to be made.

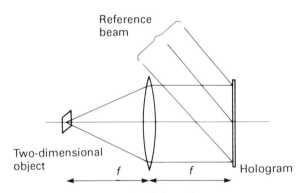

Figure 13.28 Fourier transform holography

It is seen that the image results from the reconstruction of the wavefront. If the object remains in place when the exposed hologram is returned to its original position, interference is possible between the actual object and the reconstructed image provided that both are illuminated from the same laser: thus, if the object is moved slightly or deformed, two-beam interference fringes will be clearly seen. The techniques of **interference holography** have become standard in mechanical testing and vibration analysis.

When a hologram reconstructs a complex image from a reference beam it is found that a change in the angle or the collimation of the reference beam moves and aberrates the complex image. If, instead of a complex scene, a single coherent point of light is used it is found that, after processing, the hologram acts upon the reference beam to focus it as an image of the point; in other words the hologram is acting like a lens, but using diffraction rather than refraction. It is found that the illuminating beam of Figure 13.26 can be shone directly onto the hologram plate in

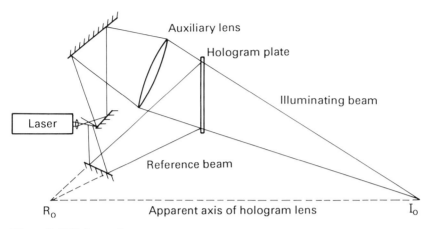

Figure 13.29 Hologram lens

the same manner as the reference beam (Figure 13.29). After processing it is found that the **hologram lens** will image a point source placed at the origin, R_o, of the reference beam so that it appears as a point source placed at the origin, I_o, of the illuminating beam, neither beam being present. The point source does not have to be coherent and so we have a perfect lens *but* only for these particular conjugates, off-axis though the lens may appear to be. Such lenses are very similar to the zone-plate lenses described in Section 13.8.

13.11 Diffraction and optical processing

We have seen that the location of the first dark ring (or fringe) in a Fraunhofer diffraction pattern is inversely proportional to the diameter (or width) of the aperture used. The pattern is centred on the image of the source so that if the aperture is very small compared with the lens, in an arrangement similar to that in Figure 13.7, the pattern will be larger (and fainter) but still centred on the image of the source. Then, a large number of equal apertures are irregularly distributed over the screen carrying the original aperture, each will form an identical pattern in the same place. Thus the intensity of the pattern will increase but, except for the central disc where the effect is modified by the spacings of the apertures, the rings will have the same diameters for a large number of circular holes as for one circular hole.

It has been shown that if the circular apertures in the opaque screen are replaced by circular opaque particles of the same size on a transparent screen, the diffraction effect is exactly the same, and examples of this phenomenon can be easily produced. On looking at a white luminous point on a dark background through a plate of glass covered with lycopodium dust, which consists of minute spheres of equal size, brilliant coloured rings will be seen around a bright centre. Similar rings will be seen through a steamy glass and also close round the sun and moon in hazy weather. In these last two cases the rings will not be as brilliant as with lycopodium because the particles of moisture causing the diffraction are not all of the same size.

These rings are known as **coronas** or **halos**, although the latter term is also applied to the larger rings often seen round the sun and moon, and which are due to refraction.

If we can measure the angle subtended by a given ring for a particular colour, we can find the radius of the particles from Equation 13.9. Thomas Young did this to measure the size of blood corpuscles, wool fibres, etc. This was an early example of what is now called **optical processing**, and for an understanding of this it is necessary to consider the concepts of Fourier transforms. When the summation of simple harmonic motion was developed in Section 9.2, it was seen that two sine waves of differing amplitude and frequency could generate a waveform of complex shape. Fourier demonstrated that any form, complex or simple, could be mathematically analysed in terms of sine and cosine wave forms of varying amplitudes and frequencies. This applies both to wave motions and to waveforms. A single perfect sine-wave motion can be described as a single frequency, as we have seen. If this wave motion is modified in any way other frequencies are required. This applies even to such simple modifications as starting and stopping the wave motion. Thus when light comprises short wave trains we find that many frequencies are required to define them and the light is described as 'white'. The longer the wave trains, the fewer frequencies are required and so we find that best interference effects occur with light having a long coherence length, which means highly monochromatic.

Two particular examples of Fourier transforms are given in Figure 13.30. Figure 13.30(a) shows that for wave motions that start and stop gradually the mathematically derived frequencies are spread about the main frequency, and this is the shape of most spectral lines. If the waveform is started and stopped abruptly the frequencies required are shown in Figure 13.30(b). Here it is seen that some frequencies are required to be negative or out of phase with the main frequency, and the shape bears a striking resemblance to the Fraunhofer diffraction patterns of Section 13.5.

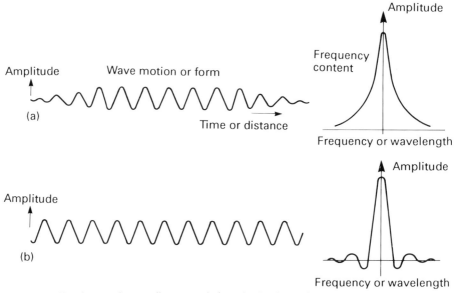

Figure 13.30 Fourier transforms – diagrammatic (see also Section 9.3)

Considering Figure 13.30 again it can be seen that for a wavefront arriving at the aperture at an angle the waveform within the aperture is a sine wave abruptly terminated at each end. It is found that Fraunhofer diffraction patterns are related to the shapes of apertures in the same way as Fourier transforms, each a unique description of the other. From this, optical systems can be devised which use diffraction patterns to perform mathematical processes or pick out particular types of object from general scenes.

Exercises

13.1 What must be the size of a circular opening in an opaque screen for it to transmit two Fresnel zones to a point 2 m away? What will be the approximate intensity of light at this point? (λ =589 nm.)

13.2 Show that, with a plane wavefront, when the wavelength is small the areas of the half-period zones for a point at a distance d from the wavefront are all very nearly equal to $\pi d\lambda$.

13.3 Find the diameters of the first four clear zones (the centre being opaque) of a zone plate that will 'focus' parallel incident light of $\lambda = 600$ nm at 50 cm from the plate.

13.4 The diffraction bands at the end of the shadow of an opaque object decrease in width from the shadow outwards; explain why this is. When using white light these bands are coloured; state and explain the order of the colours.

13.5 Explain briefly the diffraction bands produced on a screen by light from a line source passing a straight edge parallel to it. Find the approximate intensity of illumination at the geometrical shadow edge and at the first bright band and first dark band. Draw a graph to illustrate this variation of intensity.

13.6 When a thin wire is placed in the light from a slit source of monochromatic light its shadow on a screen is seen to be crossed with alternate light and dark bands parallel to the length of the wire. Explain the formation of these bands. In an experiment the distance between the bright bands was 0.7 mm when the screen was 1 m from the wire. What was the thickness of the wire? ($\lambda = 650$ nm.)

13.7 Derive the expression $\sin \theta = \lambda/b$ for the position of the first dark band in the case of diffraction by a rectangular aperture in front of a telescope objective.

Calculate the resolving power of a telescope objective of 1½ in (circular) aperture, assuming $\lambda = 560$ nm and that two stars can be resolved when the central maximum of one image falls on the first dark ring of the other.

13.8 Explain how Cornu's spiral may be used to calculate diffraction patterns. To what type of pattern is it applicable?

13.9 Explain carefully the diffraction rings observed when a small light source is observed through a diffracting screen such as a glass plate covered with lycopodium powder.

In an experiment, the angular diameter of the rings was measured using a plate containing a small central hole surrounded by a circular ring of smaller holes (Young's eriometer). Observed through lycopodium powder the distance from the diffracting screen to the hole was found to be 32.4 cm when a certain diffraction ring coincided with the ring of holes. Observed through a screen of blood corpuscles the distance for the same ring was 8.92 cm. If the mean diameter of the lycopodium grain is 0.029 mm find the diameter of the blood corpuscles.

13.10 Fraunhofer diffraction is observed using a square hole of 2 mm side. Explain qualitatively what differences can be seen when this is replaced by a circular hole 3 mm in diameter.

13.11 Explain, giving a diagram, the formation of the diffraction pattern at the focus of, say, a telescope objective. What do the dimensions of this pattern depend on and how are they associated with the resolving power of the instrument?

13.12   What is meant by the resolving power of a telescope? How may it be expressed and on what is it dependent?

Assuming that two stars can be resolved when the central maximum of one image falls upon the first dark ring of the other, calculate the resolving power of a telescope objective, the aperture of which is 3 in, and focal length 30 in. Assum $\lambda = 560$ nm.

If the limit of resolution of the eye for comfortable working is taken as 100″, find the focal length of the eyepiece necessary for clear resolution of the image when using the telescope.

13.13   Treating the eye as a perfect optical system of power 60 D, aperture 3.5 mm and refractive index of vitreous 4/3, calculate the radius of the first dark ring in the Airy diffraction pattern formed of a distant point source. ($\lambda = 560$ nm.)

Taking the diameter of the foveal cones as 0.0025 mm, comment on the eye's known capability of resolving two stars or point sources separated by less than one minute of arc.

13.14   Explain how diffraction limits the detail that can be seen with an optical instrument. Why does a microscope using ultraviolet light offer a better theoretical performance?

13.15   The angular separation of two stars is 1.5″. Find the minimum aperture that must be given to the objective of a telescope for the stars to be just resolved by the instrument. Assume $\lambda = 560$ nm.

13.16   The focal length of a telescope objective is 150 mm and the magnifying power of the whole instrument is eight. Taking the resolving power of the eye as 100″ for comfortable working, find the angular magnitude of the finest structure the eye and eyepiece are capable of resolving and hence the minimum aperture of objective that will suffice ($\lambda = 560$ nm).

Why would a larger aperture probably be used in practice?

13.17   Calculate the radius of the first dark ring in the diffraction image produced by a telescope of 28 in aperture, focal length 26 ft, $\lambda = 560$ nm.

On the assumption that two stars can be resolved when the centre of the image of one falls on the first dark ring of the other, calculate the resolving power of this objective.

By how much would the image require to be magnified by the eyepiece in order that the eye may resolve it? The limit of resolution of the eye may be taken as 1.5′ for comfortable vision.

13.18   Explain the action of the diffraction grating. State the advantages and disadvantages of using a grating to produce a spectrum as compared with the use of a prism.

13.19   Explain the production of spectra by a diffraction grating. Describe an experiment for determining the wavelength of light by means of a grating, stating clearly what quantities are measured and how the calculation is made.

13.20   What will be the angular separation of the two sodium lines ($\lambda = 589.0$ nm, $\lambda = 589.6$ nm) in the first-order spectrum produced by a diffraction grating having 14 438 lines to the inch, the light being incident normally on the grating?

13.21   Parallel light from a mercury vapour lamp falls normally on a plane grating having 10 000 lines to the inch. The diffracted light is focused on a screen by a lens of 15 in focal length. Find the distances in the first-order spectrum between the lines corresponding to the wavelengths 579.1, 577.0, 546.0 and 435.8 nm.

13.22   A grating illuminated by imaging a source on to it is just resolved by a microscope. The condensing system is now changed to a lens which collimates a point source. Explain why the grating is no longer resolved.

13.23   Explain briefly what is meant by the diffraction of light. Give an example, to be met with in everyday life, of diffraction in the case of:
(a)   Water waves.
(b)   Sound waves.
(c)   Light waves.

In what respects does the spectrum produced by a diffraction grating differ from a prismatic spectrum?

13.24 If you look through a piece of fine gauze (40 wires to 1 cm) at a narrow source, emitting light of wavelength 600 nm, placed 4 m from the gauze, what will be the linear separation between the central and the first diffracted image?

Describe the advantages and disadvantages of using a diffraction grating to produce a spectrum as compared with the use of a prism.

13.25 Explain why it is possible to study much of the theory of lenses and of optical instruments generally on the basis of a ray theory rather than a wave theory of light propagation.

In what circumstances would the neglect of the wave theory introduce serious errors?

Explain why the image of a star viewed through a telescope appears smaller as the aperture of the telescope objective is increased

13.26 Derive an expression that determines the positions of the spectral lines produced by a transmission diffraction grating, assuming normal incidence.

What influences the width of these lines? What effect has the width of each of the clear spaces on the lines?

A grating has 600 lines per mm. If the visible spectrum extends from 400 to 700 nm, find the linear separation of those wavelengths in the focal plane of a telescope objective of focal length 25 cm in the second order. Assume normal incidence, first order.

13.27 Define Fraunhofer diffraction.

A parallel beam of monochromatic light is incident on a plane diffraction grating at an angle of 30° to the normal. If the grating has 3000 lines per cm and the wavelength of the radiation is 632.8 nm, determine the angles of all the transmitted orders.

13.28 Fraunhofer diffraction at a double slit may be explained as follows: light from the two slits undergoes interference to produce fringes of the type obtained with two beams, but the intensities of these fringes are limited by the amount of light arriving at a given point on the screen by virtue of the diffraction occurring at each slit.

Explain this statement, sketch the resultant fringe pattern produced by such a double-slit experiment and comment on the importance of dimensions b (slit width) and c (slit separation).

The Fraunhofer pattern from a double slit composed of slits each 0.5 mm wide and spearated by $d = 20$ mm is observed in sodium light ($\lambda = 593$ nm) on a screen. How many fringes will occur under the central diffraction maxima?

13.29 A diffractive bifocal contact lens has a power of −6.00 D and a bifocal add of +2.50 D. If the diffractive area has a diameter of 4.8 mm calculate how many zones it contains and the diameter of the central zone assuming the lens has a design wavelength of 555 nm.

Chapter 14

Optical design: forming a good image

14.1 Sources of image defects

In most of the previous chapters the optical elements have been assumed to provide perfect optical imagery. In Chapter 7 the presence of image defects or aberrations was admitted and these were described mainly in terms of the Seidel or third-order aberrations. In Chapter 13 the effect of diffraction on image quality was described. Both these chapters assumed monochromatic light.

In Chapter 11 it was shown that the refractive index of all optical materials changes with wavelength. The extent of this change varies, with different optical materials, from less than 2 parts in 100 to more than 4 parts in 100 across the visible spectrum. This means that the actual dioptric power or focal length of the element shows a 2–4% colour spread or **chromatic aberration**. In ordinary optical elements of moderate aperture and field the monochromatic aberrations are less than this, and the diffraction effects are much less. The evaluation and correction of chromatic aberration therefore becomes the first priority in most optical systems, followed by the Seidel aberrations, followed by diffraction effects – which must be evaluated even though there is not a lot that can be done to reduce them.

Notable cases in which chromatic aberration does not need to be corrected are with monochromatic and laser systems and also, in the main, spectacle lenses. Spectacle lenses rely on the tolerance of the eye to colour effects, although it is less tolerant to lateral colour, as described in the next section, and spectacles made from high-index glass or polycarbonate, which have a large V value as described in Chapter 11, exhibit loss of image quality at the wider field angles.

The use of two different materials to correct chromatic aberration gives rise to the achromatic doublet, which then has a wider choice for the curvatures of its surfaces so that spherical aberration and coma can also be corrected. Most achromatic doublets give good imagery over an appreciable aperture but only over a very limited field of view. The conflicting requirements of aperture and field of view will be dealt with in later sections. Optical design as such is beyond the scope of this book and requires the aid of a computer and sophisticated software. Optical design programs are now available for use with personal computers. Essential to their design process is a proper description of image quality and this must relate to image valuation methods, which can be used on the finished lenses. These aspects are described in the final sections of this chapter.

14.2 Chromatism and achromatism

Chromatism or chromatic aberration arises from the inherent dispersion in optical materials as described in Chapter 8. By defining three wavelength values at the centre and near the extremes of the visible spectrum, dispersion was defined in Section 11.3 in terms of the **constringence** or **V value** of a given material. These wavelengths are reviewed from time to time and for this section the historical lines C, D and F (Table 14.1) will be used.

Table 14.1 Spectral lines used in optics (visible region)

Wavelength (nm)	Element	Designation	Colour
706.52	Helium	v	Deep red
656.28	Hydrogen	C	Red
643.85	Cadmium	C'	Red
632.8	Neon	Laser (HeNe)	Red
589.6	Sodium	D_2	Yellow
589.0	Sodium	D_1	Yellow
587.56	Helium	d	Yellow
546.07	Mercury	e	Green
543.5	Neon	Laser (GreNe)	Green
514.5	Argon (ion)	Laser	Green
486.13	Hydrogen	F	Blue
479.99	Cadmium	F'	Blue
441.6	Cadmium	Laser (HeCd)	Blue
435.84	Mercury	g	Blue
404.66	Mercury	h	Violet

For the case of refraction by a thin prism we have the approximate equation for the deviation v,

$$v = (n - 1) a \tag{14.1}$$

(from Equation 2.20), where a is the apical angle and n is some representative refractive index.

We now see that this is not good enough and the action of the prism on white light needs more information. Using the three wavelengths indicated above,

$$\left. \begin{array}{l} v_C = (n_C - 1) a \\ v_D = (n_D - 1) a \\ v_F = (n_F - 1) a \end{array} \right\} \tag{14.2}$$

where the subscript indicate the deviation and refractive index associated with that wavelength value. The actual dispersion of the light between the red C wavelength and the blue F wavelength is given by

$$\delta v = v_F - v_C = (n_F - n_C) a \tag{14.3}$$

This shows that the dispersion δv is proportional to the index difference and the apical angle. If a second prism of different material was required to give the same dispersion, the equations for the first and second prisms would be

$$\left. \begin{array}{l} \delta v = v_{F_1} - v_{C_1} = (n_{F_1} - n_{C_1}) a_1 \\ \delta v = v_{F_2} - v_{C_2} = (n_{F_2} - n_{C_2}) a_2 \end{array} \right\} \tag{14.4}$$

Although δv is the same in each case, the apical angles are different because the indices of the materials are different. If now these two prisms are arranged so that their dispersions (and deviations) are in opposite directions, it is possible to have a prism combination in which the dispersion effect is cancelled out for these two wavelengths. This is then an **achromatic prism** at these wavelengths and is normally made of materials which have very different dispersions. Crown glass with a low dispersion will need a large apical angle, and flint glass with a high dispersion will need a smaller apical angle as shown in Figure 14.1.

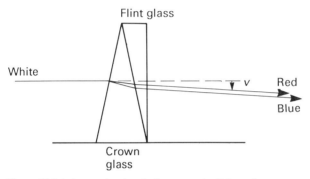

Figure 14.1 Achromatic prism (prisms to scale, light paths exaggerated)

Using Equation 14.4 we see that, for achromatism,

$$(n_{F_1} - n_{C_1}) a_1 = (n_{P_2} - n_{C_2}) a_2 \tag{14.5}$$

The residual deviation is given by

$$v_1 - v_2 = (n_{D_1} - 1) a_1 - (n_{D_2} - 1) a_2 \tag{14.6}$$

Normally materials are chosen with as large a difference as possible between n_{D_1} and n_{D_2} so that the residual deviation can be large without the component prism angles being very large as indicated in Figure 14.1, which is drawn to scale from the results of the example given below.

Worked example   A prism of Hard Crown glass has an apical angle of 25°. What apical angle should be worked on a prism of Dense Flint glass to give achromatism for the C and F lines when used as a pair, and what will be the mean deviation?
    The glass details are:

Type	n_C	n_D	n_F
Hard Crown glass	1.5150	1.5175	1.5235
Dense Flint glass	1.6176	1.6225	1.6349

(These values include that for the mean sodium D line which is not now used – see Section 14.3.1.)

For the Crown glass the dispersion is

$$\delta v_1 = v_{F_1} - v_{C_1} = (n_{F_1} - n_{C_1}) a_1 = 0.0085 \times 25°$$

For the Flint glass the dispersion is:

$$\delta v_2 = v_{F_2} - v_{C_2} = (n_{F_2} - n_{C_2}) a_2 = 0.0173 a_2$$

Because δv_2 is to be made equal to δv_1,

$$0.0173 a_2 = 0.0085 \times 25°$$

Therefore

$$a_2 = 12.28°$$

Then

$$v_{D2} = (n_{D_1} - 1) a_1 = 0.5175 \times 25 = 12.94°$$

and

$$v_{D2} = (n_{D_2} - 1) a_2 = 0.6225 \times 12.28 = 7.64°$$

The resultant mean deviation for the sodium D wavelength when the prisms are opposed is therefore given by $12.94 - 7.64 = 5.3°$.

The deviations for the red C and blue F wavelengths are now the same but not quite equal to the mean deviations, as the following calculation shows:

$$v_{C_1} = (n_{C_1} - 1) a_1 = 0.5150 \times 25 = 12.87°$$

$$v_{C_2} = (n_{C_2} - 1) a_2 = 0.6176 \times 12.28 = 7.58°$$

Residual deviation for red light $= 12.87 - 7.58 = 5.29°$.

The residual deviation for blue light will also be found to be $5.29°$. The deviation for sodium D light is $5.30°$ so there will be a small dispersion, which folds over to give a maximum deviation near the centre of the spectrum and slightly less at the red *and* blue ends. This is called the **secondary spectrum** and is very much smaller than the dispersion with a single prism giving $5.3°$ deviation.

The resultant prism combination is shown in Figure 14.1, where the prisms have been drawn to scale but the light paths have been exaggerated. Note that although the red and blue light rays now emerge with the same deviation they are slightly displaced from each other. In an imaging system where the light beam through the achromatic prism is not parallel, this can give lateral colour in the image.

The above calculation is, if course, approximate owing to the use of the *thin* prism formula, $v = (n - 1)\, a$, at these significant values of apical angles.

14.3 The achromatic doublet

14.3.1 First-order design

In the same way that a thin prism can be achromatized by combining it with another made of a different material so can a thin lens be paired with another to make an **achromatic doublet**. More often than not the curvatures of the two adjacent surfaces are made equal and opposite so that the two lenses can be cemented together. This arrangement is often called a **cemented doublet** (see Section 14.3.3).

468

Table 14.2 Representative glass types and polymers

Glass type	Mean index (n_d)	Mean dispersion ($n_F - n_C$)	V value (V_d)	404.7 (n_h)	435.8 (n_g)	486.1 (n_F)	546.1 (n_e)	656.3 (n_C)	Transmission (25 mm) (550 nm) (%)	Reflection (one surface) (%)	Density (g cm^{-3})
Glass type											
BSC 517642	1.516 80	0.008 05	64.17	1.530 24	1.526 68	1.522 38	1.518 72	1.514 32	98.3	4.3	2.51
HC 524592	1.524 00	0.008 85	59.21	1.538 96	1.534 96	1.530 15	1.526 11	1.521 30	99.1	4.4	2.55
ZC 508612	1.507 59	0.008 30	61.16	1.521 49	1.517 80	1.513 34	1.509 57	1.505 04	99.4	4.1	2.49
MBC 572577	1.572 20	0.009 91	57.74	1.589 01	1.584 51	1.579 10	1.574 56	1.569 19	99.6	4.9	3.14
DBC 620603	1.620 41	0.010 28	60.33	1.637 74	1.633 13	1.627 56	1.622 86	1.617 27	99.2	5.6	3.60
LF 581409	1.581 44	0.014 23	40.85	1.606 62	1.599 61	1.591 46	1.584 82	1.577 23	99.2	5.1	3.23
DF 620364	1.620 04	0.017 05	36.37	1.650 64	1.642 02	1.632 08	1.624 08	1.615 03	99.3	5.6	3.63
EDF 648338	1.648 31	0.019 16	33.84	1.683 07	1.673 14	1.661 87	1.652 85	1.642 71	98.5	6.0	3.74
EDF 706300	1.705 85	0.023 53	30.00	1.749 38	1.736 73	1.722 56	1.711 40	1.699 03	98.0	6.9	2.99
DEDF 755276	1.755 20	0.027 38	27.58	1.805 89	1.791 21	1.774 68	1.761 67	1.747 30	99.0	7.6	4.79
LAC 713538	1.713 00	0.013 25	53.83	1.735 43	1.729 43	1.722 22	1.716 16	1.708 97	99.6	7.0	3.81
LAF 850322	1.850 26	0.026 38	32.23	1.898 11	1.884 50	1.868 94	1.856 50	1.842 56	97.9	9.0	5.14
Polymer type											
PMMA (Perspex/Plexiglas)	1.491 76	0.008 56	57.45	1.506 61	1.502 56	1.497 76	1.493 79	1.489 20	–	3.9	–
Polystyrene	1.590 48	0.019 13	30.86	1.625 34	1.615 45	1.604 08	1.495 01	1.584 95	–	5.2	–
Polycarbonate	1.585 47	0.019 58	29.90	1.622 45	1.611 52	1.599 44	1.590 08	1.579 86	–	5.1	–

Again, achromatic doublets have equal focal lengths for two specified wavelengths. In this section the mercury e line (546.1 nm) and the cadmium F' and C' lines (480.0 and 643.8 nm) will be used, as these are a better fit to the visual response curve. A selection of glass types and optical plastics is given in Table 14.2.

For a thin lens,

$$
\left.\begin{aligned}
F_e &= (n_e - 1)(R_1 - R_2) \\
F_{F'} &= (n_{F'} - 1)(R_1 - R_2) \\
F_{C'} &= (n_{C'} - 1)(R_1 - R_2)
\end{aligned}\right\} \tag{14.7}
$$

Therefore

$$
F_{F'} - F_{C'} = (n_{F'} - n_{C'})(R_1 - R_2) = \left(\frac{n_{F'} - n_{C'}}{n_e - 1}\right) F_e \tag{14.8}
$$

Since the expression in the brackets is the reciprocal of the V value or $1/V$, and we have:

$$
F_{F'} - F_{C'} = \frac{F}{V} \tag{14.9}
$$

For an achromatic combination it will be necessary that the difference in power for the two wavelengths in one lens will be neutralized by that in the other. The power of two thin lenses *in contact*, where F_1 and F_2 are the values of F_e for each lens, is

$$
F = F_1 + F_2 \tag{14.10}
$$

the sum of the separate powers. The achromatic condition for the doublet is that the sum of the differences in power shall be zero, or

$$
\frac{F_1}{V_1} + \frac{F_2}{V_2} = 0 \tag{14.11}
$$

where V_1 and V_2 are the constringences of the two materials.

This equation makes it obvious that different materials having different V values must be used, otherwise $F_1 + F_2$ will also equal zero.

Combining Equations 14.10 and 14.11 gives

$$
F_1 = \left(\frac{V_1}{V_1 - V_2}\right) F \qquad F_2 = \left(\frac{V_2}{V_1 - V_2}\right) F \tag{14.12}
$$

where F is the mean power of the combination.

These equations may be used to find the values of F_1 and F_2, which are demanded when an achromatic doublet of power F is to be made from materials having V values V_1 and V_2. They do not define the curvatures of the surfaces.

Worked example   An achromatic lens of +3.0 D power is to be made from Hard Crown and Dense Flint glass with the following values:

	n_e	V_e
Hard Crown	1.526 11	58.95
Dense Flint	1.624 08	36.11

From Equation 14.9, for the Crown lens,

$$F_{F'} - F_{C'} = \frac{F_1}{Ve_1} = \frac{F_1}{58.95}$$

for the Flint lens,

$$F_{F'} - F_{C'} = \frac{F_2}{Ve_2} = \frac{F_2}{36.11}$$

For the two lenses to achromatize when in contact,

$$\frac{F_1}{58.95} = \frac{-F_2}{36.11}$$

or

$$F_2 = \frac{-36.11}{58.95} F_1$$

Substituting in $F_1 + F_2 = F$,

$$F_1 - \frac{36.11}{58.95} F_1 = 3.0 = \frac{22.82}{58.95} F_1$$

Therefore

$$F_1 = 7.75 \, D$$
$$F_2 = -4.75 \, D$$

The application of Equations 14.12 is seen to provide the powers of the two lenses which, when combined, give achromatism for the two wavelengths specified. They do not specify the curvatures and so these can be chosen to reduce other aberrations, particularly spherical aberration and coma. This will be discussed in Section 14.3.2. An indication of the range of choice is given in Figure 14.4.

Whatever values of curvatures are chosen, the lenses will have appreciable centre thicknesses and will not be 'thin' lenses as assumed above. This means that it is not possible to get both the principal foci and the principal points to coincide for the two colours with the same lens combination. In the case of a telescope objective or other imaging lens it is more important that foci for the different colours should coincide, that is, the back focusing distances should be equal. This is done in the next section.

In the case of eyepieces, however, it is necessary that the images in the different colours should be the same size. This means that the equivalent focal lengths of the combination for these two colours should be equal. If this is not so, images seen through the eyepiece will be fringed with colour, the defect increasing towards the edge of the field. This is **chromatic difference of magnification** or **lateral chromatic aberration**. More commonly this is called **lateral colour**.

The Huygens eyepiece (Section 6.6) is an example of a system in which this defect is corrected even by the use of two lenses of the same optical material, the blue and red rays from any point on the image emerging parallel, and the images in these two colours therefore subtending the same angle.

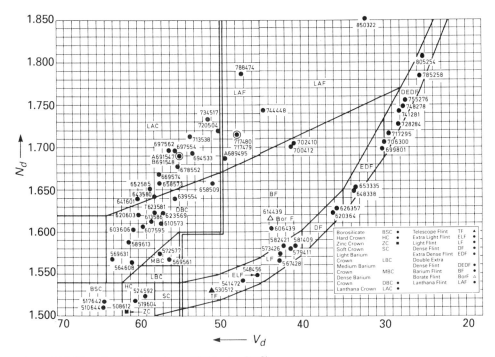

Figure 14.2 Glass diagram (Courtesy of Pilkington PLC)

For two thin separated lenses in air,

$$F = F_1 + F_2 - dF_1F_2 \tag{14.13}$$

putting

$$F_F - F_C = \delta F$$

$$F + \delta F = (F_1 + \delta F_1) + (F_2 + \delta F_2) - d(F_1 + \delta F_1)(F_2 + \delta F_2) \tag{14.14}$$

where d is the distance separating the lenses.

Subtracting Equation 14.13, from Equation 14.14 and neglecting the term involving the product of the small quantities δF_1 and δF_2, we have

$$\delta F = \delta F_1 + \delta F_2 - (F_2\delta F_1 + F_1\delta F_2)d$$

In order that $\delta F = 0$,

$$d = \frac{\delta F_1 + \delta F_2}{F_2\delta F_1 + F_1\delta F_2}$$

now

$$\delta F_1 = \frac{F_1}{V_1} \qquad \delta F_2 = \frac{F_2}{V_2}$$

$$d = \frac{V_2F_1 + V_1F_2}{(V_1 + V_2)F_1F_2} = \frac{V_1f'_1 + V_2f'_2}{V_1 + V_2}$$

If the two lenses are made of the same glass, $V_1 = V_2$. In order that there shall be no difference between F_C and F_F we have

$$d = \frac{f_1' + f_2'}{2}$$

as the condition for achromatism of magnification.

Again, there is no restriction on the curvatures of the surfaces and these can be chosen to correct other aberrations.

14.3.2 Third-order design

As was suggested in Sections 7.3 and 7.4.2, a pair of lenses can have the curvatures of their surfaces adjusted without changing their power and so allow the overall spherical aberration and coma of the combination to be neutralized without losing the achromatic condition described in the previous section. This is particularly the case with spherical aberration because the F^3 dependence of this defect means that

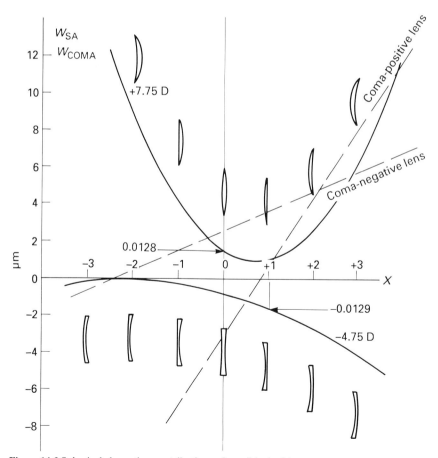

Figure 14.3 Spherical aberration contributions of possible doublet components

the negative lens will always have spherical aberration of the opposite sign to that of the positive lens.

In Section 14.3.1, an achromatic doublet of +3.0 D overall power was designed having components of +7.75 D and −4.75 D of materials with refractive indices 1.526 and 1.624 respectively. The effect of bending (that is, changing the surface curvatures without changing the power) of these component lenses can be calculated using Equation 7.21, and these have been plotted on Figure 14.3 assuming that light for a distant object is incident first on the positive component. Thus the conjugates factor (see Section 7.3), Y, is 1 for the positive lens and +2.4 for the negative lens. Equation 7.19 calculates the contribution of each lens to the spherical aberration of the wavefront. For thin lenses in contact the ray height, y, is effectively the same and could be left out of the calculation. However, for realism, a value of 10 mm (0.01 m) has been assumed and the actual shift in the wavefront is plotted.

In Figure 14.4 the influence of the F^3 term on the magnitude of the two curves is clearly seen. The doublet design will have to have a positive component somewhere near its bending for minimum spherical aberration. One of the possible combinations illustrated in Figure 14.4 is (c) where the positive component is equi-convex and the negative component almost plano-concave. Therefore the relevant X values are 0 for the positive and slightly more than 1 for the negative. As can be seen from the values indicated in the figure, third-order spherical aberration is virtually eliminated for a distant object. Clearly, other combinations can be chosen to do this. The actual choice is usually governed by extra conditions such as the need to cement the components together or the correction of fifth-order spherical aberration or third-order coma.

The value of the coma wavefront aberrations for the possible components is plotted as the dashed lines in Figure 14.3, calculated for an object 5° off-axis. It can be seen that this is not well corrected by the chosen combination and so, depending on the image quality required, the field of view of a doublet similar to that given in Figure 14.4(c) is likely to be 2° at most. For wider fields of view a combination using a meniscus negative component leading as in Figure 14.4(e) will do better. The coma of the plano-convex positive lens is small and positive, while that of the meniscus negative lens is small and negative. However, the values of the graphs of Figure 14.3 can no longer be used directly as the conjugates factors for the components are now different.

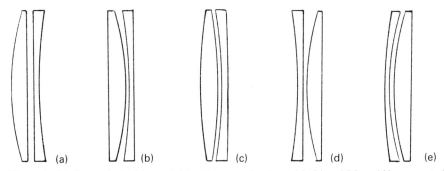

Figure 14.4 Achromatic doublets satisfying the example values. (a), (c) and (e) could be cemented doublets

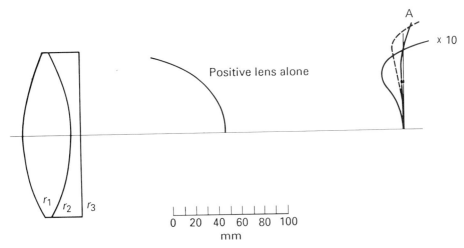

Figure 14.5 Spherical aberration correction in $F/2.5$ doublet lens (to scale). $r_1 = -r_2 = 136\,\text{mm}$; $n_1 = 1.526$, $d_1 = 44\,\text{mm}$; $r_3 = +4160\,\text{mm}$; $n_2 = 1.624$, $d_2 = 8\,\text{mm}$; EFL = 333 mm; clear aperture 140 mm; $F/2.5$

In the scale drawing of Figure 14.5, which shows the actual longitudinal aberration, curve A, calculated by ray tracing, the dramatic improvement in quality between the positive lens alone and the doublet can be clearly seen. The thicknesses of the components have been set at 44 mm and 8 mm respectively. The fifth-order contribution gives a residual longitudinal error of about 2 mm within an $f/2.5$ cone. Curve X shows this aberration exaggerated by $\times 10$ and emphasizes the rapid deterioration in aberration if the aperture is pushed too far.

The choice of a slightly different image plane can improve the image but care must be taken in interpreting graphs of this nature because the outer zones of the lens carry more energy per unit change in y and diffraction effects may become large with respect to the residual geometric aberration. When new conjugates are used with the same lens construction, the spherical aberration correction will be less good, although still generally better than with a singlet lens. If the lens in Figure 14.5 is reversed, however, the longitudinal aberration for $y = 40\,\text{mm}$ increases from 1.6 to 28 mm while that of a single lens bent for minimum aberration would be only about 7 mm for the same power and y value. Most achromatic doublets are designed for use with their more convex surface facing the incident light from a distant object.

Coma is not normally given directly for a lens. It is common practice to represent the residual coma by plotting, on the same graph as the spherical aberration, a graph of the variation of f'_m with y (see Section 7.4.2). It can be shown that, in the presence of spherical aberration, the amount of coma varies with the difference between the two curves. In Figure 14.5, the broken line shows the variation of f'_m with y value. This also indicates, by the difference between the broken and full lines, that coma is not well corrected. If the lens is stopped down the coma blur varies only with the square of the aperture, while the spherical blur reduces as the cube of the aperture. The importance of good coma correction even for small fields is thus demonstrated.

14.3.3 Practical approaches

From Sections 14.3.1 and 14.3.2 it can be seen that even when chromatic aberration and spherical aberration are to be corrected the doublet arrangement having a choice of two glass types, four surface curvatures (three if cemented) and two thicknesses has sufficient choices to allow a number of different designs. The optical designer refers to these choices as **degrees of freedom**. If a better quality of correction is needed or more aberration types must be corrected, more degrees of freedom will be required – usually by putting in more elements as extra singlets or doublets, with their attendant choices of materials, curvatures, thicknesses and spacings. This is particularly the case when a large numerical aperture and/or a large field of view is required. More degrees of freedom will also be needed if the lens must perform adequately over a range of conjugates, and even more if a zoom requirement must be met. The level of lens performance must be assessed against weight and cost of the lens as well as the problems of holding all the components in their correct location – sometimes accurate to a few micrometres. These factors make up the science of optical design, now a well developed computer-based technology. This section and those following can do no more than give a brief introduction to this technology.

A good starting point is the achromatic doublet design of Sections 14.3.1 and 14.3.2. This lens has been designed to give an image of a distant axial object which is sharp, because the chromatic and spherical aberrations have been corrected, and bright, because the numerical aperture of the lens is a reasonable $F/2.5$. No real increase in this aperture can be allowed because the spherical aberration correction rapidly deteriorates as shown in Figure 14.5. No great increase in field can be envisaged at this aperture because the coma is not well corrected and the astigmatism and curvature of field not deliberately corrected at all. If the aperture is reduced by a stop, perhaps a short distance in front of the lens, the usable field would increase – but only at the expense of brightness in the image.

Extra lens elements can, however, help in both cases. To increase the numerical aperture it is generally allowable that an aplanatic singlet lens (see Section 7.3) placed in the image space will shorten the focal length without reducing the aperture or materially affecting the spherical aberration, although some chromatic aberration will occur. Figure 14.6 shows the general arrangement. This does not, of course, increase the actual physical aperture, but the whole design can now be *scaled* so that all the radii of curvature, thickness, spacings and diameters are increased by the same factor to give the focal length of the original doublet or some other chosen value.

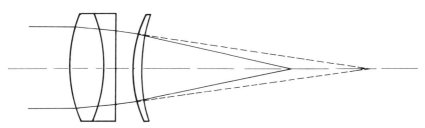

Figure 14.6 Extra meniscus lens used in the aplanatic condition to shorten the focal length of a doublet. Doublet as in Figure 14.5. Meniscus singlet: $r_1 = 108\,\text{mm}$, $n = 1.526$, $r_2 = 168\,\text{mm}$, $d = 12\,\text{mm}$, $f = 570\,\text{mm}$. System EFL = 225 mm; clear aperture 140 mm; $F/1.6$

Various manufacturers now offer cemented doublets of a well corrected design, scaled over a wide range of focal lengths. For example, a lens of 10 mm EFL and 6 mm diameter would scale to give a 250 mm EFL lens of 150 mm diameter. This latter lens would, however, be heavy and generally of larger aperture than commonly required. Most of the ranges therefore re-balance the design for the longer focal lengths to give better correction over smaller apertures; 250 mm EFL with 50 mm aperture, for example. It is with these longer focal lengths that the addition of aplanatic meniscus lenses can be most useful and, again, manufacturers offer a range of such add-on lenses.

If the original doublet is reasonably corrected for coma, it is possible to increase the usable field of view by introducing a negative plano-concave lens close to the image. Curvature of field is the aberration, as described in Section 7.4, which most rapidly affects the off-axis image after coma. The basic achromatic doublet has no deliberate correction of field curvature and Equations 7.27 and 7.28 show that lenses made from higher index materials have reduced field curvature so that the field curvature of the negative Flint lens does not entirely compensate for the extra field curvature introduced by the Crown lens being more powerful than it would need to be as a singlet. Field curvature with achromats is therefore a problem and this can be partially overcome by using a plano-concave lens as shown in Figure 14.7. As a simplistic analysis, the greater thickness of glass nearer the periphery means that the cone of rays may be directed towards the curved 'apparent' position of the image plane and be reimaged by this so-called **field-flattening lens**. Because the cones of rays here are so small, the amount of spherical aberration and coma introduced by the field flattener is very small indeed. With correct design, some correction of oblique astigmatism can also be achieved. Once again, lens manufacturers offer a range of field flattening lenses designed to work with achromatic doublets of a given focal length. As shown in Figure 14.7, the field flattener is usually placed a short distance from the image plane so that this rear surface is not at the same focus.

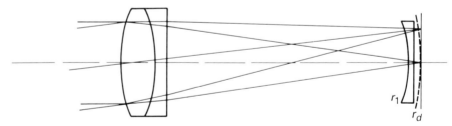

Figure 14.7 Plano-concave field flattening lens used with an achromatic doublet. Radius r_1 depends on astigmatism and field curvature of doublet. The usable field is exaggerated in the figure. Field flattener: $r_1 = -30$ to -120 mm, $n = 1.526$, $r_2 = $ plano, $d = 6.0$ mm, $r_{doublet\ image} = -200$ to 500 mm

14.4 Lens design

In the previous sections and in Chapter 7, some indication of the science and art of lens design has been given. The interrelationship between the aberrations and with the stop location is an inexact analytical tool, which the designer uses to assess the potential of a lens system, but ray tracing and computer-aided auto-design are used to finalize the design and accurately predict its performance.

These auto-design methods are entirely outside the scope of this book, but some indication of the aberration balancing act can be given. The relative importance of each aberration depends very much on the purpose of the lens being designed – particularly the field of view required. In the case of a very narrow field of view system for use as a collimator or telescope objective, the aberrations of spherical aberration and coma together with chromatic aberration are the most important. In theory the work represented by Figure 14.3 would be carried out for different wavelengths of light using different glass types in combination. This would be followed with optimization by ray tracing to balance the high-order aberrations. In practice this would all be done on the computer using design programs, which can intelligently select suitable glass types. These days a larger and larger proportion of the lens designer's work is to achieve an adequate optical performance using the least expensive glass types and the least exact mounting requirements. This last process is known as **tolerancing** the lens design.

Most telescope and binocular objectives are relatively straightforward achromatic doublets, but for a photographic lens a larger field of view is usually required. The earliest solution to this problem was the so-called **landscape lens** developed by Wollaston in 1812. (He also investigated spectacle lenses with a similar result.) By choosing a lens bending and stop location that gives low coma and negative astigmatism, it is possible to obtain moderate performance over fairly wide fields. The negative astigmatism counteracts the effect of field curvature but nothing can be done about spherical or chromatic aberration without resorting to a doublet design. Figure 14.8 shows the shape and performance of a representative singlet design. Note the considerable amount of distortion, which also varies with

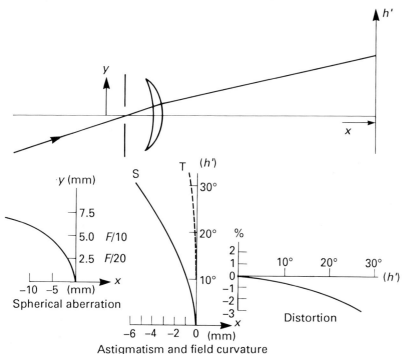

Figure 14.8 Landscape lens (EFL 100 mm)

wavelength. Although a nearly flat tangential field is obtained, considerable curvature remains with the sagittal image. The alarming amount of spherical aberration severely limits the *F*-number, as can be seen from the graph. Doublet designs known as the **Chevalier lens** were developed, but have insufficient degrees of freedom to correct all the aberrations. In recent years new glass types have become available, so that a Crown (that is, high *V* value) glass of higher index than the Flint component can give a Chevalier design with better astigmatism correction.

A landscape lens with the same balance of aberrations can be made with the stop behind the lens, but in this case the distortion is pincushion and the residual coma and lateral chromatic aberration are of the opposite sign. The **symmetrical principle** arises from this and, when two similar single lenses are arranged each side of the stop, good correction of all aberrations is possible except spherical and (longitudinal) chromatic aberration. These can be corrected with ordinary doublet lenses and the Protar lens, designed by Rudolf for the German company Zeiss, in 1890, is an example of this. Figure 14.9 shows the general symmetry, modified to some extent because of the unequal conjugates and different glass types. This was the first lens to be called an anastigmat.

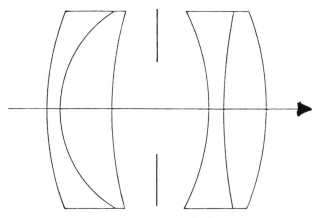

Figure 14.9 Zeiss 'Protar' lens – an example of a symmetrical photographic lens

An alternative solution was designed by H. Denis Taylor for the British company, Cooke and Sons, in 1893. Known as the **Cooke triplet**, it has the very minimum number of degrees of freedom to achieve correction of all the third-order aberrations. Figure 14.10 shows its general form. Although it is good science to match the degrees of freedom with the number of aberrations to be corrected, it is not necessarily good technology. The Cooke triplet is a difficult lens to design and manufacture because each surface seems to affect all the aberrations.

In the Protar lens, five of its six surfaces are concave to the stop, indicating the correction of field aberrations. Symmetrical lenses of five or six elements can cover fields of 60–90°, provided the aperture is not greatly faster than *F*/4. On the other hand, derivatives of the triplet are not much used above 60°. At lower fields (30°), the basic three-element design is sometimes used at apertures as fast as *F*/2.5, particularly for projection lenses. It seems that for photographic quality over a reasonable field, four lens elements are the minimum. The Tessar lens was a very popular four-element design for the first half of the twentieth century. Figure 14.11

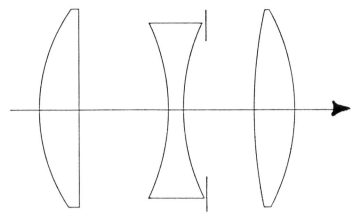

Figure 14.10 Cooke triplet – an example of an unsymmetrical photographic lens

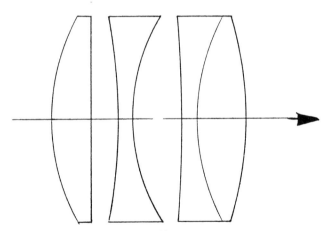

Figure 14.11 A Tessar photographic lens

shows how it can be considered to be a symmetrical lens with a separated front element or a triplet lens with a 'doubletized' rear element.

The symmetrical principle and the triplet solution were the starting points of two families of lenses developed largely during the years of this century. Generally, these later lenses give improved performance at the cost of greater complexity in the form of more elements. *Very approximately*, the *F*-number achievable by a camera lens of *N* elements is given by

$$F = \frac{\theta}{3N}$$

where θ is the total field coverage in degrees.

Whereas photographic lenses and telescope/binocular objectives commonly operate at infinity or near-infinity conjugates, microscope objectives must work at finite conjugates. They need to have very large numerical apertures but narrow fields of view. This is clearly an achromatic doublet area. However, a single

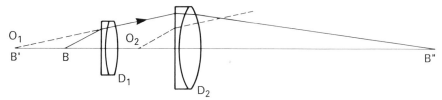

Figure 14.12 Lister doublet microscope objective

achromat will not provide a large aperture and Lister in 1830 discovered that an achromatic doublet designed for one pair of conjugates was also well corrected for another pair of conjugates, one pair having a virtual focus. Thus the same lens design could be arranged so that the object distance and the space between them ensured that both were operating at corrected conjugates. Figure 14.12 shows the general concept. The second doublet D_2 gives well corrected images for O_1 and O_2. The same design can therefore be used for D_1, which provides a virtual object for D_2 from a real object B at the same distance as O_2 from D_2.

Unfortunately, most modern off-the-shelf achromatic doublets are designed for infinity conjugates, which tend not to have any other corrected conjugates. For greater numerical apertures or short focal lengths, microscope objectives tend to retain the Lister pair of doublets and add aplanatic singlets as described in Section 14.3.3. In immersion objectives the first singlet has a flat surface, which is optically connected to the object by an oil of matching index. Figure 6.17 shows this progression of microscope objectives.

Basic eyepiece designs are described in Section 6.6. The eye ring constitutes the stop of the system and, as this is where the eye is located, it is not possible to locate lens elements on both sides of the stop as with symmetrical systems used in photographic lenses. Distortion is therefore very difficult to correct. Furthermore, the chromatic difference of distortion (and astigmatism) becomes significant. The condition for zero lateral chromatic aberration developed in Section 14.3.1 is often not fully met so that this residual lateral colour partially compensates for the chromatic difference of distortion.

Although the stop of the eyepiece is usually the image it forms of the objective in a telescope system, the rays are precisely limited by it. This is useful to the designer as it ensures that badly aberrated edge rays do not reach the eye. When an eyepiece is used to examine a real object rather than an image formed by an objective it is customary to call it a magnifier. With magnifiers, the eye may adopt any position, limited only by the diameter of the lenses of the magnifier.

This makes the design of magnifiers closer to that of the landscape photographic lens described above and used in reverse. In recent years **biocular magnifiers** have been developed with useful magnifications of ×5 and above (see Section 6.3). These are lens systems of very low F-number (large numerical aperture) so that an equivalent focal length of about 40 mm is obtained with an overall diameter of 80 mm, which permits an observer to view the magnified image with both eyes. Although the magnifier may be $F/0.5$ the small pupils of the eyes make the *used* apertures about $F/6.0$. The fact that both these small apertures are situated remotely from the lens axis changes the third-order aberrations so that, for instance, spherical aberration in the lens shows as astigmatism across the eye pupils, and coma generates an anamorphic distortion.

14.5 Image quality calculation – points and patterns

The sections in Chapter 7 dealing with the calculation of aberrations do so from the stand point of geometrical optics without considering diffraction effects. This is a valid procedure for lower quality lenses, but the effects of diffraction are of increasing importance as the aberrations are reduced. In particular the geometrical ray tracing of Section 7.6 could show the image quality continuing to improve with slight changes to curvatures, spacings, etc., when in fact diffraction effects had become the limiting factor. In the medium-quality region, diffraction effects associated with an image normally show up as extra ripples in the image predicted by geometrical ray tracing.

In the high-quality case, the lens becomes **diffraction limited** or near-diffraction limited so that the image has an intensity cross-section close to that shown in Figure 13.10. It is found that, for the nearly diffraction-limited case, the intensity of the central maximum is reduced because some of its energy is transferred to the surrounding rings, thus enlarging the image size. The ratio between the reduced central intensity and that for the zero-aberration diffraction-limited case is called the **Strehl intensity ratio** and has proved to be a useful yardstick for high-quality lens design. The profile of this non-point image (of a point object) caused by a mixture of aberrations and diffraction effects is called the **point spread function (PSF)**, which is further discussed later in this section.

The concept of Fourier transforms was introduced in a simplified way in Section 9.3 and developed a little further in Section 13.11, where it was related to regular variations in space as well as in time. In Section 10.12 the idea of contrast modulation describing a sinusoidal variation in intensity was introduced, and then in Section 13.9 a transmission diffraction grating was used as an object so that the idea of modulation *transfer* from object to image could be developed. These concepts are now brought together to provide a measure of image quality that can be used both during the computer-aided design of lenses and also for quality testing of the lenses after they have been made (Section 14.6).

A single point moved across a photographic transparency, for example, will trace out a profile of changes in light intensity due to the light and dark regions of the picture. Whatever shape this profile is, it can be represented by a host of different spatial frequencies having different modulations – its Fourier transforms as described in Section 9.3. If now this same transparency is placed in a projector or enlarger, we can find out how well the projection lens reproduces on the screen the original profile of the line across the transparency. We would expect any very sharp jumps to be blurred to a less steep profile, any very black areas to be not so black because of scattered light in the lens, and any very bright areas to be not so bright because of light loss, etc., as shown in Figure 14.13. Any sharp bright points will reflect the point spread profile of the lens as shown to the right of the figure. Clearly the image will usually be some different size than the object. Figure 14.13 ignores this, but reference will need to be made to size or spatial frequency and it must be stated whether this is in the object or in the image, or sometimes in angular terms with respect to the nodal points of the lens.

From the above, each profile will be a very special case and we can generalize the idea by using a range of sine wave gratings, each one of which represents a single spatial frequency. Due to the effects of aberrations and diffraction we expect the higher spatial frequencies to suffer more when the lens transfers them from object to image.

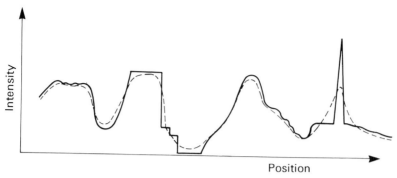

Figure 14.13 Actual and projected intensity profiles: —— object, – – – image

A real object, such as an outdoor scene or a microscopic sample, usually contains a very wide range of spatial frequencies. An image, on the other hand, is restricted to those transmitted by the lens. In the case of a photographic image it is also limited by the ability of the photographic material to respond to the image variations, and in the case of a television system by the camera tube and electronics, transmission medium, and the receiver electronics and display. The value of assessing a transfer value for a specific spatial frequency for each link in the chain is that for most systems the overall transfer value is found by multiplying together all those of the component stages. This is because, for a single frequency profile, a sine curve of simple harmonic motion, when modified in a *linear* way, each point is blurred by the same amount and the effect is to reduce its amplitude and possibly change its phase, but the result is a sine wave *of the same frequency* (ignoring any magnification).

In the optical image case, because intensities can never be negative, the expression of a sine wave intensity profile is

$$I = I_0 (1 - C_m \sin 2\pi sx) \tag{14.15}$$

where I_0 is the mean intensity and C_m the modulator as defined in Section 10.12.

$$C_m = \frac{B_t - B_b}{B_t + B_b} \tag{10.11}$$

Equation 14.15 is the same as cosine squared fringes described in Section 12.3 and shown in the central region of Plate 1 because

$$1 + \sin 2\theta = 1 + \cos 2(\theta - \pi/2) = 2\cos^2(\theta - \pi/2)$$

in which θ is quite arbitrary. In Equation 14.15, s is the spatial frequency in cycles per millimetre or similar, and x is the location across the object or image in millimetres or similar. The **principle of linearity** referred to above means that I' is some function of I that involves only the first power of I, as in the straight line equation:

$$I' = mI + c$$

It does not involve any powers of I such as I^2 or I^3.

A lens operates in a linear way because each point on the object sine wave is recreated in the image as a spread function and each point in the image is made up

Figure 14.14 Sine wave object to sine wave image via spread functions: —— object profile; – – – image profile

Figure 14.15 Object-to-image profile via spread functions: —— object profile, – – – image profile

of all the spread functions that overlap at that point. Figure 14.14 shows this effect diagrammatically in terms of the single frequency sine profile and the concept of the Fourier transforms means that if this occurs for all the component sine profiles then it is possible to get from curve O of Figure 14.15 to curve I, either by applying the spread function directly to curve O as shown or by dividing curve O into its constituent single frequencies, modifying each of these by the spread function and recombining the new single frequency image components according to the Fourier transform concepts to obtain the same image profile.

While this approach may seem tortuous, the linearity principle means that the presence or absence of particular spatial frequencies in the object has no effect on the transfer of any other single frequency. Thus, Figure 14.15 is specific to the object while Figure 14.14 is a general concept specifying how a certain lens design will transfer any object as described in Section 14.6.

For each of the single frequencies the equation of the intensity profile will be

$$I' = I_0' \left[1 + C_m' \sin \left(2\pi s' x' + \epsilon \right) \right] \tag{14.16}$$

The primed terms have the same meaning for the image profile as for the object. The important point of the linearity principle is that only one spatial frequency, s', is involved, linked to s by the magnification. As indicated by Figure 14.14, the image of a single frequency sine profile is another single frequency sine profile.

It is found that I_0'/I_0 is determined by the light gathering power and transmission of the lens and C_m'/C_m by the quality of the lens as determined by its aberrations and diffraction. The phase shift ϵ is caused by the so-called odd aberrations, such as coma and distortion, which have an asymmetrical action of the image.

The **modulation transfer function (MTF)** introduced in Section 13.9 can now be defined simply as C'_m/C_m, the ratio of the output and input modulation. In imitation of electronic systems, where the approach was first developed, it is sometimes called the **spatial frequency response** of the lens or imaging system. The phase shift, ϵ, is known as the **phase transfer function (PTF)** (the phase of the object is taken as zero) and the two transfer functions are know collectively as the **optical transfer function (OTF)**. One limitation of this approach is that it can be efficiently calculated in only one dimension. The spread functions referred to above are therefore **line spread functions (LSF)** rather than point spread functions. The line spread function is the orthogonal profile of the image of an infinitely narrow line object. This is quite different from the point spread function. The infinitely narrow line object constitutes a line of point spread functions. The profile across this line is the line spread function and this is found mathematically by integrating the effect of all the adjacent point spread functions, at each position on the crossing.

When an imaging system has more than one linear component, the modulation of a given frequency in the image can be found by multiplying the object modulation at the related frequency by the MTF values of the component parts.

$$C'_m = C_m \, \text{MTF}_1 \, \text{MTF}_2 \, \text{MTF}_3 \ldots \tag{14.17}$$

This simple approach cannot be used when component parts interact coherently as with the individual elements in a lens or two lenses used in a relay system. Then the aberrations of the overall system must be determined to give a system MTF value. It can be used when the several parts of an imaging system act independently of each other, as in the case of a distant object of contrast C_m; the atmospheric degradation, MTF_1; the camera lens, MTF_2; a photographic plate exposed and developed, MTF_3; a microscope used to study it, MTF_4; and the visual system of the observer, MTF_5. Two of the above have non-linear properties: the photographic process and the eye. Quite often, however, they are assumed to be linear and normally the inaccuracy introduced is not significant.

The calculation of the MTF of a lens design before it is manufactured is a relatively complex mathematical process. If a large number of rays are plotted through the lens design from a single object point, the density variations in the image plane allow the, usually smooth, point spread function to be calculated, which can be converted to the line spread function by numerical integration and then via a numerical Fourier transform to the MTF values across a range of spatial frequencies. More directly, it is possible to integrate the optical effect of a given width across the pupil of the lens as suggested by the Abbe theory described in Section 13.9. The amount of calculation for this can be reduced by identifying a square array of points in the lens pupil and comparing points separated by an amount that increases as the spatial frequency increases. Both these approaches are used in modern computer design programs, with many proprietary shortcuts to maximize the accuracy and/or minimize the computing time.

Programs that can display the calculated MTF, the point spread function and the line spread function give maximum choice to the designer. A further calculation is that of the edge gradient function, which looks at the image of an infinitely sharp step in the object intensity. This allows comparison with lens testing methods using sharp edges. Interference testing of lenses has never been popular because of the difficulty in interpreting the fringe pattern. Another development with optical design programs is the computation of the expected interference fringe pattern. This now allows this better analytical test to be used intelligently on the finished lens.

14.6 Lens testing

The purpose of almost every lens or lens system is to form an image, whether real or virtual, at some location other than that of the object. The quality of this image is determined by the type and size of the residual aberrations of the lens system, the accuracy of its focusing and ultimately by the diffraction effects due to the image forming light being restricted to that which gets through the lens aperture. The Airy disc described in Section 13.5 is normally the best image that can be achieved. When lens systems are being considered it is common for off-axis images to receive less light through the lens due to **vignetting**. This is the action described in Section 5.8 by which the lenses and apertures of the system combine to give a reduced lozenge-shaped aperture when viewed at an angle. Reducing the aperture increases the size of the Airy disc. Thus, the basic diffraction image quality of lenses is generally worse at the edge of the image format than at the centre.

With modern lens design it is often possible to make the residual aberrations so small that the diffraction image is only slightly modified by the aberrations. This is particularly the case when a high-quality photographic lens is used at a small aperture, say $F/16$. It is customary to measure very small residual aberrations by the number of wavelengths deviation they cause: 3λ of coma, etc.

When the aberrations of the system are large considerable blurring of the image is found. This does not necessarily mean the system is 'poor' in the widest sense of the word. It may be that the need for it to be low cost, lightweight or very wide aperture means that the optimum design exhibits large residual aberrations. The specification of lenses with larger residuals can be more difficult than those that are nearly diffraction limited, as there are more ways in which they can be right!

As an alternative to examining the image itself, Foucault allowed it to form its own pinhole camera. By placing a screen *behind* the image of a point source a patch of light is found similar in shape to the aperture of the lens system. *At the image location* a sharp edge or **knife-edge** is moved across so that it progressively cuts into the beam. If this is done at the image and it is a point image, the patch of light will darken evenly and quickly. When the image is aberrated, the patch takes on typical patterns associated with the aberration type and where the edge is located. The intensity at a particular point in the patch is given by the slope of the wavefront at the other end of that ray. **Foucault knife-edge testing**, as this is called, is very sensitive and an identical system is used, with high-quality lenses or mirrors, when we wish to measure the deformation of a wavefront caused by inhomogeneities in optical materials or by changes in refractive index of air or liquid flowing past aerofoils or obstructions. In this so-called **Schlieren test**, the optical material or fluid flow is placed close to the imaging system while the edge is placed at the image of the small source.

This test recognizes that a point image is produced by a spherical wavefront while an aberrated image must have a deformed wavefront. These deformations can be measured by interference, as described in Section 12.6, when they are not larger than a few wavelengths. However, the methods of spatial frequency response described in the previous section have a much wider application and are now routinely used on medium to high quality lenses. The original test method comprised the presentation to the lens of a pattern (grating) of a given spatial frequency and unity contrast modulation. In the specified image location a narrow slit is moved across this and the intensity variations are measured using a photomultiplier tube. It is usual to plot these on a graph with spatial frequency on the *x*-axis and relative contrast on the *y*-axis; that is, the modulation at spatial frequency compared to the modulation at the very lowest frequency. This means that the graphs in Figure 14.16 always start at $y = 1$. This so-called normalization is a useful simplification but does remove from the graph information on the veiling

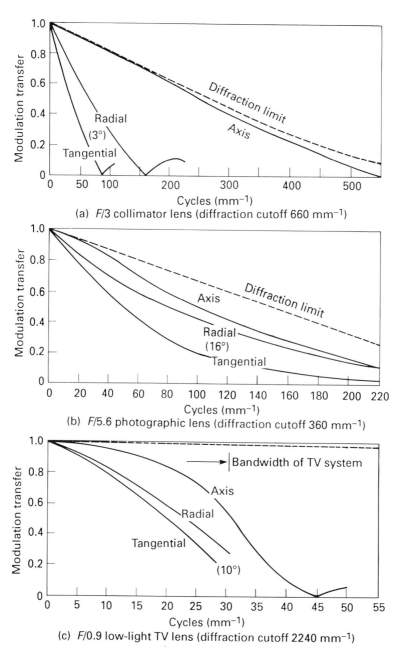

Figure 14.16 MTF curves of various lenses

glare of the lens caused by poor surface quality or dirty surfaces because this reduces the image contrast at all spatial freqencies.

For ordinary lenses under ordinary conditions the relative contrast never rises above unity. Some photographic processes can give curves above unity due to a contrast enhancing action of the developer. The eye also is less responsive at low and high spatial frequencies (Section 15.7).

The curves in Figure 14.16 show the response curves for different types of lenses. While a lens may be made diffraction limited for axial objects it is very difficult to maintain this over an appreciable field. Figure 14.16(a) shows a narrow field star tracker lens, which does not need high performance off-axis. The residual astigmatism means that the grating image has a different modulation in the meridional plane from that in the tangential plane. The frequency, s_0, at cut-off is given by

$$s_0 = \frac{2NA}{\lambda} \tag{14.18}$$

and in this case has been calculated for $F/3$ ($NA = 0.166$) and λ equal to $0.5\,\mu m$. The diffraction-limited curve may be used for other lenses by merely changing the scale on the x-axis so that s_0 satisfies Equation 14.18.

In the case of the photographic lens the performance at the centre and edges of the field are much more alike, although astigmatism still shows (Figure 14.16(b)). For a low-light TV camera lens the need is for maximum light rather than high-frequency response, as the electronic systems have a sharp cut-off. The design therefore maximizes the response at the spatial frequencies that are of use (Figure 14.16(c)).

If the lenses in Figure 14.16 were measured for other image surfaces, that is, defocused, it is possible that one or other of the off-axis curves would show an improved response, but generally they become poorer. If the photographic lens of Figure 14.16(b) were stopped down to a smaller aperture the response curves would normally show an improvement. However, the diffraction-limited curve would show a degradation, and once the actual lens performance approached the diffraction limit further aperture reduction would give a worsening response.

The methods of MTF testing originally used a periodic grating as the object and arranged for its spatial frequency to change. This takes a substantial amount of time, particularly when through focus results are needed and a number of field points tested. The advent of faster economical computers now means that the MTF of a lens or optical system can be rapidly calculated from its line spread function as described in Section 14.5. For this approach the lens test system comprises an illuminated slit which can be adjusted for width and orientation and forms the object for the lens under test. In the image plane a photodiode array is located and orientated at right angles to the slit image. This detector array has a line of small photodiodes closely spaced. A typical array comprises 256 diodes at $25\,\mu m$ intervals, thus providing an array length of $6.4\,mm$.

The output of the diode array can be captured almost instantaneously (about $20\,ms$) and provides the line spread function as shown in Figure 14.17 and expanded in Figure 14.18. An enormous practical advantage of such a system is that a real-time output of this type, displayed on the computer screen, allows the lens system under test to be accurately set up.

For chosen settings a computer program written in machine code for maximum speed calculates the MTF curve as shown in Figure 14.19. This calculation requires

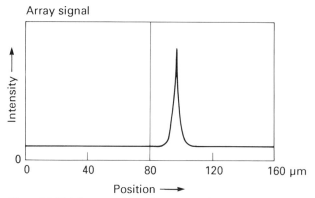

Figure 14.17 A line spread function measured using a photodiode array (From results obtained with the EROS Solid State equipment by Ealing Electro-Optics plc, Watford, UK, and reproduced with their permission)

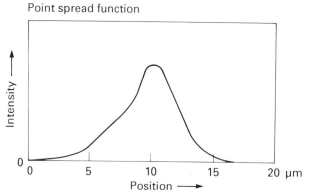

Figure 14.18 The same line spread function as in Figure 14.17, expanded to show the profile (From results obtained with the EROS Solid State equipment by Ealing Electro-Optics plc, Watford, UK, and reproduced with their permission)

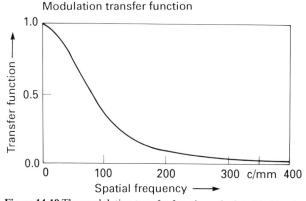

Figure 14.19 The modulation transfer function calculated by Fourier transform from the line spread function of Figure 14.18 (From results obtained with the EROS Solid State equipment by Ealing Electro-Optics plc, Watford, UK, and reproduced with their permission)

about 100 ms. The system is ultimately dependent on the fidelity of the photodiode array. However, the program does not need the LSF to be located in any specific part of the array, and test procedures can be used to show up any abnormalities. Extra lenses (diffraction limited) can be used to ensure that the LSF is large enough to cover a good number of photodiode elements so that the accuracy of the input information is high. Most MTF results are now accurate to 3–5%.

Exercises

14.1   Explain the construction of an achromatic lens showing exactly why it is necessary to use two different glasses. Why does a lens of this type not focus all colours in the same position?

14.2   Given that for a certain flint glass $(n_D - 1)/(n_F - n_C) = 36$ and $n_D = 1.6$, calculate the total curvature of a concave lens made from the glass in order that, when placed in contact with a convex lens having a difference of power for the F and C lines of 0.2167 D, the combination may be achromatic.

14.3   A positive lens is made of crown glass the refractive indices of which are as follows: red 1.5150, yellow 1.5175, blue 1.5235, the mean power of the lens being 5 D; what will be the difference in power for the red and blue? What must be the mean power of a negative lens of flint glass, the refractive indices of which are red 1.6175, yellow 1.6225, blue 1.6348, in order that when combined with the positive lens the red and blue focus at the same point? What is such a combination called?

14.4   Given glasses 1 and 2 in Table 14.3, find the curves of an achromatic lens of 25 cm focal length, the crown lens to be equi-convex and to be in exact contact with the flint.

14.5   An achromatic telescope objective is to be constructed of glasses 3 and 4 in Table 14.3. The radii of the crown lenses are to one another as 2 to −3 while the flint curve next to the crown is at the same time represented by −2.815. Determine the radii of all four surfaces on the assumption that the lens is thin and is to have a focal length of +25 cm.

14.6   Explain why two different glasses must be used in the construction of an achromatic lens. A positive achromatic lens of 30 cm focal length is to be constructed from glasses 5 and 6 in Table 14.3. The flint lens is to be cemented to the crown and its second surface is to be plane; find the radii of curvature of the surfaces.

Table 14.3

Glass	n_D	$n_F - n_C$	V
1. Crown	1.517		60.5
2. Flint	1.612		37.0
3. Crown	1.5188	0.0086	60.3
4. Flint	1.6214	0.0172	36.1
5. Crown	1.519	0.0086	
6. Flint	1.614	0.0166	

14.7   Assuming that the optical system of the eye is roughly equivalent in its action to a spherical surface of water of radius 5.1 mm and given that the refractive indices of water are for red light 1.332 and for blue-violet light 1.340, calculate the chromatic aberration as expressed by the interval between the corresponding foci. Assuming the eye pupil to be 4 mm in diameter, calculate the diameter of the least patch of confusion between these foci (where the retina may be assumed to be positioned) and comment on the result, remembering that the diameter of a foveal cone is about 0.0025 mm.

14.8   A thick lens is made of crown glass which has the constants $n_D = 1.523\,00$, $n_F = 1.529\,24$, $n_C = 1.520\,36$. The lens is of meniscus form, its radii of curvature being + 55.0 mm and + 87.16 mm, thickness 20 mm. For each of the three colours find:
(a)   The power and focal length of the lens.
(b)   The position of the principal points.
(c)   The size of image of distant object subtending 10°.

14.9   Show by diagrams how two prisms of crown and flint glass respectively may be combined to obtain:
(a)   Dispersion without deviation.
(b)   Deviation without dispersion.

14.10   Explain the principles underlying the construction of
(a)   An achromatic prism.
(b)   A direct vision prism.

14.11   A thin prism with a refracting angle of 8° is made of crown glass having refractive indices 1.527 for red, 1.530 for yellow and 1.536 for blue. A thin prism of flint glass is to be combined with the crown prism to neutralize the dispersion, the refractive indices of the flint glass being 1.630 for red, 1.635 for yellow and 1.648 for blue. What must be the angle of the flint prism and what will be the total mean deviation produced by the combination?

14.12   It is required to produce an achromatic prism combination to give a deviation of the mean ray of 1 in 70. The two component prisms are to be made of crown and flint glasses 1 and 2 in Table 14.4, respectively. Find the angles of the component prisms.

14.13   Two thin prisms are to be combined to form a thin achromatic prism combination which is required to produce a deviation of the mean ray of 1 in 100. What are the angles of the component prisms? The first component prism is made of crown glass 3 and the second of flint glass 4 in Table 14.4.

14.14   An achromatic prism is to be constructed of glasses 5 and 6 in Table 14.4. The crown prism is to have an angle of 10°. Find the angle of the flint prism and the deviation produced.

Table 14.4

Glass	n_D	n_C	n_F	$n_F - n_C$	V
1. Crown	1.5178				60.2
2. Flint	1.6190				36.2
3. Crown				0.0086	60.2
4. Flint				0.0171	36.2
5. Crown	1.530	1.527	1.536		
6. Flint	1.635	1.630	1.648		

14.15   What is meant by the term achromatism?
Prove that a system of two lenses of the same material will be approximately achromatic if the lenses are separated by a distance equal to one-half of the sum of their focal lengths. State whether the system will be converging or diverging in character.

14.16   Explain what is meant by the dispersive power and constringence of a medium.
An achromatic combiantion of overall mean power +5 D is required from crown and flint glasses of constringence values 60.5 and 37.0 and n_D values of 1.517 and 1.612 respectively. The flint lens is to be cemented by its curved surface to the crown lens with its second surface plane. Calculate the radii of curvature of all the surfaces.

14.17   List the aberrations from which lens systems may suffer, and explain how each occurs, using diagrams where applicable.

Which of the aberrations you describe are of importance in:

(a) A slide projector.

(b) A high-power microscope?

14.18 List and describe the aberrations from which optical systems may suffer. State which of these are of importance in:

(a) Telescopes.

(b) Film projectors.

14.19 An achromatic doublet has a back vertex focal length of 100 mm and a numerical aperture of 0.24. It is required to increase the numerical aperture by some 30%. Explain with a diagram how a singlet lens placed between the doublet and the image can be used to give at a shorter distance an image which is free of some aberrations. Indicate which aberration of the system is degraded.

If the surface of the singlet lens nearer to the image has a radius of curvature of 50 mm calculate the thickness of the singlet required and the radius of curvature of its front surface assuming that the new image location is 35 mm closer to the lenses than the original image position. Assume a refractive index for the singlet of 1.55.

14.20 An achromatic doublet with good coma correction comprises two components of power +33 and −25 D and index 1.50 and 1.78 respectively. Calculate the curvature of field of the image. Indicate where a negative lens may be placed to reduce the effects of field curvature and explain its action. If the lens is to be made of glass of index 1.6 with its second surface plano, calculate the radius of curvature of the first surface assuming that the best focus of the doublet lies on a surface with a curvature (due to astigmatism) of 2 × Petzval curvature.

14.21 An achromatic doublet has a total power of 12.00 D. It is known to be made from MBC 572577 and DEDF 755276. Calculate the Petzval field curvature.

Chapter 15

The eye as an optical system

15.1 Eyeball optics

In Chapter 6 a simplified description of the eye was given in terms of the refraction of light as developed in the earlier chapters. Now, using many of the topics developed in Chapter 7 onwards, it is possible to describe the optics of the eye more completely. The eye is often said to be like a camera, but this is not a good analogy. Although they both have an imaging lens, a variable aperture stop and a photosensitive surface, there are few other similarities. The poor off-axis performance of the eye is compensated by a pointing mechanism totally lacking in a photographic camera. The complex processing within the retina is entirely at variance with a simple photographic film, being approached only by the most sophisticated computer processing systems.

The treatment of the eye in this chapter is therefore not intended to be comprehensive, nor should it be as many of these topics are treated in books on visual psychophysics and physiological optics (*Handbook of Perception and Human performance,* Wiley (1986)). However, the eye is increasingly (and sometimes erroneously) evaluated in terms of instrumental and physical optics. In particular, the interaction between the eye and optical display systems has been found to need a knowledge of vision in terms that can be incorporated into the design process of the equipment. This chapter lays the basic groundwork for this but will not discuss binocular vision phenomena such as stereopsis and convergence. Furthermore, all time-varying factors have been omitted, as have colour vision phemonena.

The sections that follow give working descriptions of the eye which, when applied to real situations, will give results having a general validity provided the real situation is adequately described. Sizeable differences can occur, however, between subjects.

15.2 More schematic eyes

The Emsley schematic eye used in Section 6.2 is an example of a three-surface or **simplified schematic eye**. Many earlier examples exist due to Listing (1851), Helmholtz (1866), Tscherning (1898) and Gullstrand (1909). Details of Helmholtz, Gullstrand (no.2) and Emsley simplified schematic eyes are given in Figure 15.1. The differences between them, while not large, were significant steps in understanding the parameters of the eye, although the modifications suggested by Emsley were mainly to simplify computation.

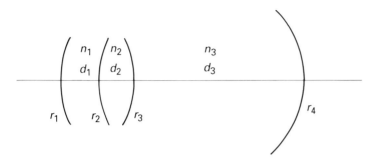

	Helmholtz (1866)	Gullstrand (1909)	Emsley (1955)
Radii of curvature (mm)			
r_1	7.83	7.80	7.80
r_2	10.00	10.00	10.00
r_3	−6.00	−6.00	−6.00
Axial distances (mm)			
d_1	3.60	3.60	3.60
d_2	3.60	3.60	3.60
d_3	15.03	16.97	16.70
Refractive indices (mm)			
n_1	1.3365	1.336	1.333
n_2	1.4371	1.413	1.416
n_3	1.3365	1.336	1.333

Figure 15.1 Simplified schematic eyes – all dimensions in millimetres

Reduced schematic eyes have also been suggested, using only one refractive surface, and these are often adequate for calculating straightforward values like retinal image sizes. The **Emsley 60 D eye** and the **Ogle 17 mm eye** are very nearly the same. Figure 15.2 gives their salient values.

There have been numerous schematic eyes using four refracting surfaces. Figure 15.3 shows the Gullstrand–Le Grand eye and the schematic eye given in the US Military Handbook 141. The cornea is represented by a lens of finite thickness having power in these cases of about 43.5 D. The variation across the population in actual corneal powers is about 38–48 D. The precision of these schematic eyes is therefore that of representing the *average* case. Comparison between these eyes and the simplified versions may be made using the procedures of Chapters 4 and 5.

The representation may be improved firstly by describing the cornea not by spherical surfaces but by elliptical surfaces. The crystalline lens is also more complex than Figure 15.3 indicates, having an increasing refractive index towards the centre in the form of a shell-like structure and also having surfaces which are aspheric. These changes affect the aberrations of the eye, which are dealt with in Section 15.5.

In terms of the first-order parameters the schematic eyes suggested differ mainly in their locations for the cardinal points. The eye given in Figure 15.3, for example, has principal points and nodal points that differ in their distances from the cornea by up to 0.3 mm compared with the Emsley eye of Section 6.2. These give rise to slightly different locations for the entrance and exit pupils of the eye. The real pupil

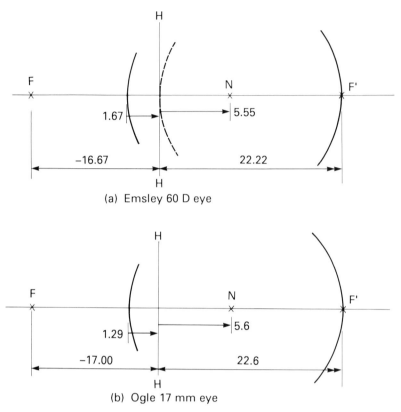

(a) Emsley 60 D eye

(b) Ogle 17 mm eye

Figure 15.2 Reduced schematic eyes – all dimensions in millimetres

is located at the anterior surface of the lens 3.6 mm from the cornea, but the power of the cornea generates an entrance pupil only 3 mm from the cornea and magnified by ×1.12. When pupil sizes are referred to in ophthalmic work it is always the entrance pupil which is meant as this is the measured pupil as seen from the outside world.

As the centre of rotation lies 13 mm behind the cornea, the entrance pupil moves as the eye fixates on different objects by a lateral distance given in millimetres by 10 tan θ, where θ is the angle of rotation. In real eyes there is some variation in this centre to cornea distance and also a wandering of the centre at different directions of gaze.

15.3 Chromatic aberration of the eye

The various media of the human eye exhibit dispersion and, in the absence of negative correcting elements, the optical system of the eye suffers from longitudinal chromatic aberration. Thus the image of a blue object is formed closer to the lens than the image of a red object. It is estimated that the average dispersion is only slightly greater than that of distilled water. For the 60 D eye of Emsley the power

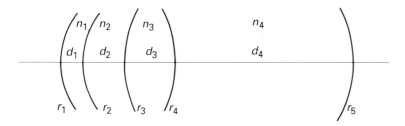

	Gullstrand–Le Grand	MIL Handbook–141
Radii of curvature (mm)		
r_1	7.8	7.98
r_2	6.5	6.22
r_3	10.2	10.20
r_4	−6.0	−6.17
r_5	−12.3	
Axial distances (mm)		
d_1	0.55	1.15
d_2	3.05	2.39
d_3	4.00	4.06
d_4	16.60	17.15
Refractive index (mm)		
n_1	1.3771	1.376
n_2	1.3374	1.336
n_3	1.420	1.420
n_4	1.336	1.337

Figure 15.3 Schematic eyes

error against wavelength can be calculated as shown in Figure 15.4, using the known refractive indices of distilled water.

In spite of nearly 2 D of chromatic aberration over the visible spectrum, the eye continues to operate without being conscious of the error. It is known that for some people a polychromatic image is needed for good accommodation. If the chromatic error is eliminated by compensating optics it is found that the ability to see low-contrast large objects is improved but there is little effect on fine detail acuity.

The extent of the chromatic aberration over the visible spectrum corresponds to an image shift of about 0.5 mm. This would be partially compensated if the receptor layers for blue light are displaced forward in the retina to be in front of those receiving red light, but this is not thought to be the case. When light is received by the eye, some of it is reflected in the regions of the retina. The sharpness of the image formed on the retina can therefore be studied. It does appear that red light penetrates deeper into the retina than blue light.

However, very careful subjective measurements of best focus for different wavelengths show a curve closely following that expected from the longitudinal chromatic aberration calculated above. The locations at which light is reflected are therefore not necessarily coincident with those at which the light is absorbed.

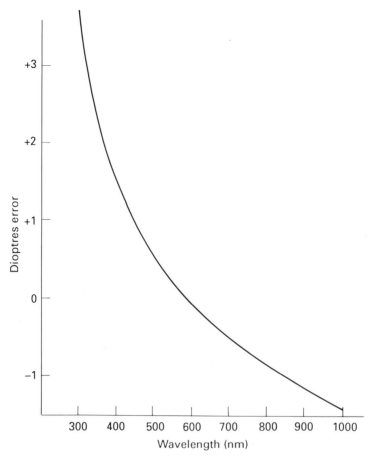

Figure 15.4 Chromatic aberration of the eye

15.4 Diffraction and the eye

Chapter 13 showed that the effects of diffraction constituted a lower limit to the size of image that an optical system could form of a point object. The eye is no exception to this. In Section 13.6, the minimum resolveable angle, w, between two images was given by

$$w = \frac{1.22\lambda}{b} \tag{13.11}$$

where λ is the wavelength of the light used (in object space) and b the diameter of the limiting circular aperture. This criterion is known as the **Rayleigh limit** and can be plotted against pupil size as shown in Figure 15.5. This graph also contains experimental values obtained by a number of workers. It can be seen that for small pupil diameters the experimental points, while following the shape of the curve, are slightly better than the theory which is, after all, based on the arbitrary figure of $\lambda/4$

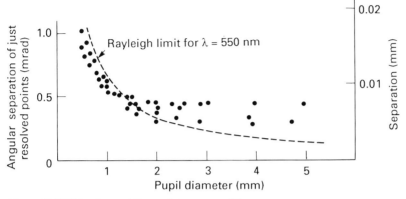

Figure 15.5 Diffraction and the resolving power of the eye

path difference. Although this pessimism has been noted for other instruments, the Rayleigh limit continues to be the criterion in general use.

For pupil diameters above 2 mm the eye is considerably worse than that predicted by diffraction alone. This is due to the aberration of the eye, which increases with increasing aperture. In general terms the eye is commonly assumed to be diffraction limited for pupil diameters of 2 mm or less. When the separation distance on the retina is calculated (using the 17 mm reduced eye) for the 2 mm pupil it is found to be commensurate with the retinal receptor spacing at the fovea.

15.5 Aberrations of the lens and cornea

The Seidel aberrations treated in Chapter 7 are applicable to the eye only in general terms as the refracting surfaces of the eye are not regular or spherical. Because the eye rotates to fixate each object of interest so that its image falls on the foveal part of the retina it might be supposed that only spherical aberration of the optical surfaces is of interest. This aberration is certainly the major interest, but the common axis of the optical surfaces does not exactly align with the fovea (Section 10.1). This means that the eye is used with its optical system off-axis by about 5–10°. Off-axis aberrations are of interest when light must be accurately focused through the optics at a large field angle for photocoagulation work on the retina (Section 9.9).

Spherical aberration of the eye may be seen by holding an opaque edge across the pupil of the eye while observing an object. When the edge is more than half-way across the eye the object will appear to move as the remaining rays are restricted to those nearer and nearer to the periphery. Care must be taken to accommodate properly for the object distance.

A more accurate method is to observe a straight line near a positive lens. By using screens one part of the line is illuminated by one small source and the other part by another small source. The distances of these sources are arranged so that they are both imaged by the lens onto the pupil of an observer's eye. Lateral movement of the sources allows one image to be central while the other is some distance off-axis. In the presence of spherical aberration the parts of the line appear displaced. An adjustment can be provided to realign them.

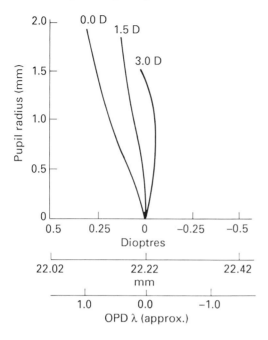

Figure 15.6 Spherical aberration of the eye across horizontal meridian (after Ivanoff, *Les Aberrations de L'Oeil* (1953))

By this method Ivanoff measured the spherical aberration at different viewing distances as shown in Figure 15.6. These are the average values over a number of subjects and defined about the 'achromatic axis of the eye', which passes through the fovea and a non-central point of the pupil such that there is no chromatic dispersion. It is seen that the aberration is undercorrect for distant viewing (periphery more powerful than centre), while it is overcorrect for near viewing. The spherical aberration of any of the schematic eyes given in this chapter may be readily calculated by using the meridional ray tracing procedures of Chapter 7. All of them overestimate the undercorrect aberration because of their spherical surfaces. The aspheric surfaces of the real eye tend to reduce spherical aberration.

However, the surfaces of the real eye are far from regular and only approximate to aspheric curves. Subjective methods of ocular aberration measurement such as those described above are very tedious to carry out and limited in their accuracy. Objective methods use light reflected from the retina. If a known mesh of apertures is placed between the eye and a point source, a defocused blur on the retina will show as a distorted pattern of spots. This method was used by Howland and Howland, who required their subjects to draw the pattern as observed. (Howland, H. C. and Howland, B. (1977) A subjective method for the measurement of monochromatic aberrations of the eye. *J. Opt. Soc. Am.*, **67**, 1508–1518.) Later, Walsh and Charman modified the method to allow a photograph to be taken of the pattern on the retina. (Walsh, G. and Charman, W. N. (1985) Measurement of the axial wavefront aberration of the human eye. *Ophthalmol. Physiol. Opt.*, **5**, 23–31.) Computer analysis of the results showed considerable irregularity of the wavefront aberration and large variations across subjects. Figure 15.7 shows the results over 6 mm pupils for ten subjects. Although for some eyes the central 2–3 mm is of good optical quality, all subjects show considerable deterioration towards the periphery. A set of pseudo-isometric representations of the same eyes

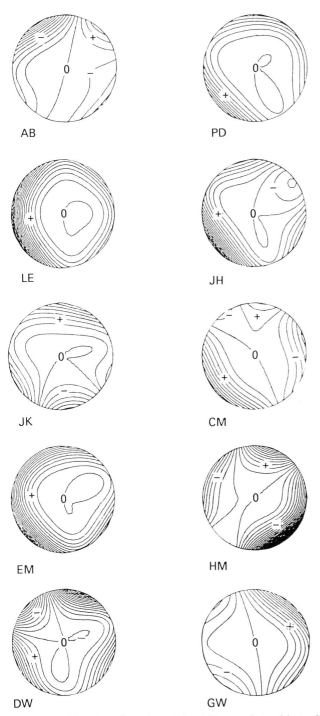

Figure 15.7 Wavefront aberrations of the right eyes of ten subjects. Contour lines at 0.25 μm intervals. Pupil size 6 mm in each case (After Walsh and Charman)

is given in Figure 15.8. The irregular and asymmetric shapes of these wavefronts cannot be corrected by spectacle lenses. Where they originate in the cornea surface some correction is obtained from rigid contacted lenses, which provide a regular surface to the main refractive power of the eye. However, a regular spherical surface also adds some spherical aberration. Some of these wavefronts have an astigmatic component and this would be corrected, at least in part, by toric spectacle lenses. Wavefront aberration data such as these allow modulation transfer functions to be calculated for these eyes as described in Section 15.6.

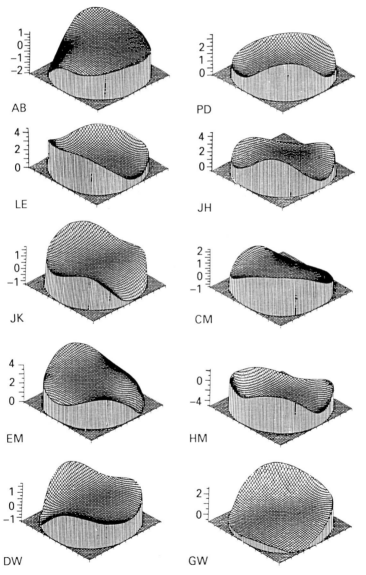

Figure 15.8 Pseudo-isometric representations of the wavefront aberrations of Figure 15.7. Scales in microns (After Walsh and Charman)

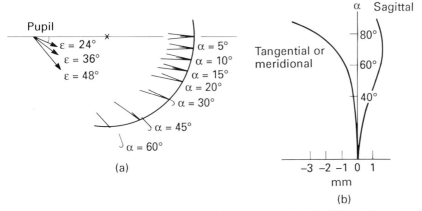

Figure 15.9 Astigmatism with schematic eyes. (a) Caustic analysis MIL-HDBK-141 eye (After Shealy and Rosenblum (1975) *Opt. Eng.*, **14**, 237–240); (b) Seidel analysis of Gullstrand–Le Grand aspherics eye (after Lotmar (1971) *JOSA*, **61**, 1522–29)

For the off-axis case, **Shealy** has calculated the caustic surfaces for the MIL-HDBK-141 schematic eye. This is shown in Figure 15.9(a). The usual shape for spherical aberration occurs on axis but as the angle is increased the two sheets separate showing the presence of considerable astigmatism. The same action can be shown by tracing rays at different field angles. Figure 15.9(b) shows the results obtained with the Gullstrand–Le Grand eye with aspheric surfaces. The x-axis shows distances along each ray with respect to the spherical retinal radius, r_5. These values are found to be less than those obtained from spherical surfaces, showing that the natural aspherics of the eye tend to reduce astigmatism. In the graphs of Figure 15.9, the angle α is the visual angle, that is the off-axis angle of the object in front of the eye. Inside the eye this angle is reduced to the angle ϵ as shown in Figure 15.9(a).

15.6 Optical performance of the eye

Even though the aberrational analysis of the eye is rendered difficult by its non-uniformity, it is still possible to obtain a measure of its optical quality using the concept of the modulation transfer function developed in Sections 14.5 and 14.6. The visual perception of modulated contrast in a scene may be considered in two parts. Firstly, the optics of the eye must form an image on the retina; secondly, the neural processes must respond to the image. This section is concerned only with the first of these – the MTF of the optics of the eye.

Methods of measurement range from imaging a fine point or line onto the retina and analysing its sharpness (or line spread function) to imaging a full sinusoidal grating on to the retina and measuring the contrast modulation of it. These so-called objective methods (which do not need a response from the subject) use light reflected from the retina and therefore use the optics of the eye twice. There is also some loss of quality in the reflection process at the retina. The double-pass effect can be allowed for by taking the square-root of the measured modulation (assuming that the retinal reflection is diffuse). The second effect is more difficult. Clearly, the reflected light has not become physiologically effective since it has not

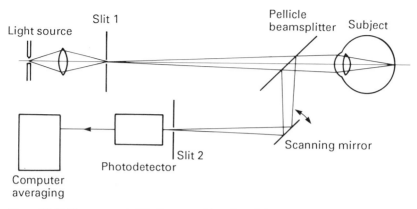

Figure 15.10 Measurement of the line spread function of the eye

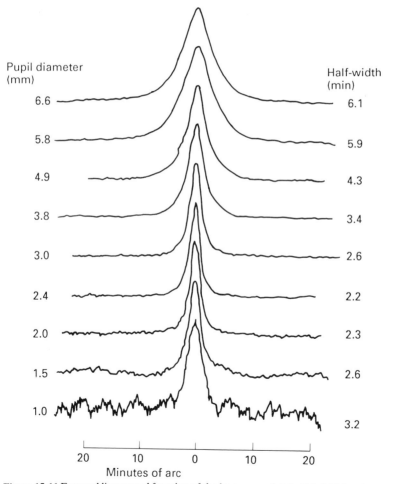

Figure 15.11 External line spread function of the human eye (white light) (After Campbell and Gubisch (1966), *J. Physiol.*, **186**, 558)

been absorbed by the receptors. However, the objective approach does allow a determination of the minimum quality of the optics of the eye. Earlier measurements were limited by the apparatus and techniques used and gave somewhat poorer values than were found subsequently.

Later determinations have used the simplest possible apparatus. Figure 15.10 shows an illuminated slit viewed by the observer via a pellicle beamsplitter. This beamsplitter directs light reflected by the retina and refocused by the optics of the eye onto a second slit via a front reflection mirror, which can scan the image across the second slit. A photodetector placed behind the second slit records the profile of the so-called external line spread function attributable to the double passage of light through the eye. The slit widths are small compared with the spread function.

Stray light must be carefully excluded from the system. A very sensitive low-noise photodetector must be used with the highest light intensity supportable on the first slit. Even so, most experimenters use a computer to average up to a thousand passes of the retinal image by the scanning mirror so that extraneous effects are reduced. The eye of the subject must be held at optimum focus during this time, as must pupil size and fixation. A beautiful series of results using white light was obtained by Campbell and Gubisch in 1966 (*J. Physiol.*, **186**, 558) and is shown in Figure 15.11, where the curves have been shifted vertically for easy comparison.

The 'half-width' of the spread (that is, the full width at half the peak value) is given for the different sizes of the artificial pupil used. Note how the noise increases as the smaller pupil returns less light. Further studies have shown that chromatic aberration and (irregular) spherical aberration contribute in roughly equal parts. Using numerical methods, these line spread functions can be converted to MTF values allowing for the double pass effect and the finite width of the slits. Figure 15.12 shows these values and, for comparison, a diffraction-limited curve for an eye of 1.5 mm pupil (for light of 550 nm). The assumption is made that the retina is a perfect diffuse reflector of the image, which is less than true at the extremes of the visible spectrum. However, these curves constitute real modulation transfer functions, being the ratio of output over input as defined in Section 14.5. Thus the modulation transfer is a dimensionless quantity.

More recently the work of Walsh and Charman described in Section 15.5 obtained wavefront aberration values from which MTF curves can be derived. Their results, obtained across ten subjects, showed considerable variation between subjects and Figure 15.13 shows the spread of MTF curves (for vertical gratings) across all these subjects with a 5 mm pupil, the dotted line giving the diffraction-limited response *at this aperture*. It is clear that for all subjects the optical performance is rather better than that derived from the line spread mesurements given in Figure 15.12. Neither of these measurements allows very high accuracies to be obtained. The wavefront approach takes no account of scattering within the eye, while the line spread function measurement may be too sensitive to this.

The MTF values derived from the wavefront aberration can be calculated for different pupil sizes and the results for one of the subjects are given in Figure 15.14. This shows a diffraction-limited response at up to 2 mm pupil size and some reduction from diffraction limit at 3 mm, although the best high-frequency response is at 3 mm. The strong loss of MTF response at wider pupil sizes is typical of the reduction occurring with aberrated systems. A further reduction occurs with defocus (Section 15.8).

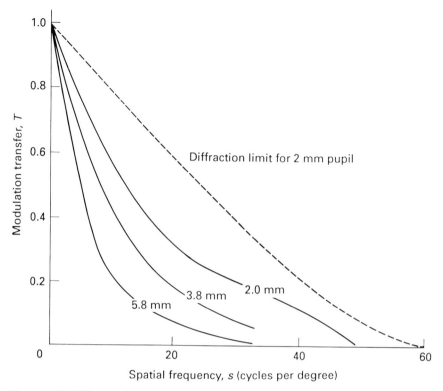

Figure 15.12 MTF values for the optics of the eye for various sizes of pupil (After Campbell and Gubisch (1966) *J. Physiol.*, **186**, 558)

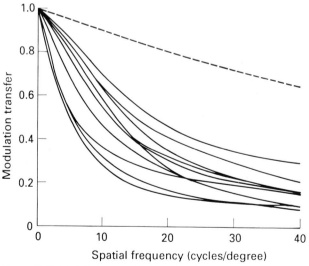

Figure 15.13 Modulation transfer functions for ten subjects, all with 5 mm diameter pupils. Vertical gratings, 590 nm. The dashed curve is that expected for a diffraction-limited eye with a 5 mm diameter pupil (after Walsh and Charman (1985) *Ophthalmol. Physiol. Opt.*, **5**, 23–31)

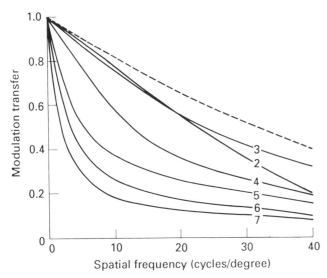

Figure 15.14 Modulation transfer functions for vertical, sinusoidal gratings and a typical subject, as deduced from the corresponding wavefront aberration data at a wavelength of 590 nm. Each curve is labelled with the corresponding pupil diameter (mm). The curve at 2 mm is essentially diffraction limited for that aperture. The dotted line gives the diffraction-limited response for a 3 mm pupil (After Walsh and Charman)

15.7 Total performance of the eye

When the response of the retinal receptors in the eye is included in an assessment of 'performance', we can no longer invoke a strictly defined modulation transfer function, as it is not possible to obtain an output value in a quantified form (although use has been made of the evoked potentials, which are measured when electrodes are attached to the scalp near the visual cortex). A further limiting feature concerns the extent to which the eye–brain system may be treated as linear. Although linearity is essential to the concepts of frequency response methods, small departures from linearity do not generate large inaccuracies and the methods may be applied, with care, to the visual response.

The description of visual performance most closely related to frequency response methods is that of the **threshold contrast**. This involves determining the minimum contrast, C_m, of a sine-wave grating of spatial frequency, s, for which the subject can detect its presence or orientation. It is found that the eye has an optimum spatial frequency at which the minimum contrast seen is lower than that at higher or lower spatial frequencies. It is also found that other parameters such as luminance, overall size of grating, viewing time, non-uniform backgrounds, etc., all affect the curve. Visual parameters such as pupil size, accommodation, peripheral or foveal viewing, monocular or binocular viewing, etc., also influence the curve. In some studies non-sinusoidal objects have been studied, leading to important advances in the understanding of the visual system. Threshold contrast curves have been obtained for almost every conceivable condition of test and subject.

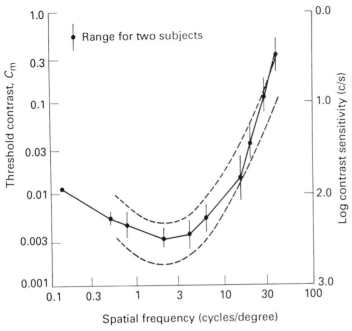

Figure 15.15 Contrast threshold under optimum conditions (After Van Meeteren and Vos (1972) *Vision Research*, **12**, 825–833). The dashed lines on these figures indicate an estimate of contrast sensitivity for the best and worst of 90% of people under optimum viewing conditions. Contrast is defined as modulation as in Section 10.12 (Reproduced from *Design Handbook for Imagery Interpretation Equipment*, Farrel and Booth, The Boeing Aerospace Company, whose permission for this use is gratefully acknowledged)

Such curves are not true MTF curves for the reasons stated above. The 'output' in this method is a 'seen' or 'not seen' response from the human subject. The curve given is therefore the boundary between combinations of contrast and frequency that are seen and combinations that are not seen. When contrast modulation is plotted on the y-axis and spatial frequency on the x-axis, using logarithmic scales on both, a J-shaped curve is found, for most conditions and subjects, similar to that shown in Figure 15.15. There is no point in normalizing this curve as it is not a ratio and cannot be used to calculate anything about points in contrast frequency space other than those on the curve. The points on the curve have no mathematical significance (as if the MTF was zero or unity at these points) other than specifying the input conditon for which the signal rises sufficiently above the noise of the eye–brain system to be detected. Sometimes the reciprocal of contrast is plotted on the y-axis. This gives the curve of greater similarity to MTF curves and to avoid confusion this practice will not be followed here (see contrast sensitivity, Section 15.9). Given a family of contrast threshold curves such as Figure 15.16 for different average scene luminances, it is permissible to interpolate between the curves to find threshold contrast values for other scene luminances, but it is notoriously difficult to extrapolate from such curves to different experimental conditions. Figure 15.16 shows that increasing scene luminance improves the performance of the eye as evidenced by the smaller threshold contrast values at each frequency.

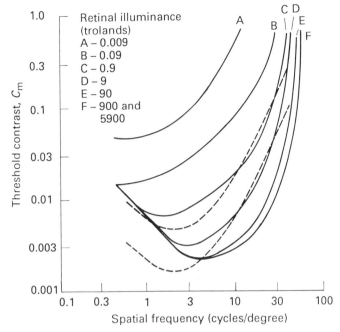

Figure 15.16 Contrast threshold variation with light level. (After Van Ness and Bouman (1966) *J. Opt. Soc. Am.*, **56**, 689–694). The dashed lines on these figures indicate an estimate of contrast sensitivity for the best and worst of 90% of people under optimum viewing conditions. Contrast is defined as modulation as in Section 10.12. (Reproduced from *Design Handbook for Imagery Interpretation Equipment*, Farrel and Booth, The Boeing Aerospace company, whose permission for this use is gratefully acknowledged)

Another family of curves may be obtained when the apparent scene luminance is kept constant but the effective pupil size adjusted. Figure 15.17 shows how a reduction in the pupil size gives a better visual performance when the retinal illumination is kept constant, as might be expected from the improving optical quality. The values given, however, show a poorer performance at lower spatial frequencies than those in Figure 15.16. Each study used a different subject and different test method, the main influence being the more restricted overall field used for the latter results. Comparison between Figures 15.12 and 15.17 shows that the general range of usable frequencies is similarly bounded for high frequencies, but the reduction in visual performance for the lower spatial frequencies is a retinal rather than an optical effect. Another more specific comparison may be made between Figure 15.14 and Figure 15.17. In the latter figure, for pupil size 2.0 mm and 3.8 mm, the contrast threshold curves run together at the higher frequencies. Although Figure 15.14 is the derived MTF curves for a single subject, the crossover of the curves for 2 and 3 mm pupils supports the contrast threshold values running together in Figure 15.17.

15.8 Variation in visual performance with focus

The ability of the eye to accommodate sets it apart from most optical instruments. However, the eye does not necessarily achieve a perfect in-focus condition at all

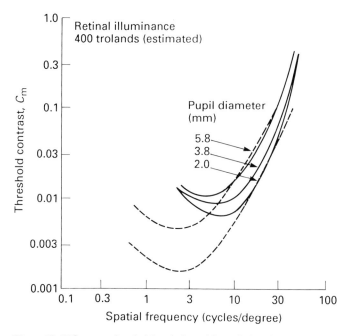

Figure 15.17 Contrast threshold variation with pupil size. (After Campbell and Green (1965) *J. Physiol.,* **181**, 576–593). The dashed lines on these figures indicate an estimate of contrast sensitivity for the best and worst of 90% of people under optimum viewing conditions. Contrast is defined as modulation as in Section 10.12. (Reproduced from *Design Handbook for Imagery Interpretation Equipment,* Farrel and Booth, The Boeing Aerospace Company, whose permission for this use is gratefully acknowledged)

times. Indeed, for the aberrated eye a single perfect in-focus position does not exist. In the first place, the considerable chromatic aberration of the eye means that an incorrect focus for one colour may be a good focus for another. It does appear that the mechanism of colour vision can mitigate the effect of poor focus by concentrating on the in-focus colour.

In the second place, the presence of spherical aberration means that the 'correct' focus is not at the paraxial focus. Indeed, it may be shown that, at intermediate pupil sizes, the best response for higher spatial frequencies may be obtained at a different focus condition than for lower spatial frequencies. The presence of irregular aberrations serves to blur things even more!

Figure 15.18 shows how various amounts of artificial defocus, introduced by lenses in front of an eye with paralysed accommodation, raise the threshold contrast values. The effect is generally symmetrical about a value of +1.5 D due to the near object and the subjects' refractive error. With a larger pupil it is possible to see an unsymmetrical effect. The curves of Figure 15.19 are for an eye with a 5 mm pupil and are derived from double-pass line spread functions on the retina. This is therefore a true MTF representation. Here again the subject shows refractive error, partly due to the use of red light. However, the peak modulation transfer value occurs at a different defocus depending on the spatial frequency considered. The shift shown is approximately commensurate with about 1 D of spherical aberration.

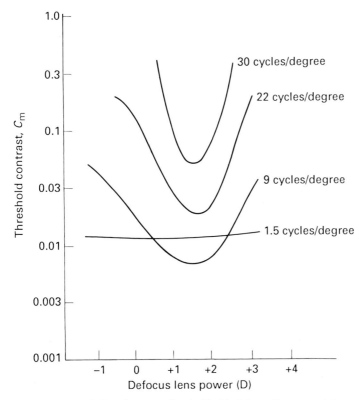

Figure 15.18 Variation of contrast threshold with defocus (2 mm pupil) (After Campbell and Green (1965) *J. Physiol.*, **181**, 576–593)

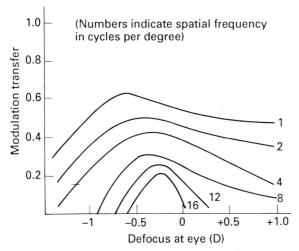

Figure 15.19 Modulation transfer of optics of eye with various amounts of defocus (5 mm pupil) (After Charman and Jennings (1976) *Brit. J. Physiol. Optics*, **31**, 119–134)

15.9 Measurement of contrast sensitivity

A graphic demonstration of the threshold effect of contrast is provided by Plate 11. This picture interacts with the eye to give an analogue of Figure 15.15. The vertical lines have a sinusoidal profile the spatial frequency of which increases from left to right in the same way as the *x*-axis of Figure 15.15. For reasons of space and photography it is not possible to provide a range much bigger than log 1.5, which means that the lines on the right are about 30 times closer than those on the left. Clearly we can make this range cover the left half of the *x*-axis by viewing the plate from a short distance or the right half by moving it 30 times further away.

The effect of the *y*-axis is achieved by making the sinusoidal profile increase in modulation, again in a logarithmic manner, with increasing value of *y*. It is difficult to be precise about the actual contrast modulation values following photographic processing and printing. However, the range is sufficient for the J-shape, as the locus of the points where each line becomes invisible, to show clearly under normal viewing conditions. As the viewing distance is varied the J-shape moves laterally without significantly changing its shape although the contrast threshold of the lower spatial frequencies is affected by the apparent size of the plate.

Although such a photograph is useful in a textbook it is not the way to obtain accurate experimental results. For this purpose a sinusoidal grating is generated on a visual display unit (VDU). A very rapid vertical scan is applied to the spot giving a vertical line, while a much slower scan moves this line horizontally to effectively fill the screen with light. This luminance level can then be modulated by applying a sine wave to the brightness grid at a frequency between the two scan rates. As this frequency is varied, vertical sine wave gratings of different spatial frequencies appear on the tube face. The apparent contrast of these gratings may be changed by adjusting the voltage amplitude of the brightness modulating sine wave. This adjustment may be controlled by the subject, who can determine when the grating is just visible. The change in this voltage with spatial frequencies will give the threshold response curve for the given viewing conditions.

A better method is to increase the modulation from zero and the subject responds when it reaches threshold. A variation in this method has been proposed for clinical assessment of contrast threshold. This approach was originally used to test our other frequency-dependent sense, hearing. In the method, proposed by Sekuler and Tynan in 1977 (*Am. J. Optometry*, **54**, 573–575), the observer sees, on pressing a button, a sinusoidal grating of high contrast and low spatial frequency, which immediately begins to reduce its contrast at the rate of 4 dB/s (about a factor of 0.4). As soon as the observer can no longer see the grating he or she releases the button. As well as recording the contrast and spatial frequency, the equipment then begins to increase the contrast and the observer presses the button as soon as the grating becomes visible again, whereupon the contrast begins to reduce. At the same time the spatial frequency of the grating is increasing logarithmically (by doubling every 50 s to a maximum of 32 cycles per degree). The values of contrast and spatial frequency are continuously recorded so that an output as in Figure 15.20 is obtained within about 6 min.

For clinical measurements and screening, 6 min is too long and naive subjects find considerable difficulty in maintaining a stable 'just seen' criterion. Various other methods have been proposed. An early approach used photographic print similar to Plate 11 but which has only one spatial frequency. The print is uncovered slowly, revealing its lowest contrast region first, and the subject indicates when the

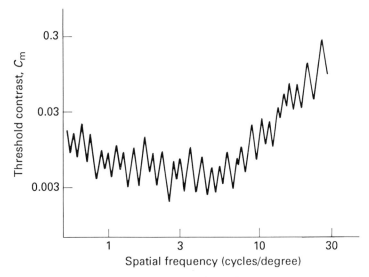

Figure 15.20 Contrast threshold curve obtained with automatic equipment (After Sekuler and Tynan)

grating is first seen. This method is also time consuming and difficult. In 1984 Ginsburg introduced the Vistech chart, in which an array of circular discs carried sine-wave gratings of increasing frequency and reducing contrast, and which were also angled slightly to the left or right of vertical. The naive subject could then be asked to state the orientation of this tilt as the criterion of grating visibility.

The contrast modulation at which this can be reliably done is the contrast threshold of the subject for that spatial frequency under the given conditions. Subjects with a visual abnormality will fail this test at higher levels of contrast modulation. Their contrast threshold is said to be *raised*. This is felt to be misleading, and clinical studies commonly plot the reciprocal of the contrast threshold and refer to it as the **contrast sensitivity**. This fits a curve as shown in Figure 15.21. As indicated in Section 15.7, this curve must not be regarded as a modulation transfer function even though it has a similar shape. It is particularly remiss in clinical studies to refer to curves such as Figure 15.21 as contrast sensitivity functions.

More recently it has been accepted that for clinical screening purposes a measurement at a single middle frequency is sufficient when used in conjunction with the measurement of visual acuity. Single frequency grating charts of reducing contrast have been introduced by Robson and others where the subject compares two areas of equal luminance; only one of these carries the grating, and the subject must indicate which. Clinical experience, however, is heavily dominated by the Snellen chart in which subjects/patients are asked to read a series of letters of reducing size but constant (100%) contrast. The decreasing letter sizes approximate to increasing spatial frequency values and the smallest letter size which can be read determines the visual acuity which for a 100% contrast value (modulation equal to unity, therefore reciprocal modulation equal to unity) gives the point VA on Figure 15.21. So-called **low contrast** Snellen charts are now available where the letters reduce in size in the same way, but have a constant low contrast in the region of

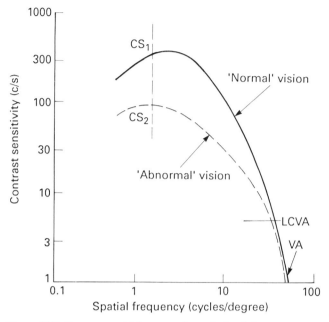

Figure 15.21 Contrast sensitivity curves for typical and abnormal observers

5–10%. These do measure a different cut-off frequency, as shown in Figure 15.21 by the line LCVA, but not one which would easily distinguish between the normal and abnormal response curves indicated.

Recently Peli and Robson have introduced a clinical chart using letters of the same size (50 mm) but reducing contrast (Plate 12). These letters have a fundamental frequency, which can be arranged to be 3 cycles per degree by viewing from a distance of 3 m (or 1 cycle per degree at 1 m). Because the letters have a square profile, they also contain higher spatial frequencies, and because they are of limited extent they also contain lower spatial frequencies. The fundamental frequency predominates and this predominance could be increased by giving them a triangular profile as suggested by Tunnacliffe. The important factor is that this frequency can be arranged to be near the peak of the visual response curve and therefore give a measurement which, as shown in Figure 15.21 as CS_1 and CS_2, is orthogonal to the cut-off frequency measured by the traditional 100% Snellen chart. If clinicians are to rely on two measurements to indicate overall visual performance these are almost certainly the best choice.

Exercises

15.1   List and describe the optical aberrations affecting vision. The fovea of the eye is 5–10° away from the axis of the lens and cornea. Which of the oblique aberrations is the most important for foveal vision? How clearly can this be seen in a normal emmetropic eye? Explain the difference between oblique astigmatism and axial astigmatism.

15.2   What is meant by the external line spread function of the eye? Why can the MTF of the visual system not be calculated from contrast threshold curves?

15.3   At pupil diameters below 2 mm the eye is diffraction limited. At pupil diameters above 2 mm the eye is aberration limited. Comment and explain.

15.4   The normal eye has a pupil diameter of 3.8 mm at a scene luminance of 80 cd/m^2. An object at 200 m approximates to a sinusoidal grating of spatial frequency 20 cycles/degree with a contrast modulation of 0.02. Will it be seen? At what distance will it just be seen assuming no change of contrast with distance. (Use Figure 15.17.)

15.5   A schematic eye formulated by Ivanoff has the following particulars:
Refractive indices:

Aqueous humour	1.3354
Crystalline lens	1.44
Vitreous humour	1.334

Radii of curvature:

Cornea (single-surface)	+8 mm
Front surface of crystalline lens	+10.2 mm
Back surface of crystalline lens	−6 mm

Axial separations:

Depth of anterior chamber	3.6 mm
Thickness of crystalline lens	4.0 mm

Determine the equivalent power of the eye and the axial length needed for emmetropia.

15.6   Discuss the aberrations of the eye, considering in particular how they differ from those of common man-made optical systems and the role of the pupil.

15.7   What are longitudinal and transverse chromatic aberration? Give details of a possible method for measuring the longitudinal chromatic aberration of the human eye.

A reduced eye consists of a single spherical refracting surface of radius of curvature 5.55 mm and has an axial length of 22.22 mm. It is emmetropic for yellow light. In red and blue light the eye is found to be 0.5 D hypermetropic and 0.5 D myopic respectively. What is the refractive index of the eye medium in red, yellow and blue light and what is its dispersive power?

15.8   Explain what is meant by threshold contrast. A sinusoidal grating is just visible to the eye. Which of the following small changes is likely to render it invisible?
(a) An increase in the spatial frequency of the grating.
(b) An increase in the contrast modulation of the grating.
(c) An increase in the illumination of the grating.
(d) An increase in the pupil size of the observer.
(e) An increase in the focus error of the observer.
(f) An increase in the viewing distance.
Comment on the 'don't know' cases.

15.9   A photo interpreter is examining an aerial reconnaissance photograph. Some very faint detail on it approximates to a sinusoidal grating of spatial frequency 60 cycles/mm. What microscope magnification will give optimum vision? Why is the reduction of veiling glare in the instrument so important?

15.10   By considering the refraction of a spherical wavefront by a spherical refracting surface of radius r, separating medii of refractive indices n and n', derive a formula relating the object and image positions for such a surface.

An eye has a cornea of radius of curvature 7.5 mm and contains a single medium of refractive index 1.33. If its length is 25 mm, how many dioptres of ametropia will the eye suffer from? What corneal radius would render the same eye emmetropic?

If the ametropia was to be corrected by a spectacle lens placed 15 mm in front of the cornea, what power of lens would be required?

15.11   The Gullstrand–Emsley schematic eye has the following specification:
Radii of curvature:

Cornea	+7.8 mm

Front surface of crystalline lens	+10.0 mm
Back surface of crystalline lens	−6.0mm

Axial separations:

Depth of anterior chamber	3.6 mm
Thickness of crystalline lens	3.6 mm

Refractive indices:

Humours	1.3333 (4/3)
Crystalline lens	1.4160

What overall axial length would render this eye myopic with its far point 200 mm in front of the corneal vertex?

15.12   Describe how the external line spread function of the human eye may be measured. How does it change as the effective pupil of the eye is reduced from 7.0 to 0.5 mm?

15.13   (a) Describe the construction and explain the principles of operation of a laser optometer.

(b) Describe the characteristics of speckle and explain what is meant by the plane of stationarity.

Answers to exercises

1.1   9° 31′; 240 ft.
1.3   Circular patch of full illumination 13 cm diameter, surrounding ring of partial illumination 8.33 cm wide.
1.4   2916 cm^2; full illumination 648 cm^2, partial illumination 4536 cm^2.
1.5   5.59, 0.349, 0.785 cm^2.
1.7   5.45 in from image.
1.8   13.5 in.
1.9   52.2 ft.
1.10 27 in diameter; penumbra.
1.17 (a) 1.205; (b) 0.8297.
1.18 (a) 1.128, (b) 1.819, (c) 0.5496, (d) 0.6060, (e) 1.10.
1.22 122° 52′.
1.25 $i_1' = 28° 45'$, $i_2' = 32° 7'$
1.28 (a) 200 mm, (b) 10 metres, (c) yes, approximately 83 mm.

2.3   100 cm × 75 cm.
2.4   12 ft.
2.5   0.55 m × 0.367 m; 1.332 m above the ground.
2.6   1° 26′.
2.9   From M$_1$, 40, 160, 240 cm. From M$_2$, 60, 140, 260 cm.
2.11 (a) $i_1' = i_2' = i_3' = 0°$; (b) $i_1' = 32° 2'$, $i_2' = 25° 53'$, $i_3' = 45°$.
2.12 $i_1' = 25° 53'$; displacement 1.82 cm; angle of emergence in oil 28° 20′.
2.13 0.29 in.
2.14 3.02 in.
2.15 (a) 41° 4′, (b) 61° 7′, (c) 74° 54′.
2.16 530 mm, 530 mm.
2.17 (a) $i_2' = 24° 28'$, $d = 18° 56'$; (b) $i_2' = 0°$, $d = 23° 8'$; (c) $i_2' = 53° 8'$, $d = 23° 8'$.
2.18 $i_1' = 24° 51'$, $i_2 = 35° 9'$, $i_2' = 61° 45'$, $d = 41° 45'$. Minimum $v - 39° 49'$.
2.20 $v = 2.19$ prism dioptres; $a = 2.5°$, $n = 1.5$.
2.21 1.577.
2.22 1.5811.

3.1   0.05 in.
3.2   1.528.
3.3   1.543.
3.4   (a) +5 D, (b) +5.26 D, (c) +1000 D, (d) −100 D.
3.5   −1⅓ D; $l' = +37.5$ cm.

515

3.6 -14.3 cm from lens; $+10.33$ D.

3.9   Virtual erect image -37.38 cm from surface, 1.64 cm high. $F = -38.46$ cm, $F' = +58.45$ cm from surface.

3.11   0.76 mm.

3.12   Centre; 3.02 cm from front surface.

3.13 -1.24 in from upper surface; $m = 1.26$.

3.14 $+22.22$ mm; -22.5 cm.

3.15 $l' = -3.05$ mm; 4.52 mm diameter.

3.16   (a) 20 cm, (b) 80 cm from curved surface.

3.17 $+53.33$ D; $+1.875$ cm from plane surface.

3.18   0.29 mm.

3.19 -3.15 D; -21.25 in; -8.85 in.

3.20 -8.53 cm from mirror.

3.21   (a) $+3.33$ cm, 1 cm; (b) $+15$ cm, 9 cm.

3.22 $l = +103.4$ mm; 0.86 mm; virtual.

3.23 -18 in.

3.24 $+5.75$ D; -2.875 D; -34.78 cm.

3.25   200 mm; 400 mm.

3.26   0.6 ft from mirror.

3.27 $m = -0.6$.

3.28   1.18 mm long; 3.92 mm behind cornea.

3.29 -53.33 cm.

3.30   12.73 cm in front of concave mirror; 0.273 cm high; inverted.

3.31 -13.85 cm.

4.1   74.2 cm, 12.37 cm.

4.2   (a) 165 mm, (b) 82.5 mm.

4.3   12.4 cm.

4.5   60 cm

4.6 -6.06 cm, virtual; -11.76 cm, real.

4.7   418.4 mm.

4.8   (a) $+25$ cm, $+4$ D; (b) $+28.57$ cm, $+3.5$ D; (c) $+33.33$ cm, $+3$ D; (d) $+50$ cm, $+2$ D; (e) ∞, 0 D; (f) -16.67 cm, -6 D.

4.9 -176.5 mm, 1.47 mm, erect.

4.10   6.98 in or 29.02 in from lamp; $m = -4$ or $-\frac{1}{4}$ (approx.).

4.11 $+15.55$ cm.

4.12   40 mm.

4.13 $+7.58$ in.

4.15   (a) $+50$ cm, 2 mm; (b) $+36.4$ cm, 0.9 mm.

4.16   Image 16 in long, one end 14 in from lens; image 13.34 in long, 16.67 in from lens.

4.17 -6.22 in from lens.

4.19   185.4 cm.

4.20   91.45 mm.

4.21   4.37 cm; 15.63 cm from the first lens.

4.23 -1.2 in; -32.8 D.

4.24   67.08 cm.

4.25   2.17 cm.

4.26   138.6 cm from lens.

4.27 $+6.4$ in.

4.29 -8 D.
4.30 $r_2 = -15$ cm; $r_1 = +7.5$ cm.
4.31 $+18.34$ cm.
4.32 $n = 1.5$; $r = 24$ cm.
4.33 -3.09 cm from front surface; -5.33 cm from front surface.

5.1 $+14.55$ D.
5.2 $+9.87$ D, $+11.12$ D, $+89.9$ mm from second surface.
5.3 $+9.64$ in from second surface; 1.33 in long.
5.4 Afocal.
5.5 -2.25 in.
5.6 $f' = +45.7$ mm; $f_V = -19.1$ mm; $f'_V = +25.7$ mm; $e = +26.6$ mm, $e' = -20$ mm.
5.7 $f' = +41.67$ cm; image $+68.73$ cm from second lens, $m = -1.25$.
5.8 $F = +6$ D; $e = -6.67$ cm; $e' = -8.33$ cm. Telephoto system.
5.9 3.513 in.
5.10 $f' = +17.2$ cm; $e = +5.2$ cm; $e' = -6.9$ cm; image $+18.2$ cm from second lens; 1.38 cm high
5.11 $F = \times 1.84$ D; $e = -34.8$ cm; $e' = -26.1$ cm; virtual image $- 47.2$ cm from second lens; 4.17 in high.
5.13 $f' = 1.429$ in, $e = +2.5$ in; $e' = -1$ in.
5.14 $f_V = -0.5$ in, $f'_V = +0.5$ in, $e = +1$ in, $e' = -1$ in.
5.15 (a) -3.28 D, (b) -3.09 D.
5.16 $+7.14$ in.
5.17 $l_2 = +178$ mm; $h' = -11.8$ mm.
5.18 $F = +4.68$ D; $+238.5$ mm from lens.
5.19 $f = -241.5$ mm; $f' = +321.3$ mm; $e = +10$ mm; $e' = -22.3$ mm; $N +49.8$ mm and $N' + 57.5$ mm from second surface.
5.20 2.18 in.
5.21 $f' = +198.8$ mm; first principal point $+13.8$ mm from first surface. Second principal point $+7.0$ mm from final surface.
5.22 $f' = +101.6$ mm; first principal point $+ 6.3$ mm from first surface. Second principal point -16.8 mm from final surface.
5.23 Entrance pupil 4.29 cm behind first lens, 2.68 cm diameter; exit pupil 2.31 cm in front of the second lens, 2.31 cm diameter; entrance port is aperture of first lens; Exit port 7.5 cm in front of 2nd lens; $m = 1.5$.
5.24 Field of full illumination 1° 52′; total field 68° 36′.
5.25 23.15 m.
5.26 11 ft 5 in to 8 ft 10.5 in.
5.27 1/7.5 s, 0.032 in.

6.1 $F/8$; $F/5.6$; $F/11$.
6.4 -77.8 mm from the lens, 18 mm.
6.6 (a) 3.75 (b) 4.0.
6.7 $\times 72$.
6.8 -16.6 mm from the objective; 75; 3.33 mm.
6.9 (a) $10\frac{2}{3}$ in and $1\frac{1}{3}$ in, (b) $13\frac{5}{7}$ in and $-1\frac{5}{7}$ in.
6.10 -36 mm, 138 mm.
6.11 $34\times$; 4.4 mm; 25.75 mm; $\pm 16.2°$.
6.12 $f' = 1.43$ in; $f_V = +1.07$ in; $f'_V = 0.428$ in; $M = 14$.

6.13 EFL is 32 mm; F' is 8 mm behind second lens; P' is 24 mm in front of second lens. System is symmetrical. Exit pupil diameter is 3.2 mm, distance from second lens is 9.7 mm.

6.15 247.5 mm.

6.17 ×11; 375 mm.

6.18 (a) Eyepiece moved inwards 7.58 mm, $M = 3.8$, (b) eyepiece moved outwards 2.57 mm, $M = 5.6$.

6.19 4.86 × 3.89 m.

6.20 0.014 mm away from the object, ×146.

6.22 51.5 mm.

7.6 $t' = +8.57$ cm; $s' = +11.57$ cm.

7.7 $l' = +14.4$ in.

7.16 $l' = +128.54$ cm.

7.17 $l' = +15.39$ cm (marginal); $+15.87$ cm (paraxial).

7.22 Initial slope $= 1.708°$, $l' = 132.819$ mm, final slope $= -6.587°$.

7.23 (a) 1.0 mm, (b) 1.2 mm.

7.24 40 cm, 2.9 cm.

7.25 (a) -70.6 mm, (b) 0; (c) -1.0; (d) -8.57 mm.

7.26 (a) 5.23 D; (b) -1.0; (c) -1.0; (d) $+8.48$ mm.

7.27 (a) $+750$ mm; (b) $+0.67$; (c) -0.56; (d) $+299$ mm; -84 mm.

8.2 (a) $+33.3$ cm from lens; 6.66 cm long, vertical. (b) Vertical line 33.3 mm long; $+166.7$ mm from lens; horizontal line 200 mm long $+1$ m from lens. Circle of least confusion 28.6 mm diameter; $+285.7$ mm from lens.

8.3 (a) 18 mm horizontal, 27 mm vertical; (b) 9 mm horizontal, 9 mm vertical; (c) 36 mm horizontal, 9 mm vertical.

8.4 $+3.0$ D cylindrical axis vertical/$+5.0$ D cylindrical, axis horizontal; toric.

8.5 52.3 mm; 87.2 mm.

8.6 0.3 in.

8.7 (a) facet location 16 mm, 20 mm; facet width 1 mm, 1 mm; facet angle 18.6°, 23.1°, facet depth 0.34 mm, 0.43 mm; (b) facet location 17.9 mm, 20 mm; facet width 0.56, 0.50; facet angle 20.7°, 23.1°; facet depth 0.21 mm, 0.21 mm.

9.2 7595 Å, 759.5 nm, 0.7595 μm, 2.99×10^{-5} in; 5894 Å, 589.4 nm, 0.5894 μm, 2.32×10^{-5} in; 4862 Å, 486.2 nm, 0.4862 μm, 1.91×10^{-5} in.

9.3 A stationary, above; B upwards, level; C downwards, below.

9.9 -8.

9.10 1.75 m/s.

9.11 (a) 2 Hz, (b) 5.0 mm, (c) 100 mm.

9.12 (a) $\pi/9.24$ rad (19.47°), (b) 120° ($2\pi/3$ rad), (c) 0.616 s.

9.13 90°.

10.1 3×10^{16} cd.

10.3 59° 48′.

10.4 4.67 lm/m^2.

10.5 1/120 lm/cm^2; no change.

10.6 3.2, 2.4, 0.267; 1:1.78:144.

10.7   17.32 cd; 80; 45.
10.8   22.5 cd.
10.10 6.37 cd/mm^2; 0.8 lm/mm^2.
10.11 110 cm and 1090 cm from 30 cd lamp.
10.12 $D = 0.699$, $T = 4\%$, $D = 1.398$.
10.13 1 m from either lamp.
10.14 (a) 50 cd, (b) 25 cd, 0.0628 lm.
10.15 3⅕ lm/mm^2.
10.16 (a) 0.0014 lm, (b) 0.0053 lm/ft^2.
10.17 11.1%; 0.95.

11.4   0.016 46; hard crown.
11.6   1° 41′.
11.8   Crown   0.0157, 0.375, 0.625, 1.125.
       Flint   0.0357, 0.296, 0.704, 1.333.
11.16 90°.
11.17 53.1°.
11.18 0.004 46k mm, where k is an odd integer.
11.20 0.03 mm.
11.21 $i_0' = 19.07°$; $i_e' = 21.8°$.

12.5   0.514 mm.
12.6   640 nm.
12.7   (a) 1.2 mm, (b) 120 mm toward film side.
12.8   658 nm.
12.9   0.0053 mm.
12.10 0.105 D.
12.11 12.25 m.
12.15 1.18 mm.
12.16 600 nm.
12.17 0.145 mm.
12.18 0.006 mm.
12.22 1′ 2″.
12.23 273 nm; (a) 537.7 nm; (b) 472 nm.
12.24 90.6 nm.
12.25 89.28 nm; 1%, 1.64%, 1.3%.

13.1   1.54 mm radius; intensity approximately zero.
13.3   1.55, 2.19, 2.68, 3.1 mm.
13.6   0.93 mm.
13.7   3.7″.
13.9   0.008 mm.
13.10 54 mm.
13.12 $w = 1.85″$; 0.56 in.
13.13 0.003 25 mm.
13.15 9.39 cm.
13.16 11.28 mm.
13.17 Radius of first dark ring 0.000 298 in. $w = 0.198″$; magnification 454.
13.20 1′ 15″.
13.21 0.0133, 0.1975, 0.6893 in.

13.24 9.6 mm.

13.26 45 mm.

13.27 $-61.6°$, $-43.62°$, $-30°$ (zero order), $-18.07°$, $-6.91°$, $3.98°$, $15.03°$, $26.69°$, $39.72°$, $55.98°$.

13.28 81, assuming d is the width of opaque strip between slits.

13.29 13 (nearly), 1.33 mm.

14.2 $R_1 - R_2 = -13.0 \, D$.

14.3 $0.0821 \, D$; $-2.96 \, D$.

14.4 $r_1 = +10.04 \, cm$, $r_2 = -10.04 \, cm$, $r_3 = -10.04 \, cm$, $r_4 = +294.1 \, cm$.

14.5 $r_1 = +8.67 \, cm$, $r_2 = -13.0 \, cm$, $r_3 = -12.2 \, cm$, $r_4 = +68.7 \, cm$.

14.6 $r_1 = +8.74 \, cm$, $r_2 = r_3 = -10.2 \, cm$, $r_4 = \infty$.

14.7 0.365 mm; 0.036 mm.

14.8

	D	F	C
F	+4.253 D	+4.312 D	+4.235 D
f'	+235.1 mm	+231.9 mm	+236.2 mm
e	−18.53 mm	−18.41 mm	−18.55 mm
e'	−29.36 mm	−29.17 mm	−29.39 mm
h'	41.46 mm	40.89 mm	41.64 mm

14.11 $4°$; $1.7°$.

14.12 Crown $3° \, 56'$, flint $1° \, 58'$.

14.13 Crown $2.77°$, flint $1.4°$.

14.14 $a = 5°$, $v = 2° \, 7.5'$.

14.16 $r_1 = 83.1 \, mm$, $r_2 = r_3 = -77.7 \, mm$.

14.19 $t = 13.6 \, mm$, $r_1 = 33.33 \, mm$.

14.20 $C = 8 \, m^{-1}$, $r_1 = 23.4 \, mm$.

14.21 $C = 8.36 \, m^{-1}$.

15.4 No. Approximately 165 m.

15.5 66.3 D, 23.09 mm.

15.7 Red   1.329 23
Yellow 1.332 93
Blue   1.336 63
Dispersive power = 0.022 22.

15.9 Between about $90\times$ and $130\times$.

15.10 +9.2 D hypermetropic.

15.11 25.85 mm.

Index

Note: numbers in **bold** refer to main entries

Abbe number, **363**
Abbe prism, **49**, 50
Abbe roof prism, **50**, 51
Aberrated images, **34**–36
Aberrations, **230**–267
  chromatic, **89**, **362**, **464**
  coma, *see* Coma
  in eye, 497–501
  fifth-order theory, 231
  lateral chromatic, 470
  oblique, 245
    see also Astigmatism
  secondary, **231**
  Seidel (primary), **231**, 258
  selected region for aberration effect, 233
  spherical, **63**, 237–244, 283–284
    aplanatic points, 239, 240
    bending factor (X), 239, 244
    caustic surfaces, 242–244
    conjugates factor, 240
    contour map, 244, 274
    correction, 64, 476
    line spread function, 242
    point spread function, 242
    possible doublet components contributions, 472
    representation, 242, 243
    single positive lens, 241, 242
    spot diagram, 242
    zero, of paraboloid mirror, 284
  stop position-distortion, 258–262
  thin lens, longitudinal, transverse, wavefront aberration relationship, 234
  third-order theory, 231
  wavefront, **231**, 232
Abney divisions of spectrum, 321
Absolute refractive index, 16–17
Absorbing optical materials, 361, 366–368
Absorption, 14–15, 366
  selective, 14, **367**
Absorption coefficient, **366**–367, **369**
Absorption filters, 367, **368**
Absorption spectra of hot gases, 317
Accommodation, **176**

Accurate conicoids, 286–289
Achromatic doublet, 466, 467–476
  first-order design, 467–472
  third-order design, 472–476
Achromatic lenses, **362**
Achromatic prism, **466**
Achromatism, 465–467
Action of cylindrical lenses, 273
Afocal systems, **126**, **194**–197
  telecentric, 195, 196
Airy disc, **437**, 485, Plate 8
Albedo, **337**
Aluminium, 14, 418
Amorphous solid glass, **375**
Amplitude, **302**, **303**
  of simple harmonic motion, 303
  of wave motion, 203
Amplitude zone plate, 450
Analyser, **376**
Angles, 65–66
  of deviation, **28**
    minimum, **43**
  of incidence, **12**, 66
  of prism, measurement, 359, 360
  of reflection, **12**, 66
  of refraction, 66
Angstrom, **314**
Anisotropic material, **375**, **376**
Anisotropic vibrations, 375
Anti-reflection coatings, 413–419
  multilayer, 416–419
  quarter wave, 414
  single layer, 413–416
Aperture stop, **147**
Apex of prism, **41**
Aphakia, **293**
Aplanatic lens, 241
Aplanatic points, **239**
Aplanatic singlet lens, 475
Apostilb, **338**
Apparent magnification, **181**–182
Apparent size, **75**
Apparent thickness of glass, **36**
Applications of lasers, *see* Lasers

Applications of polarized light, 393–394
Approximate conicoids, 282–286
Aqueous humour, **175**
Arc lamp crater colour temperature, 319
Arrowhead, meaning of, 66
Aspheric condenser lenses, 289, 293
Aspheric 'shell' concept, 285
Aspheric surfaces, 282–294
  approximate conicoids, 282–286
Astigmatic difference (Sturm's interval), **247**, 280
Astigmatic lens, 280
  image formed by, 282
Astigmatic pencil, 279, **280**
Astigmatism, 245–253
  axial, **273**
  Coddington equations, **250**–251
  coma, *see* Coma
  field, curvature of, 245–253
  oblique, 247–248
  paraboloidal mirror, 284
  Petzval condition, **246**
  Petzval surface, **245**, 253
  principal ray, **247**
  radial (sagittal) line focus, 248
  single spherical surface, 249
  Sturm's interval, **247**
  surfaces (cup and saucer diagram), 252–253
  tangential line focus, 248
  wavefront aberration, **247**–248
Astronomical telescopes, 191, 192
Asymmetrical phase plate, 451, 454
Atomic physics, 315–316
Atomic structure, 315–316
Autocollimation methods, 110–112
Axial astigmatism, **273**
Axis
  common, 65
  distances along, 65
  distances perpendicular to, 65
Axis meridian (cylinder axis), **271**

Back vertex of lens, **84**
Bandpass filter, 368, 370, **417**–418
Barium oxide, 367
Barrel distortion, 258, **259**, 260, 261, 262
Beam(s), 4–5, 7
Beam splitter, 386–389
Bending factor, 239
Bessel functions, 436
Biaxial crystals, 380
Bifocal lenses, **178**
  'flat top', 295
Binocular magnifiers, 184–185
Binoculars, 197
  eye lens, **198**
  eye relief distance, **198**
  field lens, 198–199
  magnifiers, 182, **184**, **480**
Biocular magnifiers, **184**

Birefringence, **378**
  incidences for birefringent crystals, 380
  linear, **381**
Black-body radiation, 318
Black cloth, 338
Black velvet, 338
Blazed gratings, **448**
  refractive ophthalmic prism compared, 449
Blazed zone plate, **452**–453
Blended aphakic lens, 293
'Blended' lens manufacture, 292
Bloomed lenses, **415**
Blue wavelength, 1
Brewster angle, **386**, 388, 389
Brilliance, diamond, 38
Bundled rays, **4**
  see also Pencils

Calcite crystal (Iceland spar), 378–379, 380, 382, 394
Camera, 170–174
  EFL (image distance), 172
  field of view, 172
  film format, 172
  focal plane shutter, 171
  focusing by split-field viewfinder, 219, 220
  format, 171
  inverted telephoto (retrofocus) lens, **174**
  iris diaphragm, 145, **171**
  lenses, 84, 102, 142
    telephoto, **173**
    zoom, **174**
  single lens reflex, **172**
  35mm, 172
  through the lens metering, **172**
Candela, **332**, 337, 339
Candle
  colour temperature, 319
  international, **339**
Candle-power, **332**, 339, 344–346
  mean horizontal, 346
  mean spherical, 346
Carbon dioxide laser, 326
Cardinal point, **127**, 131
Cartesian ovals, **287**
Cataracts, 293
Cathetometer, **223**
  telecentric, 222
Cauchy's formula, **364**
Caustic surfaces, **242**
Cemented doublet, **466**, 467, 476
  corrected for coma, 476
Centre of curvature, **54**
Centre thickness of lens, **84**
Centred systems, **65**, **124**
Chevalier lens, **478**
Chromatic aberration, **89**, **362**, **464**
  correction, 362
  in eye, 494–495, 496
Chromatic difference magnification, **470**

Chromatism, 465–467
CIE chromaticity, diagram, Plate 5
Circle of least confusion, **280**
Circular polarizer, **385**
Circularly polarized light, **384**
Clerk–Maxwell's colour equation, 351
Close objects, resultant intensity curves for, 441
Coddington equations, **250**–251
Coherence, 397–399
   degrees, **399**
   length, **401**
   mutual, **399**
   time, **401**
Coherent light, **321**
Collimation methods, 109–110
Collimator, 109
Colorimeters, **350**–351
   Donaldson, 350–351
   Wright, 350
Colorimetry, 349, **350**–354
Colour
   additive primaries, **353**, 354
   Clerk–Maxwell's equation, 351
   complementary, 319, **352**
   hue, **319**–320
   luminosity, **319**
   matching stimuli, 350
   mixing, 349–354
   purity, **319**, 320
   selective absorption, 353
   spectrum, 318
   subtractive primaries, **353**, 354
   temperatures, 318–319
      of moon, 319
      of sun, 319
Colour conversion filters, 348
Colour vision
   trichromatic theory, 352
Coma, 253–258, 474
   central, **258**
   correction, 474
   paraboloid mirror, 284
Comparison of photometers, 340–342
Complementary colours, 319, **352**
Complex optical systems, 164–167
Concave lenses, **86**
Concave surface, **63**
Condenser, **209**–213
   direct illumination, **211**, 213
   intermediate, 212
   for slide projectors, 209–212
   Maxwellian view, 211–213
Conic curves, 287, 288
Conicoid mirrors, 289
Conicoid surfaces, refraction of distant axial object, 290
Conicoids
   accurate, 286–289
   approximate, 282–286
Conjugate foci, **70**

Conjugate foci methods, 112
Conjugate planes, **72**
Conjugate points, **70**
Conjugates factor, **240**
Constant h, 316
Constringence, **363**, **465**
Contact lens, 136–148, 270
   diffractive bifocal, 453
   refractive power, 353
   thick lens, 136–138
Contour map, spherical surface, 244, 274
Contour projector, **223**
Contrast, 348–349
   sensitivity of eye, **511**
      measurement of, 510–512
   threshold, or eye, 348
Contrast modulation, 348
Contrast photometer, **341**
Convergence angles, 106
Convergent pencils of light, **4**, 5
Convex lenses, 86
Convex refracting surface, 69
Convex surface, **63**, **69**
Cooke triplet lens, 478, 479
Cornea, **175**
Cornu's spiral, **428**, 429–430
Corona, **460**
Corpuscular theory of light, **300**
Cosine law
   of emission (Lambert), 338
   of illumination, 334
Crimping, **270**
Critical angle, **37**
   crown glass, 38
   dense flint glass, 38
   diamond, 38
   light flint glass, 38
   polymethyl methacrylate, 38
   special flint glass, 38
   water, 38
Crossed cylinders, **274**
Crossed polarizers, **376**
Crown glass
   critical angle, 38
   hard, 365
   refractive index, 38
   spectacle, refractive index, 17
Crystalline lens, **175**
Curvature, **54**
   centre of, **54**
   radius of, **54**
   total, **278**
Cycle, **302**
Cylinder axis (axis meridian), 271
Cylindrical lenses
   action, 273
   two, at right angles, 274
   two, crossed at angle α, 276
   two, crossed at right angles, 275
   two, not at right angles, 274
   see also Cylindrical surfaces

Cylindrical surfaces, **271**–278
　lens forms, 272
　meridian, **271**
　　axis (cylinder axis), 271
　　power, 271
　　principal, 271
　see also Cylindrical lenses

Decentred lenses, **85**
Definitions, 64–66
　angle of incidence, 66
　angle of reflection, 66
　angle of refraction, 66
　distances along axis, 65
　distances perpendicular to axis, 65
　eccentricity, 258
　luminance, 337
　slope of ray, 66
Defocus, **235**
Degree of polarization, **390**
Degrees of freedom of lenses, 475
Dense flint glass, 362, 365
　critical angle, 38
　refractive index, 38
Depth of field, **162**
Depth of focus, **162**
Design of lenses, *see* Lens
Deviation of light
　by reflection, 28–29
　through prisms, 43, 46
Diamond
　brilliance, 38
　critical angle, 38
　refractive index, 17, 38
　table, 38
Diaphragm, lens system, **145**–146
Dichroic crystals, **375**
Dichroic herapathite crystals, 377
Dielectrics, 367
Diffracted light forming bright spot at centre of
　circular shadow, Plate 7
Diffraction, 4, **422**
　at straight edge, geometry/theory/resultant, 430
　effects, 425–427
　　small circular aperture, 426
　　small circular obstacle, 426–427
　eye, 496–497
　Fraunhofer, **432**–438, 442, 444, Plate 9
　Fresnel, **427**–431
　rectangular aperture/circular patterns
　　compared, 438
Diffraction grating, 405, 442–448
　grating interval (period), 446
Diffraction limited lens, **436**, **481**
Diffraction spectrum, 448
Diffractive bifocal contact lens, 453
Diffractive kinoform, 453
Diffractive optics, **22**, 448–**453**
Diffuse (irregular) reflection, **11**
Diffuse-reflection coefficient, 337

Diffused (scattered) light, **11**
Dioptre, **60**
Dioptric power (lens), **87**
Dioptric setting of magnifiers, **184**
Direct illumination, 212
Dispersion of light, **357**
　through prisms, **45**
Dispersive effect of prisms, 362
Dispersive power, **363**
Distances along axis, definition, 65
Distances perpendicular to axis, definition, 65
Distortion, 258, 259–262
　barrel, **260**
　percentage error, 260
　pincushion, **259**, 260
　thin lens, 259
Divergent pencils of light, **4**, 5
Donaldson colorimeter, 350–**351**
Double-swept contrast/frequency grating, Plate 11
Dove prism, **49**, 50
Dual nature of light, 299–301
Drysdale's method, **216**–219

Eccentricity, definition, 258
Eclipse
　of moon, 10
　solar, 10
EEL cells
　output curves, 343
　spectral response, 343
Effective point location mathematics, 233
Effective power of lenses, **101**–102
Elastic collison, 366
Electric filament lamps, 339
Electromagnetic theory, **313**–314
Electro-optics, **22**
Ellipsoid(s), 283
Ellipsoidal condensing reflector, 290
Ellipsometers, **393**
Elliptically polarized light, **384**
Emmetropia, 176
Emsley 60D eye, 177, **493**, 494–495
Entrance pupil, *see* Lens system
Equivalent focal length, **116**
　equivalent power, **116**
　measurement, 138–141
　see also Lenses
Equivalent thin lens, **116**
　erect image, **27**
Erecting telescope, *see* Telescopes
Excimer lasers, *see* Lasers
Exit pupil, *see* Lens system
Exit window (port), *see* Lens system
Extinction coefficient, **349**
Extinction ratio, **390**
Extra-axial points, **72**
Extraordinary rays (e-rays), **379**, 383
Eye, 174–180, 492–512
　aberration of lens/cornea, 497–501
　　spherical, 497–498
　　wavefront, 498–500, 503

Eye (*cont.*)
    accommodation, **176**
    aqueous humour, **175**
    chromatic aberration, 494–496
    contrast sensitivity, **511**
        measurement, 510–512
    cornea, **175**
    diffraction, 496–497
    emmetropia, **176**
    equally bright surfaces, judging, 329
    external line spread function, 502, 503
    far point, **716**
    fixation, **175**
    fovea, **175**–176
    foveal vision, **176**
    hypermetropic (hyperopic), **176**
    iris, 175
    lens, 175
    limit of resolution, **439**
    line spread function measurement, 502, 503
    modulation transfer function (MTF), 503, 504, 505
    myopic, **176**
    near point, **176**
    optical axis, **176**
    optical parameters, 143
    optical performance, 501–503, 504
    prescription, **178**
    pupil, 148
    pupillary response, **175**
    relaxed, 176
    responsiveness at low/high spatial frequencies, 487
    retina, *see* Retina
    schematic model, **176**–177, 492–494, 495
        astigmatism with, 501
        reduced, **493**
        simplified, **492**
    threshold contrast, **348**, **505**–506, 507, 508
    total performance, 505–507
    variation in visual performance with focus, 507–508, 509
    visual axis, **176**
    vitreous humour, **175**
Eye relief distance, **198**
Eye ring, **188**
Eyepieces, **186**, 188–189, 199–208, 480
    Huygens, 200, 210–204
        Ramsden eyepiece compared, 206–207
    Ramsden, 201, 204–207
        Huygens eyepiece compared, 206–207

f-number of lens, **161**
Fabry–Perot etalon, **406**
Far point, **176**
Fechner's fraction, **331**
Fermat's principles of least time, **20**, 21
Fibre-like lens, *see* Lens
Field-flattening lens, **476**
Field lens, **167**, **198**

Field lens microscope, 198–199
Field lens telescope, *see* Telescope
Field of view, **153**
    binocular, **185**
    monocular, **185**
    of plane mirrors, **27**
Field points, **72**
Field stop, **152**–153
Fifth-order theory of aberrations, **231**
Filters, 348
    absorption, 367, **368**
    bandpass, 368, 370, **417**–418
    colour conversion, 370
    continuously changing in index throughout thickness, 373
    edge, **417**–418
        cold mirror, **418**
        hot mirror, **418**
    heat absorbing, **370**–371
    neutral density, **368**
    photochromic, 371
    selectivity absorbing, 370
Finite velocity of light, 299
First focal length, **71**
First nodal point, **130**
First-order design, 467–472
First-order optics, **64**–65
First principal focus, 70, **78**, **90**
First principal plane, **116**
First principal point, **116**
First vertex focal length, **116**
First vertex focal power, **116**
Fixation, eye, 175
Fizeau fringes, **412**
Flat-top segment, **296**
Flint glass, 365
    dense, 365
        critical angle, 38
        refractive index, 38
    light
        critical angle, 38
        refractive index, 38
    special
        critical angle, 38
        refractive index, 38
Fluorescence, **14**
Fluorescent lamps, 326
    colour temperature, 319
Focal length
    calculation of for thick lenses, 117, 138–141
    equivalent, **116**
    first, **71**
    measurement, 138–141
Focal power (lens), **87**
Focal plane shutter, **171**
Focal planes, **71**
Focal properties
    thick lenses, 115–121
        see also Thick lenses
    thin lenses, 89–91
        see also Thin lenses

Focimeter, **137**, 215–216
   projection, 216
Foco-collimator, 139–140
Focus, **4**
   depth (geometric), 160–164
   optical length of rays coming to, 440
Folded Abbé prism system, **49**, 50
Foolscap paper, 338
Foot-lambert, 338
Format of camera, **171**
Forms (bending) for a lens, **85**
Foucault knife-edge testing, **485**
Fourier transforms, 310–312, 460–461, 481
   hologram, **458**
Fovea, **175**–176
Foveal vision, **176**
Fraunhofer diffraction, **432**–438, 442, 444, Plate 9
Fraunhofer hologram, **458**
Fraunhofer test plates, **412**–413
Frequency, **1**, **302**
   of heat waves, 2
   of light, 1, 2
   of radio waves, 2
   of red light, 2
   of ultra-violet light, 2
   of wave motion, 302
   of X-rays, 2
   of yellow light, 2
Fresnel bi-prism, **400**
Fresnel diffraction, **426**–441, Plates, 3, 4
Fresnel double-mirrors, **400**
Fresnel equations, 12, **387**
Fresnel holograms, **458**
Fresnel holography, 457
Fresnel integrals, **429**
Fresnel lenses, **294**–297
Fresnel zone plate, 296
Fresnel zones, **424**–425
Fringes
   degrees of coherence, **399**
   of equal thickness, **412**
   Fizeau, **412**
   mutual coherence, **399**
   optical, **398**
Front-reflection mirrors, 419
Front vertex of lens, **84**
Frosted glass, 15
Fundamental paraxial equation, **69**
Fused quartz crystals, 394
Fused silica, 394

Galilean telescopes, 192
Gallium arsenide, 324
Gas lenses, 373
Gauss, J. K. F. (1777–1855), 64
Gaussian optics, **64**–65
Geometrical blur, **163**
Geometrical depth of focus, **163**
Geometrical optics, **20**, 22, **299**
Glan–Foucault prism, **383**

Glass, 364–366
   amorphous solid, 275
   apparent thickness, 36
   dense flint, 36, 365
   diagram, 471
   extra dense fluid, 365
   flint, 365
   frosted, 15
   ground, 15
   ground opal, 338
   hard crown, 362, 365
   Jena, 364
   optical, 364
      transmission, 364–365
   photochromic, 371, 372
   refraction from, into air, 35
   representative types, 468
   striae, 373
Glass mirror, 12
Gold, 14
Gradient index lens
   meridional ray travelling through, 374
   skew ray travelling through, 374
Graphical constructions, 61–63
Graphical methods for use of thin lenses, 96–98
   see also Thin lenses
Graticules, 207–208
Grating
   images, 454–456
      quality curves, 455
   interval, **446**
   spatial frquency, **455**
Grating vector addition, 446, 447
Grease spot, 12
Green glass, 15
Grid polarizer, 377, 378
GRIN (gradex index; gradient index), **373**–374
Ground glass, 15
Guild, **341**
Guild flicker photometer, 341–342
Gullstrand–Le Grand schematic eye, 493, 495, 501

Haidinger's brush, 393–394
Half-period zones, **424**–425
Half-speed rotators, **49**
Half-wave plate, **384**
Halo, **460**
Harmonic oscillators, **300**–301
Hazards to vision of lasers, see Lasers
Head-up display, 28, **184**
Heat absorbing filters, **370**–371
Heat waves
   frequency, 2
   velocity, 1, 2
   wavelength, 2
Helmholtz schematic eye, 493
Hertz, **302**
Higher-Chance V-block refractometer, 360–361
Holographic gratings, **445**

Holography, **422**, 456–459
    focused hologram, **458**
    Fourier-transform hologram, **458**
    Fraunhofer hologram, **458**
    Fresnel, 457, **458**
    hologram lens, **459**
    interference, **458**
Homogeneous materials, **357**
Hot mirror, **418**
Huygens, C. (1629–1695), 2, 300
Huygens' construction, double refraction, 381,
    382
Huygens' construction, ordinary/extraordinary
    wavefronts/ray directions, 381
Huygens' eyepiece, 200, 201–204, 470
    Ramsden eyepiece compared, 206–207
Huygens' principle, **2**, 3, 13–14, 16, 25, 26
Hydrogen atom, emission spectrum, 315, 316
Hyperboloids, 283
Hyperfocal distance of lenses, **164**
Hypermetropia (hyperopia), **176**

Ice crystals, 380, 382
Iceland spar (calcite crystal), 378–379, 380, 382,
    394
Illuminance, **331**, **333**–336
    of surface, **331**
Illumination, **331**, **333**–336
    cosine law, 334
    measurement, 347–348
    of surface, 334–335
Images (imaging), **4**
    aberrated, 34–36
    by reflection, 25–27
    by refraction, 32–36
    by spherical surface reflection, 80–81
    by spherical surface refraction, *see* Spherical
        surface refraction
    erect, **27**
    inverted, 8
    with thick lenses, 127–136
    with thin lenses, 91–96
        see also Thin lenses
    real, **25**
    reverted, **27**
    virtual, **25**
Image brightness calculation, 161–162
Image curvature, 245
Image defects sources, 464
Image quality calculation, 481–484
Incandescent lamps, **319**–320
Incoherence, 397–399
Incoherent light, **321**
Infinity adjustment, **191**
Inhomogeneous optical materials, 373–374
Integrating sphere, **346**
Intensity profiles, actual/projected, 481, 482
Interaction of light with matter, 357–394

Interference, 397–419
    fringes, 398, 402–405
        cosine squared, 404, 405
        equal thickness, **412**
        Fizeau, **412**
        multiple beam, Plate 10
        two-beam, intensity profile, 404
        viewed via Fresnel bi-prism, Plate 1
        width, 402
    incoherent/coherent sources, 399–401
    laser, 401
    multiple beam, 405–406
        intensity profile, 406
    partial reflection, 407–408
    thin film, 409–411
Interference holography, **458**
Interferometer, non-contact, 413
Interleaved micrometer, 244–245
Intermediate condenser, 211
Internal transmittance, **366**
International candle, **339**
Interval of Sturm, **247**, 280
Intraocular lens, 293
Inverted image, **8**
Inverted population of electronic energy levels,
    321
Inverting prisms, 49–50, 51
Iris, 148, 175
Iris diaphragm, **171**
Irregular (diffuse) reflection, **11**
Isotropic materials, **357**

Javal–Schiotz ophthalmometer, 383
Jena glass, 364
Jolie's proposal, 285–286
Jones' vectors, **393**

Keratometer, **255**–256
Kinoform, **453**
    diffractive, 453
Knife-edge, 485

Lagrange invariant, **74**, **106**
Lambert, **338**
Lambert's cosine law of emission, 338
Lambert's law, **369**
Landscape lens, **477**–478
Lanthanum, 365
Large distances measurement, 227
Lasers, **301**, 322–326
    application, 325–326
    continuous wave carbon dioxide, 326
    excimer, **325**
    interference with, 401
    interferometer systems, 401
    light emitting diodes, 324
    neodymium, 324
    operating, 325

Lasers (*cont.*)
  photometry, 324–325
  practical types, 322–324
  pulsed, 325
  ruby, 323–324
  traversing goldfish eye, 374
  vision hazards, 324–325
Lateral chromatic aberration, **470**
Lateral colour, 470
Lateral displacement by diffraction, 39
Lateral magnification, **73**, **93**
Lateral size measurement, 222
Lathes, computer-controlled, 270
Law of inverse squares, 333
Law of reflection, 12–14
Laws of refraction, 15, 18–20
Lead oxide, 365
Lens
  achromatic, **362**
  aplanatic, 241
  back vertex of, 84
  centre thickness, **84**
  Chevalier, **478**
  Cooke triplet, 478, 479
  crystalline, **175**
  decentred, **85**
  degrees of freedom, 475
  design, 476–480
    aberration balancing act, 477
    tolerancing, 477
  diffraction limited, **436**, **481**
  dioptric power of, **87**
  effective power, **101**–102
  equivalent power/focal length measurement,
    116, 138–141
    dual magnification method, 139
    Newton's method, 139
    rotation method: nodal points, 140–141
    single magnification method (foco-
      colimeter), 139–140
  eye, 175
  f-number, 161
  fibre-like, 373
  field, 167
  field-flattening, 476
  Fresnel, **294**–297
  front vertex of, **84**
  gas, 373
  hologram, **459**
  hyperfocal distance, **164**
  intraocular, **293**
  inverted telephoto, **174**
  landscape, **477**–478
  microscope, eye lens, 198
  modulation transfer function, **455**, **483**
  narrow aperture, 163
  neutralization of, **107**
  objective, **186**
  optical centre of, **84**
  prismatic effect, **101**
  progressive, **294**–295

Lens (*cont.*)
  projector, 102
  Protar, 478
  retrofocus, **174**
  segmented, 294–296
  spectacle, *see* Spectacle lenses
  spherical, 279
  system, *see* Lens system
  telephoto, **173**
  Tessar, 478–479
  thin, *see* Thin lenses
  toric, 279
  Zeiss 'Protar', 478
  zoom, 174
Lens clock, **56**, 57
Lens elements, *see* Lens system
Lens pair relay system, 165–167
Lens power measurement, 107
  neutralization, 107
  see also Optic bench
Lens system, **123**, 141–164
  aperture stop, 146, **147**
  construction rays/actual rays, 145
  depth of field, 162
  depth of focus, 162
  diaphragm (stop), **145**–146
  entrance pupil, **147**, 148
  entrance window (port), **155**, 159
  exit pupil, **149**–150
  exit window (port), **155**
  field of view, 153
    full illumination, 153
    plane mirror, 27–28
    total, 153
  field stop, 152–153
  focal depth, geometric, 160–164
  lens elements, **141**
  marginal rays, 149, 150
  principal ray, 149, 150
  reciprocity failure, 161
  telecentric stop, 150
  usual conditions, 232
Light
  amount: visual response, 329–331
  circularly polarized, **384**
  deviation by reflection, 28–29
  differing phase superposed, 311
  diffraction, **422**
  dispersion through prisms, 45
  elliptically polarized, **384**
  interaction with matter, 357–394
  linearly polarized, **375**, **384**
  pencils, *see* Pencils
  rectilinear propagation, 2–4
  scattering, 354, **385**
  sine wave, 311
  smallest perceptible difference, 329–331
  spontaneous emission, 315
  standard sources, 339
  unpolarized, **376**
Light-emitting diodes, **324**

Light flint glass
  critical angle, 38
  refractive index, 38
Light quanta, *see* Photons
Light transmitting optical fibres, 361
Limit of resolution of eye, **439**
Line spread function, **242**, **484**
  measurment of, in eye, 502, 503
Linear birefringence, **381**
Linearly polarized light, **375**, 384
Lister doublet microscope objective, 480
Listin schematic eye, 492
Lloyd's mirror, **400**, 403
Log-log scale, 369
Long-focus microscope, 225
Longitudinal magnification, **94**
Loupe (simple magnifier), **180**–184
Lucite, *see* Polymethyl methacrylate
Lumen, **332**
Lumen per square foot, 333
Lumen per square metre (lux), 333
Luminance, **331**, **336**–339
  definition, 337
  measurement, 347–348
  moon, 338
  expression, 337
  sun, 337
Luminescent sources, 1, 326–327
Luminosity of colour, 319
Luminous flux, **331**–333
Luminous intensity, **331**–333
Luminous sources, **1**
Lummer–Brodhun photometer, **341**
Lux (lumen per square metre), **333**

Magnesium fluoride, 419
Magnesium oxide, 338
Magnification
  apparent, **181**
  lateral, **93**
  longitudinal, **94**
Magnifiers, 180–185
  binocular, 184–185, **480**
  biocular, **184**
  dioptric setting, **184**
  magnifying power (apparent magnification),
    181–182
  simple (loupe), **180**–184
Magnifying power, **181**–182
Maiman ruby laser, 323
Major principal transmittance, **390**
Marginal rays, 149, **150**
Martin Marietta Black, 338
Matching stimuli, 350
Matt white celluloid, 338
Maxwellian view of condensers, **211**–212
Maxwell's discs, 350
Maxwell's equations, 300
Mean horizontal candle-power, **346**

Mean refractive index, **362**
Mean spherical candle-power, 346
Measurement
  of angles of prism, 359, 360
  of luminance, 347–348
Measuring instruments, 212–227
  large distances, 227
  photo-electronic, 227
  small objects, 222–225
Mechanical tube length, **187**
Meniscus lens, 475
Meridian of cylindrical surfaces, **271**
Meridians, principal, 271, 279
Meridional plane, **236**
Meridional rays, **262**–264
Meteorological range (visibility), **349**
Mica crystals, 382
Michelson interferometer, 407–408
  Twyman–Green modification, 408
Micro reciprocal degrees (mireds), 370
Micrometre, **314**
Microscope, compound, 185–191
  eye lens, 198
  eye relief distance, **198**
  eyepiece, **186**, 188–189
    given focal length, 189
  field lens, 198–199
  limit of resolution (resolving power), 439,
    440–441
  long-focus, 223, 225
  measuring, 222
  mechanical tube length, **187**
  objective, 186
  optical tube length, **186**, 187
  parfocality, **188**
  travelling, **223**
  working distance, **189**
MIL-HDBK-141 schematic eye, 501
Millilambert, 338
Minimum angle of deviation, 43, 45
  see also Prisms
Minor principal transmittance, **390**
Mireds (micro reciprocal degrees), **370**
Mirror
  coincident methods for, 111, 112
  glass, 12
  front-reflection, **419**
  hot, 418
  plane, 27–28
    applications, 28
    field of view, 27–28
  reflectance, 12
  reflection of light, 399
  rotating, 29
  steel, 12
Modulation transfer function (MTF), **455**,
    483–489
  collimator lens, 486
  of eye, 503, 504, 505
  photographic lens, 486
  TV lens, 486

Moon, 10
    colour temperature, 319
    eclipse, 10
    luminance, 338
    surface, incident light reflected, 338
3M's Nextel paint, 338
Multiple beam interferences, 405–406
    intensity profile, 406
Myopia, **176**

Nanometre, **314**
Nanotechnology, **397**
Narrow aperture lenses, 163
Near point, **176**
Neodymium lasers, *see* Lasers
Neutral density filters, **368**
Neutral materials, **14**
Neutralization of a lens, **107**
Newton's corpuscular theory of light, 300
Newton's equation, **94**
Newton's relation, **74**, 81
Newton's rings, 402, **411–412**, Plate 2
Newton's scale of colours, **412**
Nicol prism, **383**
Nit, 337
No-parallax condition, **111**
Nodal points, **72**, 130, 134
Nodal slide, **140–141**
Non-spheroidal optical surfaces, 270–297
Normal spectrum, **448**
Normal to surface at point of incidence, 12
Numerical aperture (NA), **161–162**, **441**

Object-to-image via point spread functions, 483
Objective lens, **186**
Oblique aberrations, 245
Offence against the sine condition (OSC), 257
Ogle 17 mm eye, **493**, 494
Opal glass, ground, 338
Opaque materials, **14**
Operation of lasers, *see* Lasers
Ophthalmic prisms, *see* Prisms
Optical activity of quartz, **394**
Optical axis, **65**, **84**, **176**, **376**
Optical bench, **107–112**, 139
    coincident methods for mirrors, 112
    conjugate foci methods, 112
    positive lenses, 108–112
        autocollimation methods, 110–112
        collimation methods, 109–110
        distant object, 108–109
        telescope methods, 110
Optical coatings, 397
Optical contact, **409**
Optical density, **368**
Optical fibres, 14
    light transmitting, 361
Optical fringes, **398**

Optical glass, 364
    refractive index, 17
    transmission, 364–365
Optical length, **20**
    see also Optical thickness
Optical media, 2
Optical metrology, **397**
Optical parameters of eye, 143
Optical performance of eye, 501–503, 504
Optical plastics, refractive index, 365–366
Optical processing, **460**
Optical pumping, **321**
Optical rangefinder, **220–221**
Optical surfaces, 11
Optical thickness (length), 20, **403**
    apparent, **403**
Optical thin films, **397**
Optical transfer function, **455**, **484**
    see also Modulation transfer function
Optical tube length, **186**
Optician's lens measure, **56**, 57
Optics
    diffractive, 22
    electro-, 22
    geometrical, **20**, 22, **299**
    paraxial (Gaussian), 64–65
    physical, **20**, **299**
Optimization program, **262**
Opto-electronics, **22**
Ordinary rays (o-rays), **379**, 383
Organic polymers, 365–366

p (parallel), 386
Paraboloid, comparison with spherical curve, 283
Paraboloid mirror, 284
    off-axis, 284
    zero spherical aberration, 284
Parallel pencils of light, 5
Parallel plate micrometer, **41**
Parallel plates, refraction through, 38–41
Paraxial approximation, 64
Paraxial equation, fundamental, 69
Paraxial optics, 64–65
Paraxial rays, **64**
Paraxial reflection at spherical surfaces, 75–78
    all-positive diagram, 77
    positive reflecting surface; principal foci, 79
    ray path change, 75–76
    vergence change, 76–78
Paraxial region, **64**
Parfocality, **188**
Partial dispersion values, **364**
Partially polarized light, 376
Pelli–Robson Sensitivity Chart, Plate 12
Pencils of light (bundles of rays), **4–8**
    aperture, 6
    convergent, 4, 5
    divergent, 4, 5
    parallel, 5
    principal (chief) ray, **4**, **72**, 149, **150**, 160, **247**
    width, 7

Penumbra, **9**
Perfect point imagery, required surface for, 287
Period/periodic time, **302**
Perspex, *see* Polymethyl methacrylate
Petzval condition, 246
Petzval surface, **245**
Phase, **302, 303**
Phase plate (phase zone plate), **427**, 450–451
    asymmetrical, 451, 454
    symmetrical, 451, 452, 454
Phase transfer function, **484**
Phosphors, **326**
Photochromic filters, 371
Photochromic glass, 371, 372
Photocopier, desk top, 373
Photoelasticity, 393
Photoelectric photometers, 342–344
Photoelectronic measuring instruments, 227
Photometers, 340–344
    comparison, 340–342
    contrast, **341**
    Guild flicker, 341–342
    Lummer–Brodhun, **341**
    Spectra–Pritchard, **347**
Photometry, **329**–349
    integrating spheres, 346–347
    Lummer–Brodhun, 341
    photoelectric, 342–344
    principle, 340
    screens, 344
    sources, 347
Photons (light quanta), **301**, 315, 366, 367
    stimulated emission, 321
Photopic curve (V curve), 329
Physical optics, 20, 299
Pincushion distortion, **259**, 260
Pinhole camera, 8–9
Planck's theory of light, 300–301
Plane mirror, *see under* Mirror
Plane of incidence, 12
Plane of vibration, **375**
Plane surface, 54
    reflectance at, 13–14
    reflection at, 25–38
Plaster of Paris, 338
Plastics (organic polymers), 365–366
Platinum, 339
Poincaré sphere, **393**
Point images, 438–442
Point spread function (PSF), **242, 481**
Polarizance, **390**
Polarization, 375–385
    applications of polarized light, 393–394
    by reflection, 385–389, 392
    by scattering, 385–389
    circular polarizer, 384, **385**
    circularly polarized light, **384**, 385
    degree of, **390**
    elliptically polarized light, **384**
    extinction ratio, 390
    linearly polarized light, **375, 384**

Polarization (*cont.*)
    major principal transmittance, **390**
    minor principal transmittance, **390**
    partial, **376**
    passage of polarized light through retarder, 384
    in spectacle lenses, 375, 387
Polarizer, **376**
    circular, **385**
    crossed, **376**
    transmission axis, **390**
Polarizing angle, **386**, 388, 389
Polarizing sunglasses, 375, 387
Polaroid, 377
    E-sheet, 377
Pole of surface, **65**
Polished surfaces, 12
Polymers, representative types, 468
Polymethyl methacrylate
    critical angle, 38
    refractive index, 38
Porro prism, **49**–50, 51
    telescope, 193
    variant, 51
Power meridian, **271**
Practical types of laser, *see* Lasers
Presbyopia, **178, 293**
Prescription eyes, 178
Pressure broadening, 316
Principal planes, **93**
Principal ray, 4, 72, 149, 160, **247**
Principal section, **382**
    of prisms, **41**
Principal transmittance ratio, 390
Principle of linearity, **482**
Principle of reversibility of optical path, **20**
Principle of superposition, **303**
Principles of photometry, 340
Prismatic effect of lenses, **100**
Prisms, 41–50
    Abbe, **49**, 50
    Abbe roof, **50**, 51
    achromatic, **466**
    apex of, **41**
    deviation, 43, 46
    dispersive effect, **362**
    Dove, **49**, 50
    erecting system for telescopes, 192–193
    folded Abbé system, **49**, 50
    Glan–Foucault, **383**
    image doubling in measuring instruments, 383
    inverting, 49–50, 51
    light dispersion, 45
    measuremnt of angle, 359, 360
    minimal angle of deviation, 43
        measurement, 360
    Nicol, **383**
    ophthalmic, 45–48
        crossline appearance, 47
        refraction through, 47
    Porro, *see* Porro prism
    principal section, **41**

Prisms (*cont.*)
  reflecting, 48–50
  refracting, 41–45
  refracting edge (apex), 41
  refraction through, 42
  refractive index determination, 359–360
  roof edge, **50**
  viewing objects through, 46
  Wollaston double-image, **383**
Prism dioptre, **47**
Progressive lens, **293**–294
Projection focimeter, 217
Projectors, 208–209, 211
  lenses, 102
  movie, 213
  slide, condensers for, 210
  slide projection system, 211
  throw, **208**
Protar lenses, 478
Pulsed lasers, *see* Lasers
Pupil
  entrance, **147**, 148
  exit, **149**–150
  eye, 148
  response, **175**

Quanta, light, *see* Photons
Quantum theory of light, 366
Quarter-wave coatings, **414**
Quarter-wave plate, **384**
Quartz crystals, 301, 315, 380, 382
  fused (fused silica), 394
  indices, 382
  optical activity, 394

Radian measure of angles, 54
Radical line focus, **248**
Radio waves
  frequency, 2
  velocity, 1
  wavelength, 2
Radiometry, **331**
Radius of curvature, **54**, 55
Radiuscope (Drysdale's method), **216**–219
Ramsden eyepiece, 204–206
  Huygens' eyepiece compared, 206–207
Range finder, 219–221
  optical, **220**–221
Ray, 4
  bundles of rays, *see* Pencils of light
  construction/actual, 145
  marginal, 149, **150**
  principal, 149, **150**, 160
  slope, 66
Ray tracing, **230**, 262–267
  meridional rays, **262**–264
  skew rays, 264–267
Rayleigh limit, 496, 497
Reactolite Rapide, 371–373

Real images, **25**
Reciprocity failure, **161**
Rectilinear propagation of light, 2–4
Red light
  frequency, 2
  velocity, 2
  wavelength, 1, 2
Reflectance
  of mirrors, 12
  at a plane surface, 13–14
Reflecting prisms, 48–50
Reflection, 11–14
  amount of light reflected, 11
  angle of deviation, 28
  angle of reflection, 12
  curvature of change, 61
  deviation of light by, 28–29
  imaging by, 25–27
  irregular (diffuse), 11
  law, 12–14
  multiple images, 30–32
  near-hemispherical surface, 63
  normal to surface at point of incidence line, 12
  plane of incidence, 12
  plane surfaces, 25–38
  regular (specular), 11, 12, 13
  selective, 12
  spherical surface, 75, 77
  thin film, 412
  total (total internal), 37–38
  two mirrors, 29–32
Refracting edge of prism, 41
Refracting prisms, 41–45
Refraction, 15–17
  curvature change, 59
  glass into air, 211
  imaging by, 32–36
  lateral displacement, 39
  laws, 18–20
  near hemispherical surface, 62
  negative refracting surfaces, 71
  plane surfaces, graphical construction, 33–34
  positive refracting surface, 70
  spherical surface, 66–67
  through parallel plates, 38–41
  through prisms, 42
Refraction equation, **106**, **151**
Refractive index
  absolute, 16–17
  aluminium oxide, 414
  crown glass, 38
  dense flint glass, 38
  'high index' spectacle glass, 17
  light flint glass, 38
  optical glass, 17
  magnesium fluoride, 414, 415
  optical plastics, 17
  polymethyl methacrylate, 38
  prisms, 359–361
  silicon dioxide, 414
  silicon monoxide, 414

Refractive index (*cont.*)
  special flint glass, 38
  spectacle crown glass, 17
  titanium dioxide, 414
  water, 17, 38
  zinc sulphide, 414
Refractive power of contact lenses, 353
Refractometer, **315**, **357**, 358–361
  Higher-Chance V-block, 361
Regular reflection, **11**, 12, 13
Relative dispersion, 363
Relative refractive index, 17
Relaxed eye, 176
Relay systems, **165**–167
  lens pair relay, 165, 166
  single lens, 165
  two-fold relay, 165–167
Representative types of glass, 468
Resolving power, **439**
  of eye, 439
  of telescopes, 439
Resonant cavity, **321**–322
Resultant intensity curves for close objects, 441
Retarder, 383
Retina, **175**, 396
  fovea, 175–176
  image intensity, 337
Retinal illuminance, 339
Retrofocus lens, **174**
Reversibility of optic path principle, 20, 160
Reverted image, **27**
Römer's observations, **299**
Roof edge prism, **50**
Rotating mirrors, 29
Ruby lasers, *see* Lasers
Russell angles, **346**

s (senkrecht), 386
Sag, 55, 57–58
Sagittal line focus, **248**
Scattering of light, 11, 354, 366, **385**
Schematic model of eye, 176, 495
  see also Eye
Schlieren test, **485**
Schmidt plate, 291, 292
Screen
  gain, **344**, 345
  photometry of, 344
Second focal length, **70**, **90**
Second nodal point, **130**
Second principal focus, **70**, **79**, **89**, **116**
Second principal plane, **116**
Second principal point, **116**
Second vertex focal length, **116**
Second vertex focal power, **116**
Secondary aberrations, **231**
Secondary spectrum, 467
Segmented lenses, **294**–296
Segmented optical surface, 270–297
  see also Non-spherical optical surfaces
Seidel aberrations, **231**, 258

Seidel coefficients for displaced stop, 258–259
Selective absorption, **14**, **367**
  of colour, 353
Selective reflection, 12
Selective transmission, **14**
Selectively absorbing filters, 370
Selenium barrier-layer photo cell, **344**
Seven-layer polarizing beam splitter, 386–387
Shadows, 9–10
Short focus telescopes, *see* Telescopes
Sign convention, 64–65
  optical axis, 65
  pole (vertex), 65
Silica, fused (fused quartz), 394
Silicon dioxide, refractive index, 414
Silicon monoxide, refractive index, 414
Silver halide, 371
Simple harmonic motion, **302**–308
  amplitude, 303
  equal amplitude/equal frequency, composition
    at right angles, 308
  four, equal frequency/amplitude, progressively
    out of phase, 305
  phase, 303
  principle of superposition, 303
  two, executed in directions at right angles, 307
  two, of different frequences, 304
  two, of same frequency, 204–205
Simple magnifiers (loupe), **180**–184
Sinc function, **434**
Sine condition, **255**
  offence against, **257**
Sine wave, 311
Sine wave object to sine wave image via point
    spread functions, 483
Single lens reflex (SLR) camera, **172**
Single lens relay systems, 165
Single ray refraction, 99
Size measurement by doubling, 224
Skew rays, 264–267
Sky blueness, colour temperature, 319
Slope of ray, definition, 66
Small objects measurement, 222–225
Smallest perceptible difference of light, 329–331
Snellen charts, low contrast, 511–512
Snell's law, **19**, 43, 238, 262, 287, 388
Sodium nitrate crystals, 382
Sodium vapour, 17
Solar eclipse, 10
Sources of interference, 399–401
Spatial frequency of grating, **454**
Spatial frequency response, **483**
Special flint glass
  critical angle, 38
  refractive index, 38
Spectacle lenses, 84, 178
  aspheric shell, 285
  bifocal, **178**
  'flat top' bifocal, 294, 296
  Jolie's proposal, 285–286
  polarizing, 375, 387

Spectra-Pritchard photometer, **347**–348
Spectral lines used in optics, 465
Spectral response of EEL cells, 343
Spectrometers, 358–361
Spectrophotometers, **315**, **357**, **358**, 419
Spectroscope, **315**
Spectrum, 320
  Abney division, 320
  colour, 318
  normal, **448**
  secondary, 467
  Tilton's equal-line bands, 320
Specular reflectance, transparent surfaces, 12
Speed of light, 1, 2
  in air/water compared, 16
  in space (vacuum), 16
Spherical aberrations, *see under* Aberrations
Spherical curvature, 54–58
  centre, 54
  comparison with paraboloid, 283
  definition, 54–55
  distortion into toric surface, 271
  measurement, 55–56
  radius, 54–55
  sag, 55, 57–58
Spherical lens, 279
Spherical mirror, spherical aberration, 284
Spherical reflecting surfaces
  first principal focus, 78
  focal properties, 78–80
  reflection at, 75, 77
  second principal focus, 78
Spherical surface reflection, imaging by, 80–81
Spherical surface refraction, imaging by, 72–75
  apparent size, 75
  conjugate planes, 72
  conjugate points (foci), 70
  distant object, 74
  extra-axial point, 72
  field point, 72
  first focal length, 71
  first principal focus, 70
  focal planes, 71
  focal properties, 70–72
  Lagrange invariant, 74, 106
  lateral magnification, 73
  negative, 71
  Newton's relation, 74, 81
  nodal point, 72
  paraxial region, 73
  positive, 70
  principal ray, 72
  second principal focus, 70
Spherical waves, 422–425
Spherometer, **56**, 57
  precision, 58
Spontaneous emission of light, **315**
Spot diagram, 242
Standard sources of light, 339
Steel mirrors, 12
Steradian, **332**

Stimulated emission of light, **319**
Stimulated emission of photons, 321–322
Stokes' vectors, **393**
Stops, **145**
Strehl intensity ratio, **481**
Stress-optical coefficient, **393**
Sturm's conicoid, 280
Sturm's interval, **247**, **280**
Substrates, 416
Sugar solutions, 394
Sun
  absorption spectra of hot gases, 316
  colour temperature, 319
  eclipse, 10
  increasing redness, 354
  luminance, 337
Sunglasses, polarized, 375
Superposition principle, 303
Surface illumination, 334–335
Surface power, **69**
Symmetrical phase plate, 451, 452, 454
Symmetrical planes, **93**
Symmetrical principle, 261, **478**
Symmetrical system, **65**

Tangential line focus, **248**
Telecentric cathetometer, 221
Telecentric principle, **110**, 150
Telecentric stop, 150
Telecentric systems, **150**
Telecentric telescope, 222
Telephoto lens, **173**
  inverted (retrofocus) lens, **174**
Telephoto power, **173**
Telescopes, **191–199**
  afocal systems, **194**
    telecentric, 195, 196
  astronomical, 191, 192
  erecting, 192, 197, **199**
  eye lens, **198**
  eye relief distance, **198**
  eyepiece system, 197
  field lens, 198–199
  focus inaccuracy, 213–214
  Galilean, 192
  infinity adjustment, 191
  limit of resolution, 439
  methods, **110**
  Porro prism system, 192–193
  prism erecting system, 192–193
  short focus, 223
  telecentric, 222
Television picture tubes, 325
Tessar lens, 478–479
Test plates, Fraunhofer's, **412–413**
Theodolite, **191**, 226, **227**
Thick lenses, 115–141
  cardinal points/distances, 127, 131
  contact lens as, 136–138
  distant image, 117

Thick lenses (*cont.*)
  distant object, 116
  focal length calculation, 117, 138–141
  focal properties, 115–121
    first principal plane, **116**
    first principal point, **116**
    first vertex focal length, **116**
    first vertex focal power, **116**
    second principal focus, **116**
    second principal plane, **116**
    second principal point, **116**
    second vertex focal length, **116**
    second vertex focal power, **116**
  imaging with, 127–136
  negative, image formation, 133
  nodal points, 130, 134
  object/image distances from principal planes,
    132
  positive, image formation, 128, 129
  unequal indices effect, 125
  unit planes, 128
Thin film, 409–410
  interference, 411
  reflection, 410
Thin lens pairs, separated, 122–136
  cardinal points/distances, 124, 127, 131
  equivalent power of lens system in terms of
    power/focal length of first lens, 126
  focal properties, 122–126
  image formation, 128
  imaging with, 127–136
  nodal points, 130, 134
  principal planes action, 127
  unit planes, 128
  with glass block, 123
Thin lenses, 84–107
  back vertex, 84
  concave, 86
  convex, 86
  decentred, 85
  distortion, 259
  effective power, 101–102
  equivalent, **116**
  focal power (dioptric power), 87–88
  focal properties, 89–91
    first principal focus, 90
    second focal length, 90
    second principal focus, 89–90
  forms (bending), 85
  front vertex, 84
  graphical methods, 96–98
    negative lens: virtual object/real image
      (erect), 98
    negative lens: virtual object/virtual image
      (inverted), 98
    positive lens: object/image position, 96
    positive lens: real object/virtual image, 97
    positive lens, virtual object/real image, 97
  imaging with, 91–96
    longitudinal magnification, 94
    negative lens, 92

Thin lenses (*cont.*)
  imaging with (*cont.*)
    positive lens, 92
    positive lens: distant object, 94
    principal planes, 93–94
    symmetrical planes, 93
  longitudinal, transverse, wavefront aberration
    relationship, 234
  more than one, 102–107
    convergence angle ray tracing, 105
    Γ diagram, 104
    ray tracing through two positive lenses, 103
  negative lens: distant object, 95
  optical axis, 84, 85
  optical centre, 84
  pairs, *see* Thin lens pairs
  paraxial ray tracing through, 151
  positive lens: object/image positions, 96
  prismatic effect, 100–101
  refraction equation, 106
  series of thin prisms, 99
  single ray refraction, 99
  transfer equation, 106–107
Third-order theory, 231–236
Three-colour printing, 353–354
Threshold contrast, **348**, **505**–506, 507, 508
Through-focus aberrated images of distant point
    source, Plate 6
Through the lens (TTL) metering, **172**
Tilton's equal-hue bands, 320
Titanium dioxide, refractive index, 416
Toric lens, 279
Toric surfaces, 270, 271
  barrel, 272
  capstan, 272
  two line image formation, 274
Toric wavefront, 273
Toroidal (astigmatic) lens, image formed by, 282
  total curvature, **278**
Total (internal) reflection, **37**–38
Total field of view, **153**
Tourmaline crystals, 376, 380, 382
Transfer equation, 106–107, 151
Translucent materials, **14**
Transmission axis of polarizers, **390**
Transmittance, **366**
  internal, **366**
  of polarization, 390
Transmitting optical materials, 361–366
Transparent materials, **14**
Travelling microscope, **233**
Traversing goldfish eye lasers, *see* Lasers
Trial case lenses, 84
Trichromatic coefficients, 352
Trichromatic theory of colour vision, **352**
Trichromatic units, **351**
Tritium gas-filled lamps, 327
Trolands, **339**
Tscherming schematic eye, 492
Tungsten filament lamp, 317
  colour temperature, 319

Two-fold relay systems, 166, 167
Tyndall effect, **354**

Ultraviolet light
    frequency, 2
    velocity, 2
    wavelength, 2
Umbra, **9**
Uniaxial crystals, **379**
    linear birefringence, 381
    wavelet, 380
Unit planes, **128**
Unit trichromatic equation, **351**–352
Unpolarized light, **376**

V curve, **329**
V value, **363, 465**
Vector addition of amplitudes, 425, 428
Velocity, **1**
    of heat waves, 1, 2
    of radio waves, 2
    of red light, 2
    of ultraviolet light, 2
    of X-rays, 2
    of yellow light, 2
Vergences, **60**
Vernier acuity, **219**
Vernon Harcourt lamp, 339
Vertex of surface, **65**
Vibrating cord, 301, 302
View finders, 219–221
Vignetting, **153**, 154, **485**
Virtual images, **25**
Visibility (meteorological range), **349**
Visual angle, **182**
Visual axis, **176**
Vitreous humour, **175**

Water
    critical angle, 38
    refractive index, 38
    waves, **1**
Waveforms, complex, 308–313
    frequency spectra, 312
Wavefront aberrations, **231**, 232
Wavefronts, **2**, 310
    curvature, 59–60
    optics, 422

Wavelength, **1, 302**
    of heat waves, 2
    of radio waves, 2
    of red light, 1, 2
    of ultraviolet light, 2
    of wave motion, 302
    of X-rays, 2
    of yellow light, 2
Wavelets, 2, 380, 422
    biaxial crystals, 380
    uniaxial crystals, 380
Wave motion, 301–302
    amplitude, 203
    cycle, 302
    period (periodic time), 302
    phase, 302
    wavelength, 302
Wave superposition, 309–310
Wave theory, **300**
Wave trains, 308–309
Weber's law, 329
White blotting paper, 338
White light, 3, 318, 319, 353
Width of a light pencil, 7
Wollaston double-image prism, 383
Wright colorimeter, 350

X-rays
    frequency, 2
    velocity, 2
    wavelength, 2

Yellow light, 17, 230
    frequency, 2
    velocity, 2
    wavelength, 1, 2
Young construction, 61–62
Young–Helmholtz theory, 352

Zeiss 'Protar' lens, 478
Zinc sulphide, refractive index, 414
Zone plate, 427, 450
    amplitude, 450, 454
    blazed, **452**–453
Zoom lens, **174**